D1943063

Theories of Molecular Reaction Dynamics

Preface

This book focuses on the basic concepts in molecular reaction dynamics, which is the microscopic atomic-level description of chemical reactions, in contrast to the macroscopic phenomenological description known from chemical kinetics. It is a very extensive field and we have obviously not been able, or even tried, to make a comprehensive treatment of all contributions to this field. Instead, we limited ourselves to give a reasonable coherent and systematic presentation of what we find to be central and important theoretical concepts and developments, which should be useful for students at the graduate or senior undergraduate level and for researchers who want to enter the field.

The purpose of the book is to bring about a deeper understanding of the atomic processes involved in chemical reactions and to show how rate constants may be determined from first principles. For example, we show how the thermally-averaged rate constant $k(T)$, known from chemical kinetics, for a bimolecular gas-phase reaction may be calculated as proper averages of rate constants for processes that are highly specified in terms of the quantum states of reactants and products, and how these state-to-state rate constants can be related to the underlying molecular dynamics. The entire spectrum of elementary reactions, from isolated gas-phase reactions, such as in molecular-beam experiments, to condensed-phase reactions, are considered. Although the emphasis has been on the development of analytical theories and results that describe essential features in a chemical reaction, we have also included some aspects of computational and numerical techniques that are used when the simpler analytical results are no longer accurate enough.

We have tried, without being overly formalistic, to develop the subject in a systematic manner with attention to basic concepts and clarity of derivations. The reader is assumed to be familiar with the basic concepts of classical mechanics, quantum mechanics, and chemical kinetics. In addition, some knowledge of statistical mechanics is required and, since not all potential readers may have that, we have included an appendix that summarizes the most important results of relevance. The book is reasonably self-contained such that a standard background in mathematics, physics, and physical chemistry should be sufficient and make it possible for the students to follow and understand the derivations and developments in the book. A few sections may be a little more demanding, in particular some of the sections on quantum dynamics and stochastic dynamics.

Earlier versions of the book have been used in our course on advanced physical chemistry and we thank the students for many useful comments. We also thank our colleagues, in particular Dr Klaus B. Møller for making valuable contributions and comments.

The book is divided into three parts. Chapters 2–8 are on gas-phase reactions, Chapters 9–11 on condensed-phase reactions, and Appendices A–I contain details about concepts and derivations that were not included in the main body of the text. We have put a frame around equations that express central results to make it easier for the reader to navigate among the many equations in the text.

In Chapter 2 we develop the connection between the microscopic description of isolated bimolecular collisions and the macroscopic rate constant. That is, the *reaction cross-sections* that can be measured in molecular-beam experiments are defined and the relation to $k(T)$ is established. Chapters 3 and 4 continue with the theoretical microscopic description of isolated bimolecular collisions. Chapter 3 has a description of *potential energy surfaces*, i.e., the energy landscapes for the nuclear dynamics. Potential energy surfaces are first discussed on a qualitative level. The more quantitative description of the energetics of bond breaking and bond making is considered, where this is possible without extensive numerical calculations, leading to a semi-analytical result in the form of the London equation. These considerations cannot, of course, replace the extensive numerical calculations that are required in order to obtain high quality potential energy surfaces. Chapter 4 is the longest chapter of the book with the focus on the key issue of the *nuclear dynamics of bimolecular reactions*. The dynamics is described by the *quasi-classical* approach as well as by exact *quantum mechanics*, with emphasis on the relation between the dynamics and the reaction cross-sections.

In Chapter 5, attention is directed toward the direct calculation of $k(T)$, i.e., a method that bypasses the detailed state-to-state reaction cross-sections. In this approach the rate constant is calculated from the *reactive flux* of population across a dividing surface on the potential energy surface, an approach that also prepares for subsequent applications to condensed-phase reaction dynamics. In Chapter 6, we continue with the direct calculation of $k(T)$ and the whole chapter is devoted to the approximate but very important approach of *transition-state theory*. The underlying assumptions of this theory imply that rate constants can be obtained from a stationary equilibrium flux without any explicit consideration of the reaction dynamics.

In Chapter 7 we turn to the other basic type of elementary reaction, i.e., *unimolecular reactions*, and discuss detailed reaction dynamics as well as transition-state theory for unimolecular reactions. In this chapter we also touch upon the question of the atomic-level *detection* and *control* of molecular dynamics. In the final chapter dealing with gas-phase reactions, Chapter 8, we consider unimolecular as well as bimolecular reactions and summarize the insights obtained concerning the microscopic interpretation of the Arrhenius parameters, i.e., the pre-exponential factor and the activation energy of the Arrhenius equation.

Chapters 9–11 deal with elementary reactions in condensed phases. Chapter 9 is on the energetics of *solvation* and, for bimolecular reactions, the important interplay between *diffusion and chemical reaction*. Chapter 10 is on the calculation of reaction rates according to *transition-state theory*, including static solvent effects that are taken into account via the so-called potential-of-mean force. Finally, in Chapter 11, we describe how dynamical effects of the solvent may influence the rate constant, starting with *Kramers* theory and continuing with the more recent *Grote–Hynes* theory for

$k(T)$. Both theories are based on a stochastic dynamical description of the influence of the solvent molecules on the reaction dynamics.

We have added several appendices that give a short introduction to important disciplines such as statistical mechanics and stochastic dynamics, as well as developing more technical aspects like various coordinate transformations. Furthermore, examples and end-of-chapter problems illustrate the theory and its connection to chemical problems.

Contents

1 Introduction

Chemical reactions, the transformation of matter at the atomic level, are distinctive features of chemistry. They include a series of basic processes from the transfer of single electrons or protons to the transfer of groups of nuclei and electrons between molecules, that is, the breaking and formation of chemical bonds. These processes are of fundamental importance to all aspects of life in the sense that they determine the function and evolution in chemical and biological systems.

The transformation from reactants to products can be described at either a phenomenological level, as in classical **chemical kinetics**, or at a detailed molecular level, as in **molecular reaction dynamics**.[1] The former description is based on experimental observation and, combined with chemical intuition, rate laws are proposed to enable a calculation of the rate of the reaction. It does not provide direct insight into the process at a microscopic molecular level. The aim of molecular reaction dynamics is to provide such insight as well as to deduce rate laws and calculate rate constants from basic molecular properties and dynamics. Dynamics is in this context the description of atomic motion under the influence of a force or, equivalently, a potential.

The main objectives of molecular reaction dynamics may be briefly summarized by the following points:

- the microscopic foundation of chemical kinetics;
- state-to-state chemistry and chemistry in real time;
- control of chemical reactions at the microscopic level.

Before we go on and discuss these objectives in more detail, it might be appropriate to consider the relation between molecular reaction dynamics and the science of *physical chemistry*. Normally, physical chemistry is divided into four major branches, as sketched in the figure below (each of these areas are based on fundamental axioms). At the macroscopic level, we have the old disciplines: 'thermodynamics' and 'kinetics'. At the microscopic level we have 'quantum mechanics', and the connection between the two levels is provided by 'statistical mechanics'. Molecular reaction dynamics encompasses (as sketched by the oval) the central branches of physical chemistry, with the exception of thermodynamics.

A few concepts from classical chemical kinetics should be recalled [1]. Chemical change is represented by a reaction scheme. For example,

[1]The roots of molecular reaction dynamics go back to a famous paper by H. Eyring and M. Polanyi, *Z. Phys. Chem.* **B12**, 279 (1931).

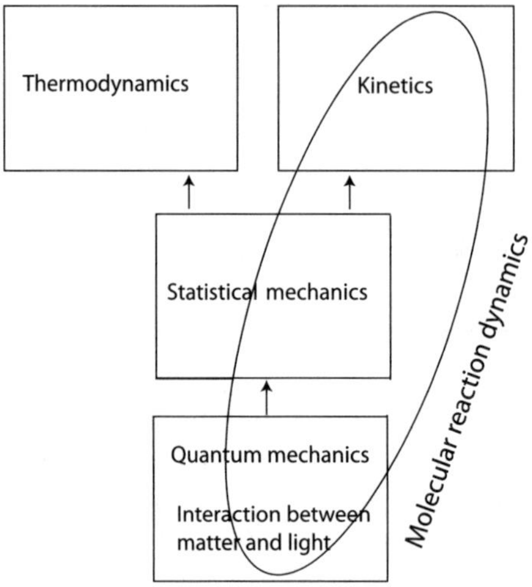

$$2H_2 + O_2 \rightarrow 2H_2O$$

The rate of reaction, R, is the rate of change in the concentration of one of the reactants or products, such as $R = -d[H_2]/dt$, and the rate law giving the relation between the rate and the concentrations can be established experimentally.

This reaction scheme represents, apparently, a simple reaction but it does not proceed as written. That is, the oxidation of hydrogen does not happen in a collision between two H_2 molecules and one O_2 molecule. This is also clear when it is remembered that all the stoichiometric coefficients in such a scheme can be multiplied by an arbitrary constant without changing the content of the reaction scheme. Thus, most reaction schemes show merely the overall transformation from reactants to products without specifying the path taken. The actual path of the reaction involves the formation of intermediate species and includes several elementary steps. These steps are known as *elementary reactions* and together they constitute what is called the *reaction mechanism* of the reaction. It is a great challenge in chemical kinetics to discover the reaction mechanism, that is, to unravel which elementary reactions are involved.

Elementary reactions are reactions that directly express basic chemical events, that is, the making or breaking of chemical bonds. In the gas phase, there are only two types of elementary reactions:[2]

- *unimolecular reaction* (e.g., due to the absorption of electromagnetic radiation);
- *bimolecular reaction* (due to a collision between two molecules);

and in condensed phases, in addition, a third type:

- *bimolecular association/recombination reaction.*

[2]The existence of *trimolecular* reactions is sometimes suggested. For example, $H + OH + M \rightarrow H_2O + M$, where M is a third body. However, the reaction probably proceeds by a two-step mechanism, i.e., (1) $H + OH \rightarrow H_2O^*$, and (2) $H_2O^* + M \rightarrow H_2O + M$, where H_2O^* is an energy-rich water molecule with an energy that exceeds the dissociation limit, and the function of M is to take away the energy. That is, the reaction actually proceeds via bimolecular collisions.

The reaction schemes of elementary reactions are to be taken literally. For example, one of the elementary reactions in the reaction between hydrogen and oxygen is a simple atom transfer:

$$\mathrm{H + O_2 \rightarrow OH + O}$$

In this bimolecular reaction the stoichiometric coefficients are equal to one, meaning that one hydrogen atom collides with one oxygen molecule. Once the reaction mechanism and all the rate constants for the elementary reactions are known, the reaction rates for all species are given by a simple set of coupled first-order differential equations. These equations can be solved quite easily on a computer, and give the concentrations of all species as a function of time. These results may then be compared with experimental results.

From the discussion above, it follows that: **elementary reactions are at the heart of chemistry**. The study of these reactions is the main subject of this book.

The rearrangement of nuclei in an elementary chemical reaction takes place over a distance of a few ångström (1 ångström $= 10^{-10}$ m) and within a time of about 10–100 femtoseconds (1 femtosecond $= 10^{-15}$ s; a femtosecond is to a second what one second is to 32 million years!), equivalent to atomic speeds of the order of 1 km/s. The challenges in molecular reaction dynamics are: (i) to understand and follow in real time the detailed atomic dynamics involved in the elementary processes, (ii) to use this knowledge in the control of these reactions at the microscopic level, e.g., by means of external laser fields, and (iii) to establish the relation between such microscopic processes and macroscopic quantities like the rate constants of the elementary processes.

We consider the detailed evolution of isolated *elementary reactions*[3] in the **gas phase**, for example,

$$\mathrm{A + BC}(n) \longrightarrow \mathrm{AB}(m) + \mathrm{C}$$

At the fundamental level the course of such a reaction between an atom A and a diatomic molecule BC is governed by quantum mechanics. Thus, within this theoretical framework the reaction dynamics at a given collision energy can be analyzed for reactants in a given quantum state (denoted by the quantum number n) and one can extract the transition probability for the formation of products in various quantum states (denoted by the quantum number m). At this level one considers the 'state-to-state' dynamics of the reaction.

When we consider elementary reactions, it should be realized that the outcome of a bimolecular collision can also be non-reactive. Thus,

$$\mathrm{A + BC}(n) \longrightarrow \mathrm{BC}(m) + \mathrm{A}$$

and we distinguish between an *elastic collision* process, if quantum states n and m are identical, and otherwise an *inelastic collision* process. Note that inelastic collisions

[3] An elementary reaction is defined as a reaction that takes place as written in the reaction scheme. We will here distinguish between a *truly* elementary reaction, where the reaction takes place in isolation without any secondary collisions, and the traditional definition of an elementary reaction, where inelastic collisions among the molecules in the reaction scheme (or with container walls) can take place.

correspond to energy transfer between molecules—in the present case, for example, the transfer of relative translational energy between A and BC to vibrational energy in BC.

The realization of an isolated elementary reaction is experimentally difficult. The closest realization is achieved under the highly specialized laboratory conditions of an ultra-high vacuum molecular-beam experiment. Most often collisions between molecules in the gas phase occur, making it impossible to obtain state-to-state specific information because of the energy exchange in such collisions. Instead, thermally-averaged rate constants may be obtained. Thus, energy transfer, that is, inelastic collisions among the reactants, implies that an equilibrium Boltzmann distribution is established for the collision energies and over the internal quantum states of the reactants. A parameter in the equilibrium Boltzmann distribution is the macroscopic temperature T. Under such conditions the well-known rate constant $k(T)$ of chemical kinetics can be defined and evaluated based on the underlying detailed dynamics of the reaction.

The macroscopic rate of reaction is, typically, much slower than the rate that can be inferred from the time it takes to cross the transition states (that is, all the intermediate configurations between reactants and products) because the fraction of reactants with sufficient energy to react is very small at typical temperatures.

Reactions in a **condensed phase** are never isolated but under strong influence of the surrounding solvent molecules. The solvent will modify the interaction between the reactants, and it can act as an energy source or sink. Under such conditions the state-to-state dynamics described above cannot be studied, and the focus is then turned to the evaluation of the rate constant $k(T)$ for elementary reactions. The *elementary reactions* in a solvent include both unimolecular and bimolecular reactions as in the gas phase and, in addition, *bimolecular association/recombination reactions.* That is, an elementary reaction of the type A + BC $\rightarrow$ ABC, which can take place because the products may not fly apart as they do in the gas phase. This happens when the products are not able to escape from the solvent 'cage' and the ABC molecule is stabilized due to energy transfer to the solvent.[4] Note that one sometimes distinguishes between association as an outcome of a bimolecular reaction and recombination as the inverse of unimolecular fragmentation.

On the *experimental* side, the chemical dynamics on the state-to-state level is being studied via molecular-beam and laser techniques [2]. Alternative, and complementary, techniques have been developed in order to study the real-time evolution of elementary reactions [3]. Thus, the time resolution in the observation of chemical reactions has increased dramatically over the last decades. The 'race against time' has recently reached the ultimate femtosecond resolution with the direct observation of chemical reactions as they proceed along the reaction path via transition states from reactants to products. This spectacular achievement was made possible by the development of femtosecond lasers, that is, laser pulses with a duration as short as a few femtoseconds. In a typical experiment two laser pulses are used, a 'pump pulse' and a 'probe

[4]Association/recombination can, under special conditions, also take place in the gas phase (in a single elementary reaction step), e.g., in the form of so-called radiative recombination; see Section 6.5.

pulse'. The first femtosecond pulse initiates a chemical reaction, say the breaking of a chemical bond in a unimolecular reaction, and a second time-delayed femtosecond pulse probe this process. The ultrashort duration of the pump pulse implies that the zero of time is well defined. The probe pulse is, for example, tuned to be in resonance with a particular transition in one of the fragments and, when it is fired at a series of time delays relative to the pump pulse, one can directly observe the formation of the fragment. This type of real-time chemistry is called *femtosecond chemistry* (or simply, femtochemistry). Another interesting aspect of femtosecond chemistry concerns the challenging objective of using femtosecond lasers to control the outcome of chemical reactions, say to break a *particular* bond in a large molecule. This type of control at the molecular level is much more selective than traditional methods for control where only macroscopic parameters like the temperature can be varied. In short, femtochemistry is about the detection and control of transition states, that is, the intermediate short-lived states on the path from reactants to products.

On the *theoretical* side, advances have also been made both in methodology and in concepts. For example, new and powerful techniques for the solution of the time-dependent Schrödinger equation (see Section 1.1) have been developed. New concepts for laser control of chemical reactions have been introduced where, for example, one laser pulse can create a non-stationary nuclear state that can be intercepted or redirected with a second laser pulse at a precisely timed delay.

The theoretical foundation for reaction dynamics is quantum mechanics and statistical mechanics. In addition, in the description of nuclear motion, concepts from classical mechanics play an important role. A few results of molecular quantum mechanics and statistical mechanics are summarized in the next two sections. In the second part of the book, we will return to concepts and results of particular relevance to condensed-phase dynamics.

1.1 Nuclear dynamics: the Schrödinger equation

The reader is assumed to be familiar with some of the basic concepts of quantum mechanics. At this point we will therefore just briefly consider a few central concepts, including the *time-dependent Schrödinger equation* for nuclear dynamics. This equation allows us to focus on the nuclear motion associated with a chemical reaction.

We consider a system of K electrons and N nuclei, interacting through Coulomb forces. The basic equation of motion in quantum mechanics, the *time-dependent Schrödinger equation*, can be written in the form

$$i\hbar\frac{\partial\Psi(\boldsymbol{r}_{\text{lab}},\boldsymbol{R}_{\text{lab}},t)}{\partial t} = (\hat{T}_{\text{nuc}} + \hat{H}_e)\Psi(\boldsymbol{r}_{\text{lab}},\boldsymbol{R}_{\text{lab}},t) \tag{1.1}$$

where i is the imaginary unit, $\hbar = h/(2\pi)$ is the Planck constant divided by 2π, and the wave function depends on $\boldsymbol{r}_{\text{lab}} = (\boldsymbol{r}_1, \boldsymbol{r}_2, \ldots, \boldsymbol{r}_K)$ and $\boldsymbol{R}_{\text{lab}} = (\boldsymbol{R}_1, \boldsymbol{R}_2, \ldots, \boldsymbol{R}_N)$, which denote all electron and nuclear coordinates, respectively, measured relative to a fixed laboratory coordinate system. The operators are

$$\hat{T}_{\text{nuc}} = \sum_{g=1}^{N} \frac{\hat{P}_g^2}{2M_g} \tag{1.2}$$

which is the kinetic energy operator of the nuclei, where $\hat{P}_g = -i\hbar\nabla_g$ is the momentum operator and M_g the mass of the gth nucleus, and $\hat{H}_e$ is the so-called electronic Hamiltonian *including* the internuclear repulsion,

$$\hat{H}_e = \sum_{i=1}^{K} \frac{\hat{p}_i^2}{2m_e} - \sum_{g=1}^{N}\sum_{i=1}^{K} \frac{Z_g e^2}{4\pi\epsilon_0 r_{ig}} + \sum_{i=1}^{K}\sum_{j>i}^{K} \frac{e^2}{4\pi\epsilon_0 r_{ij}} + \sum_{g=1}^{N}\sum_{h>g}^{N} \frac{Z_g Z_h e^2}{4\pi\epsilon_0 r_{gh}} \tag{1.3}$$

where the first term represents the kinetic energy of the electrons, the second term the attraction between electrons and nuclei, the third term the electron–electron repulsion, and the last term the internuclear repulsion. $\hat{p}_i = -i\hbar\nabla_i$ is the momentum operator of the ith electron and m_e its mass, r_{ig} is the distance between electron i and nucleus g, the other distances r_{ij} and r_{gh} have a similar meaning, and $Z_g e$ is the electric charge of the gth nucleus, where Z_g is the atomic number. The Hamiltonian is written in its non-relativistic form, i.e., spin-orbit terms, etc. are neglected. Note that the electronic Hamiltonian does not depend on the absolute positions of the nuclei but only on internuclear distances and the distances between electrons and nuclei.

The translational motion of the particles as a whole (i.e., the center-of-mass motion) can be separated out. This is done by a change of variables from $\boldsymbol{r}_{\text{lab}}, \boldsymbol{R}_{\text{lab}}$ to $\boldsymbol{R}_{\text{CM}}$ and $\boldsymbol{r}, \boldsymbol{R}$, where $\boldsymbol{R}_{\text{CM}}$ gives the position of the center of mass and $\boldsymbol{r}, \boldsymbol{R}$ are internal coordinates that describe the relative position of the electrons with respect to the nuclei and the relative position of the nuclei, respectively. This coordinate transformation implies

$$\Psi(\boldsymbol{r}_{\text{lab}}, \boldsymbol{R}_{\text{lab}}, t) = \Psi(\boldsymbol{R}_{\text{CM}}, t)\Psi(\boldsymbol{r}, \boldsymbol{R}, t) \tag{1.4}$$

where $\Psi(\boldsymbol{R}_{\text{CM}}, t)$ is the wave function associated with the free translational motion of the center of mass, and $\Psi(\boldsymbol{r}, \boldsymbol{R}, t)$ describes the internal motion, given by a time-dependent Schrödinger equation similar to Eq. (1.1). The kinetic energy operators expressed in the internal coordinates take, however, a more complicated form than specified above, which will be described in a subsequent chapter.[5]

Fortunately, a direct solution of Eq. (1.1) is normally not necessary. The electrons are very light particles whereas the nuclei are, at least, about three orders of magnitude heavier. From the point of view of the electronic state, the nuclear positions can be considered as slowly changing external parameters, which means that the electrons experience a slowly changing potential. When the electrons are in a given quantum state (say the electronic ground state) it can be shown that the electronic quantum number, in the following indicated by the subscript i, is unchanged as long as the nuclear motion can be considered as being slow. Thus, no transitions among the electronic states will take place under these conditions. This is the physical basis for the so-called *adiabatic approximation*, which can be written in the form

$$\Psi(\boldsymbol{r}, \boldsymbol{R}, t) = \chi(\boldsymbol{R}, t)\psi_i(\boldsymbol{r}; \boldsymbol{R}) \tag{1.5}$$

[5]Normally, three approximations are introduced in this context: (i) the center of mass is taken to be identical to the center of mass of the nuclei; (ii) the kinetic energy operators of the electrons are taken to be identical to the expression given above, which again means that the nuclei are considered to be infinitely heavy compared to the electrons; and (iii) coupling terms between the kinetic energy operators of the electrons and nuclei, introduced by the transformation, are neglected.

where

$$\hat{H}_e\psi_i(\boldsymbol{r};\boldsymbol{R}) = E_i(\boldsymbol{R})\psi_i(\boldsymbol{r};\boldsymbol{R}) \tag{1.6}$$

$\psi_i(\boldsymbol{r};\boldsymbol{R})$ is the usual stationary electronic wave function, $E_i(\boldsymbol{R})$ is the corresponding *electronic energy* (including internuclear repulsion) which is a function of the nuclear geometry, and $\chi(\boldsymbol{R},t)$ is the wave function for the nuclear motion. Equation (1.6) is solved at fixed values of the nuclear coordinates, as indicated by the ';' in the electronic wave function. Note that the electronic energy is invariant to isotope substitution within the adiabatic approximation, since the electronic Hamiltonian is independent of nuclear masses.

The solution to Eq. (1.6) for the electronic energy in, e.g., a diatomic molecule is well known. In this case there is only one internuclear coordinate and the electronic energy, $E_i(R)$, is consequently represented by a curve as a function of the internuclear distance. For small displacements around the equilibrium bond length $R = R_0$, the curve can be represented by a quadratic function. Thus, when we expand to second order around a minimum at $R = R_0$,

$$\begin{aligned} E(R) &= E(R_0) + \left(\frac{\partial E}{\partial R}\right)_0 (R-R_0) + (1/2)\left(\frac{\partial^2 E}{\partial R^2}\right)_0 (R-R_0)^2 + \cdots \\ &= E(R_0) + (1/2)k(R-R_0)^2 + \cdots \end{aligned} \tag{1.7}$$

where the subscript indicates that the derivatives are evaluated at the minimum, and

$$k = \left(\frac{\partial^2 E}{\partial R^2}\right)_0 \tag{1.8}$$

is the *force constant.* However, when we consider chemical reactions, where chemical bonds are formed and broken, the electronic energy for all internuclear distances is important. The description of the simultaneous making and breaking of chemical bonds leads to multidimensional *potential energy surfaces* that are discussed in Chapter 3.

Substituting Eq. (1.5) into Eq. (1.1), we obtain

$$i\hbar\frac{\partial\chi(\boldsymbol{R},t)}{\partial t} = (\hat{T}_{\rm nuc} + E_i(\boldsymbol{R}) + \langle\psi_i|\hat{T}_{\rm nuc}|\psi_i\rangle_0)\chi(\boldsymbol{R},t) \tag{1.9}$$

where we have used that $\langle\psi_i|\nabla_g|\psi_i\rangle = \nabla_g\langle\psi_i|\psi_i\rangle/2 = 0$, when $\psi_i(\boldsymbol{r};\boldsymbol{R})$ is real, and that the electronic wave function is normalized. The subscript on the matrix element implies that $\hat{T}_{\rm nuc}$ acts only on ψ_i and the matrix element involves an integration over electron coordinates.

The term $\langle\psi_i|\hat{T}_{\rm nuc}|\psi_i\rangle_0$ is normally very small compared to the electronic energy, and may consequently be dropped (the resulting approximation is often referred to as the *Born–Oppenheimer approximation*):

$$\boxed{i\hbar\frac{\partial\chi(\boldsymbol{R},t)}{\partial t} = [\hat{T}_{\rm nuc} + E_i(\boldsymbol{R})]\chi(\boldsymbol{R},t)} \tag{1.10}$$

Equation (1.10) is the fundamental equation of motion within the adiabatic approximation. We see that: *the nuclei move on a potential energy surface given by* *the electronic energy*. Thus, one must first solve for the electronic energy, Eq. (1.6), and subsequently solve the time-dependent Schrödinger equation for the nuclear motion, Eq. (1.10).

The physical implication of the adiabatic approximation is that the electrons remain in a given electronic eigenstate during the nuclear motion. The electrons follow the nuclei, for example, as a reaction proceeds from reactants to products, such that the electronic state 'deforms' in a continuous way without electronic transitions. From a more practical point of view, the approximation implies that we can separate the solutions to the electronic and nuclear motion.

The probability density, $|\chi(\boldsymbol{R}, t)|^2$, may be used to calculate the *reaction probability*. The probability density associated with the nuclear motion of a chemical reaction is illustrated in Fig. 1.1.1. The reaction probability may be evaluated from $P = \int_{\boldsymbol{R}\in\text{Prod}} |\chi(\boldsymbol{R}, t \sim \infty)|^2 d\boldsymbol{R}$, where the integration is restricted to configurations representing the products (Prod), in the example for large B–C distances, R_{BC}. The limit $t \sim \infty$ implies that the probability density obtained long after reaction is used in the integral. In this limit, where the reaction is completed, there is a negligible probability density in the region where R_{AB} as well as R_{BC} are small. In practice, 'long after' is identified as the time where the reaction probability becomes independent of time and, typically, this situation is established after a few hundred femtoseconds.

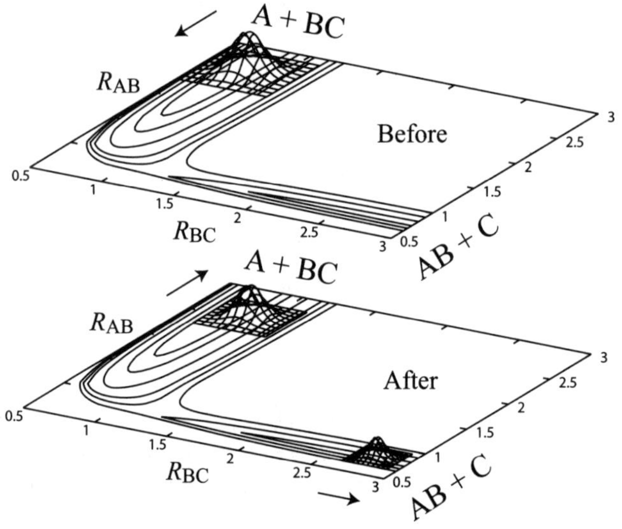

Fig. 1.1.1 Schematic illustration of the probability density, $|\chi(\boldsymbol{R}, t)|^2$, associated with a chemical reaction, A + BC $\rightarrow$ AB + C. The contour lines represent the potential energy surface (see Chapter 3), and the probability density is shown at two times: before the reaction where only reactants are present, and after the reaction where products as well as reactants are present. The arrows indicate the direction of motion associated with the relative motion of reactants and products. (Note that, due to the finite uncertainty in the A–B distance, R_{AB}, there is some uncertainty in the initial relative translational energy of A + BC.)

The general time-dependent solutions to Eq. (1.10) are denoted as *non-stationary states*. They can be expanded in terms of the eigenstates $\phi_{\boldsymbol{n}}(\boldsymbol{R})$ of the Hamiltonian

$$\hat{H} = \hat{T}_{\text{nuc}} + E_i(\boldsymbol{R}) \tag{1.11}$$

with $\phi_{\boldsymbol{n}}(\boldsymbol{R})$ given by

$$\hat{H}\phi_{\boldsymbol{n}}(\boldsymbol{R}) = \mathcal{E}_{\boldsymbol{n}}\phi_{\boldsymbol{n}}(\boldsymbol{R}) \tag{1.12}$$

where $\boldsymbol{n}$ is a set of quantum numbers that fully specify the eigenstates of the nuclear Hamiltonian. The general solution can then be written in the form[6] (as can be checked by direct substitution into Eq. (1.10))

$$\begin{aligned} \chi(\boldsymbol{R}, t) &= \exp(-i\hat{H}t/\hbar)\chi(\boldsymbol{R}, 0) \\ &= \exp(-i\hat{H}t/\hbar)\sum_{\boldsymbol{n}} c_{\boldsymbol{n}}\phi_{\boldsymbol{n}}(\boldsymbol{R}) \\ &= \sum_{\boldsymbol{n}} c_{\boldsymbol{n}}\phi_{\boldsymbol{n}}(\boldsymbol{R})e^{-i\mathcal{E}_{\boldsymbol{n}}t/\hbar} \end{aligned} \tag{1.13}$$

Each state in the sum, $\Phi_{\boldsymbol{n}}(\boldsymbol{R}, t) = \phi_{\boldsymbol{n}}(\boldsymbol{R})e^{-i\mathcal{E}_{\boldsymbol{n}}t/\hbar}$, is denoted as a *stationary state*, because all expectation values $\langle\Phi_{\boldsymbol{n}}(t)|\hat{A}|\Phi_{\boldsymbol{n}}(t)\rangle$ (e.g., for the operator representing the position) are independent of time. That is, there is no observable time dependence associated with a single stationary state. Equation (1.13) shows that the non-stationary time-dependent solutions can be written as a superposition of the stationary solutions, with coefficients that are independent of time. The coefficients are determined by the way the system was prepared at $t = 0$.

The eigenstates of $\hat{H}$ are well known, for *non-interacting* molecules, say the reactants A + BC (an atom and a diatomic molecule), giving quantized vibrational and rotational energy levels. Within the so-called rigid-rotor approximation where couplings between rotation and vibration are neglected, Eq. (1.11) can for non-interacting molecules be written in the form

$$\hat{H}^0 = \hat{H}_{\text{trans}} + \hat{H}_{\text{vib}} + \hat{H}_{\text{rot}} \tag{1.14}$$

where $\hat{H}_{\text{trans}}$ represents the free relative motion of A and BC, and $\hat{H}_{\text{vib}}$ and $\hat{H}_{\text{rot}}$ correspond to the vibration and rotation of BC, respectively. This form of the Hamiltonian, with a sum of independent terms, implies that the eigenstates take the form

$$\phi_{\boldsymbol{n}}^0(\boldsymbol{R}) = \phi_{\text{trans}}(\boldsymbol{R}_{\text{rel}})\phi_{\text{vib}}^n(R)\phi_{\text{rot}}^J(\theta, \phi) \tag{1.15}$$

where the functions in the product are eigenfunctions corresponding to translation, vibration, and rotation. The eigenvalues are

$$\mathcal{E}_{\boldsymbol{n}}^0 = E_{\text{tr}} + E_n + E_J \tag{1.16}$$

That is, the total energy is the sum of the energies associated with translation, vibration, and rotation. The translational energy is continuous (as in classical mechanics).

[6] A function of an operator is defined through its (Taylor) power series. The summation sign should really be understood as a summation over discrete quantum numbers and an integration over continuous labels corresponding to translational motion.

For a one-dimensional *harmonic oscillator*, with the potential in Eq. (1.7), the Hamiltonian is

$$\hat{H}_{\text{vib}} = -\frac{\hbar^2}{2\mu}\frac{\partial^2}{\partial R^2} + (1/2)k(R - R_0)^2 \tag{1.17}$$

where μ is the reduced mass of the two nuclei in BC. This Hamiltonian has the well-known eigenvalues

$$E_n = \hbar\omega(n + 1/2)\,, \ n = 0, 1, 2, \ldots \tag{1.18}$$

where $\omega = 2\pi\nu = \sqrt{k/\mu}$, and ν is identical to the frequency of the corresponding classical harmonic motion. The vibrational energy levels of the one-dimensional harmonic oscillator are illustrated in Fig. 1.1.2. Note, for example, that the vibrational zero-point energy changes under isotope substitution since the reduced mass will change. We shall see, later on, that this purely quantum mechanical effect can change the magnitude of the macroscopic rate constant $k(T)$.

The *rigid-rotor* Hamiltonian for a diatomic molecule with the moment of inertia $I = \mu R_0^2$ is

$$\hat{H}_{\text{rot}} = \frac{\hat{L}^2}{2I} \tag{1.19}$$

where $\hat{L}$ is the angular momentum operator, giving the well-known eigenvalues

$$E_J = \hbar^2 J(J + 1)/(2I)\,, \ J = 0, 1, 2, \ldots \tag{1.20}$$

with the degeneracy $\omega_J = 2J + 1$.

Typically, the energy spacing between rotational energy levels is much smaller than the energy spacing between vibrational energy levels which, in turn, is much smaller than the energy spacing between electronic energy levels:

$$\Delta E_{\text{rot}} \ll \Delta E_{\text{vib}} \ll \Delta E_{\text{elec}} \tag{1.21}$$

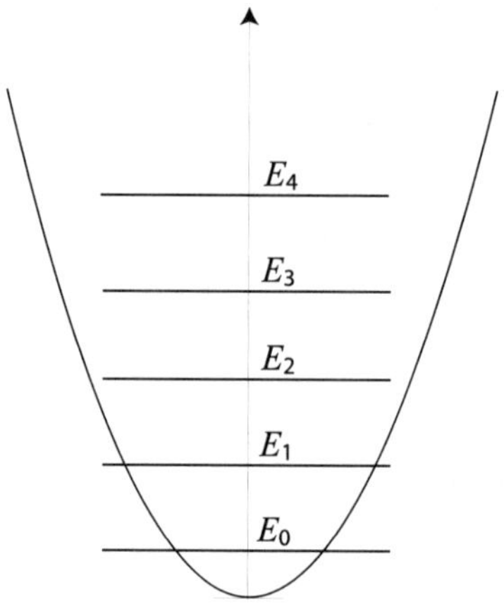

Fig. 1.1.2 The energy levels of a one-dimensional harmonic oscillator. The zero-point energy $E_0 = \hbar\omega/2$, where for a diatomic molecule $\omega = \sqrt{k/\mu}$, k is the force constant, and μ is the reduced mass.

where the energy spacing is defined as the energy difference between adjacent energy levels.

It is possible to solve Eq. (1.10) numerically for the nuclear motion associated with chemical reactions and to calculate the reaction probability including detailed *state-to-state* reaction probabilities (see Section 4.2). However, with the present computer technology such an approach is in practice limited to systems with a small number of degrees of freedom.

For practical reasons, a quasi-classical approximation to the quantum dynamics described by Eq. (1.10) is often sought. In the quasi-classical trajectory approach (discussed in Section 4.1) only one aspect of the quantum nature of the process is incorporated in the calculation: the initial conditions for the trajectories are sampled in accord with the quantized vibrational and rotational energy levels of the reactants.

Obviously, purely quantum mechanical effects cannot be described when one replaces the time evolution by classical mechanics. Thus, the quasi-classical trajectory approach exhibits, e.g., the following deficiencies: (i) zero-point energies are not conserved properly (they can, e.g., be converted to translational energy), (ii) quantum mechanical tunneling cannot be described.

Finally, it should be noted that the motion of the nuclei is not always confined to a single electronic state (as assumed in Eq. (1.5)). This situation can, e.g., occur when two potential energy surfaces come close together for some nuclear geometry. The dynamics of such processes are referred to as *non-adiabatic*. When several electronic states are in play, Eq. (1.10) must be replaced by a matrix equation with a dimension given by the number of electronic states (see Section 4.2). The equation contains coupling terms between the electronic states, implying that the nuclear motion in all the electronic states is coupled.

1.2 Thermal equilibrium: the Boltzmann distribution

Statistical mechanics gives the relation between microscopic information such as quantum mechanical energy levels and macroscopic properties. Some important statistical mechanical concepts and results are summarized in Appendix A. Here we will briefly review one central result: the *Boltzmann distribution* for thermal equilibrium.

For reactants in complete thermal equilibrium, the probability of finding a BC molecule in a specific quantum state, n, is given by the Boltzmann distribution (see Appendix A.1). Thus, in the special case of non-interacting molecules the probability, $p_{\mathrm{BC}(n)}$, of finding a BC molecule in the internal (electronic, vibrational, and rotational) quantum states with energy E_n is

$$p_{\mathrm{BC}(n)} = \frac{\omega_n}{Q_{\mathrm{BC}}} \exp(-E_n/k_B T) \tag{1.22}$$

where ω_n is the degeneracy of the nth quantum level (i.e., the number of states with the same energy E_n) and Q_{BC} is the 'internal' partition function of the BC molecule where center-of-mass motion is excluded, given by

$$Q_{\mathrm{BC}} = \sum_n \omega_n \exp(-E_n/k_B T) \tag{1.23}$$

i.e., a weighted sum over all energy levels, where the weights are proportional to the occupation probabilities of each level.

The distribution depends on the temperature; only the lowest energy level is populated at $T = 0$. When the temperature is raised, higher energy levels will also be populated. The probability of populating high energy levels decreases exponentially with the energy.

The Boltzmann distribution is illustrated in Fig. 1.2.1 for the vibrational states of a one-dimensional harmonic oscillator with the frequency $\omega = 2\pi\nu$, where the energy levels are given by Eq. (1.18), and in Fig. 1.2.2 for the rotational states of a linear molecule with the moment of inertia I, where the energy levels are given by Eq. (1.20) with the degeneracy $\omega_J = 2J + 1$.

The Boltzmann distribution for free translational motion takes a special form (see Appendix A.2.1), since the energy is continuous in this case. The probability of finding a translational energy in the range $E_{\rm tr}, E_{\rm tr} + dE_{\rm tr}$ is given by

$$P(E_{\rm tr})dE_{\rm tr} = 2\pi \left(\frac{1}{\pi k_B T}\right)^{3/2} \sqrt{E_{\rm tr}} \exp(-E_{\rm tr}/k_B T) dE_{\rm tr} \tag{1.24}$$

This function is shown in Fig. 1.2.3.

These distribution functions show that, at a given temperature, many energy levels will be populated when the energy spacing is small (which is the case for translational and rotational degrees of freedom), whereas only a few of the lowest energy levels will have a substantial population when the energy spacing is large (vibrational and electronic degrees of freedom). Furthermore, it is often found that the degeneracy

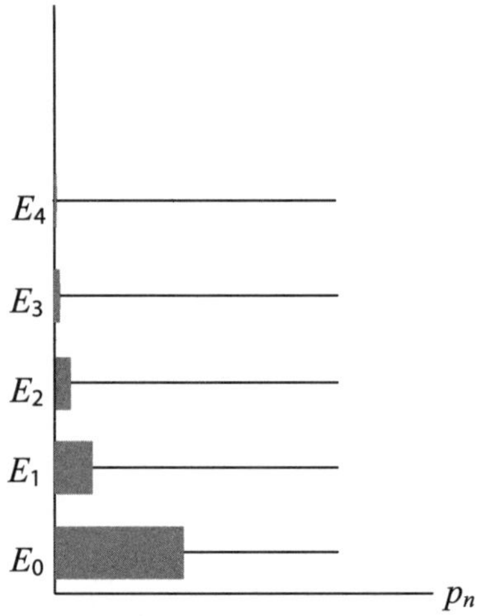

Fig. 1.2.1 The Boltzmann distribution for a system with equally-spaced energy levels E_n and identical degeneracy ω_n of all levels ($T > 0$). This figure gives the population of states at the temperature T for a harmonic oscillator.

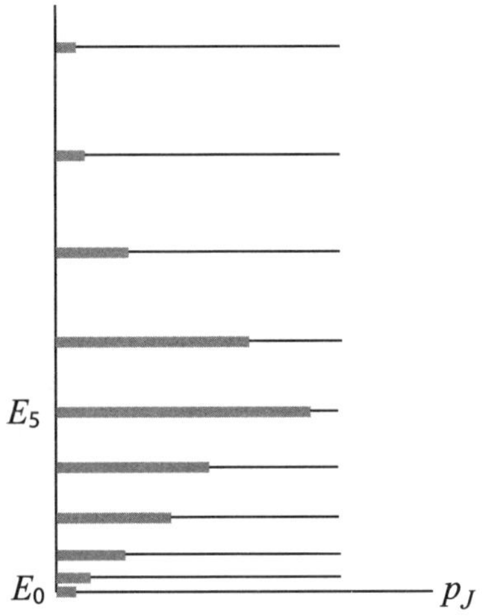

Fig. 1.2.2 The Boltzmann distribution for the rotational energy of a linear molecule ($T > 0$). The maximum is at $J_{\text{max}} = \sqrt{Ik_BT}/\hbar - 1/2$ (rounded off to the closest integer).

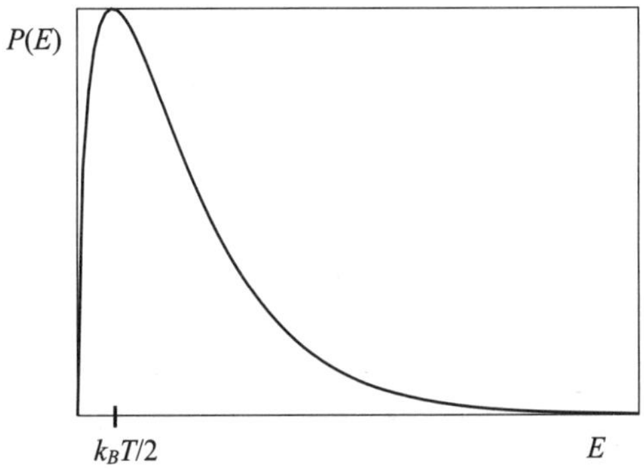

Fig. 1.2.3 The Boltzmann distribution for free translational motion. The maximum is at $E_{\text{max}} = k_BT/2$.

of the energy levels is increasing as a function of the energy. The maxima of the distributions will therefore often be found for energies above the ground-state energy. Thus, for the translational energy, the maximum is at the energy $k_BT/2$.

An *elementary reaction* is in classical chemical kinetics defined under conditions where energy transfer among the molecules in the reaction scheme or with surrounding solvent molecules can take place. In this case, we write

$$\text{A} + \text{BC} \longrightarrow \text{AB} + \text{C}$$

for an elementary reaction. We have deleted the quantum numbers associated with the molecules, and it is understood that the states are populated according to the

Boltzmann distribution. Furthermore, when the reaction takes place, we will normally assume that the thermal equilibrium among the reactants can be maintained at all times.

Further reading/references

[1] J.I. Steinfeld, J.S. Francisco, and W.L. Hase, *Chemical kinetics and dynamics*, second edition (Prentice Hall, 1999)

[2] D.R. Herschbach, *Angewandte Chemie-international edition in English* **26**, 1221 (1987). Y.T. Lee, *Science* **236**, 793 (1987). J.C. Polanyi, *Science* **236**, 680 (1987)

[3] A.H. Zewail, *Scientific American*, Dec. 1990, page 40. A.H. Zewail, *J. Phys. Chem.* **104**, 5660 (2000)

[4] R.D. Levine, *Molecular reaction dynamics* (Cambridge University Press, 2005)

Problems

1.1 Show that Eq. (1.13) is a solution to Eq. (1.10), and that Eqs (1.15) and (1.16) are eigenstates and eigenvalues of the Hamiltonian in Eq. (1.14).

1.2 Show that the maximum in the Boltzmann distribution for the rotational energy of a linear molecule is at $J_{\rm max} = \sqrt{Ik_BT}/\hbar - 1/2$ (when J is considered as a continuous variable).

1.3 Consider the Boltzmann distribution for free translational motion, and calculate the probability of finding translational energies that exceed $E = E^*$. Compare with the expression $\exp[-E^*/(k_BT)]$.
Use the integral: $\frac{4}{\sqrt{\pi}}\int_x^\infty u^2 e^{-u^2}du = 1 - {\rm erf}(x) + \frac{2x}{\sqrt{\pi}}e^{-x^2}$,
where the error function is defined as: ${\rm erf}(x) = \frac{2}{\sqrt{\pi}}\int_0^x e^{-u^2}du$.
Note that ${\rm erf}(0) = 0$, ${\rm erf}(0.5) = 0.5205$, ${\rm erf}(1.0) = 0.8427$, ${\rm erf}(2.0) = 0.9953$, and ${\rm erf}(\infty) = 1$.

1.4 Elementary concepts of probability and statistics play an important role in this book. Thus, these concepts are an integral part of, e.g., quantum mechanics and statistical mechanics. The probability that some continuous variable x lies between x and $x + dx$ is denoted by $P(x)dx$. Often we refer to $P(x)$ as the probability distribution for x (although $P(x)$ strictly speaking is a probability density). The *average value* or *mean value* of a variable x, which can take any value between $-\infty$ and ∞, is defined by

$$\langle x\rangle = \int_{-\infty}^{\infty} xP(x)dx$$

(a) Calculate the average translational energy using Eq. (1.24).
The *variance* of x is defined by

$$\sigma_x^2 = \langle (x - \langle x\rangle)^2\rangle$$

where σ_x is called the *standard deviation*. It is a measure of the spread of the distribution about its mean value.

(b) Show that $\sigma_x^2 = \langle x^2 \rangle - \langle x \rangle^2$, where $\langle x^2 \rangle$ is the average value of x^2.

(c) Calculate the standard deviation associated with the Boltzmann distribution for translational motion, Eq. (1.24).

In connection with the evaluation of the integrals the Gamma function $\Gamma(n)$ is useful. It is defined as

$$\Gamma(n) = \int_0^\infty x^{n-1} \exp(-x) dx, \quad n > 0$$

with the special values

$\Gamma(n+1) = n!$, where $n = 1, 2, \ldots,$

$\Gamma(n + 1/2) = \dfrac{1 \cdot 3 \cdot 5 \cdots (2n-1)}{2^n} \sqrt{\pi}$, where $n = 1, 2, \ldots.$

Part I

Gas-phase dynamics

2

From microscopic to macroscopic descriptions

Key ideas and results

In this chapter we consider bimolecular reactions from both a microscopic and a macroscopic point of view and thereby derive a theoretical expression for the macroscopic phenomenological rate constant. That is, a relation between molecular reaction dynamics and chemical kinetics is established.

The outcome of an *isolated* (microscopic) reactive scattering event can be specified in terms of an intrinsic fundamental quantity: the *reaction cross-section*. The cross-section is an effective *area* that the reactants present to each other in the scattering process. It depends on the quantum states of the molecules as well as the relative speed of the reactants, and it can be calculated from the collision dynamics (to be described in Chapter 4).

In this chapter, we define the cross-section and derive its relation to the rate constant. We show the following.

- The macroscopic rate constant is related to the relative speed of the reactants and the reaction cross-section, and the expression contains a weighted average over all possible quantum states and velocities of the reactants, and sums and integrals over all possible quantum states and velocities of the products.
- Specialized to thermal equilibrium, the velocity distributions for the molecules are the Maxwell–Boltzmann distribution (a special case of the general Boltzmann distribution law). The expression for the rate constant at temperature T, $k(T)$, can be reduced to an integral over the relative speed of the reactants. Also, as a consequence of the time-reversal symmetry of the Schrödinger equation, the ratio of the rate constants for the forward and the reverse reaction is equal to the equilibrium constant (detailed balance).

In chemical kinetics, we learn that an elementary bimolecular reaction,

$$\mathrm{A} + \mathrm{B} \longrightarrow \text{products} \tag{2.1}$$

obeys a second-order rate law, given by

$$-\frac{d[\mathrm{A}]}{dt} = k[\mathrm{A}][\mathrm{B}] \tag{2.2}$$

where $k \equiv k(T)$ is the temperature-dependent bimolecular rate constant. The purpose of the following chapters (Chapters 2–6) is to obtain an in-depth understanding of this relation and the factors that determine $k(T)$.

2.1 Cross-sections and rate constants

We begin by establishing the relation between the so-called reaction cross-section σ_R and the bimolecular rate constant. Let us consider an elementary *gas-phase* reaction,

$$\mathrm{A}(i, \boldsymbol{v}_\mathrm{A}) + \mathrm{B}(j, \boldsymbol{v}_\mathrm{B}) \rightarrow \mathrm{C}(l, \boldsymbol{v}_\mathrm{C}) + \mathrm{D}(m, \boldsymbol{v}_\mathrm{D}) \tag{2.3}$$

where an A and a B molecule collide,[1] and a C and D molecule are formed. The reactant molecule A is prior to the collision in a given internal quantum state i, which specifies a set of quantum numbers corresponding to the rotational, vibrational, and electronic state of the molecule, and moves with velocity $\boldsymbol{v}_\mathrm{A}$ relative to some laboratory fixed coordinate system (the velocity is specified by a vector with a given direction and length, $|\boldsymbol{v}_\mathrm{A}|$, which is the speed). Reactant molecule B is likewise in a given internal quantum state j and moves with velocity $\boldsymbol{v}_\mathrm{B}$. The product molecules move with velocities $\boldsymbol{v}_\mathrm{C}$ and $\boldsymbol{v}_\mathrm{D}$, and end up in internal quantum states as specified by the quantum numbers l and m, respectively. These conditions are readily specified in a theoretical calculation of the reaction but difficult to realize in an experiment, because inelastic molecular collisions will upset the detailed specification of the molecular states. The requirements of an experimental set-up for the investigation of the chemical reaction in Eq. (2.3) may be summarized in the following way.

- Establishment and maintenance of two molecular beams, where the molecules move in a specified direction with a specified speed and are in a specified internal quantum state.
- Detection of internal quantum states, direction of motion, and speed of product molecules after the collision.
- Single-collision conditions, that is, there is one and just one collision in the reaction zone defined as the zone where the beams cross, and no collisions prior to or after this collision.

These requirements can be met in a so-called crossed *molecular-beam experiment*, which is sketched in Fig. 2.1.1. Here we can generate beams of molecules with well-defined velocities and it is possible to determine the speed of the product molecules, e.g., $v_\mathrm{C} = |\boldsymbol{v}_\mathrm{C}|$, by the so-called *time-of-flight* technique. The elimination of *multiple scattering* in the reaction zone and collisions in the beams are obtained by doing the experiments in high vacuum, that is, at very low pressures.

In an experiment, we can monitor the number of product molecules, C or D, emerging in a space angle $d\Omega$ around the direction Ω; $d\Omega$ is given by the physical design of the detector and Ω by its position (Ω is conveniently specified by the two polar angles θ and ϕ). This is the simplest analysis of a scattering process, where we just count the number of product molecules independent of their internal state

[1]The word 'collision' should not be taken too literally, since molecules are not, say, hard spheres where it is straightforward to count the 'hits'. Thus, a 'collision' should really be interpreted as the broader term 'a scattering event'.

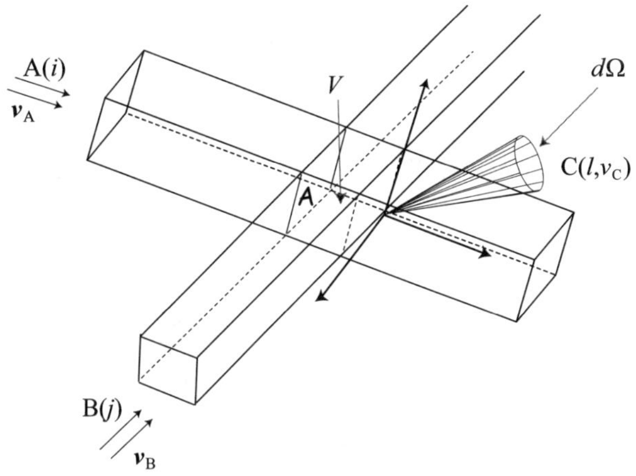

Fig. 2.1.1 Idealized molecular-beam experiment for the reaction $A(i, \boldsymbol{v}_A) + B(j, \boldsymbol{v}_B) \rightarrow C(l, \boldsymbol{v}_C) + D(m, \boldsymbol{v}_D)$. The coordinate system is fixed in the laboratory. The reactants move with the relative speed $v = |\boldsymbol{v}_A - \boldsymbol{v}_B|$.

and speed. In a more advanced analysis, one may use the time-of-flight technique to analyze the speed of product molecules, and only monitor products with a certain speed, that is, C molecules with a speed in the range $v_C, v_C + dv_C$ and D molecules with a speed in the range $v_D, v_D + dv_D$. A still more advanced detection method also allows for the detection of the quantum states of the product molecules. This is the ultimate degree of specification of a scattering experiment.

At sufficiently low pressure (as in the beam experiment) where an A molecule only collides with one single B molecule in the reaction zone, it will hold that the number of product molecules is proportional to the number of collisions between A and B molecules. Clearly, that number depends on the relative speed of the two molecules, $v = |\boldsymbol{v}_A - \boldsymbol{v}_B|$, the time interval dt, and the number of B molecules. Therefore, if we assume that the number density (number/m^3) of B molecules in quantum state j and with velocity $\boldsymbol{v}_B$ is $n_{B(j,\boldsymbol{v}_B)}$, and that the flux density of A molecules relative to the B molecules is $J_{A(i,v)}$ (number/(m^2 s)), then the number of collisions between A and B in the time interval dt is proportional to $n_{B(j,\boldsymbol{v}_B)} V J_{A(i,v)} \mathcal{A} dt$. Here V is the volume of the reaction zone (see Fig. 2.1.1), and $\mathcal{A}$ the cross-sectional area of the beam of A molecules.

In the experiment sketched in Fig. 2.1.1, we monitor the number of product molecules, $C(l, v_C)$, emerging in the space angle $d\Omega$ around the direction Ω, with the speed in the range $v_C, v_C + dv_C$, and in the internal quantum state specified by l. Furthermore, let us for the moment assume that we can also detect the state of the product molecule D, $m, \boldsymbol{v}_D$, as specified in Eq. (2.3). The number of product molecules, $dN_{C(l,v_C,v_C+dv_C)}(\Omega, \Omega + d\Omega, t, t + dt)\big|_{m,\boldsymbol{v}_D}$, registered in the detector in the time interval dt, given that D is in the state $m, \boldsymbol{v}_D$, may therefore be written as

$$
\begin{aligned}
dN_{C(l,v_C,v_C+dv_C)}(\Omega,\Omega+d\Omega,t,t+dt)\big|_{m,\boldsymbol{v}_D} &\equiv \left.\frac{d^3N_{C(l,v_C)}(\Omega,t)}{dv_C d\Omega dt}\right|_{m,\boldsymbol{v}_D} dv_C d\Omega dt \\
&= \mathcal{P}(ij,v|ml,v_C,\Omega,\boldsymbol{v}_D;t)n_{B(j,\boldsymbol{v}_B)}J_{A(i,v)}V\mathcal{A}\,dv_C d\Omega dt
\end{aligned}
\tag{2.4}
$$

where we have introduced the factor $\mathcal{P}(ij,v|ml,v_C,\Omega,\boldsymbol{v}_D;t)$, which is proportional to the *probability* that a product molecule is formed and found in the specified state. Note that on the left-hand side of the vertical bar in the argument list we have written the quantum numbers and the relative speed that specifies the state of the reactants, whereas the quantum numbers and velocities on the right-hand side specify the state of the products. The notation $|_{m,\boldsymbol{v}_D}$ implies that the number of C molecules in the specified state is counted only when D is in the state $m,\boldsymbol{v}_D$.

The complete degree of specification of a scattering experiment is rarely realized in an actual experiment and, normally, we will just monitor the number of C molecules in the specified state, irrespective of the quantum state and velocity of D. In order to obtain that quantity, we integrate over $\boldsymbol{v}_D$ and sum over m in Eq. (2.4). Thus,

$$
\frac{d^3N_{C(l,v_C)}(\Omega,t)}{dv_C d\Omega dt}dv_C d\Omega dt = \mathcal{P}(ij,v|l,v_C,\Omega;t)n_{B(j,\boldsymbol{v}_B)}J_{A(i,v)}V\mathcal{A}\,dv_C d\Omega dt \tag{2.5}
$$

where $\mathcal{P}(ij,v|l,v_C,\Omega;t) = \sum_m \int_{\text{all }\boldsymbol{v}_D} \mathcal{P}(ij,v|ml,v_C,\Omega,\boldsymbol{v}_D;t)d\boldsymbol{v}_D$. By division with $V dv_C d\Omega dt$, we obtain an expression for the number of product molecules per reaction zone volume, per space angle, per time unit, and per unit speed, $d^3n_{C(l,v_C)}/(dv_C d\Omega dt)$, where $n_{C(l,v_C)} = N_{C(l,v_C)}/V$ is the number density of product molecules in the reaction zone. We find

$$
\frac{d^3n_{C(l,v_C)}(\Omega,t)}{dv_C d\Omega dt} = \mathcal{P}(ij,v|l,v_C,\Omega;t)\mathcal{A}\,n_{B(j,\boldsymbol{v}_B)}J_{A(i,v)} \tag{2.6}
$$

The time dependence of the number of particles in the third-order differential and in the probability can be dropped, since the experiments are typically conducted under stationary conditions. Thus,

$$
\boxed{
\begin{aligned}
\frac{d^3n_{C(l,v_C)}(\Omega)}{dv_C d\Omega dt} &= \mathcal{P}(ij,v|l,v_C,\Omega)\mathcal{A}\,n_{B(j,\boldsymbol{v}_B)}J_{A(i,v)} \\
&\equiv \left(\frac{d^2\sigma_R}{dv_C d\Omega}\right)(ij,v|l,v_C,\Omega)n_{B(j,\boldsymbol{v}_B)}J_{A(i,v)}
\end{aligned}
}
\tag{2.7}
$$

where we have replaced $\mathcal{P}\mathcal{A}$ by $\left(\frac{d^2\sigma_R}{dv_C d\Omega}\right)(ij,v|l,v_C,\Omega)$, which defines the *differential reactive scattering cross-section.* It has the dimension of an *area* per unit speed, per unit space angle, because $\mathcal{P}$ is a probability density, i.e., $\mathcal{P}dv_C d\Omega$ is dimensionless. Since $n_{B(j,\boldsymbol{v}_B)}$, $J_{A(i,v)}$, and $d^3n_{C(l)}/(dv_C d\Omega dt)$ are intensive properties (i.e., independent of the size of the system), $\mathcal{P}\mathcal{A}$ and $d^2\sigma_R/(dv_C d\Omega)$ must therefore also be intensive properties independent of the beam geometries. The differential cross-section is a function of the quantum states (ijl), the relative speed $v = |\boldsymbol{v}_A - \boldsymbol{v}_B|$ of the reactants, and the *continuous* velocity of the product, specified by the space angle Ω and the speed v_C. Since it is an intensive property of the chemical reaction, it is often used to report the results of scattering experiments. Physically, the cross-section represents an effective area that the reactants present to each other in connection with a scattering process.

We now need expressions for the relative flux of A molecules with regard to the B molecules, and the number of B molecules. Let us introduce the following notation that allows for any distribution of velocities in the molecular beams:

number density of A(i) with velocity in the range $\boldsymbol{v}_{\rm A}$, $\boldsymbol{v}_{\rm A} + d\boldsymbol{v}_{\rm A}$: $n_{{\rm A}(i)} f_{{\rm A}(i)}(\boldsymbol{v}_{\rm A}) d\boldsymbol{v}_{\rm A}$

number density of B(j) with velocity in the range $\boldsymbol{v}_{\rm B}$, $\boldsymbol{v}_{\rm B} + d\boldsymbol{v}_{\rm B}$: $n_{{\rm B}(j)} f_{{\rm B}(j)}(\boldsymbol{v}_{\rm B}) d\boldsymbol{v}_{\rm B}$

where $n_{{\rm A}(i)}$ is the number density (number/m^3) of A in quantum state i, irrespective of their velocity, and $f_{{\rm A}(i)}(\boldsymbol{v}_{\rm A}) d\boldsymbol{v}_{\rm A}$ is the normalized velocity probability distribution of A(i), that is, the probability of finding an A(i) with velocity in the range from $\boldsymbol{v}_{\rm A}$ to $\boldsymbol{v}_{\rm A} + d\boldsymbol{v}_{\rm A}$, and $n_{{\rm B}(j)}$ is the number density (number/m^3) of B molecules in quantum state j. The product of the incoming flux, $J_{{\rm A}(i,v)}$, and the number of B molecules may then be expressed as

$$J_{{\rm A}(i,v)} n_{{\rm B}(j,\boldsymbol{v}_{\rm B})} = v n_{{\rm A}(i)} f_{{\rm A}(i)}(\boldsymbol{v}_{\rm A}) d\boldsymbol{v}_{\rm A} \times n_{{\rm B}(j)} f_{{\rm B}(j)}(\boldsymbol{v}_{\rm B}) d\boldsymbol{v}_{\rm B} \tag{2.8}$$

and we obtain from Eq. (2.7) the following expression:

$$\frac{d^3 n_{{\rm C}(l,v_{\rm C})}(\Omega)}{dv_{\rm C} d\Omega dt} = \left(\frac{d^2\sigma_R}{dv_{\rm C} d\Omega}\right)(ij, v|l, v_{\rm C}, \Omega)\, v n_{{\rm A}(i)} f_{{\rm A}(i)}(\boldsymbol{v}_{\rm A}) n_{{\rm B}(j)} f_{{\rm B}(j)}(\boldsymbol{v}_{\rm B}) d\boldsymbol{v}_{\rm A} d\boldsymbol{v}_{\rm B} \tag{2.9}$$

Note that the idealized beam experiment corresponds to 'sharp' distributions in the velocities, which mathematically can be expressed in terms of the so-called delta function[2] $\delta(x)$. For example, if the velocity of all the A molecules is $\boldsymbol{v}^0_{\rm A}$, we have $f_{{\rm A}(i)}(\boldsymbol{v}_{\rm A}) = \delta(\boldsymbol{v}_{\rm A} - \boldsymbol{v}^0_{\rm A})$.

Often we are not interested in the distribution of speeds of the product molecules, so we may accordingly multiply both sides of the equation by $dv_{\rm C}$ and integrate over the speed:

$$\frac{d^2 n_{{\rm C}(l)}(\Omega)}{d\Omega dt} = \left(\frac{d\sigma_R}{d\Omega}\right)(ij, v|l, \Omega)\, v n_{{\rm A}(i)} f_{{\rm A}(i)}(\boldsymbol{v}_{\rm A}) n_{{\rm B}(j)} f_{{\rm B}(j)}(\boldsymbol{v}_{\rm B}) d\boldsymbol{v}_{\rm A} d\boldsymbol{v}_{\rm B} \tag{2.10}$$

where $\left(\frac{d\sigma_R}{d\Omega}\right) = \int_0^\infty \left(\frac{d^2\sigma_R}{dv_{\rm C} d\Omega}\right) dv_{\rm C}$.

Likewise, if we multiply both sides of the equation by $d\Omega$ and integrate over all Ω, we get an expression for the total reaction rate of the *state-to-state* reaction as specified by the internal quantum numbers and the relative speed of the reactants:

$$\frac{dn_{{\rm C}(l)}}{dt} = \sigma_R(ij, v|l) v f_{{\rm A}(i)}(\boldsymbol{v}_{\rm A}) f_{{\rm B}(j)}(\boldsymbol{v}_{\rm B}) d\boldsymbol{v}_{\rm A} d\boldsymbol{v}_{\rm B} n_{{\rm A}(i)} n_{{\rm B}(j)} \tag{2.11}$$

where

[2]The delta function has the property $\int_{-\infty}^{\infty} \delta(x-x') f(x) dx = f(x')$, where $f(x)$ is an arbitrary function. Additional properties of the delta function are described, e.g., in many textbooks on quantum mechanics.

$$\sigma_R(ij, v|l) = \int_0^{4\pi} \left(\frac{d\sigma_R}{d\Omega}\right)(ij, v|l, \Omega)d\Omega \tag{2.12}$$

is the integrated cross-section, often referred to as the *total cross-section* for the state-to-state reaction as specified by the internal quantum numbers.[3] Since the cross-section is still resolved with respect to the quantum states, it is also referred to as a *partial cross-section*. The various cross-sections are summarized in the following table:

	Reaction cross-section	Dimension
Differential	$\left(\frac{d^2\sigma_R}{dv_C d\Omega}\right)(ij, v\|l, v_C, \Omega)$	area/(speed × space angle)
Differential	$\left(\frac{d\sigma_R}{d\Omega}\right)(ij, v\|l, \Omega)$	area/(space angle)
Partial	$\sigma_R(ij, v\|l)$	area

Equation (2.11) has the form of a normal rate equation for the isolated bimolecular reaction given by (compare with Eq. (2.3))

$$\mathrm{A}(i, \boldsymbol{v}_\mathrm{A}) + \mathrm{B}(j, \boldsymbol{v}_\mathrm{B}) \rightarrow \mathrm{C}(l) + \mathrm{D} \tag{2.13}$$

This rate, at the state-to-state level, can be measured provided that all inelastic collisions are eliminated. Note that the rate constant (with the proper unit for a bimolecular reaction, volume/time) can be identified as the cross-section times the relative speed $\sigma_R(ij, v|l)v$.

At this point, let us consider how to interpret the cross-section. Since the partial cross-section is an intensive property, it is clear from Eqs (2.7) and (2.10) that $\mathcal{P}$ is an extensive variable given by

$$\mathcal{P}(ij, v|l, \Omega) = \frac{d\sigma_R(ij, v|l, \Omega)/d\Omega}{\mathcal{A}} \tag{2.14}$$

or, after integration over $d\Omega$, $\mathcal{P}_\mathrm{tot} = \mathcal{P}(ij, v|l) = \sigma_R(ij, v|l)/\mathcal{A}$.

That is, the reaction probability is proportional to the ratio of the reaction cross-section and the area of the beam ($\mathcal{A}$); see Fig. 2.1.2.

Example 2.1: Molecular-beam studies, some experimental data

Many elementary chemical reactions have been investigated via molecular-beam techniques. An example is the reaction

$$\mathrm{F} + \mathrm{H_2} \rightarrow \mathrm{HF} + \mathrm{H}$$

and its variant with D_2 [see D.M. Neumark, A.M. Wodtke, G.N. Robinson, C.C. Hayden, and Y.T. Lee, *J. Chem. Phys.* **82**, 3045 (1985) and M. Faubel, L. Rusin, S. Schlemmer, F. Sondermann, U. Tappe, and J.P. Toennies, *J. Chem. Phys.* **101**, 2106 (1994)]. It is found that the total reaction cross-section increases with collision

[3]The 4π indicates that integration is over the full unit sphere, $d\Omega = \sin\theta d\theta d\phi$, with $\theta \in [0, \pi]$ and $\phi \in [0, 2\pi]$.

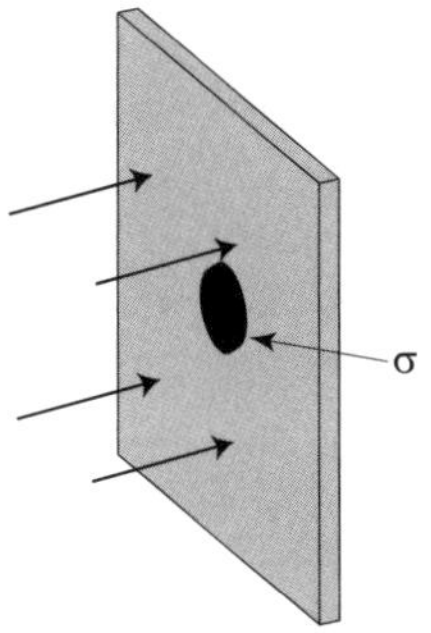

Fig. 2.1.2 A beam of molecules incident on a rectangle of area $\mathcal{A}$. The ratio of the total cross-section $(\sigma_R(ij, v|l))$ to the area of the rectangle $(\mathcal{A})$ is, according to Eq. (2.14), related to the fraction of molecules that are undergoing reaction.

energy. Differential cross-sections associated with angular distributions of products were resolved with respect to the different vibrational states of HF(n). The angular distributions of HF(n) depend on the vibrational state and were all found to be non-isotropic.

The formation of the stronger HF bond, with a bond dissociation energy that is about 130 kJ/mol higher than for H_2, implies a substantial drop in potential energy and hence a large release of kinetic energy that can be distributed among the translational, vibrational, and rotational degrees of freedom of the products. At collision energies from 2.9 to 14.2 kJ/mol, it is found that the HF vibrational distribution is highly inverted, with most of the population in $n = 2$ and $n = 3$.

Another reaction studied via molecular-beam techniques is the (S_N2) reaction:

$$Cl^- + CH_3Br \to ClCH_3 + Br^-$$

where the total cross-section as a function of the relative collision energy has been determined [L.A. Angel and K.M. Ervin, *J. Am. Chem. Soc.* **125**, 1014 (2003)]. A special feature of this (ion–molecule) reaction is found at low collision energies. Thus, with increasing collision energies over the range 0.06–0.6 eV, the cross-section declines from 1.3×10^{-16} cm^2 to 0.08×10^{-16} cm^2.

The total rate of reaction, $dn_C/dt = -dn_A/dt$, is obtained from Eq. (2.11) by summing over all possible quantum states of reactants and products and all possible velocities $\boldsymbol{v}_A$ and $\boldsymbol{v}_B$. We find

$$-dn_A/dt = \sum_{ijl} \int_{\text{all } \boldsymbol{v}_A} \int_{\text{all } \boldsymbol{v}_B} v\sigma_R(ij, v|l) f_{A(i)}(\boldsymbol{v}_A) f_{B(j)}(\boldsymbol{v}_B) d\boldsymbol{v}_A d\boldsymbol{v}_B n_{A(i)} n_{B(j)} \quad (2.15)$$

where the integration is over the three velocity components of A and B, respectively.

If we now write the number density of A(i) as

$$n_{A(i)} = n_A p_{A(i)} \quad (2.16)$$

where $n_{\rm A}$ is the number density of species A and $p_{{\rm A}(i)}$ is the probability of finding A in quantum state i, with a similar expression for the B molecules $n_{{\rm B}(j)} = n_{\rm B}p_{{\rm B}(j)}$, a rate expression equivalent to the well-known phenomenological macroscopic rate expression is obtained. Thus,

$$-dn_{\rm A}/dt = k_\sigma n_{\rm A} n_{\rm B} \tag{2.17}$$

where the rate constant, k_σ, is given by

$$\boxed{\begin{aligned} k_\sigma &= \sum_{ijl} p_{{\rm A}(i)} p_{{\rm B}(j)} \int_{{\rm all}\,\boldsymbol{v}_{\rm A}} \int_{{\rm all}\,\boldsymbol{v}_{\rm B}} v\sigma_R(ij, v|l) f_{{\rm A}(i)}(\boldsymbol{v}_{\rm A}) f_{{\rm B}(j)}(\boldsymbol{v}_{\rm B}) d\boldsymbol{v}_{\rm A} d\boldsymbol{v}_{\rm B} \\ &\equiv \sum_{ijl} p_{{\rm A}(i)} p_{{\rm B}(j)} k_\sigma(ij|l) \end{aligned}} \tag{2.18}$$

This equation relates the bimolecular rate constant to the *state-to-state rate constant* $k_\sigma(ij|l)$ and ultimately to $v\sigma_R(ij, v|l)$. Note that the rate constant is simply the average value of $v\sigma_R(ij, v|l)$. Thus, in a short-hand notation we have $k_\sigma = \langle v\sigma_R(ij, v|l)\rangle$. The average is taken over all the microscopic states including the appropriate probability distributions, which are the velocity distributions $f_{{\rm A}(i)}(\boldsymbol{v}_{\rm A})$ and $f_{{\rm B}(i)}(\boldsymbol{v}_{\rm B})$ in the experiment and the given distributions over the internal quantum states of the reactants.

Outside high vacuum systems we will have an ensemble of molecules that will exchange energy. Typically, thermal equilibrium will be maintained during chemical reaction. There are, though, important exceptions such as chemical reactions in flames and in explosions, as well as reactions that take place at very low pressures.

2.2 Thermal equilibrium

We now proceed to develop a specific expression for the rate constant for reactants where the velocity distributions $f_{{\rm A}(i)}(\boldsymbol{v}_{\rm A})$ and $f_{{\rm B}(j)}(\boldsymbol{v}_{\rm B})$ for the translational motion are independent of the internal quantum state (i and j) and correspond to thermal equilibrium.[4] Then, according to the kinetic theory of gases or statistical mechanics, see Appendix A.2.1, Eq. (A.65), the velocity distributions associated with the center-of-mass motion of molecules are the Maxwell–Boltzmann distribution, a special case of the general Boltzmann distribution law:

$$\begin{aligned} f(\boldsymbol{v}_{\rm A}) d\boldsymbol{v}_{\rm A} &= f(v_{xA}, v_{yA}, v_{zA}) d\boldsymbol{v}_{\rm A} \\ &= f(v_{xA}) f(v_{yA}) f(v_{zA}) d\boldsymbol{v}_{\rm A} \\ &= \left(\frac{m_{\rm A}}{2\pi k_B T}\right)^{3/2} \exp\left\{-\frac{m_{\rm A} v_{xA}^2}{2k_B T}\right\} \exp\left\{-\frac{m_{\rm A} v_{yA}^2}{2k_B T}\right\} \exp\left\{-\frac{m_{\rm A} v_{zA}^2}{2k_B T}\right\} d\boldsymbol{v}_{\rm A} \end{aligned} \tag{2.19}$$

where $d\boldsymbol{v}_{\rm A} = dv_{xA} dv_{yA} dv_{zA}$, k_B is the Boltzmann constant, and a similar expression holds for molecule B.

[4] We assume that the mean-free path is much larger than the molecular dimensions; see Section 9.2. At very high pressures this 'assumption of free flight' is not valid and the overall reaction rate is controlled by the diffusional motion of the reactants.

Since the relative speed v appears in the integrand in Eq. (2.18), it will be convenient to change to the center-of-mass velocity $\boldsymbol{V}$ and the relative velocity $\boldsymbol{v}$ (where $v = |\boldsymbol{v}|$). We find

$$\begin{aligned} \boldsymbol{v} &= \boldsymbol{v}_{\rm A} - \boldsymbol{v}_{\rm B} \\ \boldsymbol{V} &= (m_{\rm A}\boldsymbol{v}_{\rm A} + m_{\rm B}\boldsymbol{v}_{\rm B})/M \end{aligned} \tag{2.20}$$

where $M = m_{\rm A} + m_{\rm B}$. The velocities of A and B expressed in terms of the center-of-mass velocity and the relative velocity are then given by

$$\begin{aligned} \boldsymbol{v}_{\rm A} &= \boldsymbol{V} + \frac{m_{\rm B}}{M}\boldsymbol{v} \\ \boldsymbol{v}_{\rm B} &= \boldsymbol{V} - \frac{m_{\rm A}}{M}\boldsymbol{v} \end{aligned} \tag{2.21}$$

Since the Jacobi determinant for this substitution is equal to one, we have

$$d\boldsymbol{v}_{\rm A} d\boldsymbol{v}_{\rm B} = d\boldsymbol{V} d\boldsymbol{v} \tag{2.22}$$

In order to simplify the notation, we consider only the x-component (the derivation is easily generalized to include all three velocity components). The fraction of A molecules (mass $m_{\rm A}$) with velocity in the range from v_{xA} to $v_{xA} + dv_{xA}$, at the temperature T, is

$$f(v_{xA})dv_{xA} = \sqrt{\frac{m_{\rm A}}{2\pi k_B T}} \exp\left\{-\frac{m_{\rm A}v_{xA}^2}{2k_B T}\right\} dv_{xA} \tag{2.23}$$

and the corresponding expression for the B molecules (mass $m_{\rm B}$) is

$$f(v_{xB})dv_{xB} = \sqrt{\frac{m_{\rm B}}{2\pi k_B T}} \exp\left\{-\frac{m_{\rm B}v_{xB}^2}{2k_B T}\right\} dv_{xB} \tag{2.24}$$

We need to evaluate the following product:

$$f(v_{xA})f(v_{xB})dv_{xA}dv_{xB}$$

appearing in Eq. (2.18). This is the probability of finding an A molecule with velocity in the range from v_{xA} to $v_{xA} + dv_{xA}$ *and* a B molecule with velocity in the range from v_{xB} to $v_{xB} + dv_{xB}$.

We introduce new variables according to Eq. (2.20):

$$\begin{aligned} v_x &= v_{xA} - v_{xB} \\ V_x &= v_{xA}m_{\rm A}/M + v_{xB}m_{\rm B}/M \end{aligned}$$

where v_x and V_x are the x-component of the relative velocity $\boldsymbol{v}$ and the center-of-mass velocity $\boldsymbol{V}$, respectively. The following relation is easily established (see, for example, Appendix D.1):

$$m_{\rm A}v_{xA}^2/2 + m_{\rm B}v_{xB}^2/2 = \mu v_x^2/2 + MV_x^2/2$$

where

$$\mu = m_{\rm A}m_{\rm B}/M \tag{2.25}$$

is the *reduced* mass and M is the total mass. The total kinetic energy is, accordingly, equal to the kinetic energy for the center-of-mass motion plus the kinetic energy of the relative motion. We obtain now

$$f(v_{xA})f(v_{xB})dv_{xA}dv_{xB} = \sqrt{\frac{m_\mathrm{A}m_\mathrm{B}}{(2\pi k_BT)^2}}\exp\left\{-\frac{\mu v_x^2}{2k_BT}\right\}\exp\left\{-\frac{MV_x^2}{2k_BT}\right\}dv_xdV_x$$

That is, the Maxwell–Boltzmann distribution for the two molecules can be written as a product of two terms, where the terms are related to the relative motion and the center-of-mass motion, respectively. After substitution into Eq. (2.18) we can perform the integration over the center-of-mass velocity V_x. This gives the factor $\sqrt{2\pi k_BT/M}$ ($\int_{-\infty}^{\infty}\exp(-ax^2)dx = \sqrt{\pi/a}$) and, from the equation above, we obtain the probability distribution for the relative velocity, irrespective of the center-of-mass motion.

The probability distribution for the relative velocity in the x-direction is accordingly

$$f_1(v_x)dv_x = \left(\frac{\mu}{2\pi k_BT}\right)^{1/2}\exp\left\{-\frac{\mu v_x^2}{2k_BT}\right\}dv_x \tag{2.26}$$

Note that this equation is identical to the Maxwell–Boltzmann velocity distribution of a single particle with a mass given by the reduced mass.

Due to the simple product form of the Maxwell–Boltzmann distribution, the derivations given above are easily generalized to the expression for the relative velocity in three dimensions. Since the integrand in Eq. (2.18) (besides the Maxwell–Boltzmann distribution) depends only on the relative speed, we can simplify the expression in Eq. (2.18) further by integrating over the orientation of the relative velocity. This is done by introducing polar coordinates for the relative velocity. The full three-dimensional probability distribution for the *relative speed* is

$$f_3(v)dv = 4\pi\left(\frac{\mu}{2\pi k_BT}\right)^{3/2}v^2\exp\left\{-\frac{\mu v^2}{2k_BT}\right\}dv \tag{2.27}$$

where the speed is $v = \sqrt{v_x^2+v_y^2+v_z^2}$, i.e., the length of the velocity vector, and the factor $4\pi v^2$ comes from the integration over the orientation of the velocity.

It is useful to determine the corresponding energy distribution, that is, the fraction of molecules with relative kinetic energy in the range from E_tr to $E_\mathrm{tr}+dE_\mathrm{tr}$. We note that the relative kinetic energy is $E_\mathrm{tr} = \mu v^2/2$, which implies that $dE_\mathrm{tr} = \mu v dv$ or $dv = dE_\mathrm{tr}/\sqrt{2\mu E_\mathrm{tr}}$. The energy distribution for motion in three dimensions is then obtained from Eq. (2.27):

$$f_3(E_\mathrm{tr})dE_\mathrm{tr} = 2\pi\left(\frac{1}{\pi k_BT}\right)^{3/2}\sqrt{E_\mathrm{tr}}\exp\left\{-\frac{E_\mathrm{tr}}{k_BT}\right\}dE_\mathrm{tr} \tag{2.28}$$

This relation was illustrated in Fig. 1.2.3.

We can now collect the results and obtain an expression for Eq. (2.18) in the case where the translational motion of the reactants is in thermal equilibrium:

$$\boxed{\begin{aligned} k(T) &= \int_0^\infty v\sigma_R(v)f_3(v)dv \\ &= \int_0^\infty \sqrt{2E_{\text{tr}}/\mu}\,\sigma_R(E_{\text{tr}})f_3(E_{\text{tr}})\,dE_{\text{tr}} \\ &= \frac{1}{k_BT}\left(\frac{8}{\pi\mu k_BT}\right)^{1/2}\int_0^\infty \sigma_R(E_{\text{tr}})E_{\text{tr}}\exp\left\{-\frac{E_{\text{tr}}}{k_BT}\right\}dE_{\text{tr}} \end{aligned}} \tag{2.29}$$

where

$$\sigma_R(v) = \sum_{ijl} p_{\text{A}(i)}p_{\text{B}(j)}\sigma_R(ij, v|l) \tag{2.30}$$

We can also write Eq. (2.29) in the form

$$\begin{aligned} k(T) &= \sum_{ijl} p_{\text{A}(i)}p_{\text{B}(j)}\frac{1}{k_BT}\left(\frac{8}{\pi\mu k_BT}\right)^{1/2}\int_0^\infty \sigma_R(ij, E_{\text{tr}}|l)E_{\text{tr}}\exp\left\{-\frac{E_{\text{tr}}}{k_BT}\right\}dE_{\text{tr}} \\ &\equiv \sum_{ijl} p_{\text{A}(i)}p_{\text{B}(j)}k_{(ij|l)}(T) \end{aligned} \tag{2.31}$$

which defines the *state-to-state rate constant* $k_{(ij|l)}(T)$, that is, the rate constant associated with reactants and products in the specified quantum states $(ij|l)$ with the translational motion of the reactants in thermal equilibrium.

For reactants in complete internal thermal equilibrium, $p_{\text{A}(i)}$ and $p_{\text{B}(j)}$ are Boltzmann distributions (see Appendix A.1), e.g.,

$$p_{\text{A}(i)} = \frac{\omega_i}{Q_\text{A}}\exp(-E_i/k_BT) \tag{2.32}$$

where Q_A is the internal partition function of molecule A,

$$Q_\text{A} = \sum_i \omega_i \exp(-E_i/k_BT) \tag{2.33}$$

and E_i and ω_i are the energy and the degeneracy of the ith quantum level.

2.2.1 Principle of detailed balance

An additional important general result for $k(T)$ can be derived in the case of complete thermal equilibrium, that is, translational as well as complete internal thermal equilibrium: the relationship between rate constants for the forward, $k_f(T)$, and the reverse, $k_r(T)$, reaction

$$\text{A} + \text{B} \underset{k_r}{\overset{k_f}{\rightleftarrows}} \text{C} + \text{D}$$

at complete thermal equilibrium is known as the principle of *detailed balance*,[5] and given by

[5]This relation between rate constants and the equilibrium constant can, of course, also be derived within classical chemical kinetics using the equilibrium condition $dn_\text{A}/dt = 0$.

$$\boxed{\frac{k_f(T)}{k_r(T)} = \left(\frac{\mu_{\rm CD}}{\mu_{\rm AB}}\right)^{3/2} \left(\frac{Q_{\rm C}Q_{\rm D}}{Q_{\rm A}Q_{\rm B}}\right)_{\rm int} \exp(-\Delta E_0/k_B T) \equiv K(T)} \tag{2.34}$$

where $\mu_{\rm CD}$ and $\mu_{\rm AB}$ are the reduced masses of the products and the reactants, respectively, $\Delta E_0 = E_{0,p} - E_{0,r}$, $E_{0,p}$ and $E_{0,r}$ are the zero-point energies of the products and the reactants, respectively, the partition functions refer to the internal (subscript 'int' for internal) motion of the molecules, and $K(T)$ is the equilibrium constant of the reaction. This relation is very useful for obtaining information about reverse reactions once the forward rate constants or cross-sections are known. This principle is a consequence of *microscopic reversibility*, that is, the time-reversal symmetry of the equation of motion—the time-dependent Schrödinger equation (or the classical equations of motion). This result is proved in Appendix B.

Example 2.2: Vibrational relaxation, Boltzmann distribution

This example illustrates several important points.

(i) Inelastic collisions change the internal energy distribution of molecules.

(ii) Inelastic collisions are the physical mechanism behind thermal equilibration, i.e., a Boltzmann distribution is obtained in the long-time limit of vibrational relaxation.

(iii) The relaxation time is inversely proportional to the pressure.

To that end, consider the following inelastic collision:

$$\mathrm{He} + \mathrm{N_2}(n=1) \underset{k_{01}}{\overset{k_{10}}{\rightleftarrows}} \mathrm{He} + \mathrm{N_2}(n=0)$$

where we assume that N_2 can only exist in the two vibrational states indicated above. The bimolecular rate constant of the forward process is k_{10}, and the rate constant of the reverse process is k_{01} (note that the formalism of this chapter also applies to inelastic collisions).

We assume that thermal equilibrium, at the temperature T, has been established in the *translational* degrees of freedom. Then, according to the principle of detailed balance (Eq. (B.37)),

$$k_{10}/k_{01} = e^{-(E_0-E_1)/k_B T}$$

where E_0 and E_1 are the energies associated with the vibrational energy levels. This equation is similar to Eq. (2.34), except that thermal equilibrium in the vibrational degree of freedom has not been assumed. The rate constant k_{10} has been determined at $T = 700$ K and $k_{10} = 3.0 \times 10^4$ liter/(mol s). The initial conditions ($t = 0$) are $[\mathrm{N_2}(n=1)] = [\mathrm{N_2}(n=1)]_0$ and $[\mathrm{N_2}(n=0)] = 0$.

The rate law takes the form

$$
\begin{aligned}
-\frac{d[N_2(n=1)]}{dt} &= k_{10}[He][N_2(n=1)] - k_{01}[He][N_2(n=0)] \\
&= k_{10}[He][N_2(n=1)] - k_{01}[He]([N_2(n=1)]_0 - [N_2(n=1)]) \\
&= (k_f + k_r)[N_2(n=1)] - k_r[N_2(n=1)]_0
\end{aligned}
$$

where $k_f = k_{10}[He]$ and $k_r = k_{01}[He]$. This is an inhomogeneous first-order differential equation, with the solution

$$
[N_2(n=1)] = [N_2(n=1)]_0 \left(k_r + k_f e^{-(k_f+k_r)t}\right) / (k_f + k_r)
$$

The concentration in the vibrational ground state is

$$
\begin{aligned}
[N_2(n=0)] &= [N_2(n=1)]_0 - [N_2(n=1)] \\
&= [N_2(n=1)]_0 (1 - [N_2(n=1)]/[N_2(n=1)]_0) \\
&= [N_2(n=1)]_0 \left(1 - e^{-(k_f+k_r)t}\right) k_f/(k_f + k_r)
\end{aligned}
$$

In the long-time limit, the fraction of molecules in the vibrational ground state becomes

$$
\begin{aligned}
[N_2(n=0)]_\infty/[N_2(n=1)]_0 &= \frac{k_f}{k_f + k_r} \\
&= \frac{k_{10}}{k_{10} + k_{01}} \\
&= \frac{k_{10}/k_{01}}{1 + k_{10}/k_{01}} \\
&= \frac{e^{-(E_0-E_1)/k_BT}}{1 + e^{-(E_0-E_1)/k_BT}}
\end{aligned}
$$

According to the Boltzmann distribution, the probability of observing the N_2 molecule in the vibrational ground state is

$$
\begin{aligned}
p_0 &= \frac{e^{-E_0/k_BT}}{Q} \\
&= \frac{e^{-E_0/k_BT}}{e^{-E_0/k_BT} + e^{-E_1/k_BT}} \\
&= \frac{e^{-(E_0-E_1)/k_BT}}{1 + e^{-(E_0-E_1)/k_BT}}
\end{aligned}
$$

that is, *identical* to the probability derived from the kinetic approach.

We define the vibrational relaxation time, t_{relax}, as the time it takes to reach 90% of the final concentration in the vibrational ground state. That is,

$$
\begin{aligned}
[N_2(n=0)] &= 0.9[N_2(n=0)]_\infty \\
&= 0.9[N_2(n=1)]_0 \frac{k_f}{k_f+k_r}
\end{aligned}
$$

which implies that $1-\exp(-(k_f+k_r)t_{\text{relax}})=0.9$, and

$$
\begin{aligned}
t_{\text{relax}} &= \frac{\ln 10}{[\text{He}](k_{10}+k_{01})} \\
&= \frac{RT\ln 10}{p_{\text{He}}k_{10}(1+k_{10}/k_{01})}
\end{aligned}
$$

where the ideal gas equation, $p_{\text{He}}=[\text{He}]RT$, was used in the second line. Thus, this equation shows that the relaxation time is inversely proportional to the pressure. Now for N_2, we have $E_0-E_1=-0.29$ eV and $t_{\text{relax}}=4.4\times 10^{-3}$ s, at $p_{\text{He}}=1$ atm. Thus, the vibrational relaxation is relatively slow, especially at low pressures. Rotational and translation relaxation is much faster.

Further reading/references

[1] J.C. Light, J. Ross, and K.E. Shuler, in *Kinetic processes in gases and plasmas* (Academic Press, 1969) Chapter 8

Problems

2.1 Using Eq. (2.18), what is the expression for the rate constant if all distributions associated with internal quantum states and velocities of the reactants are 'sharp'? Mathematically, this condition can be expressed in the following way: $p_{A(i)}=\delta_{ij}$ and $p_{B(j)}=\delta_{jk}$, where the so-called Kronecker delta is defined by $\delta_{mn}=1$ if $m=n$, and $\delta_{mn}=0$ if $m\neq n$, that is, $\sum_m \delta_{mn}g_m=g_n$. For the continuous velocity we assume that the distributions are peaked at $\boldsymbol{v}_A^0$ and $\boldsymbol{v}_B^0$, respectively, that is, $f_{A(i)}(\boldsymbol{v}_A)=\delta(\boldsymbol{v}_A-\boldsymbol{v}_A^0)$ and $f_{B(j)}(\boldsymbol{v}_B)=\delta(\boldsymbol{v}_B-\boldsymbol{v}_B^0)$, where the so-called delta function is defined by $\delta(\boldsymbol{r}-\boldsymbol{r}')=\delta(x-x')\delta(y-y')\delta(z-z')$, where for an arbitrary function $g(x)$, $\int_{-\infty}^{\infty}\delta(x-x')g(x)dx=g(x')$, with equivalent expressions in y and z.

2.2 The equilibrium probability distributions for the *relative speed* and the *relative kinetic energy* are frequently used in various one- or two-dimensional models.

(a) When motion is restricted to two dimensions, show that the probability distribution for the relative speed is

$$
f_2(v)dv = 2\pi\left(\frac{\mu}{2\pi k_B T}\right) v\exp\left\{-\frac{\mu v^2}{2k_B T}\right\}dv
$$

where the speed is $v=\sqrt{v_x^2+v_y^2}$.

(b) When motion is restricted to one dimension, show that the energy distribution is

$$f_1(E_{tr})dE_{tr} = \frac{1}{\sqrt{\pi E_{tr} k_B T}} \exp\left\{-\frac{E_{tr}}{k_B T}\right\} dE_{tr}$$

Note that the speed is the length of the velocity vector, i.e., $v = \sqrt{v_x^2}$, and use that the velocity distribution must be multiplied by 2, when the speed is considered. When motion is restricted to two dimensions, show that the energy distribution is

$$f_2(E_{tr})dE_{tr} = \frac{1}{k_B T} \exp\left\{-\frac{E_{tr}}{k_B T}\right\} dE_{tr}$$

2.3 (a) For a reaction *without* an energy threshold, it was found that the total reaction cross-section could be represented as $\sigma_R(E_{tr}) = A/\sqrt{E_{tr}}$, where A is a constant. Calculate the thermal rate constant $k(T)$ for the reaction.

In connection with the evaluation of Eq. (2.29), the Gamma function $\Gamma(n)$ can sometimes be useful. It is defined as

$$\Gamma(n) = \int_0^\infty x^{n-1} \exp(-x)dx, \quad n > 0$$

with the special values
$\Gamma(n+1) = n!$, where $n = 1, 2, \ldots$,
$\Gamma(n + 1/2) = \dfrac{1 \cdot 3 \cdot 5 \cdots (2n-1)}{2^n}\sqrt{\pi}$, where $n = 1, 2, \ldots$.

(b) For a reaction *with* an energy threshold, it was found that the total reaction cross-section could be represented as

$$\sigma_R(E_{tr}) = \begin{cases} 0 & \text{for} \quad E_{tr} < E_0 \\ f(E_{tr} - E_0) & \text{for} \quad E_{tr} \geqslant E_0 \end{cases}$$

Show that the thermal rate constant $k(T)$ for the reaction can be written in the form $k(T) = A(T)e^{-E_0/k_B T}$, and specify the connection between $A(T)$ and $f(E_{tr} - E_0)$.

2.4 We consider, in the gas phase, the (S_N2) reaction

$$Cl^- + CH_3Br \rightarrow ClCH_3 + Br^-$$

Reaction cross-sections have been determined in a molecular-beam experiment [*J. Am. Chem. Soc.* **125**, 1014 (2003)]. The total cross-section at the relative translational energy $E_{tr} = 0.06$ eV is $\sigma_R(E_{tr}) = 1.3 \times 10^{-16}$ cm^2.

(a) Calculate the thermally-averaged rate constant $k(T)$ at $T = 300$ K, under the assumption that in the relevant energy range determined by the Boltzmann distribution the cross-section is independent of the relative translational energy.

(b) Assume that the differential reaction cross-section, for small deflection angles θ, can be described by the equation $d\sigma_R/d\Omega = AE_{tr}^{-1/2}\theta^{-3/2}$, where $d\Omega =$

$\sin\theta d\theta d\phi$ and A is a constant. What is the ratio between the number of product molecules in the two angular regions $\theta \in [0.2, 0.3]$ radians and $\theta \in [0.3, 0.366]$ radians, respectively, when $\phi \in [0.1, 0.2]$ radians? (Use that, at small deflection angles $\sin\theta \sim \theta$.)

3

Potential energy surfaces

Key ideas and results

A potential energy surface is the electronic energy as a function of the internuclear coordinates, and it is obtained as a solution to the electronic Schrödinger equation, Eq. (1.6), for a set of values of the internuclear coordinates. We consider, in particular, the general topology of such energy surfaces for unimolecular and bimolecular reactions, as well as some approximate analytical solutions to the electronic Schrödinger equation associated with bond breaking and bond making.

- For direct reactions a single saddle point is found on the path from reactants to products. In indirect reactions wells and one or more saddle points are found along the reaction path. The electronic energy at the saddle point relative to the electronic energy of the reactants is the barrier height. The height depends on the particular reaction. In special cases no barriers exist.
- Barrier heights in bimolecular reactions depend on the approach angle. For example, in $D + H_2 \rightarrow H + HD$ (and its isotopic variants), the lowest barrier is found when D attacks along the bond axis of H_2, that is, collinearly.
- A barrier that occurs in the entrance channel while the reactants are approaching each other is denoted as an 'early' barrier, whereas a 'late' barrier occurs in the exit channel as the products are separating.
- A simple description of the electronic energy of three interacting hydrogen atoms is given by the London equation. A significant point is that the energy is *not* equal to the sum of H–H pair energies.
- A semi-empirical extension of the London equation—the LEPS method—allows for a simple but somewhat crude construction of potential energy surfaces.

The electronic Schrödinger equation is, according to Eq. (1.6), given by

$$\hat{H}_e \psi_i(\boldsymbol{r}; \boldsymbol{R}) = E_i(\boldsymbol{R}) \psi_i(\boldsymbol{r}; \boldsymbol{R}) \tag{3.1}$$

where $\hat{H}_e$ is the electronic Hamiltonian, $\psi_i(\boldsymbol{r}; \boldsymbol{R})$ is the stationary electronic wave function, and $E_i(\boldsymbol{R})$ is the corresponding *electronic energy* (including internuclear repulsion) which is a function of the nuclear geometry. The geometry is specified by coordinates which are denoted by $\boldsymbol{R}$. Actually, the electronic energy depends only on a subset of the nuclear coordinates (which we refer to as internuclear coordinates, to be specified below) since the electronic Hamiltonian in Eq. (3.1) is invariant to overall translation as well as overall rotation.

In order to generate an electronic potential energy surface, Eq. (3.1) must be solved at a set of *fixed* values of the nuclear coordinates $\boldsymbol{R}$, as indicated by ';' in the electronic wave function.

First, let us consider what is meant by *internuclear coordinates* and, in particular, how many of these coordinates are needed in order to specify the electronic energy. We consider a collection of N atomic nuclei, which in this context are considered as point particles. In the following, we will for convenience refer to any collection of nuclei and electrons as a 'molecule'. The atomic nuclei and the electrons may form one or more stable molecules but this is of no relevance to the following argument. The internuclear coordinates are defined as coordinates that are invariant to overall translation and rotation. These coordinates can, for example, be chosen as internuclear distances and bond angles.

$3N$ coordinates are needed in order to completely specify the position of the nuclei. Three coordinates are needed in order to specify the position of the center of mass.[1] Thus, $3N-3$ coordinates account for the internal degrees of freedom, that is, overall orientation and internuclear coordinates. The overall orientation can be specified by two coordinates for a linear molecule, say by the two polar angles (θ, ϕ). For a non-linear molecule three coordinates are needed in order to specify the orientation. These coordinates are often chosen as the so-called Euler angles. Thus, for a molecule with N atomic nuclei,

$$\text{number of internuclear coordinates} = \begin{cases} 3N-5 & \text{for a linear molecule} \\ 3N-6 & \text{for a non-linear molecule} \end{cases} \tag{3.2}$$

Note that in a non-linear molecule, one of the vibrational modes of the linear molecule has been replaced by a rotational coordinate. As an illustration, let us consider two examples. For the stable linear triatomic molecule CO_2, there are $3 \times 3 - 5 = 4$ internuclear coordinates, which corresponds to the vibrational degrees of freedom, namely the symmetric and antisymmetric stretch and two (degenerate) bending modes (see Appendix E). For the three atoms in the reaction $D + H - H \to D - H + H$, there are $3 \times 3 - 6 = 3$ internuclear coordinates. These coordinates can, for example, be chosen as a D–H distance, the H–H distance, and the D–H–H angle.

In this chapter, we will first discuss the general topology of potential energy surfaces that have been calculated for elementary chemical reactions. Second, we will consider the solution of the electronic Schrödinger equation with a focus on analytical results and elementary concepts rather than detailed computational procedures.

3.1 The general topology of potential energy surfaces

For *bimolecular* elementary chemical reactions there are, roughly speaking, two general categories of potential energy surfaces: (1) surfaces with a single potential energy barrier and no wells along the reaction path, and (2) surfaces with wells and one or more barriers along the reaction path. A bimolecular reaction is in the former case denoted as a *direct reaction* and in the latter case as an *indirect reaction* (or complex mode

[1]These three coordinates have in fact already been separated out of Eq. (1.6).

reaction). Typically, for direct reactions the transformation from reactants to products occurs within a vibrational period, whereas for indirect reactions the transformation can take several vibrational periods.

For direct reactions a single saddle point is found on the path from reactants to products, with potential energy valleys extending in the directions of separate reactants and products; see Fig. 3.1.1.

A *saddle point* is a stationary point on the multidimensional potential energy surface. It is a stable point in all dimensions except one, where the second-order derivative of the potential is negative (see Appendix E). The *classical* energy threshold E_{cl} or *barrier height* of the reaction corresponds to the electronic energy at the saddle point relative to the electronic energy of the reactants.

Example 3.1: Data for simple reactions

Highly accurate potential energy surfaces have been calculated for simple direct reactions like $D + H_2 \rightarrow HD + H$ and $F + H_2 \rightarrow HF + H$. The lowest classical barrier height for $D + H_2$ (and its isotopic variants) has been calculated to be 40.50 ± 0.50 kJ/mol [1,2]. The lowest barrier is found at the collinear geometry where all atoms are along the same line. For the $F + H_2$ reaction, the classical barrier height at the collinear geometry is 7.66 ± 1.0 kJ/mol. Recent calculations [3] show, however, that the lowest barrier is 6.07 ± 1.0 kJ/mol, which is found for a bent geometry corresponding to an F–H–H angle of about 119°. The barrier heights of these reactions are much smaller than the bond dissociation energy of H_2. Thus, the saddle point is on the path from reactants to products, at a stage where the H–H bond is partly broken and a new bond is simultaneously being formed.

The energy threshold of a reaction corresponds to the minimum relative translational energy that must be supplied to the reactants in order to produce products. This energy threshold will differ from the classical barrier height, even when the reac-

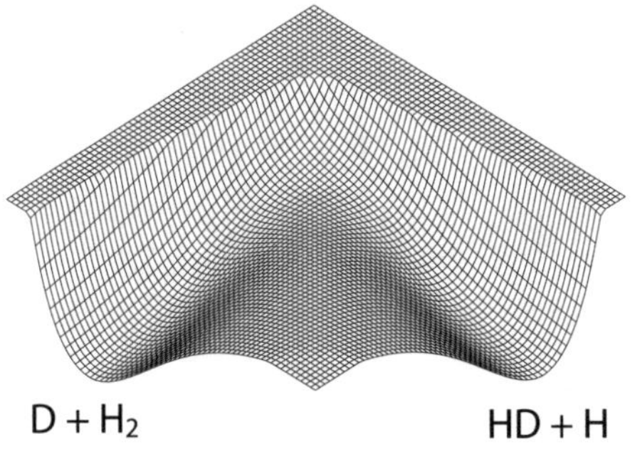

Fig. 3.1.1 A potential energy surface for a direct bimolecular reaction. The surface corresponds to a reaction like $D + H - H \rightarrow D - H + H$ at a fixed approach angle, say in a collinear configuration specified by the D–H and H–H distances. These distances are measured along the two perpendicular axes. (Note that in this figure all energies above a fixed cut-off value E_{max} have been replaced by E_{max}.)

tants are in their ground states, due to vibrational zero-point energies and quantum mechanical tunneling. The relation between the energy threshold and the activation energy of chemical kinetics is discussed in Chapter 8.

Although the entire potential energy surface is important, it is often useful to focus on particular features. For example, the *minimum-energy path* or the *reaction coordinate*, which corresponds to points on a line through the saddle point along the deepest part of the two valleys on each side of the saddle point. A chemical reaction that follows this path will, at all times, experience the lowest possible potential energy. The energy along the reaction coordinate is shown in Fig. 3.1.2 for two cases: first for a case where the electronic energy of reactants and products is identical, as in Fig. 3.1.1, and for an exothermic reaction like $F+H_2$, where the electronic energy of the products is lower than the electronic energy of the reactants. This is the definition of an exothermic reaction at the microscopic level, and the implications for the dynamics are a large release of energy into translational, vibrational, and rotational degrees of freedom in the products.

As mentioned above, the barrier height depends on the approach angle, as illustrated in Fig. 3.1.3. Another important feature is the location of the barrier along the reaction coordinate [4]. An *'early' barrier* is a barrier that occurs in the entrance channel while the reactants are approaching each other. In Fig. 3.1.4 the potential energy surface for such a situation is shown as a contour plot, that is, all the points on a contour line correspond to a fixed value of the energy. The implications for the dynamics are that translational energy is most effective for passage across the barrier and that the products will show up with high vibrational excitation. The $F + H_2$ reaction is an example of a reaction with an early barrier. A *'late' barrier* is a barrier that occurs in

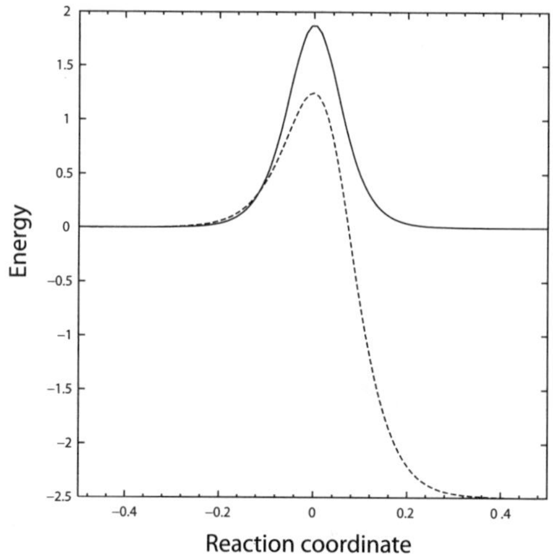

Fig. 3.1.2 Energy along reaction coordinate (arbitrary units). The solid line corresponds to a symmetric reaction like $H + H_2$ and the dashed line corresponds to an exothermic reaction like $F + H_2$.

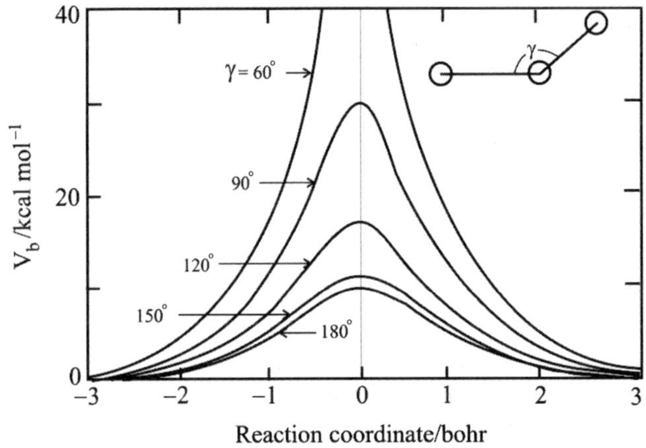

Fig. 3.1.3 Energy along the reaction coordinate for the reaction $D + H - H \rightarrow D - H + H$ (and its isotopic variants), as a function of the approach angle. Note that the lowest barrier is found for the collinear approach. [Adapted from R.D. Levine and R.B. Bernstein, *Molecular reaction dynamics and chemical reactivity* (Oxford University Press, 1987).]

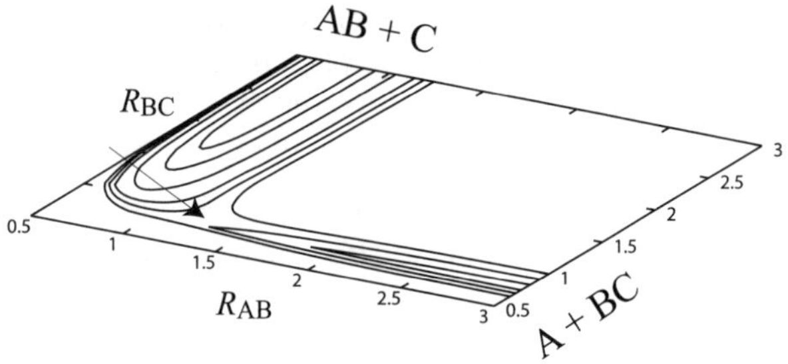

Fig. 3.1.4 Contour plot of a potential energy surface for the reaction A + BC → AB + C. The surface is shown as a function of the two internuclear distances R_{AB} and R_{BC} at a fixed approach angle. The barrier (marked with an arrow) occurs in the entrance channel, i.e., an 'early' barrier.

the exit channel where the products are separating; in Fig. 3.1.4 such a barrier exists for the reverse reaction, i.e., AB + C → A + BC. The implications for the dynamics are that vibrational energy is most effective for passage across the barrier and that the products will show up with low vibrational excitation.[2] Note that for a symmetric reaction, as in Fig. 3.1.1, the barrier always occurs at the 45° line.

So far, we have considered potential energy surfaces without any (local) minima along the reaction coordinate. However, wells along the reaction coordinate can occur.

[2]These conclusions are based on the assumption that there is no interchange of translational and vibrational energy prior to the crossing of the barrier. The validity of the assumption depends on the exact form of the potential energy surface as well as on the masses of the particles. In order to get a quick estimate of the mass effects, a mass-weighted coordinate system can be used, see Appendix D.

For example, for a bimolecular nucleophilic substitution (S_N2) reaction like $Cl^- + CH_3Br \to ClCH_3 + Br^-$, the potential energy has a double-well shape, i.e., two minima separated by a central barrier. The minima for this reaction reflects the stability (an effect that is also well known within classical electrostatics) of the ion–dipole complexes $Cl^- \cdots CH_3Br$ and $ClCH_3 \cdots Br^-$. For other indirect (or complex mode) reactions one finds two saddle points separated by a well on the path from reactants to products. The existence of a well along the reaction path implies that the collision may be 'sticky', and a long-lived intermediate complex can be formed before the products show up. Examples of complex mode reactions are $H + O_2 \to OH + O$ (with the intermediate HO_2), $H^+ + D_2$ and $KCl + NaBr$.

In *unimolecular* reactions, initially, there is only a single stable molecule. This configuration corresponds to a minimum, that is, a well on a multidimensional potential energy surface; see Figs 3.1.5 and 3.1.6.

One can again consider two general categories: *direct* reactions and *complex* mode reactions. Saddle points are found for some, but not all unimolecular reactions. Thus, for the unimolecular dissociation of H_2O in its electronic ground state no saddle point is found (see Figs 3.1.5 and 3.1.6). For an isomerization like $HCN \to HNC$, a saddle point does exist.

Above, we have focused on surfaces associated with the electronic ground state. However, sometimes *excited electronic states* play an important role. This is the case for unimolecular photo-activated reactions, as well as for some bimolecular reactions where non-adiabatic effects play a role. We return to both cases in subsequent chapters.

The techniques for the solution of the electronic Schrödinger equation are highly developed. Computer programs that solve this equation and that can locate saddle

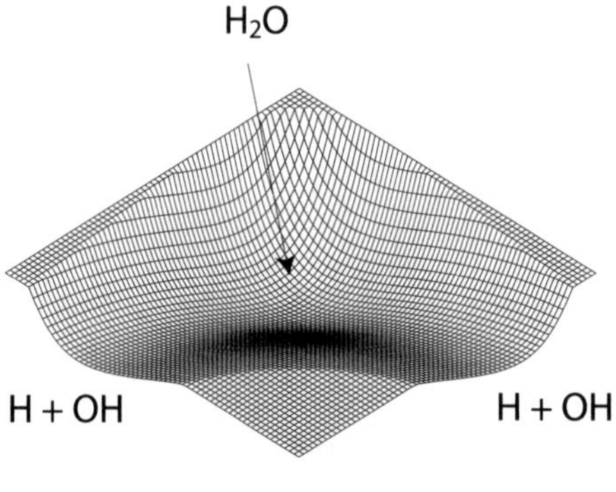

Fig. 3.1.5 A potential energy surface for a direct unimolecular reaction without a saddle point. The surface corresponds to a reaction like $H_2O \to H + OH$ for dissociation along a fixed bond angle, where only two internuclear coordinates are required in order to specify the configuration. (Note that in this figure all energies above a fixed cut-off value E_{max} have been replaced by E_{max}.)

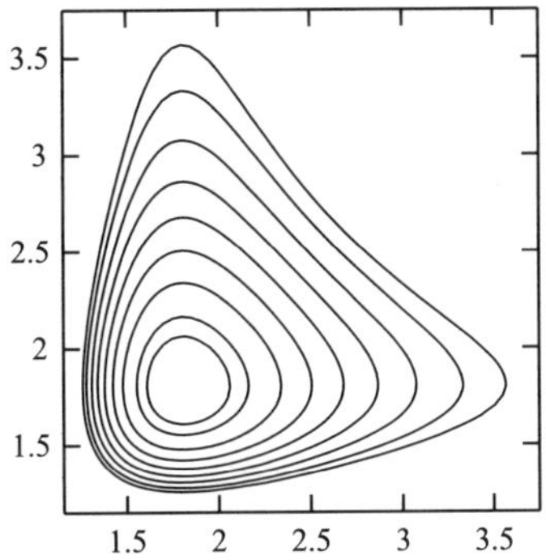

Fig. 3.1.6 Contour plot of the potential energy surface of Fig. 3.1.5. H_2O is at its equilibrium bond angle of 104.5° and the inner contours correspond to the lowest energies. The minimum is at an OH distance of 1.81 a_0, where 1 a_0 = 0.529 Å.

points, etc., are available today and in widespread use.[3] These programs can run on standard personal computers. When the molecules contain a large number of electrons it is, however, difficult to obtain a highly accurate potential energy surface. This is, in part, because the total electronic energy that is obtained from the electronic Schrödinger equation is very large compared to the small changes in the energy that are observed for different internuclear configurations. Thus, at the saddle point the energy is typically of the order of 0.1 to 1 eV[4] higher than the electronic energy of the reactants. If the energy of the reactants is thousands of eV or more, a highly accurate scheme is, obviously, required in order to determine an accurate potential energy surface, including an accurate barrier height.

3.2 Molecular electronic energies, analytical results

We consider in this section some approximate analytical solutions to the electronic Schrödinger equation, in order to provide some basic insight into the energetics of the making and breaking of chemical bonds. Since most of the results are well known from quantum mechanics/chemistry, we will only present the key points and sometimes omit detailed proofs.

3.2.1 The variational principle

In order to obtain the potential energy surfaces associated with chemical reactions we, typically, need the lowest eigenvalue of the electronic Hamiltonian. Unlike systems such as a harmonic oscillator and the hydrogen atom, most problems in quantum mechanics cannot be solved exactly. There are, however, approximate methods that can be used to obtain solutions to almost any degree of accuracy. One such method is the *variational method*. This method is based on the variational principle, which says

[3]Several commercial quantum chemical programs like 'Gaussian' (by Gaussian, Inc., `www.gaussian.com`) and 'Spartan' (by Wavefunction, Inc., `www.wavefun.com`) are available for the solution of the electronic Schrödinger equation.

[4]1 eV = 96.485 kJ/mol.

that the ground-state energy that is calculated using an arbitrary (trial) wave function is always equal to or greater than the energy associated with the exact ground-state wave function.

The exact ground-state wave function ψ_0 and the associated energy E_0 satisfy the Schrödinger equation

$$\hat{H}_e \psi_0 = E_0 \psi_0 \tag{3.3}$$

Thus,

$$E_0 = \frac{\int \psi_0 \hat{H}_e \psi_0 d\tau}{\int \psi_0 \psi_0 d\tau} \tag{3.4}$$

where $d\tau$ represents the appropriate volume element (note that the denominator is equal to one if the wave function is normalized). Then, according to the variational principle, if we substitute *any* other function for ψ_0 in Eq. (3.4) and calculate

$$E_\phi = \frac{\int \phi \hat{H}_e \phi d\tau}{\int \phi \phi d\tau} \tag{3.5}$$

then

$$E_\phi \geqslant E_0 \tag{3.6}$$

Thus, if we use a trial function (ϕ) that depends on some parameters, we can vary these parameters in order to minimize E_ϕ, and we will always obtain an energy that is larger than or equal to (if we happen to obtain the exact ground-state wave function) the exact ground-state energy.

If we, for example, have a trial function that depends linearly on the variational parameters (c_n), i.e.,

$$\phi = \sum_{n=1}^{N} c_n \phi_n \tag{3.7}$$

then minimization of the energy with respect to c_n leads to the following so-called *secular equation* associated with the wave function (assuming for simplicity normalized real basis functions ϕ_n):

$$\begin{vmatrix} H_{11} - E & H_{12} - ES_{12} & \cdots & H_{1N} - ES_{1N} \\ H_{21} - ES_{21} & H_{22} - E & \cdots & H_{2N} - ES_{2N} \\ \vdots & \vdots & & \vdots \\ H_{N1} - ES_{N1} & H_{N2} - ES_{N2} & \cdots & H_{NN} - E \end{vmatrix} = 0 \tag{3.8}$$

where

$$\begin{aligned} H_{ij} &= \int \phi_i \hat{H}_e \phi_j d\tau \\ S_{ij} &= \int \phi_i \phi_j d\tau \end{aligned} \tag{3.9}$$

Since $\hat{H}_e$ is Hermitian the matrix is a real symmetric matrix. This $N \times N$ determinant gives an Nth-order polynomial in E. The lowest root is the best approximation to the ground-state energy within the framework of the trial function in Eq. (3.7).

3.2.2 Chemical bonding in H_2^+ and H_2

We consider first, at a qualitative level, the chemical bonding in the simplest molecule H_2^+. This summarizes some important concepts and, at the same time, introduces the notation to be used later on.

The electronic Hamiltonian of H_2^+ (see Eq. (1.3)) is

$$\hat{H}_e = -\frac{\hbar^2}{2m_e}\nabla^2 - \frac{e^2}{4\pi\epsilon_0 r_A} - \frac{e^2}{4\pi\epsilon_0 r_B} + \frac{e^2}{4\pi\epsilon_0 R} \tag{3.10}$$

where r_A is the distance of the electron from nucleus A, r_B is the distance of the electron from nucleus B, and R is the internuclear separation,[5] see Fig. 3.2.1.

As a trial function for the wave function (molecular orbital, MO) of the electron, consider the linear combination

$$\psi = c_1 1s_A + c_2 1s_B \tag{3.11}$$

where $\psi \equiv \psi(\boldsymbol{r})$ with $\boldsymbol{r}$ denoting the three coordinates of the electron, and where $1s_A$ and $1s_B$ denote a normalized hydrogenic $1s$ orbital centered on nuclei A and B, respectively,

$$\begin{aligned} 1s_A &= \sqrt{\frac{1}{\pi a_0^3}} e^{-r_A/a_0} \\ 1s_B &= \sqrt{\frac{1}{\pi a_0^3}} e^{-r_B/a_0} \end{aligned} \tag{3.12}$$

where a_0 is the Bohr radius. The secular equation, Eq. (3.8), associated with the wave function is

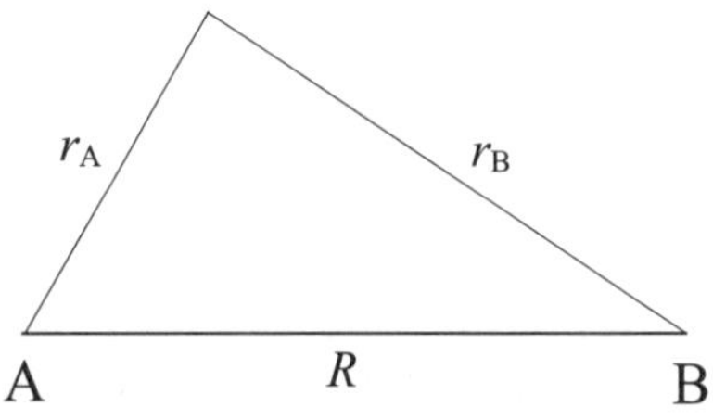

Fig. 3.2.1 Definition of the distances involved in the electronic Hamiltonian of H_2^+.

[5]In Section 1.1, we mentioned that this form of the Hamiltonian refers to electronic coordinates specified in a laboratory fixed coordinate system. For H_2^+, after the proper transformation from laboratory fixed coordinates to center-of-mass coordinates and internal coordinates of the three-particle system, one will obtain an electronic Hamiltonian that is very close to the form given above. When so-called Jacobi coordinates (to be introduced in Section 4.1.4) are used as internal coordinates, the coordinates of the electron are measured relative to the center of mass of the two protons, and the electronic Hamiltonian can be written as above, except that the electron mass m_e is replaced by $\mu = m_e 2m_p/(m_e + 2m_p) = 9.1069 \times 10^{-31}$ kg $\sim 0.9997 m_e$, where m_p is the proton mass. This small change of effective mass will only introduce a very small change in the energy. Thus, when we consider the change in the ground-state energy of the free hydrogen atom introduced by such a change in mass, it will correspond to 0.003 eV, which is indeed a very small change.

$$\begin{vmatrix} H_{\mathrm{AA}}(R) - E(R) & H_{\mathrm{AB}}(R) - E(R)S(R) \\ H_{\mathrm{BA}}(R) - E(R)S(R) & H_{\mathrm{BB}}(R) - E(R) \end{vmatrix} = 0 \tag{3.13}$$

where

$$\begin{aligned} H_{\mathrm{AA}}(R) &= \int d\boldsymbol{r} 1s_{\mathrm{A}} \hat{H}_e 1s_{\mathrm{A}} \\ H_{\mathrm{AB}}(R) &= \int d\boldsymbol{r} 1s_{\mathrm{A}} \hat{H}_e 1s_{\mathrm{B}} \\ S(R) &= \int d\boldsymbol{r} 1s_{\mathrm{A}} 1s_{\mathrm{B}} \end{aligned} \tag{3.14}$$

where $d\boldsymbol{r}$ is the volume element associated with the three coordinates of the electron, $S(R)$ is the *overlap integral*, $H_{\mathrm{AA}} = H_{\mathrm{BB}}$, and $H_{\mathrm{AB}} = H_{\mathrm{BA}}$. Thus,

$$[H_{\mathrm{AA}}(R) - E(R)]^2 - [H_{\mathrm{AB}}(R) - E(R)S(R)]^2 = 0 \tag{3.15}$$

which implies

$$E_{\pm}(R) = \frac{H_{\mathrm{AA}}(R) \pm H_{\mathrm{AB}}(R)}{1 \pm S(R)} \tag{3.16}$$

with the corresponding wave functions

$$\psi_{\pm} = \frac{1}{\sqrt{2(1 \pm S)}}(1s_{\mathrm{A}} \pm 1s_{\mathrm{B}}) \tag{3.17}$$

The matrix elements can be expressed in the form

$$H_{\mathrm{AA}}(R) = E_1 + J'(R) \tag{3.18}$$

where E_1 is the ground-state energy of the free hydrogen atom, i.e.,

$$\left(-\frac{\hbar^2}{2m_e}\nabla^2 - \frac{e^2}{4\pi\epsilon_0 r_{\mathrm{A}}}\right) 1s_{\mathrm{A}} = E_1 1s_{\mathrm{A}} \tag{3.19}$$

and

$$J'(R) = \frac{e^2}{4\pi\epsilon_0 R} - \int d\boldsymbol{r} [1s_{\mathrm{A}}]^2 \frac{e^2}{4\pi\epsilon_0 r_{\mathrm{B}}} \tag{3.20}$$

where the integral is called a *Coulomb integral*. The first term in J' is the nuclear–nuclear Coulombic repulsion. Since $d\boldsymbol{r}[1s_{\mathrm{A}}]^2$ is the probability of finding the electron in the volume element $d\boldsymbol{r}$ at the position $\boldsymbol{r}$ around A, the second term can be interpreted as the charge cloud of the electron around nucleus A interacting with nucleus B via the Coulomb potential. The off-diagonal matrix element can be written as

$$H_{\mathrm{AB}}(R) = E_1 S(R) + K'(R) \tag{3.21}$$

where E_1 is again the ground-state energy of the free hydrogen atom:

$$\left(-\frac{\hbar^2}{2m_e}\nabla^2 - \frac{e^2}{4\pi\epsilon_0 r_{\mathrm{B}}}\right) 1s_{\mathrm{B}} = E_1 1s_{\mathrm{B}} \tag{3.22}$$

and

$$K'(R) = \frac{S(R)e^2}{4\pi\epsilon_0 R} - \int d\boldsymbol{r}\, 1s_{\rm A} \frac{e^2}{4\pi\epsilon_0 r_{\rm A}} 1s_{\rm B} \tag{3.23}$$

where the integral is called an *exchange integral.* This does not lend itself to the same type of 'classical' interpretation as discussed for J'. K' is a strictly quantum mechanical quantity. Equation (3.16) now takes the form

$$\boxed{E_\pm(R) = E_1 + \frac{J'(R) \pm K'(R)}{1 \pm S(R)}} \tag{3.24}$$

where the last term gives the energy of H_2^+ relative to a separated proton and a hydrogen atom (E_1). The integrals can be evaluated analytically. J' is positive while K' is negative.

The positive value of the Coulomb integral in H_2^+ implies that the internuclear repulsion is always larger than the attraction between the electron and a proton, see Eq. (3.20). These integrals as well as the energies (relative to a separated proton and a hydrogen atom) of the two states are shown in Fig. 3.2.2. The state corresponding to the energy E_+ describes a *stable molecule* with dissociation energy 1.77 eV and equilibrium bond length 1.32 Å (= 2.5 a_0) (the experimental values are 2.78 eV and 1.06 Å, respectively). Because the exchange integral is a strictly quantum mechanical quantity, the existence of the chemical bond is a quantum mechanical effect.

In the description above only two atomic orbitals were used in the trial function. The way to improve the accuracy is to include more atomic orbitals in the trial function, such as

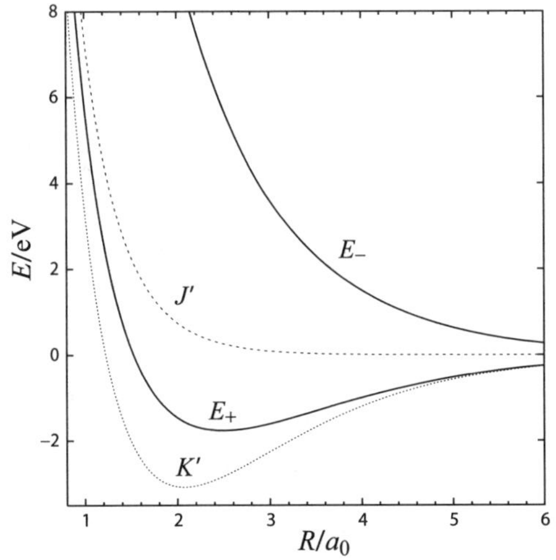

Fig. 3.2.2 The energies of the bonding (E_+) and antibonding (E_-) states in a simple molecular orbital description of H_2^+. The Coulomb (J') and exchange (K') integrals are also shown. All quantities are shown as a function of the internuclear separation in units of the Bohr radius.

$$\psi = c_1 1s_{\mathrm{A}} + c_2 2s_{\mathrm{A}} + c_3 2p_{z\mathrm{A}} + c_4 1s_{\mathrm{B}} + c_5 2s_{\mathrm{B}} + c_6 2p_{z\mathrm{B}} \tag{3.25}$$

This way of constructing a molecular orbital (MO) is called linear combination of atomic orbitals–molecular orbital (LCAO–MO).

Next we consider the hydrogen molecule. The electronic Hamiltonian of H_2 is

$$\begin{aligned}\hat{H}_e = -\frac{\hbar^2}{2m_e}(\nabla_1^2 + \nabla_2^2) - \frac{e^2}{4\pi\epsilon_0 r_{1\mathrm{A}}} - \frac{e^2}{4\pi\epsilon_0 r_{1\mathrm{B}}} \\ - \frac{e^2}{4\pi\epsilon_0 r_{2\mathrm{A}}} - \frac{e^2}{4\pi\epsilon_0 r_{2\mathrm{B}}} + \frac{e^2}{4\pi\epsilon_0 r_{12}} + \frac{e^2}{4\pi\epsilon_0 R}\end{aligned} \tag{3.26}$$

where $r_{1\mathrm{A}}$ is the distance from electron '1' to nucleus A, r_{12} is the distance between electron '1' and '2', etc. Electrons are indistinguishable from each other and cannot really be labeled. All electronic wave functions must be *antisymmetric* (i.e., change sign) under the interchange of any two electrons:

$$\Psi(1,2) = -\Psi(2,1) \tag{3.27}$$

where '1' and '2' here is a shorthand notation for the coordinates associated with the two electrons.

If the electron–electron repulsion was absent in the Hamiltonian, the wave function could be represented exactly as a product of independent one-electron wave functions, i.e., $\Psi(1,2) = \phi_1(1)\phi_2(2)$. Now, working within such a description and taking into account Eq. (3.27), the wave function can be expressed as a *Slater determinant*:

$$\psi_{\mathrm{MO}}(1,2) = \frac{1}{\sqrt{2!}}\begin{vmatrix}\psi_+\alpha(1) & \psi_+\alpha(2) \\ \psi_+\beta(1) & \psi_+\beta(2)\end{vmatrix} \tag{3.28}$$

where we have included the two possible spin states of the electron (α and β), and placed the two electrons with opposite spins in the molecular orbital ψ_+ of Eq. (3.17). The (unnormalized) wave function can be written in the form

$$\begin{aligned}\psi_{\mathrm{MO}}(1,2) &= [1s_{\mathrm{A}}(1)1s_{\mathrm{B}}(2) + 1s_{\mathrm{B}}(1)1s_{\mathrm{A}}(2)]\,\frac{1}{\sqrt{2}}\,[\alpha(1)\beta(2) - \alpha(2)\beta(1)] + \text{ionic terms} \\ &\equiv \psi_{\mathrm{VB}}^{+}(1,2) + \text{ionic terms}\end{aligned} \tag{3.29}$$

where the 'ionic terms' (excluding spin) are $1s_{\mathrm{A}}(1)1s_{\mathrm{A}}(2) + 1s_{\mathrm{B}}(1)1s_{\mathrm{B}}(2)$. The energy corresponding to ψ_{MO} can be calculated analytically and explains, qualitatively, chemical bonding in H_2 but, as expected, without complete quantitative agreement with experimental data.

When the ionic terms are neglected, one gets the so-called valence-bond (VB) description. The (unnormalized) valence-bond wave function can also be written in the form

$$\psi_{\mathrm{VB}}^{\pm}(1,2) = \frac{1}{\sqrt{2!}}\begin{vmatrix}1s_{\mathrm{A}}\alpha(1) & 1s_{\mathrm{A}}\alpha(2) \\ 1s_{\mathrm{B}}\beta(1) & 1s_{\mathrm{B}}\beta(2)\end{vmatrix} \mp \frac{1}{\sqrt{2!}}\begin{vmatrix}1s_{\mathrm{A}}\beta(1) & 1s_{\mathrm{A}}\beta(2) \\ 1s_{\mathrm{B}}\alpha(1) & 1s_{\mathrm{B}}\alpha(2)\end{vmatrix} \tag{3.30}$$

The energy corresponding to this state is given by the Heitler–London equation:

$$\boxed{E_\pm(R) = 2E_1 + \frac{J_{\rm AB}(R) \pm K_{\rm AB}(R)}{1 \pm S(R)^2}} \tag{3.31}$$

where the last term is the energy of H_2 relative to two isolated ground-state hydrogen atoms, and

$$\begin{aligned} J_{\rm AB}(R) &= \iint d\boldsymbol{r}_1 d\boldsymbol{r}_2 1s_{\rm A}(1)1s_{\rm B}(2) \\ &\quad \times \left(\frac{e^2}{4\pi\epsilon_0 R} - \frac{e^2}{4\pi\epsilon_0 r_{1\rm B}} - \frac{e^2}{4\pi\epsilon_0 r_{2\rm A}} + \frac{e^2}{4\pi\epsilon_0 r_{12}} \right) 1s_{\rm A}(1)1s_{\rm B}(2) \\ &= \frac{e^2}{4\pi\epsilon_0 R} - \int \frac{d\boldsymbol{r}_1[1s_{\rm A}(1)]^2 e^2}{4\pi\epsilon_0 r_{1\rm B}} \\ &\quad - \int \frac{d\boldsymbol{r}_2[1s_{\rm B}(2)]^2 e^2}{4\pi\epsilon_0 r_{2\rm A}} + \iint \frac{d\boldsymbol{r}_1 d\boldsymbol{r}_2[1s_{\rm A}(1)1s_{\rm B}(2)]^2 e^2}{4\pi\epsilon_0 r_{12}} \end{aligned} \tag{3.32}$$

is a *Coulomb integral* (similar to Eq. (3.20)) and

$$\begin{aligned} K_{\rm AB}(R) &= \iint d\boldsymbol{r}_1 d\boldsymbol{r}_2 1s_{\rm A}(1)1s_{\rm B}(2) \\ &\quad \times \left(\frac{e^2}{4\pi\epsilon_0 R} - \frac{e^2}{4\pi\epsilon_0 r_{1\rm A}} - \frac{e^2}{4\pi\epsilon_0 r_{2\rm B}} + \frac{e^2}{4\pi\epsilon_0 r_{12}} \right) 1s_{\rm B}(1)1s_{\rm A}(2) \end{aligned} \tag{3.33}$$

is an *exchange integral* (similar to Eq. (3.23)). The integrals can be evaluated analytically. The two energy levels in the Heitler–London equation correspond to a *bonding* (E_+) and an *antibonding* state (E_-), similar to the energy expression for H_2^+, Eq. (3.24). The exchange integral gives the overwhelming contribution to the stability of the bonding state of H_2, giving the dissociation energy 3.15 eV and equilibrium bond length 0.87 Å. When compared to the experimental values 4.75 eV and 0.74 Å, respectively, it is clear that the quantitative predictions are not too accurate.

There are two ways to improve the accuracy in order to obtain solutions to almost any degree of accuracy. The first is via the so-called self-consistent field–Hartree–Fock (SCF–HF) method, which is a method based on the variational principle that gives the optimal one-electron wave functions of the Slater determinant. Electron correlation is, however, still neglected (due to the assumed product of one-electron wave functions). In order to obtain highly accurate results, this approximation must also be eliminated.[6] This is done via the so-called configuration interaction (CI) method. The CI method is again a variational calculation that involves several Slater determinants.

3.2.3 $\mathbf{H + H_2 \to H_2 + H}$, the London equation

The generalization of the Heitler–London equation, Eq. (3.31), to three hydrogen atoms was also considered in the early days of quantum mechanics (around 1930). This description contains the essence of the energetics associated with bond breaking and bond making.

[6] For the $F + H_2$ reaction high levels of theory are required. The Hartree–Fock limit for the classical barrier height is about 67 kJ/mol [*J. Phys. Chem.* **89**, 5336 (1985)], that is, almost ten times larger than the exact value!

It is straightforward to write down the electronic Hamiltonian of three interacting hydrogen atoms according to Eq. (1.3). In the London description, a $1s$ orbital is centered on each of the three hydrogen atoms denoted by A, B, and C, respectively. This gives Slater determinants like

$$\psi_{\rm VB}(1,2,3) = \frac{1}{\sqrt{3!}} \begin{vmatrix} 1s_{\rm A}\alpha(1) & 1s_{\rm A}\alpha(2) & 1s_{\rm A}\alpha(3) \\ 1s_{\rm B}\alpha(1) & 1s_{\rm B}\alpha(2) & 1s_{\rm B}\alpha(3) \\ 1s_{\rm C}\alpha(1) & 1s_{\rm C}\alpha(2) & 1s_{\rm C}\alpha(3) \end{vmatrix} \tag{3.34}$$

where all three electrons have identical spin in this example. There are $2^3 = 8$ such Slater determinants, corresponding to the two possible spin states of each electron. The London equation for the three hydrogen atoms is based on a variational calculation that involves these eight electronic (spin) configurations. It is a simple example of a CI calculation. The essential approximations in the derivation are: (i) only one atomic orbital on each atom (here s orbitals); (ii) overlap integrals between the atomic orbitals are ignored; and (iii) multiple-exchange integrals, i.e., terms arising from permutations of more than two electrons are neglected in the matrix elements. The ground-state[7] energy is given by the London equation:

$$\boxed{\begin{aligned} E(R_{\rm AB}, R_{\rm AC}, R_{\rm BC}) &= 3E_1 + J_{\rm AB} + J_{\rm AC} + J_{\rm BC} \\ &\quad - \sqrt{(K_{\rm AB} - K_{\rm BC})^2/2 + (K_{\rm AB} - K_{\rm AC})^2/2 + (K_{\rm BC} - K_{\rm AC})^2/2} \end{aligned}} \tag{3.35}$$

where $R_{\rm AB}$, $R_{\rm AC}$, and $R_{\rm BC}$ are the internuclear distances, and $J_{\rm AB}$ and $K_{\rm AB}$ are Coulomb and exchange integrals as defined in Eqs (3.32) and (3.33) that depend on the distance $R_{\rm AB}$. The other symbols are defined in an equivalent way.

The London equation reduces to the Heitler–London equation, Eq. (3.31) for $S = 0$, if one of the atoms is removed to infinity. For example, if C is removed to infinity,

$$\begin{aligned} E &\to 3E_1 + J_{\rm AB} - \sqrt{(1/2)(K_{\rm AB})^2 + (1/2)(K_{\rm AB})^2} \\ &= 3E_1 + J_{\rm AB} - |K_{\rm AB}| \\ &= 3E_1 + J_{\rm AB} + K_{\rm AB} \end{aligned} \tag{3.36}$$

where, in the last line, the fact that the exchange integral is negative has been used. Thus, as expected, the description is on the same level as the valence-bond description for H_2.

When the Coulombic and exchange integrals J and K for H_2 are calculated and introduced into the London equation, the right general form of the potential energy surface is obtained. Thus, bond breaking is assisted by bond making, i.e., barrier energies are much less than the bond energy of H_2. However, the entrance and exit valleys do not meet at a single saddle point as in Fig. 3.1.1. There is a potential energy well (basin) corresponding to the symmetrical H–H–H configuration. As mentioned in Section 3.1, it is known (from highly accurate CI calculations) that such an energy well does not exist and that the classical barrier height associated with the saddle point

[7] The associated wave function, including spin functions, is a *doublet* state corresponding to two paired electrons and one unpaired electron.

is ~ 0.4 eV. These problems with the London equation are not highly surprising. We have seen that the London equation is related to the Heitler–London equation and, as shown above, the errors associated with this equation are clearly bigger than the value of the true barrier height.

Although the underlying approximations are too crude to obtain an accurate potential energy surface, another very important observation can be made when the London equation is compared to the energy expression for H_2: the total energy is *not* equal to the sum of pairwise H–H interactions. Thus, $E(R_{\text{AB}}, R_{\text{AC}}, R_{\text{BC}}) \neq E_{\text{AB}} + E_{\text{AC}} + E_{\text{BC}}$, where E_{AB} corresponds to E_+ of Eq. (3.31), and E_{AC} and E_{BC} are given by similar expressions. The simple summation of pairwise H–H interactions only holds for the Coulomb integrals!

3.2.4 A semi-empirical method: the LEPS surface

The London–Eyring–Polanyi–Sato (LEPS) method is a semi-empirical method.[8] It is based on the London equation, but the calculated Coulombic and exchange integrals are replaced by experimental data. That is, some experimental input is used in the construction of the potential energy surface. The LEPS approach can, partly, be justified for $H + H_2$ and other reactions involving three atoms, as long as the basic approximations behind the London equation are reasonable.

The potential energy curve corresponding to the bonding state is often well described by a *Morse potential*:

$$\begin{aligned} E_+(R) &= J_{\text{AB}} + K_{\text{AB}} \\ &\sim D(1 - e^{-\beta(R-R_e)})^2 - D \\ &= D(e^{-2\beta(R-R_e)} - 2e^{-\beta(R-R_e)}) \end{aligned} \tag{3.37}$$

where the parameters of the Morse potential are D (the classical dissociation energy), β, and R_e (the equilibrium bond length). These parameters can be extracted from experimental (spectroscopic) data. In the LEPS method, the potential energy curve corresponding to the antibonding state is described by a modified Morse potential:

$$\begin{aligned} E_-(R) &= J_{\text{AB}} - K_{\text{AB}} \\ &\sim (D/2)(e^{-2\beta(R-R_e)} + 2e^{-\beta(R-R_e)}) \end{aligned} \tag{3.38}$$

which contains the same parameters as the Morse potential. We now have two equations for the two unknowns, that is, J_{AB} and K_{AB} can be determined as a function of the internuclear distance. For $H + H_2$ all pairs (AB, AC, and BC) are equivalent and the London equation can be evaluated. For $H + H_2$ a surface free of wells is obtained.

The LEPS method is also used for general triatomic systems. Here equations that are equivalent to Eqs (3.37) and (3.38) are written down and solved for the other diatomic pairs (AC and BC). The LEPS method provides a quick but somewhat crude estimate of potential energy surfaces.

Further reading/references

[1] P. Siegbahn and B. Liu, *J. Chem. Phys.* **68**, 2457 (1978)

[8] This approach was suggested by H. Eyring and M. Polanyi in *Z. Phys. Chem.* **B12**, 279 (1931).

[2] D.L. Diedrich and J.B. Anderson, *Science* **258**, 786 (1992)
[3] K. Stark and H.-J. Werner, *J. Chem. Phys.* **104**, 6515 (1996)
[4] J.C. Polanyi, *Acc. Chem. Res.* **5**, 161 (1972)
[5] D.R. Yarkony (ed.), *Modern electronic structure theory* Part I (World Scientific, Singapore, 1995)

Problems

3.1 From the force constants for the bonds C–H (460 N/m), C–C (440 N/m), and C=O (1300 N/m), calculate the corresponding harmonic frequencies and vibrational periods in femtoseconds (using classical mechanics).

3.2 Derive an expression for the (quadratic) force constant of a Morse potential

$$V(r) = D_e[1 - \exp\{-\beta(r - r_e)\}]^2$$

in terms of D_e and β.

3.3 A Lennard–Jones (12–6) potential

$$V(r) = 4\epsilon \left[\left(\frac{r_0}{r}\right)^{12} - \left(\frac{r_0}{r}\right)^{6} \right]$$

is often used to describe an intermolecular interaction that does not involve the formation of a covalent bond (for example, between noble gas atoms).

(a) Determine the internuclear distance at the potential energy minimum.

(b) Using this value, derive an expression for the (quadratic) force constant.

3.4 Assume that the results from a calculation of the electronic energy of an ABC molecule (at a fixed bond angle) close to the equilibrium bond lengths ($r^0_{\rm AB}, r^0_{\rm BC}$) are well represented by the expression

$$E(r_{\rm AB}, r_{\rm BC}) = (1/2)k_1(r_{\rm AB} - r^0_{\rm AB})^2 + (1/2)k_2(r_{\rm BC} - r^0_{\rm BC})^2$$

Make a contour plot of the potential.

3.5 Consider the calculation of electronic energies for the simplest bimolecular reaction

$$\rm H + H_2 \to (H \cdots H \cdots H)^{\ddagger} \to H_2 + H$$

using an electronic structure program available to you. First, choose the Hartree–Fock (HF) method and the (LCAO) basis set denoted by 6-311G**.

(a) Calculate the electronic energy of H_2 using 'geometry optimization' with a reasonable initial guess for the H–H distance. Note that the multiplicity of the electronic state is 'singlet'.

(b) Calculate the total electronic energy of the reactants at infinite distance, $H + H_2$, using the analytical result for the ground-state energy of the free hydrogen atom.

(c) Calculate the electronic energy at the saddle point, which corresponds to a linear symmetric configuration, denoted by $(\mathrm{H}\cdots\mathrm{H}\cdots\mathrm{H})^{\ddagger}$, with $\mathrm{H}\cdots\mathrm{H}$ distances of 0.93 Å. For this problem with three electrons, use the unrestricted Hartree–Fock (UHF) method and that the multiplicity of the electronic state is 'doublet'. Submit the calculation as 'single point energy'. Calculate the classical barrier height and compare to the exact barrier height reported in Example 3.1.

(d) With the program available to you, can you improve the agreement?

4

Bimolecular reactions, dynamics of collisions

Key ideas and results

We consider the dynamics of bimolecular collisions within the framework of (quasi-)classical mechanics and quantum mechanics, and show the following.

- The reaction cross-section σ_R is related to the reaction probability, which can be calculated theoretically from the collision dynamics based on classical mechanics.
- Models based on physically reasonable assumptions for the reaction probability lead to simple analytical expressions for the rate constant $k(T)$.
- As an introduction to reactive scattering, we consider classical two-body scattering and describe analytically how the deflection angle can be evaluated as a function of the impact parameter.
- Three-body (and many-body) quasi-classical scattering is formulated and the numerical evaluation of the reaction probability is described.
- The full quantum mechanical evaluation of the reaction cross-section σ_R is described.

The relation between the key quantities (the rate constant $k(T)$, the cross-section σ, and the reaction probability P) and various approaches to the description of the nuclear dynamics are illustrated below.

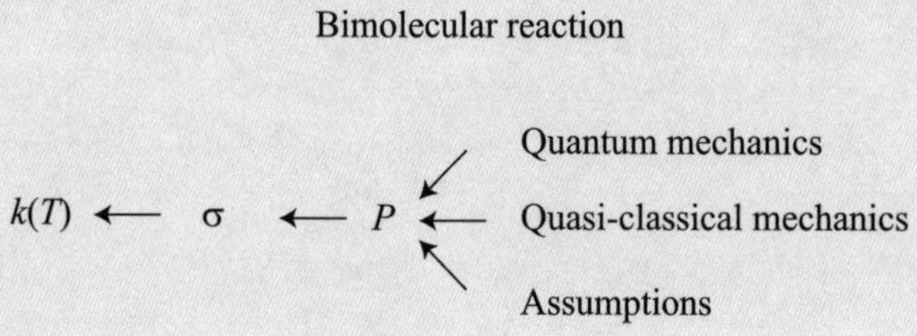

4.1 Quasi-classical dynamics

Essential features of the nuclear motion associated with chemical reactions can be described by classical mechanics. The special features of quantum mechanics cannot, of course, be properly described but some aspects like quantization can, in part, be taken into account by a simple procedure which basically amounts to a proper assignment of the trajectories to quantum states. Thus, the quantized states are naturally connected to a swarm of trajectories with the common feature of having the same energy. In short, this is the quasi-classical trajectory approach which has turned out to be quite successful. This approach is the subject of the present section.

4.1.1 Cross-sections from classical mechanics

A theoretical determination of the rate constant for a chemical reaction requires a calculation of the reaction cross-section based on the dynamics of the collision process between the reactant molecules. We shall develop a general relation, based on classical dynamics, between reaction probabilities that can be extracted from the dynamics of the collision process and the phenomenological reaction cross-section introduced in Chapter 2. That is, we give a recipe for how to calculate the reaction cross-section in accord with the general definition in Eq. (2.7).

We consider again the bimolecular collision process

$$\mathrm{A}(i, \boldsymbol{v}_\mathrm{A}) + \mathrm{B}(j, \boldsymbol{v}_\mathrm{B}) \rightarrow \mathrm{C}(l, \boldsymbol{v}_\mathrm{C}) + \mathrm{D}(m, \boldsymbol{v}_\mathrm{D}) \tag{4.1}$$

which is *reactive* when products C and D are different from reactants A and B, and *non-reactive* if they are identical. In the latter case we distinguish between an *elastic collision* process, if quantum states l and m are identical with i and j, and otherwise an *inelastic collision* process.

Since the outcome of the collision only depends on the relative motion of the reactant molecules, we begin with an elimination of the center-of-mass motion of the system. From classical mechanics it is known that the relative translational motion of two atoms may be described as the motion of one 'pseudo-atom', with the reduced mass $\mu = m_\mathrm{A} m_\mathrm{B}/(m_\mathrm{A} + m_\mathrm{B})$, relative to a fixed center of force. This result can be generalized to molecules by introducing proper relative coordinates, to be described in detail in Section 4.1.4.

The internal (vibrational and rotational) motion of molecule A is the same as that of the pseudo-molecule, while the center of mass of molecule B is at the fixed center of force. The force from the center of force on the pseudo-molecule is determined as the force between A and B, with A at the position of the pseudo-molecule and B at the position of the center of force. The scattering geometry is illustrated in Fig. 4.1.1. The pseudo-molecule moves with a velocity $\boldsymbol{v} = \boldsymbol{v}_\mathrm{A} - \boldsymbol{v}_\mathrm{B}$ relative to the fixed center of force. We have drawn a line through the force center parallel to $\boldsymbol{v}$ that will be convenient to use as a reference in the specification of the scattering geometry. In addition to the internal quantum states of the pseudo-molecule and velocity $\boldsymbol{v}$, the impact parameter b and angle ϕ are used to specify the motion of the molecule.

Several classical trajectories may result from such a collision process, as sketched in the figure. What makes the manifold of trajectories possible are the internal states i and j of the colliding molecules. To make that evident, let us first consider a situation where there are no internal states of the molecules and where the interaction potential only depends on the distance between the molecules, like for two hard spheres. Then there will only be one trajectory possible for a given b, ϕ, $\boldsymbol{v}$, because the initial conditions for the *deterministic* classical equations of motion are completely specified. This will not be the case when the molecules have internal degrees of freedom, even if

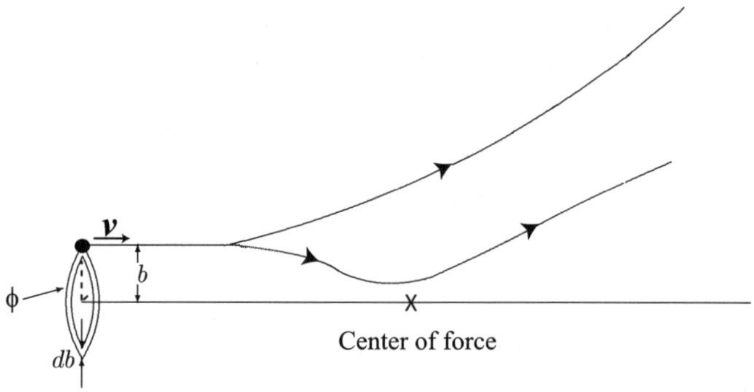

Fig. 4.1.1 Trajectories corresponding to the relative motion of two molecules. The filled circle represents the 'pseudo-molecule' and the cross the fixed center of force. Two trajectories are shown, for the same quantum states of the reactants, but with different 'phases' of the internal degrees of freedom, as explained in the text.

the internal states are completely specified by the appropriate quantum numbers, like i and j in our case.

In classical mechanics there is no quantization of states, so in order to represent a quantum state one resorts to an artifice which is often referred to as a *quasi-classical* approach (see Section 4.1.4 for an account of the computational details). The quantum state, like a vibrational or rotational state, has a given energy and we choose the energy of the classical system to be identical to the quantum energy. This does not, however, fully specify the coordinates and momenta since there will be an infinite number of choices that will correspond to a given energy. To explain the meaning of that, let us consider a diatomic molecule. Classically, the atoms perform at a given energy, an oscillatory motion corresponding to a vibrating chemical bond. From quantum mechanics, we know that the molecule may be in one of a discrete set of stationary vibrational quantum states with certain energies. Within the framework of classical mechanics, the specification of the quantum state does not, however, inform us about the 'phase' of the vibrator (see Fig. 4.1.2), that is the position q and momentum p, and thereby the potential energy and kinetic energy of the oscillator, at a given time. In the same manner, the molecule may be in one of the rotational quantum states with a certain energy, but we do not know the 'phase' of the rotator at a given time, that is, the orientation of the bond.

Therefore, the trajectory calculations with the same initial states, i, j, b, ϕ, $\boldsymbol{v}$, may lead to different trajectories, as shown in Fig. 4.1.1, when the phases of the internal degrees of freedom are different. It therefore makes sense to speak about a *transition probability* from an initial state to a final state, even in a case where the dynamics is governed by a deterministic equation, because of the indefiniteness of the phase of the internal degrees of freedom. In order to determine this probability, we need to make many trajectory calculations, with the given initial state, for different choices of the phases of the internal degrees of freedom. We may report the results of the trajectory calculation in Fig. 4.1.1, by counting the number, $dN_{\mathrm{C}(l,v_{\mathrm{C}},v_{\mathrm{C}}+dv_{\mathrm{C}})}(\Omega, \Omega + d\Omega)$, of C

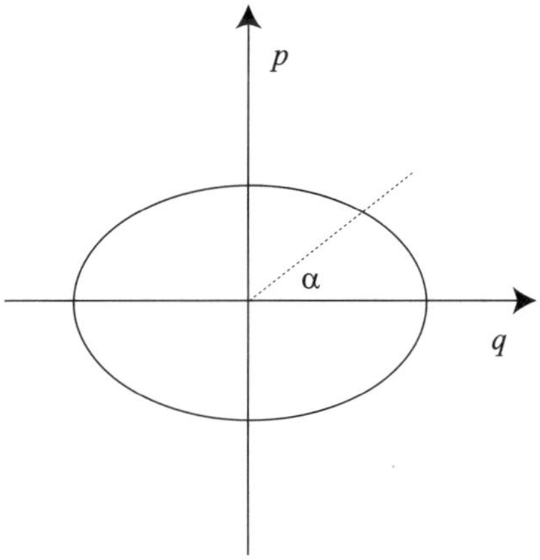

Fig. 4.1.2 Harmonic oscillator with the energy $E = p^2/(2m) + (1/2)kq^2$ (which is the equation for an ellipse in the (q,p)-space). In the quasi-classical trajectory approach, E is chosen as one of the quantum energies, and all points on the ellipse may be chosen as initial conditions in a calculation, i.e., corresponding to all phases $\alpha \in [0, 2\pi]$.

molecules in quantum state l and with the speed $v_C, v_C + dv_C$ in the direction $\Omega, \Omega + d\Omega$. This number may be expressed in terms of the transition probability P, according to

$$\begin{aligned} dN_{C(l,v_C,v_C+dv_C)}(\Omega, \Omega + d\Omega)\big|_{b,\phi} &\equiv \left.\frac{d^2N_{C(l,v_C)}(\Omega)}{dv_C d\Omega}\right|_{b,\phi} dv_C d\Omega \\ &= I_{\text{traj}} P(ij, v, b, \phi | l, v_C, \Omega) dv_C d\Omega \end{aligned} \tag{4.2}$$

Here I_{traj} is the total number of trajectories in the calculation corresponding to different phases of the internal states, and $P dv_C d\Omega$ is the fraction of trajectories leading to the desired result. The $|_{b,\phi}$ notation implies that the results refer to the given impact point b, ϕ.

Let us now relate this expression to the phenomenological expressions for the beam experiment in Chapter 2. First, we choose $d\Omega$ and dv_C in Eq. (4.2) to be identical to the ones in Chapter 2, which are determined by the detector. Then we realize the fundamental difference between dN_C in Eq. (4.2) and dN_C in Eq. (2.4). The former is given for a specific impact point b, ϕ, whereas the latter is averaged over all impact points, since they cannot be controlled in an experiment. So to make contact between experiment and theory, we have to average over all impact points b, ϕ in Eq. (4.2). I_{traj} was the number of trajectories in the simulation and therefore equivalent to the number of pseudo-molecules A colliding with B molecules in the experiment. Then considering the scattering geometry in Fig. 4.1.1, it will be natural to express I_{traj} in terms of a flux density $J_{A(i,v)}$, the area element $bdbd\phi$ associated with the b, ϕ point, and a time t:

$$I_{\text{traj}} = J_{A(i,v)} bdbd\phi\, t \tag{4.3}$$

We now choose $J_{A(i,v)}$ and t to be identical to the values of, respectively, the flux density and the duration of the experiment. From Eqs (4.2) and (4.3), we have

$$\left.\frac{d^2N_{C(l,v_C)}(\Omega)}{dv_C d\Omega}\right|_{b,\phi} dv_C d\Omega = J_{A(i,v)} t\, b db d\phi\, P(ij, v, b, \phi|l, v_C, \Omega) dv_C d\Omega \tag{4.4}$$

which, after averaging over all impact points, takes the form

$$\frac{d^2N_{C(l,v_C)}(\Omega)}{dv_C d\Omega} dv_C d\Omega = J_{A(i,v)} t\, dv_C d\Omega \int_0^{2\pi} \int_0^{\infty} P(ij, v, b, \phi|l, v_C, \Omega) b db d\phi \tag{4.5}$$

where we have removed $|_{b,\phi}$ on the left-hand side, because we now include all impact points.

This equation is now compared to Eq. (2.7), which after multiplication by the volume V of the reaction zone takes the form

$$\frac{d^3N_{C(l,v_C)}(\Omega)}{dv_C d\Omega dt} = \left(\frac{d^2\sigma_R}{dv_C d\Omega}\right)(ij, v|l, v_C, \Omega) J_{A(i,v)} N_{B(j,\boldsymbol{v}_B)} \tag{4.6}$$

It is important to note that the scattering angle in this equation refers to the angle between the incident and final directions of $\boldsymbol{v}_A$ and $\boldsymbol{v}_C$, respectively, viewed in the fixed laboratory coordinate system. The scattering in the theoretical analysis is viewed in the center-of-mass coordinate system and concerns the relative motion of molecules. The relation between the scattering angles and differential cross-sections in the two coordinate systems is discussed in Section 4.1.3. In the simulations there is only one scattering center, so we set $N_{B(j,\boldsymbol{v}_B)} = 1$ in the equation. Then we integrate over time t, in order to get the same differential number of C molecules on the left-hand sides of Eqs (4.6) and (4.5). Thus, we multiply both sides of Eq. (4.6) by dt, and integrate over time from 0 to t:

$$\frac{d^2N_{C(l,v_C)}(\Omega)}{dv_C d\Omega} = \left(\frac{d^2\sigma_R}{dv_C d\Omega}\right)(ij, v|l, v_C, \Omega) J_{A(i,v)} t \tag{4.7}$$

where

$$\frac{d^2N_{C(l,v_C)}(\Omega)}{dv_C d\Omega} = \int_0^{\infty} \frac{d^3N_{C(l,v_C)}(\Omega)}{dv_C d\Omega dt} dt$$

When compared to Eq. (4.5), we get

$$\left(\frac{d^2\sigma_R}{dv_C d\Omega}\right)(ij, v|l, v_C, \Omega) = \int_0^{2\pi} \int_0^{\infty} P(ij, v, b, \phi|l, v_C, \Omega) b db d\phi \tag{4.8}$$

When the speed of the product is not resolved, we integrate over v_C, and get

$$\left(\frac{d\sigma_R}{d\Omega}\right)(ij, v|l, \Omega) = \int_0^{2\pi} \int_0^{\infty} P(ij, v, b, \phi|l, \Omega) b db d\phi \tag{4.9}$$

where $P(ij, v, b, \phi|l, \Omega) = \int_0^{\infty} P(ij, v, b, \phi|l, v_C, \Omega) dv_C$. Likewise, when the angular distribution of the product is not resolved, the relevant cross-section is obtained, after integration over Ω, and we get the important relation between the *transition (reaction) probability* P, which may be calculated theoretically from the dynamics, and the *cross-section* related directly to experiments:

$$\boxed{\sigma_R(ij, v|l) = \int_0^{2\pi} \int_0^{\infty} P(ij, v, b, \phi|l) b db d\phi} \tag{4.10}$$

where $P(ij, v, b, \phi|l) = \int P(ij, v, b, \phi|l, \Omega) d\Omega$.

4.1.2 Simple models for chemical reactions

We consider in this section some crude approximations to actual reactive collisions. They provide some insight into the more complex real situations and, most importantly, simple analytical expressions for the rate constant are obtained.

It is assumed that the molecules are structureless hard spheres. Thus, A and B are described by *hard-sphere potentials* with diameters d_A and d_B, respectively. The interaction potential depends accordingly only on the distance between A and B, and

$$U(r) = \begin{cases} \infty & r \leqslant d = (d_A + d_B)/2 \\ 0 & r > d \end{cases} \tag{4.11}$$

In addition, assumptions for the reaction probability of the hard spheres—which strictly speaking cannot react—are introduced. That is, the reaction probability is not calculated from the actual potentials or dynamics of the collisions but simply postulated based on physical intuition. Note that the assumption of a spherically symmetric (hard-sphere) interaction potential implies that the reaction probability P cannot depend on ϕ (see Fig. 4.1.1), since there will be a cylindrical symmetry around the direction of the relative velocity. In addition, the assumption of structureless particles implies that the quantum numbers that specify the internal excitation cannot be defined within the present model.

Model 1, the collision frequency

A collision between A and B will occur if the impact parameter b fulfills the condition $b \leqslant d$, as seen from Fig. 4.1.3.

The following form for the reaction probability is assumed:

$$P(b) = \begin{cases} 1 & \text{for} \quad b \leqslant d \\ 0 & \text{for} \quad b > d \end{cases} \tag{4.12}$$

Thus, it is assumed that a reaction between the two particles occurs with unit probability when they collide, irrespective of their relative speed. The cross-section can be calculated from Eq. (4.10):

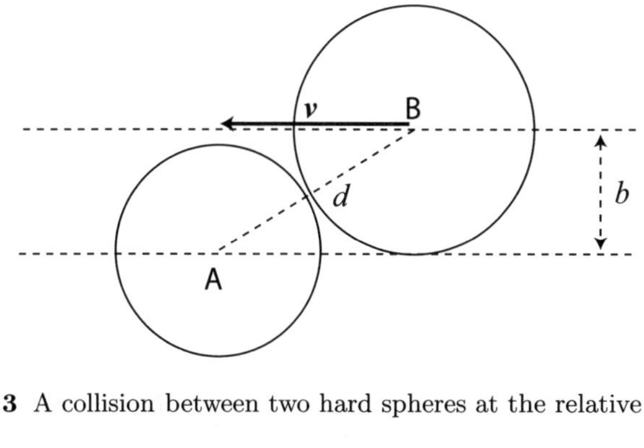

Fig. 4.1.3 A collision between two hard spheres at the relative velocity $\boldsymbol{v}$.

$$\sigma_R = 2\pi \int_0^\infty P(b) b db$$
$$= 2\pi \int_0^d b db$$
$$= \pi d^2 \tag{4.13}$$

that is, the cross-section is constant and independent of the relative kinetic energy. It is obvious that σ_R has the unit of an area, and the cross-section is the area that the reactants present to each other in the collision. Then, from Eq. (2.29), we find

$$k(T) = \pi d^2 \int_0^\infty v f_3(v) dv$$
$$= \pi d^2 \langle v \rangle \tag{4.14}$$

where

$$\langle v \rangle = \left(\frac{8 k_B T}{\pi \mu} \right)^{1/2} \tag{4.15}$$

is the average value of the relative speed of A and B. Typical values of $\langle v \rangle$ are of the order of 1000 m/s.

This result has a simple physical interpretation. The collision frequency between two hard spheres is given by[1]

$$Z_{\mathrm{AB}} = \pi d^2 \langle v \rangle [\mathrm{A}][\mathrm{B}] \tag{4.16}$$

that is, the number of collisions between A and B per second and per unit volume. Thus, the reaction rate predicted by this model is simply the collision frequency.

When compared to experimental data, this model predicts, typically, rate constants that are too large—by many orders of magnitude. In addition, the predicted temperature dependence is, usually, not in agreement with experimental observations, where often a dependence in agreement with the Arrhenius equation is found: $k(T) = A \exp(-E_a/k_B T)$.

Model 2, the energy threshold

In order to make a more realistic model, a refinement of the assumptions for the reaction probability is needed. That is, it is not realistic to assume that every collision will lead to a reaction. First, a reaction will occur only if the relative kinetic energy exceeds a certain critical value and, second, a head-on collision ($b = 0$) is more likely to lead to a reaction than a collision where the two hard spheres barely touch ($b \sim d$). These ideas are invoked in the following model.

[1] First, consider one A molecule that moves with the speed v relative to B. The number of collisions with B molecules within the time Δt is: $\pi d^2 v \Delta t$[B], since [B] is the number of moles of B molecules per unit volume and $\pi d^2 v \Delta t$ is the volume within which A will collide with B. It is a broken cylinder, since A changes direction after each elastic collision with a B molecule. Second, multiply by the number of moles of A molecules, that is [A]V, average over the distribution of speeds, and Eq. (4.16) is obtained after division by $V \Delta t$.

It is assumed that for a reaction to occur, the relative kinetic energy E_c along the line of centers shall exceed a critical value E^*. Thus,

$$P(E_c) = \begin{cases} 1 & \text{for} \quad E_c \geqslant E^* \\ 0 & \text{for} \quad E_c < E^* \end{cases} \tag{4.17}$$

This condition can be re-expressed as a function of the impact parameter b. From Fig. 4.1.4, we observe $|\boldsymbol{v}_c| = |\boldsymbol{v}| \cos\theta$ and

$$\begin{aligned} E_c/E_{\text{tr}} &= \frac{1/2\mu|\boldsymbol{v}_c|^2}{1/2\mu|\boldsymbol{v}|^2} \\ &= \cos^2\theta \\ &= 1 - \sin^2\theta \\ &= 1 - (b/d)^2 \end{aligned} \tag{4.18}$$

where $b \leqslant d$. That is, $(b/d)^2 = 1 - E_c/E_{\text{tr}}$ and when b increases the ratio E_c/E_{tr} decreases. The smallest value of E_c/E_{tr} that leads to reaction is E^*/E_{tr}. We can now determine the impact parameters that will lead to a reaction,

$$(b/d)^2 \leqslant 1 - E^*/E_{\text{tr}} \tag{4.19}$$

that is, the maximum value of the impact parameter b_{max} that will lead to a reaction is

$$b_{\text{max}} = d(1 - E^*/E_{\text{tr}})^{1/2} \tag{4.20}$$

The reaction probability as a function of the impact parameter b takes the form

$$P(b) = \begin{cases} 1 & \text{for} \quad b \leqslant b_{\text{max}} \\ 0 & \text{for} \quad b > b_{\text{max}} \end{cases} \tag{4.21}$$

The calculation of the cross-section is now equivalent to Eq. (4.13), that is,

$$\begin{aligned} \sigma_R(E_{\text{tr}}) &= \pi b_{\text{max}}^2 \\ &= \pi d^2(1 - E^*/E_{\text{tr}}) \end{aligned} \tag{4.22}$$

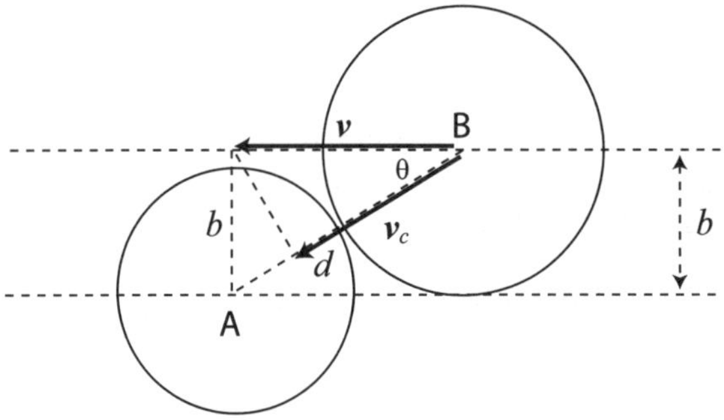

Fig. 4.1.4 The relative velocity $\boldsymbol{v}$ and its projection $\boldsymbol{v}_c$ along the line of centers; otherwise same as Fig. 4.1.3.

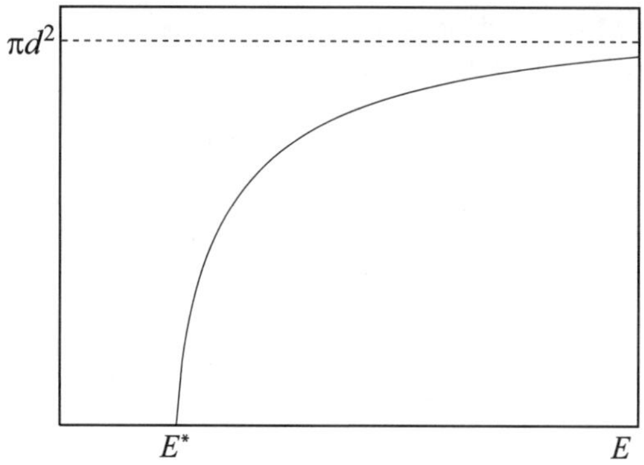

Fig. 4.1.5 Energy dependence of the reaction cross-section in 'model 2'.

for $E_{\rm tr} \geqslant E^*$ and 0 otherwise. The *reaction cross-section* as a function of energy is shown in Fig. 4.1.5.

The rate constant is evaluated by applying Eq. (2.29):

$$
\begin{aligned}
k(T) &= \int_0^\infty \sqrt{2E_{\rm tr}/\mu}\,\sigma_R(E_{\rm tr}) f_3(E_{\rm tr}) dE_{\rm tr} \\
&= \frac{1}{k_B T}\left(\frac{8}{\pi\mu k_B T}\right)^{1/2} \int_{E^*}^\infty \sigma_R(E_{\rm tr}) E_{\rm tr} e^{-E_{\rm tr}/k_B T} dE_{\rm tr} \\
&= \frac{1}{k_B T}\left(\frac{8}{\pi\mu k_B T}\right)^{1/2} \pi d^2 \int_{E^*}^\infty (E_{\rm tr} - E^*) e^{-E_{\rm tr}/k_B T} dE_{\rm tr} \\
&= \pi d^2 \left(\frac{8 k_B T}{\pi\mu}\right)^{1/2} e^{-E^*/k_B T}
\end{aligned}
\qquad (4.23)
$$

where the last integral is evaluated by partial integration. The temperature dependence predicted by this model is, typically, in good agreement with experimental observations. If, however, the pre-exponential factor in Eq. (4.23) is compared with experimental data, as in Example 4.1 below, it is often found to be 10–100 times too big!

The physical interpretation of this result is, relatively, simple. The reaction rate predicted by the model is equal to the collision frequency, Eq. (4.16), times the factor $\exp(-E^*/k_B T)$. This factor is clearly related to the Boltzmann distribution.[2] To that end, let us evaluate the probability of finding a relative velocity, irrespective of its direction, corresponding to a free translational energy $E_{\rm tr} = (1/2)\mu v^2$ that exceeds $E_{\rm tr} = E^*$ (see Problem 1.3):

[2]The interpretation of the factor $\exp(-E^*/k_B T)$ is, in many physical chemistry textbooks, somewhat inaccurate.

$$
\begin{aligned}
P_3(E_{\rm tr} > E^*) &= \int_{E^*}^{\infty} f_3(E_{\rm tr}) dE_{\rm tr} \\
&= 1 - \mathrm{erf}\left[\sqrt{\frac{E^*}{k_B T}}\right] + \frac{2}{\sqrt{\pi}}\sqrt{\frac{E^*}{k_B T}}\, e^{-E^*/k_B T}
\end{aligned}
\tag{4.24}
$$

where erf denotes the error function. Thus, this expression is not equal to $\exp(-E^*/k_B T)$, although this factor plays a dominating role; see Fig. 4.1.6. Actually, from the first line in Eq. (4.23), we observe that the rate constant will be proportional to the probability factor of Eq. (4.24), only when the cross-section $\sigma_R(E_{\rm tr})$ is 0 for $E_{\rm tr} < E^*$ and $A/\sqrt{E_{\rm tr}}$ for $E_{\rm tr} \geqslant E^*$, where A is a constant. That is, it is *not* possible to interpret Eq. (4.23) as a collision frequency times a probability factor, related to the relevant Boltzmann distribution for free translational motion in three dimensions.

When we consider the Boltzmann distribution associated with free translational motion in two dimensions (Problem 2.2) and the associated probability of finding relative translational energies that exceed E^*, we get

$$
\begin{aligned}
P_2(E_{\rm tr} > E^*) &= \int_{E^*}^{\infty} e^{-E_{\rm tr}/k_B T}/(k_B T) dE_{\rm tr} \\
&= e^{-E^*/k_B T}
\end{aligned}
\tag{4.25}
$$

which happens to be identical to the probability factor in Eq. (4.23). Furthermore, this factor is contained in the expression for the rate constant, whenever there is an energy threshold in the cross-section (see Problem 2.3).

Thus, we have only a rough identification of the factor $\exp(-E^*/k_B T)$ with the probability of finding relative translational energies that exceed E^*. If one ignores the different density of energy states for free translational motion in two and three dimensions, one can claim that $\exp(-E^*/k_B T)$ is the probability of finding relative translational energies that exceed E^*.

The *activation energy* E_a is defined from the Arrhenius equation, that is, $k(T) = A\exp(-E_a/k_B T)$, where A is a constant. According to this equation, we can extract

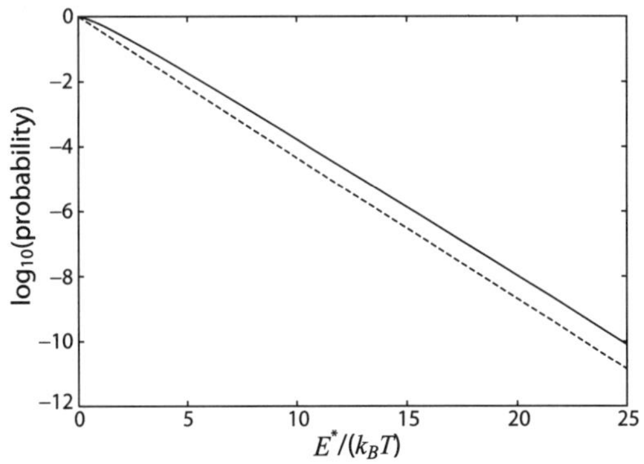

Fig. 4.1.6 The probabilities $P(E_{\rm tr} > E^*)$, Eq. (4.24) (solid line) and Eq. (4.25) (dashed line), of finding relative translational energies between reactants that exceed $E_{\rm tr} = E^*$.

the activation energy from a plot of $\ln[k(T)]$ versus $1/T$, which implies

$$E_a = -k_B \frac{d \ln k(T)}{d(1/T)} = k_B T^2 \frac{d \ln k(T)}{dT} \tag{4.26}$$

This equation is used as a general definition of the activation energy whether or not A is actually a constant. For the model above we get

$$E_a = k_B T^2 \left(\frac{1}{2T} + \frac{E^*}{k_B T^2} \right) = E^* + k_B T/2 \tag{4.27}$$

Since typically $E^* \gg k_B T/2$, the predicted activation energy is in practice independent of the temperature.

Example 4.1: Comparison of model and experimental data

We can use Eq. (4.23) to calculate the magnitude of the pre-exponential factor and compare it to experimental data. Molecular diameters d_{A} (or d_{B}) may, e.g., be estimated from gas viscosity data based on the standard model of the kinetic theory of gases, or from an analysis of the elastic scattering of molecules (see Section 4.1.3). Some values are given in Table 4.1 and are typically around 2–3 Å. Table 4.1 also lists a collision frequency; here as the number of collisions that a *single* molecule undergoes per second. According to Eq. (4.16), the number of collisions that a single A molecule undergoes per second with all the B molecules is $Z = \pi d^2 \sqrt{8k_B T/(\pi\mu)} N_A[\mathrm{B}]$, where Avogadro's constant has been introduced in order to convert from moles to numbers. $N_A[\mathrm{B}] = p_{\mathrm{B}}/(k_B T)$, $R = N_A k_B$, according to the ideal gas law, and $d = (d_{\mathrm{A}} + d_{\mathrm{B}})/2 = d_{\mathrm{B}}$, since A $\equiv$ B in this case. Thus the number of collisions per second is very large, of the order of $\sim 10^{10}$ at 1 atm and 300 K. Table 4.2 compares calculated (A_{th}) and experimental[3] (A_{exp}) pre-exponential factors for several reactions. Typically, the calculated values of the pre-exponential factors are of the order of 10^{11} dm^3 mol^{-1} s^{-1}.

Table 4.1 Collision parameters. The collision frequency is calculated at 1 atm and 300 K.

Species	Molecular diameter, d_{B}/Å	Collision frequency, Z/s^{-1}
H_2	2.4	1.1×10^{10}
CO	3.2	5.3×10^{9}
O_2	3.6	6.3×10^{9}
F_2	3.6	5.8×10^{9}
CH_4	3.8	9.9×10^{9}

[3] A good source of experimental data is the NIST kinetics database `http://kinetics.nist.gov`.

Table 4.2 Comparison of collision model and experimental data. Pre-exponential factors are given in units of $dm^3\ mol^{-1}\ s^{-1}$. The experimental data are from *J. Chem. Phys.* **92**, 4811 (1980); *J. Phys. Chem. Ref. Data* **15**, 1087 (1986); and *J. Phys. Chem. A* **106**, 6060 (2002), respectively. Note that the third reaction is a bimolecular association reaction. For this reaction, the experimental data are derived in the high-pressure limit.

Reaction	T/K	E_a/(kJ mol^{-1})	A_{th}	A_{exp}	A_{exp}/A_{th}
$F + H_2 \rightarrow HF + H$	300	3.6	1.5×10^{11}	6.0×10^{10}	0.4
$CO + O_2 \rightarrow CO_2 + O$	300	200	1.4×10^{11}	2.5×10^{9}	0.018
$CH_3 + CH_3 \rightarrow C_2H_6$	300	0	2.5×10^{11}	1.5×10^{10}	0.06

In order to obtain agreement with experimental data, a realistic interaction potential must be used and it is necessary to calculate the reaction probability from the basic equations of motion.

4.1.3 Two-body classical scattering

We consider in this section the collision between two (structureless) particles that interact via a spherically symmetric potential $U(r)$, where r is the distance between the particles [1]. Since there are no internal degrees of freedom that may exchange energy with the translational degrees of freedom, the collision is elastic, i.e., the total kinetic energy of the two particles before and after the collision is conserved (as mentioned earlier, we can also have elastic collisions when molecules collide, which implies that no energy is exchanged with internal degrees of freedom). Thus, the present section describes the dynamics of atomic collisions with the relevant potential $U(r)$ given by the electronic energy.

The dynamics of the two-particle problem can be separated into center-of-mass motion and relative motion with the reduced mass $\mu = m_A m_B/(m_A + m_B)$, of the two particles. The kinetic energy of the relative motion is a conserved quantity. The outcome of the elastic collision is described by the deflection angle of the trajectory, and this is the main quantity to be determined in the following. The *deflection angle*, χ, gives the deviation from the incident straight line trajectory due to attractive and repulsive forces. Thus, χ is the angle between the final and initial directions of the relative velocity vector for the two particles.

The spherical symmetry of the interaction implies, in particular, that the angular momentum of the relative motion is conserved. That is, since the angular momentum is a vector, both direction and magnitude are conserved quantities. The collision process will, accordingly, take place in the plane defined by the initial values of the radius vector and the momentum vector. This implies that only two coordinates are required in order to describe the relative motion. These coordinates are chosen as the polar coordinates in the plane (r, θ). The scattering in the center-of-mass coordinate system is shown in Fig. 4.1.7.

The total energy associated with the relative motion can be written in the form (see Problem 4.2)

$$E = (1/2)\mu v^2 + \frac{L^2}{2\mu r^2} + U(r) \tag{4.28}$$

where

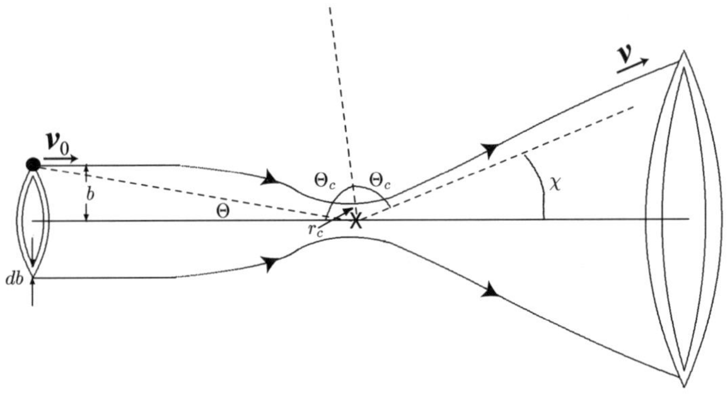

Fig. 4.1.7 Two-body classical scattering in a spherically symmetric potential. The relative motion of two atoms may be described as the motion of one 'pseudo-atom', with the reduced mass $\mu = m_A m_B/(m_A+m_B)$, relative to a fixed center of force. Two trajectories are shown; for the first trajectory, the final and initial relative velocity vector and the associated deflection angle χ are shown. This trajectory corresponds to the impact point $(b, \phi = 0)$, whereas the second trajectory corresponds to the impact point $(b, \phi = \pi)$.

$$\begin{aligned} v &= dr/dt \\ L &= \mu r^2 d\theta/dt \end{aligned} \tag{4.29}$$

The energy (E) and the magnitude of the angular momentum (L) are constants of motion and their magnitudes are determined as follows. Long before the collision ($t \to -\infty$) we have $r \to \infty$ and $\lim_{r\to\infty} U(r) = 0$, and from Eq. (4.28) we obtain $E = (1/2)\mu v_0^2$, where v_0 is the relative speed prior to the collision. The angular momentum is given by $\boldsymbol{L} = \boldsymbol{r} \times \mu \boldsymbol{v}_0$ and the magnitude is therefore given by $L = \mu r v_0 \sin\theta$, where θ is the angle between $\boldsymbol{r}$ and $\boldsymbol{v}_0$. The *impact parameter* is defined by $b = r \sin\theta$, and

$$\begin{aligned} L &= \mu v_0 b \\ &= b\sqrt{2\mu E} \end{aligned} \tag{4.30}$$

Thus, the angular momentum can be expressed in terms of the energy and the impact parameter.[4]

We may find the distance of closest approach $r = r_c$ between the two particles by solving the equation $dr/dt = 0$. From Eqs (4.28), (4.29), and (4.30), we find

$$\begin{aligned} \left(\frac{dr}{dt}\right)^2 &= \frac{2}{\mu}\left(E - \frac{L^2}{2\mu r^2} - U(r)\right) \\ &= \frac{2}{\mu}\left(E - \frac{Eb^2}{r^2} - U(r)\right) \\ &= \frac{2}{\mu}(E - V_{\text{eff}}(r)) \end{aligned} \tag{4.31}$$

[4]Note that from Eqs (4.28) and (4.30) we have the relation $(1/2)\mu v^2 = E(1 - b^2/r^2) - U(r)$, which is equivalent to Eq. (4.18) when $U(r)$ is a hard-sphere potential.

where the effective potential is defined by

$$V_{\text{eff}}(r) \equiv U(r) + \frac{Eb^2}{r^2} = U(r) + \frac{L^2}{2\mu r^2} \tag{4.32}$$

and see that if $E = V_{\text{eff}}(r)$ has a real-valued positive solution there will be a distance of closest approach, also referred to as a *turning point* for the two particles, at the total energy E. Obviously, such a solution will always exist for potentials with a repulsive part, as in the present case where the potentials are repulsive at small internuclear distances. Turning points may also exist in the attractive part of a potential where $U(r) < 0$ provided that $b > 0$. As illustrated in Fig. 4.1.8 where $U(r)$ is chosen as a Lennard–Jones potential, see Eq. (4.46), the distance of closest approach depends on the impact parameter b.

When $b > 0$, a barrier can show up in the effective potential, due to the centrifugal energy. In Fig. 4.1.8(a), the total energy E_1 is below the centrifugal barrier, whereas

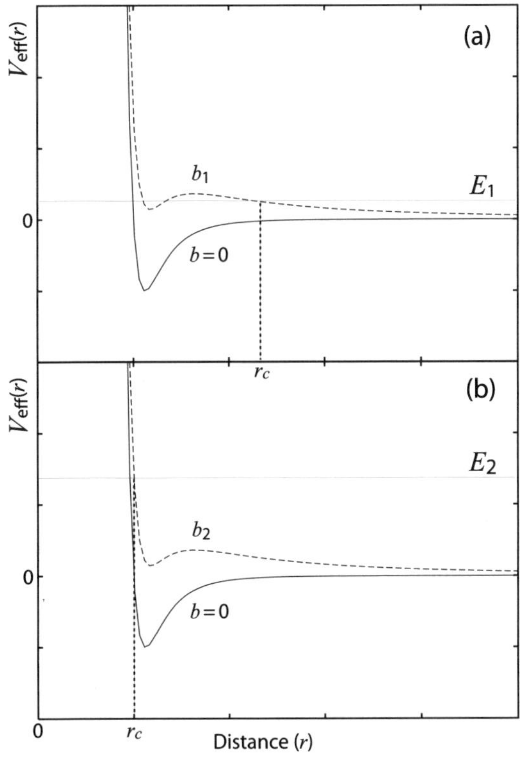

Fig. 4.1.8 Two-body classical scattering in a spherically symmetric potential. The effective potential $V_{\text{eff}}(r)$, defined in Eq. (4.32), is shown for two different impact parameters, $b = 0$ where $V_{\text{eff}}(r) = U(r)$ and $b > 0$. The upper panel (a) is for a fixed total energy E_1, and the lower panel (b) for another total energy E_2. The intersection with $V_{\text{eff}}(r)$ gives the distance of closest approach r_c.

the total energy in Fig. 4.1.8(b) is above this barrier.

In order to obtain an expression for the deflection angle, we get from Eq. (4.29)

$$d\theta = \frac{L}{\mu r^2} dt \tag{4.33}$$

Integration gives

$$\theta(t) = \int_{-\infty}^{t} \frac{L}{\mu r^2} dt \tag{4.34}$$

since $\theta(t \to -\infty) = 0$. We may eliminate time from Eq. (4.34) and derive an equation between θ and r by using Eq. (4.31):

$$dt = -dr \Big/ \sqrt{\frac{2}{\mu}\left(E - \frac{L^2}{2\mu r^2} - U(r)\right)} \tag{4.35}$$

where the solution with the minus sign in front of the square root is chosen because r decreases when t increases. Equation (4.35) is now introduced into Eq. (4.34), and we get

$$\begin{aligned} \theta(r) &= -\int_{\infty}^{r} \frac{L/(\mu\tilde{r}^2)}{\sqrt{2\left(E - L^2/(2\mu\tilde{r}^2) - U(\tilde{r})\right)/\mu}} d\tilde{r} \\ &= -b\int_{\infty}^{r} \frac{1}{\tilde{r}^2\sqrt{1 - b^2/\tilde{r}^2 - U(\tilde{r})/E}} d\tilde{r} \end{aligned} \tag{4.36}$$

where Eq. (4.30) was used in the last line.

As shown above, for potentials with a repulsive part, a closest approach distance $r = r_c$ will exist, and this distance corresponds, according to the expression above, to an angle $\theta = \theta_c$ (see Fig. 4.1.7). When the collision is over, we have $t \to \infty$ and $r \to \infty$. It is easily verified that the same angle, θ_c, is found if we had started the integration in Eq. (4.34) from this limit.[5] Thus, the trajectory is symmetric around a line at $\theta = \theta_c$. A (straight) trajectory that is not deflected corresponds to $\theta = \pi$. The deflection angle is, accordingly, defined as

$$\boxed{\begin{aligned} \chi(E, b) &= \pi - 2\theta_c \\ &= \pi - 2b\int_{r_c}^{\infty} \frac{1}{r^2\sqrt{1 - b^2/r^2 - U(r)/E}} dr \end{aligned}} \tag{4.37}$$

This expression shows that, in general, the deflection angle depends on both the impact parameter and the total energy which is equal to the relative kinetic energy prior to the collision.

[5]The limits in Eq. (4.34) are now from ∞ to t. The minus sign in Eq. (4.35) is changed to plus and the limits for the integration in $\tilde{r}$ are the same as in Eq. (4.36). Thus, the absolute value of the angles is the same.

Let us derive a relation between the deflection angle and the scattering cross-section. From the definition of the cross-section, Eq. (4.6), specialized to classical elastic two-body scattering with one scattering center, we get

$$\frac{d^2 N_{\mathrm{A}}(\Omega)}{d\Omega dt} d\Omega = \left(\frac{d\sigma_R}{d\Omega}\right)(v|\Omega) J_{\mathrm{A}(v)} d\Omega \tag{4.38}$$

for the number of A particles emerging, per unit time, in the space angle $d\Omega$ around the direction Ω. The experimentally observed deflection angle is the angle between the final and the incident directions of the scattered particles in a laboratory fixed coordinate system. The deflection angle determined above is viewed in the center-of-mass (c.m.) coordinate system and refers to the relative motion as shown in Fig. 4.1.7. The relation between the scattering angles and differential cross-sections in the center-of-mass coordinate system and in the fixed laboratory coordinate system is discussed, shortly, in the following subsection.

As seen from Fig. 4.1.7, a trajectory that starts in the area element $bdbd\phi$ will end up in the solid angle $d\Omega_{\mathrm{c.m.}}$ with scattering angles between $\Omega_{\mathrm{c.m.}}$ and $\Omega_{\mathrm{c.m.}} + d\Omega_{\mathrm{c.m.}}$, where $d\Omega_{\mathrm{c.m.}} = \sin\chi d\chi d\phi$. The number of scattered particles, per unit time, is accordingly also identical to the product of the flux density and the area element $bdbd\phi$:

$$J_{\mathrm{A}(v)} bdbd\phi = \left(\frac{d\sigma_R}{d\Omega}\right)_{\mathrm{c.m.}} (v|\Omega) J_{\mathrm{A}(v)} \sin\chi d\chi d\phi \tag{4.39}$$

Specializing further to scattering in a spherically symmetric potential, the cross-section will be independent of the angle ϕ, and

$$bdb = \left(\frac{d\sigma_R}{d\Omega}\right)_{\mathrm{c.m.}} (\chi, E) \sin\chi d\chi \tag{4.40}$$

where we have replaced the relative speed v with the equivalent total relative energy E of the collision. The *differential scattering cross-section* for the elastic scattering is then

$$\left(\frac{d\sigma}{d\Omega}\right)_{\mathrm{c.m.}} (\chi, E) = \frac{b}{|d\chi/db| \sin\chi} \tag{4.41}$$

where absolute signs are introduced because the cross-section must always be positive while b and χ can vary in opposite directions. Note that only one impact parameter b contributes to this differential cross-section, and that b can be expressed in terms of χ and E via Eq. (4.37).

Let us consider a few examples. We begin with a *hard-sphere potential*

$$U(r) = \begin{cases} \infty & \text{for } r \leqslant d \\ 0 & \text{for } r > d \end{cases} \tag{4.42}$$

For $b \leqslant d$, the closest approach distance is $r_c = d$, and for $r > d$ we have $U(r) = 0$, that is,[6]

[6]Note that $d[\cos^{-1} u(x)]/dx = -1/[\sqrt{1-u^2}]du(x)/dx$, where $\cos^{-1}(x) \equiv \arccos(x)$.

$$\begin{aligned}\chi(E,b) &= \pi - 2b\int_d^\infty \frac{1}{r^2\sqrt{1-b^2/r^2}}dr \\ &= \pi - 2b[(1/b)\cos^{-1}(b/r)]_{r=d}^{r\to\infty} \\ &= \pi - 2[\cos^{-1}(0) - \cos^{-1}(b/d)] \\ &= 2\cos^{-1}(b/d) \end{aligned} \tag{4.43}$$

Note that for a head-on collision $\chi(b = 0) = \pi$, that is, in this case the outcome is a backward scattering, whereas for $b = d$ the deflection angle is $\chi(b = d) = 0$. Furthermore, the result is always independent of the energy. The inverse cosine is defined only for $b \leqslant d$. However, if $b > d$ the closest approach distance is $r_c = b$, and when d is replaced by b in the equation above, we get $\chi(b > d) = 0$ as expected. This relation is illustrated in Fig. 4.1.9.

The differential scattering cross-section takes (after a little algebra) the form

$$\left(\frac{d\sigma}{d\Omega}\right)_{\text{c.m.}}(\chi, E) = d^2/4 \tag{4.44}$$

That is, the scattering is isotropic, i.e., it is independent of the deflection angle (which is, in general, not the case for more realistic potentials). The *total cross-section* takes the form

$$\begin{aligned}\sigma &= \int \left(\frac{d\sigma}{d\Omega}\right)(\chi, E)d\Omega \\ &= (d^2/4)\int_0^\pi 2\pi \sin\chi d\chi \\ &= \pi d^2 \end{aligned} \tag{4.45}$$

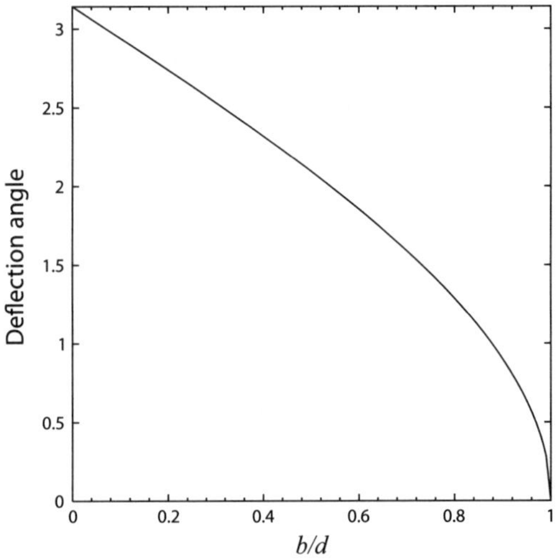

Fig. 4.1.9 The deflection angle χ (in radians) as a function of the impact parameter b, for the collision between two hard spheres with the average diameter d.

a result that we also encountered in Eq. (4.13). The cross-section is an effective area of the target in the scattering process; see Fig. 2.1.2. This result confirms our intuition concerning scattering of hard spheres.

As an example of a more realistic potential, we consider the *Lennard–Jones potential*

$$U(r) = 4\epsilon\left[\left(\frac{r_0}{r}\right)^{12} - \left(\frac{r_0}{r}\right)^{6}\right] \tag{4.46}$$

This potential has an attractive part as well as a repulsive part. The minimum is at $r_m = 2^{1/6}r_0$ where $U(r_m) = -\epsilon$. Equation (4.37) may also be evaluated analytically in this case, but here we will only present a short outline of the results. When b is small the deflection function is very similar to that for hard-sphere collisions, that is, $\chi \to \pi$. When b is large the deflection function is again very similar to the hard-sphere result, that is $\chi \to 0$ corresponding to no scattering. This limit is reached more rapidly when the relative energy E/ϵ is large. In the intermediate region of b values, a new type of behavior is observed due to the attractive part of the potential. Thus, at selected energies one can, e.g., find trajectories corresponding to *orbiting*. As illustrated in Fig. 4.1.8, we have identified a turning point as the point where $E = V_{\text{eff}}(r)$, since this implies that $dr/dt = 0$. If this point happens to be close to the top of the barrier of the effective potential, where $dV_{\text{eff}}(r)/dr = 0$, the motion away from the turning point will be very slow, and in the meantime the associated angular motion might have gone through angles larger than 2π.

The differential cross-section depends, in general, on the energy, E, as well as the angle, χ. Two types of singularities can show up in the classical expression for the differential cross-section, Eq. (4.41). The first type of singularity occurs when $\sin\chi = 0$, i.e., for $\chi = 0$ and $\chi = \pi$, provided $b/|d\chi/db|$ is finite at these angles. This singularity arises due to the vanishing space angle $d\Omega$ at these two deflection angles. The second type of singularity occurs if $d\chi/db = 0$, i.e., at an extremum in the deflection function. The two types of singularities are referred to as *glory scattering* and *rainbow scattering*, respectively (named in analogy to corresponding phenomena in optics).

Scattering angles and differential cross-section in various frames

We have considered above the deflection of the trajectory in the center-of-mass coordinate system. The deflection angle χ is the angle between the final and initial directions of the relative (velocity) vector between the two particles. Experimental observations normally take place in a coordinate system that is fixed in the laboratory, and the scattering angle Θ measured here is the angle between the final and the incident directions of the scattered particle. These two angles would be the same only if the second particle had an infinite mass. Thus, we need a relation between the angles in the two coordinate systems in order to be able to compare calculations with experiments.

The relation between the *deflection angles* in the two coordinate systems is derived below, in the special case where the target atom is at rest before the collision. This case represents of course not the typical situation in a crossed molecular-beam experiment. However, it greatly simplifies the relation and the derivation displays the essence of the problem. The general case is considered in Appendix C.

Particle A is the incident particle and B the target. The position of particle A in the *laboratory coordinate system* $\boldsymbol{r}_{\mathrm{A}}$ can be expressed in terms of the position of particle A in the *center-of-mass coordinate system* $\boldsymbol{r}'_{\mathrm{A}}$, and the position of the center-of-mass in the laboratory coordinate system $\boldsymbol{R}$:

$$\boldsymbol{r}_{\mathrm{A}} = \boldsymbol{R} + \boldsymbol{r}'_{\mathrm{A}} \tag{4.47}$$

and the corresponding relation for the velocities is

$$\boldsymbol{v}_{\mathrm{A}} = \boldsymbol{V} + \boldsymbol{v}'_{\mathrm{A}} \tag{4.48}$$

which is obtained from Eq. (4.47) by differentiation with respect to time. Note that a velocity vector is tangent to the path specified by the position vectors.

The center-of-mass velocity $\boldsymbol{V}$ is a constant of motion, i.e.,

$$\boldsymbol{V} = (m_{\mathrm{A}}\boldsymbol{v}_{\mathrm{A}} + m_{\mathrm{B}}\boldsymbol{v}_{\mathrm{B}})/M = (m_{\mathrm{A}}\boldsymbol{v}^0_{\mathrm{A}} + m_{\mathrm{B}}\boldsymbol{v}^0_{\mathrm{B}})/M = (m_{\mathrm{A}}/M)\boldsymbol{v}^0_{\mathrm{A}} \tag{4.49}$$

where $\boldsymbol{v}^0_{\mathrm{A}}$ and $\boldsymbol{v}^0_{\mathrm{B}}$ are the start velocities before the scattering event, and $\boldsymbol{v}^0_{\mathrm{B}} = 0$ since B is assumed to be at rest before the collision. We choose the laboratory coordinate system such that $\boldsymbol{V}$ is parallel to the x-axis. The scattering angle Θ in the laboratory coordinate system, i.e., the angle between $\boldsymbol{v}_{\mathrm{A}}$ and $\boldsymbol{v}^0_{\mathrm{A}}$, is then as illustrated in Fig. 4.1.10.

From Eq. (4.49) and the definition of the relative velocity

$$\boldsymbol{v} = \boldsymbol{v}_{\mathrm{A}} - \boldsymbol{v}_{\mathrm{B}} \tag{4.50}$$

we get $\boldsymbol{v}_{\mathrm{A}} = \boldsymbol{V} + (m_{\mathrm{B}}/M)\boldsymbol{v}$, and from Eq. (4.48)

$$\boldsymbol{v}'_{\mathrm{A}} = (m_{\mathrm{B}}/M)\boldsymbol{v} \tag{4.51}$$

that is, $\boldsymbol{v}'_{\mathrm{A}}$ and the relative velocity $\boldsymbol{v}$ have the same direction. Before the scattering event $\boldsymbol{v}^0 = \boldsymbol{v}^0_{\mathrm{A}} - \boldsymbol{v}^0_{\mathrm{B}} = \boldsymbol{v}^0_{\mathrm{A}}$, which was parallel to $\boldsymbol{V}$ and the x-axis. Thus, $\boldsymbol{V}$, $\boldsymbol{v}^0_{\mathrm{A}}$, and $\boldsymbol{v}^0$ are collinear. The scattering angle χ, i.e., the angle between $\boldsymbol{v}$ and $\boldsymbol{v}^0$, is then as illustrated in Fig. 4.1.10.

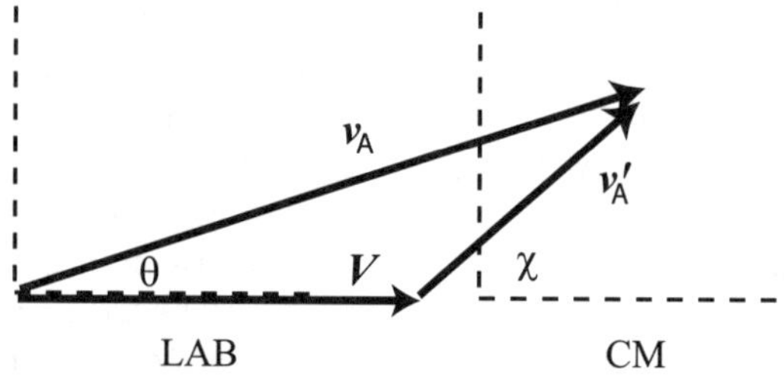

Fig. 4.1.10 The relation in Eq. (4.48) after the scattering has been completed. The laboratory coordinate system (LAB) is chosen such that $\boldsymbol{V}$ is parallel to the x-axis and the deflection angles Θ and χ with respect to, respectively, this system and the center-of-mass coordinate system (CM) are defined.

We consider Eq. (4.48) after the scattering has been completed. The angles of $\boldsymbol{v}_{\rm A}$ and $\boldsymbol{v}'_{\rm A}$ with respect to the x-axis are now the scattering angles defined above. The components of the two velocities along the y-axis are

$$v_{\rm A}\sin\Theta = v'_{\rm A}\sin\chi \tag{4.52}$$

and along the x-axis

$$v_{\rm A}\cos\Theta = v'_{\rm A}\cos\chi + V \tag{4.53}$$

The ratio between these two relations gives

$$\tan\Theta = \frac{\sin\chi}{\cos\chi + (V/v'_{\rm A})} \tag{4.54}$$

Now the magnitude of the center-of-mass velocity is, according to Eq. (4.49), $V = (m_{\rm A}v_{\rm A} + m_{\rm B}v_{\rm B})/M = (m_{\rm A}/M)v^0_{\rm A} = (m_{\rm A}/M)v^0$, where $v^0_{\rm A}$ is the initial speed of particle A. The magnitude of the velocity of A with respect to the center-of-mass system is obtained from Eq. (4.51), $v'_{\rm A} = (m_{\rm B}/M)v = (m_{\rm B}/M)v^0$, since the relative speed v is constant for an elastic collision. Thus, the relation between the *scattering angles in the two frames* is

$$\tan\Theta = \frac{\sin\chi}{\cos\chi + (m_{\rm A}/m_{\rm B})} \tag{4.55}$$

This relation is illustrated in Fig. 4.1.11, where the impact parameter has been introduced via Eq. (4.43). The impact parameter cannot, of course, be precisely controlled for objects of atomic size. When $m_{\rm B} \to \infty$, corresponding to a stationary target, we observe that the two angles coincide.

The *differential cross-section* $d\sigma/d\Omega$ is not invariant when we change the description from one coordinate system to another. Clearly, due to the relation in Eq. (4.55) a change in χ will not lead to the same change in Θ and the space angle $d\Omega_{\rm c.m.} = \sin\chi d\chi d\phi$ is not identical to the space angle $d\Omega = \sin\Theta d\Theta d\phi$ in the laboratory frame. Thus,

$$\left(\frac{d\sigma}{d\Omega}\right) = \left(\frac{d\sigma}{d\Omega}\right)_{\rm c.m.}\left|\frac{d\Omega_{\rm c.m.}}{d\Omega}\right| \tag{4.56}$$

where $d\sigma/d\Omega$ is the cross-section associated with experimental observations, and the absolute value is taken in the last factor because the differential cross-sections are positive, and an increment in χ will not always give rise to an increment in Θ. It can now be shown that (see Appendix C)

$$\left(\frac{d\sigma}{d\Omega}\right) = \left(\frac{d\sigma}{d\Omega}\right)_{\rm c.m.} \frac{\left(1 + \left(\frac{m_{\rm A}}{m_{\rm B}}\right)^2 + 2\left(\frac{m_{\rm A}}{m_{\rm B}}\right)\cos\chi\right)^{3/2}}{\left|1 + \left(\frac{m_{\rm A}}{m_{\rm B}}\right)\cos\chi\right|} \tag{4.57}$$

This result is valid for elastic scattering in a spherically symmetric potential under the assumption that the target is at rest prior to the collision.

The relations discussed above can be generalized to elastic, inelastic, and reactive scattering of two molecules for any initial conditions. A detailed discussion of these results is presented in Appendix C.

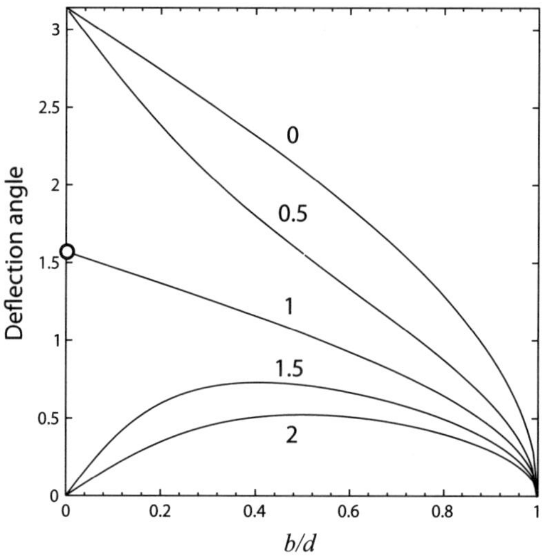

Fig. 4.1.11 The deflection angle Θ (in radians) of the scattered particle in a laboratory fixed coordinate system. The angle is given as a function of the impact parameter b, for the collision between two hard spheres A and B with the average diameter d. Particle B is initially at rest, and the five curves correspond to the mass ratios (m_A/m_B) 0, 0.5, 1, 1.5, and 2. The $\circ$ at $b = 0$ and $m_A/m_B = 1$ indicates that the deflection angle is undefined in this case, since $v_A = 0$, that is, A is at rest after the collision.

4.1.4 Many-body classical scattering

The two-body dynamics described in the preceding section has been useful in introducing a number of important concepts, and we have obtained valuable insights concerning the angular distribution of scattered particles. However, there is obviously no way to faithfully describe a chemical reaction in terms of only two interacting particles; at least three particles are required. Unfortunately, the three-body problem is one for which no analytic solution is known. Accordingly, we must use numerical analysis and computers to solve this problem.[7]

Different formulations of Newton's equation of motion

There exist three different formulations of Newton's equation of motion [1]. They all lead to exactly the same motions of the particles in the system but one form may be more convenient to use than another, dependent on both the system and properties studied. It will therefore be useful to summarize the three forms very briefly and without proof, so they will be familiar on an operational basis.

First, there is the well-known *vectorial* form

$$m\frac{d^2\boldsymbol{r}}{dt^2} = \boldsymbol{F} \tag{4.58}$$

[7]The first classical trajectory—computed by hand—for the reaction H + $H_2 \rightarrow H_2$ + H can be found in *J. Chem. Phys.* **4**, 170 (1936).

which is based on an evaluation of the force $\boldsymbol{F}$ in the system.

A quite different formulation is based on the evaluation of the *scalar* kinetic energy and potential energy functions. This is the so-called *Lagrangian* formulation that is based on a variational principle by d'Alembert and Lagrange. From expressions for the kinetic energy T_{kin} and potential energy V_{pot}, the Lagrange function $L(q,\dot{q})$ is introduced as

$$L(q,\dot{q}) = T_{\mathrm{kin}}(q,\dot{q}) - V_{\mathrm{pot}}(q) \tag{4.59}$$

where q is a set of coordinates and $\dot{q} \equiv dq/dt$ is a set of velocities. When Cartesian coordinates are used, the kinetic energy is given by the well-known expression $T_{\mathrm{kin}}(\dot{q}) = \sum_i (1/2) m_i \dot{q}_i^2$. If, for example, polar coordinates are used the kinetic energy can also be a function of the coordinates. The equation of motion for each coordinate can be shown to be given by

$$\frac{d}{dt}\left(\frac{\partial L}{\partial \dot{q}_i}\right) = \left(\frac{\partial L}{\partial q_i}\right) \qquad i = 1, 3N \tag{4.60}$$

where N is the number of particles.

In the third formulation, the so-called *Hamiltonian* formulation, the velocities $\dot{q}_i$ in the Lagrangian form are replaced by the so-called generalized momenta p_i via a Legendre transformation. The *generalized momentum* p_i, conjugate to the coordinate q_i, is defined as

$$p_i = \left(\frac{\partial L}{\partial \dot{q}_i}\right) \tag{4.61}$$

When Cartesian coordinates are used, the generalized momenta in Eq. (4.61) are equal to the well-known expression $p_i = m_i \dot{q}_i$. If, for example, polar coordinates are used, the momenta can take a form that is not equal to mass times velocity. For the systems we shall treat, the kinetic energy can always be written in the form $T_{\mathrm{kin}}(q,\dot{q}) = \sum_i a_i(q)\dot{q}_i^2$, where a_i are coefficients that can depend on the coordinates. The Legendre transform of the Lagrange function L is the Hamilton function H

$$H(p,q) = T_{\mathrm{kin}}(q,p) + V_{\mathrm{pot}}(q) \tag{4.62}$$

which is seen to be the total energy of the system. Note that in the Hamiltonian formulation the kinetic energy is a function of the generalized momenta p and not the velocities $\dot{q}$ as in the Lagrange formulation. Equation (4.61) is used to substitute the velocities in the expression for the kinetic energy by the momenta. Then the equations of motion for p and q are

$$\begin{aligned} \dot{p}_i &= -\left(\frac{\partial H}{\partial q_i}\right) \\ \dot{q}_i &= \left(\frac{\partial H}{\partial p_i}\right) \end{aligned} \tag{4.63}$$

These equations are called Hamilton's equations of motion. It is easy to show that these various forms of Newton's equation of motion are equivalent.

The Hamiltonian formulation plays an important role in connection with quantum mechanics. The Hamilton operator of quantum mechanics $\hat{H}$ is constructed from the Hamilton function of classical mechanics H by replacing the momenta by operators. If Cartesian coordinates are used, these operators are given by $\hat{p}_i = -i\hbar\partial/\partial q_i$.

Quasi-classical trajectory calculations

A determination of the rate constant for a given chemical reaction involves, as has been described previously, the following three steps.

(i) Determination of the reaction probability $P_R(ij, v, b, \phi|l)$, as will be described in this section.

(ii) Determination of the reaction cross-section $\sigma_R(ij, v|l)$ from the reaction probability $P_R(ij, v, b, \phi|l)$ using Eq. (4.10).

(iii) Determination of the rate constant from the reaction cross-section $\sigma_R(ij, v|l)$. Equation (2.29) is used if the translational motion of the molecules is thermalized (the internal motions may or may not be thermalized), or more generally the rate constant is determined via Eq. (2.18), where thermal equilibration in the translational motion is not assumed.

In this section, we shall consider how the solution of the classical equations of motion for more than two atoms may be used to find reaction probabilities and cross-sections for chemical reactions. Although the treatment is based on classical mechanics, it is termed *quasi-classical* because quantization of vibrational and rotational energy levels is accounted for.

Thus, the initial conditions are sampled in accord with the quantized energy levels of the reactants, and when the scattering is over, the vibrational and rotational quantum numbers of the products are assigned by binning product states possessing suitable ranges of internal energies into groups according to the known quantized vibrational and rotational energies of the products.

Let us, before we describe how the reaction probability is determined in a simulation, show and discuss the results of such a calculation on the simple reaction $H + H_2 \rightarrow H_2 + H$, where the potential energy surface is well known. In Fig. 4.1.12 the total reaction cross-section is shown as a function of the total energy E_{tot} for different initial vibrational states $n = 0, 1, 2$. The initial rotational quantum state of H_2 is in all cases the $J = 0$ state, that is, the molecule is not rotating before the reaction. The energies $E_{n,J=0}$ of the molecule in vibrational state n and rotational state $J = 0$ are indicated by arrows. The total reaction cross-section $\sigma_R(ij, v)$ is determined from the reaction probability $P_R(ij, v, b, \phi)$ for the reaction no matter the state of the product. We see that the relative kinetic energy necessary for a reaction, that is, the energy difference between the energy at which $\sigma_R > 0$ and $E_{n,J=0}$, decreases when n increases. Also, we notice how the reaction cross-section rises more steeply for larger n. This means that the reactivity of the H_2 molecule increases when the molecule is vibrationally excited.[8]

Fig. 4.1.13(a–c) shows partial cross-sections for reactions with the reactant molecules in vibrational quantum states $n = 0, 1, 2$ and rotational quantum state $J = 0$ and products in vibrational states $n' = 0, 1, 2$, respectively, and any rotational quantum state. Note that the abscissa axis in this plot is the translational energy and not the total energy as in Fig. 4.1.12. The translational energy is found in the latter plot by subtracting the molecular energy $E_{n,J=0}$ from the total energy. If that is done, we

[8] Note that the energy dependence of the reaction cross-section predicted by the model in Section 4.2.2 (Fig. 4.1.5) is in rough agreement with the quasi-classical calculations.

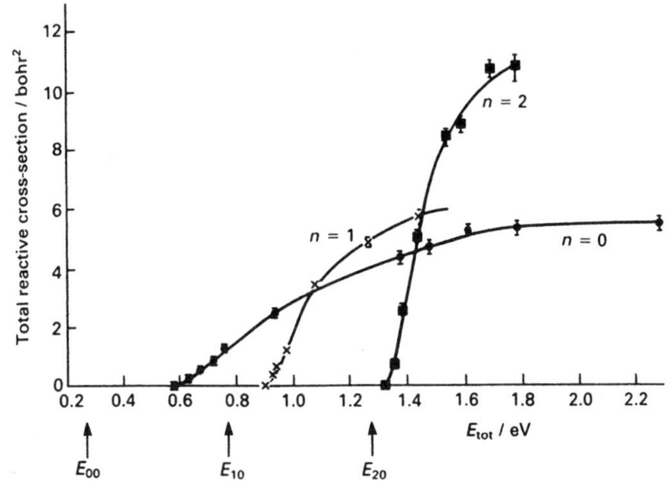

Fig. 4.1.12 Total reaction cross-sections for the reaction $H + H_2(n, J = 0) \rightarrow H_2 + H$ as functions of the total energy E_{tot} for initial vibrational states $n = 0, 1, 2$: $\bullet\, n = 0$, $\times\, n = 1$, $\square\, n = 2$. The energies of the $E_{n,J=0}$ levels of H_2 are indicated by arrows. [Adapted from Barg *et al.*, *J. Chem. Phys.* **74**, 1017 (1981).]

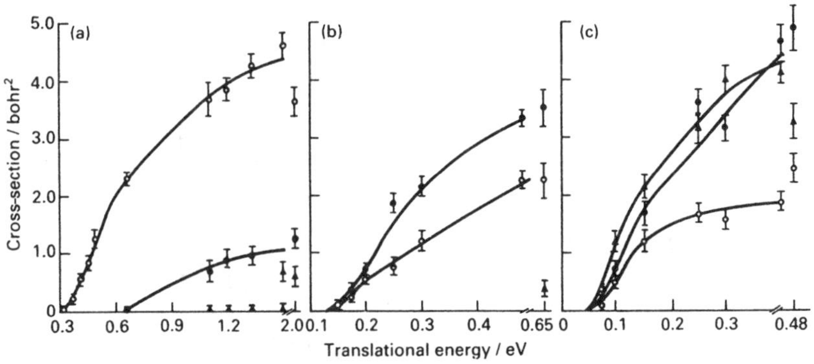

Fig. 4.1.13 Reaction cross-sections for the reaction $H + H_2(n, J = 0) \rightarrow H_2(n', \text{all } J') + H$ as functions of translational energy for (a) $n = 0$, (b) $n = 1$, and (c) $n = 2$: $\circ\, n' = 0$, $\bullet\, n' = 1$, $\triangle\, n' = 2$. [Adapted from Barg *et al.*, *J. Chem. Phys.* **74**, 1017 (1981).]

may compare the two figures and find that the partial cross-sections add up to the total cross-section, as they should, at a given translational energy.

Fig. 4.1.14 shows another example of the calculation of partial cross-sections for product molecules in specified rotational quantum states J' and any vibrational quantum state.

Finally, Fig. 4.1.15 shows the differential cross-section $d\sigma_R/d\theta$, which gives the angular distribution of product molecules independently of their quantum state. The differential cross-section shows a peak for backward scattering ($\theta = 180^0$) with the

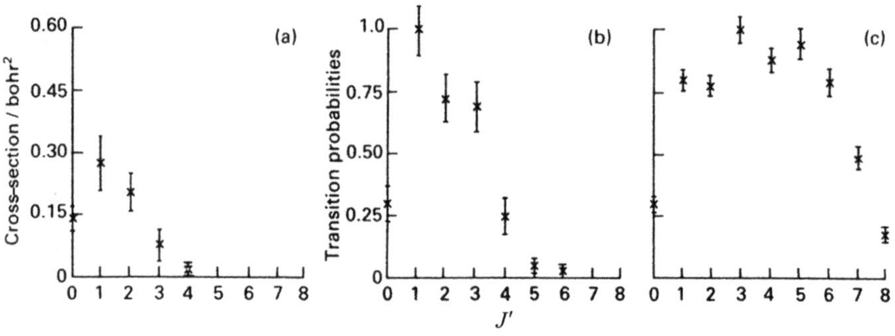

Fig. 4.1.14 Reaction cross-sections or transition probabilities for the reaction H + $H_2(n, J = 0) \rightarrow H_2(\text{all } n', J') + H$ as a function of J' for (a) $n = 0$, $E_{\text{trans}} = 0.428$ eV, (b) $n = 1$, $E_{\text{trans}} = 0.25$ eV, and (c) $n = 2$, $E_{\text{trans}} = 0.25$ eV. [Adapted from Barg *et al.*, *J. Chem. Phys.* **74**, 1017 (1981).]

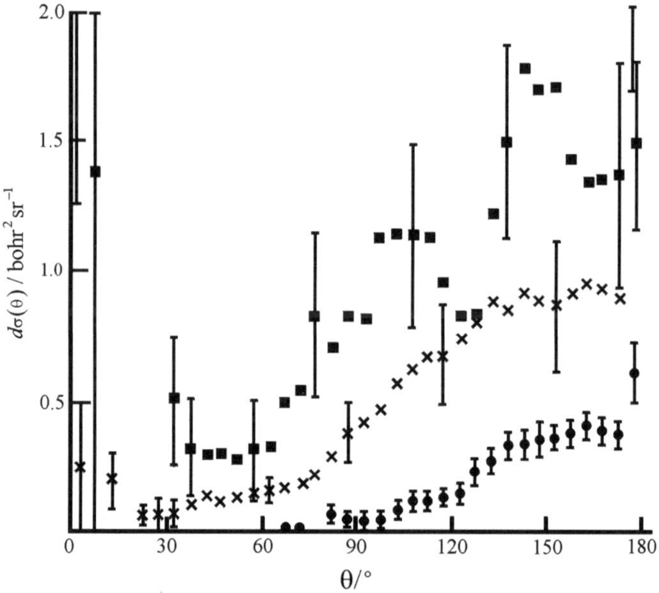

Fig. 4.1.15 Differential reaction cross-sections for the reaction $H + H_2(n, J = 0) \rightarrow H_2 + H$ at E_{trans}= 0.48 eV: • $n = 0$, × $n = 1$, □ $n = 2$. [Adapted from Barg *et al.*, *J. Chem. Phys.* **74**, 1017 (1981).]

distributions becoming broader as n increases from 0 to 2.

Tests of quasi-classical calculations against full quantum dynamics have shown surprisingly good agreement [2]. Several so-called *semi-classical methods* have been developed in order to obtain even better agreement. These approximations to quantum dynamics cover a range of more ingenious hybrids of classical and quantum mechanics. These methods are also interesting from the point of view of providing deeper insights into the nature of quantum dynamics.

Let us then go on and describe how the quasi-classical calculations leading to these results are done. We begin with the Hamiltonian for a system of N atoms:

$$H(\boldsymbol{p},\boldsymbol{r}) = T_{\text{kin}}(\boldsymbol{p}) + E(\boldsymbol{r}) = \sum_{i=1}^{N} \frac{\boldsymbol{p}_i^2}{2m_i} + E(\boldsymbol{r}_1,\ldots,\boldsymbol{r}_N) \tag{4.64}$$

and assume that there is only one potential energy surface $E(\boldsymbol{r}_1,\ldots,\boldsymbol{r}_N)$, which is supposed to be known. $\boldsymbol{r}_i$ is the position vector of atom i in the laboratory fixed Cartesian coordinate system and $\boldsymbol{p}_i$ the momentum of atom i.

For simplicity, let us consider a reaction

$$\text{A} + \text{BC}(n,J) \rightarrow \text{AB} + \text{C} \tag{4.65}$$

where only three atoms A, B, and C are involved, that is, $N = 3$, and the BC molecule is in a well-defined vibrational and rotational quantum state, denoted by the quantum numbers n, J. All important features in the classical treatment of the dynamics will be displayed in this simple example and it will be obvious how to extend the procedure to any number of atoms.

In a collision process, it is the relative position of the atoms that matters, not the absolute positions, when external fields are excluded, and the potential energy E will depend on the distances between atoms rather than on the absolute positions. It will therefore be natural to change from absolute Cartesian position coordinates to a set that describes the overall motion of the system (e.g., the center-of-mass motion for the entire system) and the relative motions of the atoms in a laboratory fixed coordinate system. This can be done in many ways as described in Appendix D, but often the so-called *Jacobi coordinates* are chosen in reactive scattering calculations because they are convenient to use. The details about their definition are described in Appendix D. The salient feature of these coordinates is that the kinetic energy remains diagonal in the momenta conjugated to the Jacobi coordinates, as it is when absolute position coordinates are used.

Following the procedure in Appendix D.1.1, we start by choosing the distance between the two atoms in molecule BC as the first distance vector. Then, the distance between A and the center of mass of molecule BC is chosen as the second distance vector, and finally the center-of-mass position vector of all three atoms is chosen as the third position vector; see Fig. 4.1.16.

This may be written in matrix form as

$$\begin{pmatrix} Q_1 \\ Q_4 \\ Q_7 \end{pmatrix} = \begin{pmatrix} 0 & -1 & 1 \\ 1 & -\frac{m_\text{B}}{m_\text{B}+m_\text{C}} & -\frac{m_\text{C}}{m_\text{B}+m_\text{C}} \\ \frac{m_\text{A}}{m_\text{A}+m_\text{B}+m_\text{C}} & \frac{m_\text{B}}{m_\text{A}+m_\text{B}+m_\text{C}} & \frac{m_\text{C}}{m_\text{A}+m_\text{B}+m_\text{C}} \end{pmatrix} \begin{pmatrix} r_{\text{A},x} \\ r_{\text{B},x} \\ r_{\text{C},x} \end{pmatrix} \tag{4.66}$$

Q_1, Q_4, and Q_7 are the x-components of the Jacobi coordinates and similar equations may be written for the y-components Q_2, Q_5, and Q_8 and the z-components Q_3, Q_6, and Q_9 by choosing, respectively, the y- and z-components of the Cartesian position vectors arranged in the column vector on the right-hand side of the equation. We see

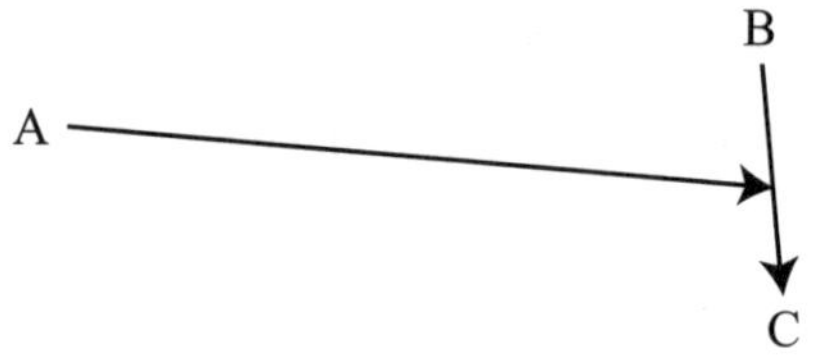

Fig. 4.1.16 Schematic illustration of Jacobi coordinates. The Cartesian coordinates of the BC distance vector are (Q_1, Q_2, Q_3), and the coordinates of the distance vector from A to the center of mass of BC are (Q_4, Q_5, Q_6).

from Appendix D.1.1 that α_1 and α_2 have both been set equal to one. Having defined the (3×3) A matrix in Eq. (4.66), we see from Eq. (D.4) that the relation between the absolute Cartesian momenta p_i and the momenta P_i conjugated to the Jacobi coordinates is given by the transposed A matrix:

$$\begin{pmatrix} p_{\mathrm{A},x} \\ p_{\mathrm{B},x} \\ p_{\mathrm{C},x} \end{pmatrix} = \begin{pmatrix} 0 & 1 & \frac{m_\mathrm{A}}{m_\mathrm{A}+m_\mathrm{B}+m_\mathrm{C}} \\ -1 & -\frac{m_\mathrm{B}}{m_\mathrm{B}+m_\mathrm{C}} & \frac{m_\mathrm{B}}{m_\mathrm{A}+m_\mathrm{B}+m_\mathrm{C}} \\ 1 & -\frac{m_\mathrm{C}}{m_\mathrm{B}+m_\mathrm{C}} & \frac{m_\mathrm{C}}{m_\mathrm{A}+m_\mathrm{B}+m_\mathrm{C}} \end{pmatrix} \begin{pmatrix} P_1 \\ P_4 \\ P_7 \end{pmatrix} \tag{4.67}$$

and equivalently for the y- and z-components. With coordinates Q_i and conjugated momenta P_i, we see from Appendix D.1.1 that the Hamiltonian may be written

$$H = \frac{1}{2\mu_{\mathrm{BC}}} \sum_{i=1}^{3} P_i^2 + \frac{1}{2\mu_{\mathrm{A,BC}}} \sum_{i=4}^{6} P_i^2 + \frac{1}{2M} \sum_{i=7}^{9} P_i^2 + E(Q_1, \ldots, Q_6) \tag{4.68}$$

where

$$\begin{aligned} \frac{1}{\mu_{\mathrm{BC}}} &= \frac{1}{m_\mathrm{B}} + \frac{1}{m_\mathrm{C}} \\ \frac{1}{\mu_{\mathrm{A,BC}}} &= \frac{1}{m_\mathrm{A}} + \frac{1}{m_\mathrm{B} + m_\mathrm{C}} \\ M &= m_\mathrm{A} + m_\mathrm{B} + m_\mathrm{C} \end{aligned} \tag{4.69}$$

The argument list in the potential energy E includes only the relative coordinates, since an external field is absent. The equations of motion are then, according to Eqs (4.63) and (4.68), given by

$$\begin{aligned} \dot{Q}_i &= \frac{P_i}{\mu_{\mathrm{BC}}} \quad (i = 1, 2, 3) \\ \dot{Q}_i &= \frac{P_i}{\mu_{\mathrm{A,BC}}} \quad (i = 4, 5, 6) \\ \dot{Q}_i &= \frac{P_i}{M} \quad (i = 7, 8, 9) \\ \dot{P}_i &= -\frac{\partial E}{\partial Q_i} \quad (i = 1, 2, 3, 4, 5, 6) \\ \dot{P}_i &= 0 \quad (i = 7, 8, 9) \end{aligned} \tag{4.70}$$

There are $6N$ or 18 coupled first-order differential equations. The six differential equations associated with the center of mass ($i = 7, 8, 9$) are, however, trivial. They correspond to free motion; thus $Q_i(t) = Q_i(t_0) + (P_i(t_0)/M)t$ and $P_i(t) = P_i(t_0)$. The remaining twelve differential equations may be integrated using some numerical integration algorithm such as, for example, a Runge–Kutta scheme. However, before we can integrate the equations and determine the coordinates and momenta as a function of time, often referred to as *the classical trajectory*, we need to choose values for all coordinates and momenta at some initial time t_0.

A schematic illustration of a classical trajectory associated with a chemical reaction is shown in Fig. 4.1.17. A detailed description of the choice of initial coordinates and momenta as well as the analysis of the trajectory results is given in the following subsections.

Initial coordinates and momenta

We are free to choose any initial condition for the center-of-mass motion, since it is of no importance for the reaction. We therefore choose to set $P_i(t_0) = P_i^0 = Q_i(t_0) = Q_i^0 = 0$ for $i = 7, 8, 9$. The equations of motion for the center of mass then show that $P_i(t) = P_i(t_0) = 0$ and $Q_i(t) = Q_i(t_0) = 0$ at all times t, that is, no center-of-mass motion.

The major problem in implementing classical trajectories is that the typical experiment samples a wide distribution of initial conditions. Since each trajectory is run for a specified set of initial coordinates and momenta, a large number of trajectories must be run with a wide range of initial conditions in order to simulate a particular experimental situation. An efficient choice of initial conditions is essential to getting useful information from trajectory calculations.

Here we will set up a trajectory calculation that is designed to simulate a molecular-beam experiment, where the velocity of A relative to the center of mass of BC is well defined. Also, we assume that the BC molecule is in a well-defined vibrational and rotational quantum state, denoted by the quantum numbers n, J.

In addition, we need to specify the parameters $\rho, b, \theta, \phi, \eta$, and ξ, as defined in Fig. 4.1.18. In the following we describe how the initial values of the coordinates and momenta are chosen given these parameters. The procedure for choosing the parameters will be described in the following subsection.

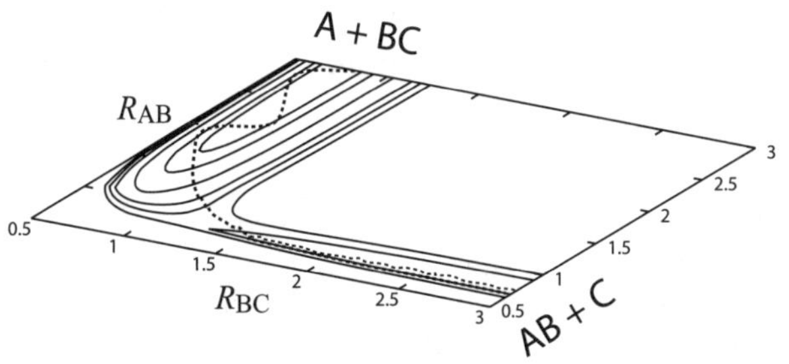

Fig. 4.1.17 Schematic illustration of a classical trajectory (dotted line) superimposed on the potential energy surface for the reaction A + BC → AB + C. Note the vibrational energy in BC, and that this particular trajectory leads to chemical reaction.

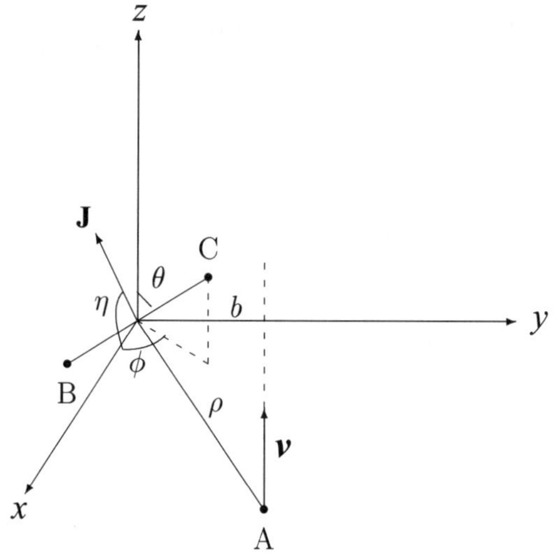

Fig. 4.1.18 Initial conditions for the A + BC collision.

First, we consider the coordinates and momenta associated with the relative motion between A and BC, that is, (Q_4^0, Q_5^0, Q_6^0) and (P_4^0, P_5^0, P_6^0).

- The y–z plane in Fig. 4.1.18 is defined by the position of A, the center-of-mass position of BC, and the velocity $\boldsymbol{v}$ of A relative to BC with $\boldsymbol{v}$ being parallel to the z-axis. The origin of the coordinate system is put at the center-of-mass position of BC. This leads to the following initial values of selected coordinates and momenta:
  - $Q_4^0 = 0$: no x-component since we are in the y–z plane;
  - $P_4^0 = 0$: no x-component since we are in the y–z plane;
  - $Q_5^0 = b$: y-coordinate of distance vector between A and the center of mass of BC; b is the impact parameter;
  - $P_5^0 = 0$: no motion in the y-direction since $\boldsymbol{v}$ is parallel to the z-axis;
  - $Q_6^0 = -\sqrt{\rho^2 - b^2}$: z-coordinate of A with BC at the origin;
  - $P_6^0 = \mu_{\text{A,BC}}\, v$: z-component of momentum of A with respect to BC.
- ρ is chosen large enough such that the interaction energy between A and BC is negligible.

Second, we consider the coordinates and momenta associated with the motion of BC, that is, (Q_1^0, Q_2^0, Q_3^0) and (P_1^0, P_2^0, P_3^0).

The phase-space curve for a one-dimensional harmonic oscillator with a given energy E_{vib} is an ellipse; see Fig. 4.1.2, and for an anharmonic oscillator it is a distorted ellipse. The turning points for the oscillator q_- and q_+ may be determined as the solutions for q $(= \sqrt{Q_1^2 + Q_2^2 + Q_3^2})$ to

$$E_{\text{vib}} = V_{\text{BC}}(q) \tag{4.71}$$

where the total energy of the oscillator is equated to the potential energy $V_{\rm BC}$ of B–C. For a Morse potential, for example,

$$V_{\rm Morse}(q) = D[1 - \exp(-b(q - q_{\rm eq}))]^2 = V_{\rm BC} \tag{4.72}$$

it is easy to solve the equation for q.

- We find the turning points

$$\begin{aligned} q_- &= q_{\rm eq} - \frac{1}{b}\ln\left[1 + \sqrt{\frac{E_{\rm vib}}{D}}\right] \\ q_+ &= q_{\rm eq} - \frac{1}{b}\ln\left[1 - \sqrt{\frac{E_{\rm vib}}{D}}\right] \end{aligned} \tag{4.73}$$

$q_{\rm eq}$ is the equilibrium distance between the two atoms. Alternatively, one may determine the turning points numerically. The vibrational energy is one of the quantized energy levels of the Morse potential, $E_{\rm vib} = E_n$, where

$$E_n = \hbar\omega[(n + 1/2) - x_e(n + 1/2)^2] \quad (n = 0, 1, \ldots) \tag{4.74}$$

with $\omega = \sqrt{2Db^2/\mu_{\rm BC}}$ and $x_e = \sqrt{b^2\hbar^2/(8D\mu_{\rm BC})}$. To set up the initial conditions for the BC molecule, we begin by choosing a random number ξ between 0 and 1. If this number is smaller than 0.5, we start out with the oscillator in the turning point q_-, otherwise in the turning point q_+.

The kinetic energy of the oscillator at the turning points is zero, so the kinetic energy of the molecule will be rotational energy alone. With the molecule in a rotational state with quantum number J, the kinetic energy may be written as

$$\frac{P_{\rm rot}^2}{2\mu_{\rm BC}} = \hbar^2 J(J+1)\frac{1}{2\mu_{\rm BC}q_\pm^2} \quad (J = 0, 1, \ldots) \tag{4.75}$$

The expression on the right-hand side of the equation is the quantum expression for the rotational energy of a diatomic rotor in quantum state J and with the moment of inertia $\mu_{\rm BC}q_\pm^2$.

- The magnitude of the rotational momentum $P_{\rm rot}$ is then

$$P_{\rm rot} = \frac{\hbar}{q_\pm}\sqrt{J(J+1)} \equiv \frac{j_r}{q_\pm} \tag{4.76}$$

- The spherical angles θ and ϕ give the orientation of the B–C chemical bond of length $q_\pm$, leading to the following Cartesian coordinates:

$$\begin{aligned} Q_1'' &= q_\pm \sin\theta\cos\phi \\ Q_2'' &= q_\pm \sin\theta\sin\phi \\ Q_3'' &= q_\pm \cos\theta \end{aligned} \tag{4.77}$$

- The η angle gives the angle of rotation of the $\boldsymbol{J}$ vector around the BC bond. The Cartesian components of the rotational momentum vector are given by the three Eulerian angles θ, ϕ, and η as

$$\begin{aligned}P_1'' &= \frac{j_r}{q_\pm}[\sin\phi\cos\eta - \cos\theta\cos\phi\sin\eta] \\ P_2'' &= -\frac{j_r}{q_\pm}[\cos\phi\cos\eta + \cos\theta\sin\phi\sin\eta] \\ P_3'' &= \frac{j_r}{q_\pm}\sin\theta\sin\eta\end{aligned} \tag{4.78}$$

- To determine the initial phase of the oscillator, that is, the vibrational momentum and coordinate, we need to know the time τ for one oscillation. For a harmonic oscillator $\tau = 1/\nu$, where ν is the frequency. In general τ is determined numerically by an integration of the equation of motion for the one-dimensional non-rotating ($J = 0$) vibrator. We start out with the oscillator in one of the turning points, q_- or q_+, where the momentum is zero and integrate the equations of motion for the oscillator ($i = 1, 2, 3$ and $J = 0$) until the initial state is recovered. The elapsed time is the time for one oscillation. The equations of motion for the rotating BC molecule ($i = 1, 2, 3$) are then integrated for a time $\xi\tau$ with P_1'', P_2'', and P_3'' as initial momenta and Q_1'', Q_2'', and Q_3'' as initial coordinates; the resulting values of the momenta and coordinates at time $\xi\tau$ are taken as the initial values (P_1^0, P_2^0, P_3^0) and (Q_1^0, Q_2^0, Q_3^0) for the integration of the entire set of equations in Eq. (4.70).

Sampling of initial conditions

In an experiment, the parameters θ, ϕ, η, ξ, and b used above to specify the initial coordinates and momenta cannot be controlled. The observed result therefore represents an average over the various parameters. Assume, for a moment, that the impact parameter b is under experimental control, and consider the probability of reaction $P_R(b; v, n, J)$ with the molecule in the (n, J) quantum state, relative speed v, and impact parameter b. The experimental value of the reaction probability is an average over the parameters θ, ϕ, η, and ξ. Since all orientations of the BC molecule, as given by the space angle $d\Omega = \sin\theta d\theta d\phi$, and all angles η between 0 and 2π are equally probable, like all values of ξ between 0 and 1, the average may be written as

$$\langle P_R(b; v, n, J)\rangle = \frac{1}{2(2\pi)^2}\int_{\theta=0}^{\pi}\int_{\phi=0}^{2\pi}\int_{\eta=0}^{2\pi}\int_{\xi=0}^{1} P_R(b, \theta, \phi, \eta, \xi; v, n, J)\sin\theta d\theta d\phi d\eta d\xi \tag{4.79}$$

where the factor in front of the integrals contains the normalization constant (4π) for the space angle $\sin\theta d\theta d\phi$, the normalization constant 2π for the η angle, and the normalization constant 1 for the ξ variable.

We do not know the integrand, except that it is either 0 when no reaction occurred or 1 when the reaction took place, so we cannot formally perform the integration. Instead, we use the so-called *Monte Carlo method* to evaluate the integral (Appendix I). The idea is that we should try to convert the integrand

$$\frac{1}{2}\frac{1}{2\pi}\frac{1}{2\pi}P_R \sin\theta\, d\theta\, d\phi\, d\eta\, d\xi \equiv P_R f_\theta f_\phi f_\eta f_\xi d\theta\, d\phi\, d\eta\, d\xi$$
$$= P_R d\theta'\, d\phi'\, d\eta'\, d\xi' \tag{4.80}$$

to a form such that the integral, like in Eq. (I.2), may be determined as the average value of the integrand evaluated at uniformly distributed values of the primed variables between 0 and 1.

In Eq. (4.80) we see that if we set

$$\begin{aligned} f_\theta &= \frac{\sin\theta}{2} \\ f_\phi &= \frac{1}{2\pi} \\ f_\eta &= \frac{1}{2\pi} \\ f_\xi &= 1 \end{aligned} \tag{4.81}$$

then all the f functions represent a normalized probability density in the respective variables as seen by integration over the respective intervals as given in Eq. (4.79). They therefore play the role of the normalized probability function $w(x)$ in Eq. (I.7), and the primed quantities play the role of the u variable in Eq. (I.7), that is,

$$\begin{aligned} d\theta' &= \frac{\sin\theta}{2} d\theta \\ d\phi' &= \frac{1}{2\pi} d\phi \\ d\eta' &= \frac{1}{2\pi} d\eta \\ d\xi' &= d\xi \end{aligned} \tag{4.82}$$

Since the reaction probability P_R in the integral is given as a function of the original variables, and not of the primed variables, we need to express the old variables in terms of the primed variables. This is achieved by an integration of the relations in Eq. (4.82):

$$\begin{aligned} \theta' &= -\frac{1}{2}\cos\theta + \text{const}_\theta = -\frac{1}{2}\cos\theta + \frac{1}{2} \\ \phi' &= \frac{1}{2\pi}\phi + \text{const}_\phi = \frac{1}{2\pi}\phi \\ \eta' &= \frac{1}{2\pi}\eta + \text{const}_\eta = \frac{1}{2\pi}\eta \\ \xi' &= \xi + \text{const}_\xi = \xi \end{aligned} \tag{4.83}$$

The integration constants, const_α, are chosen such that all the primed variables vary between 0 and 1, as may be seen from the second relation in each of the lines in Eq. (4.83). The relations are easily inverted and we find

$$\begin{aligned} \theta &= \arccos(1 - 2\theta') \\ \phi &= 2\pi\phi' \\ \eta &= 2\pi\eta' \\ \xi &= \xi' \end{aligned} \tag{4.84}$$

The integrand in Eq. (4.79) is now multiplied and divided by the probability functions f, following the procedure given in Eq. (I.8), and we find

$$\begin{aligned} &\langle P_R(b; v, n, J)\rangle \\ &= \frac{1}{2(2\pi)^2}\int_{\theta=0}^{\pi}\int_{\phi=0}^{2\pi}\int_{\eta=0}^{2\pi}\int_{\xi=0}^{1} P_R(b,\theta,\phi,\eta,\xi; v,n,J) \\ &\qquad\qquad \times \frac{(\sin\theta/2)(1/(2\pi))(1/(2\pi))}{(\sin\theta/2)(1/(2\pi))(1/(2\pi))}\sin\theta\, d\theta\, d\phi\, d\eta\, d\xi \\ &= \frac{1}{2(2\pi)^2}\int_{\theta'=0}^{1}\int_{\phi'=0}^{1}\int_{\eta'=0}^{1}\int_{\xi'=0}^{1} P_R(b,\theta(\theta'),\phi(\phi'),\eta(\eta'),\xi(\xi'); v,n,J) \\ &\qquad\qquad \times \frac{(\sin\theta/2)(1/(2\pi))(1/(2\pi))}{(\sin\theta/2)(1/(2\pi))(1/(2\pi))} 2\, d\theta'\, 2\pi\, d\phi'\, 2\pi\, d\eta'\, d\xi' \\ &= \int_{\theta'=0}^{1}\int_{\phi'=0}^{1}\int_{\eta'=0}^{1}\int_{\xi=0}^{1} P_R(b,\theta(\theta'),\phi(\phi'),\eta(\eta'),\xi(\xi'); v,n,J)\, d\theta'\, d\phi'\, d\eta'\, d\xi' \\ &= \frac{1}{N}\sum_{i=1}^{N} P_R(b,\theta(\theta_i'),\phi(\phi_i'),\eta(\eta_i'),\xi(\xi_i'); v,n,J) \\ &= \frac{N_R(b; v,n,J)}{N(b; v,n,J)} \end{aligned} \tag{4.85}$$

We have used Eq. (4.84) to introduce the primed variables and emphasized that the unprimed variables are functions of the primed variables, as indicated in the argument list to the reaction probability P_R. The primed variables are now chosen at random from a uniform distribution of numbers between 0 and 1, and used to determine the value of the unprimed variables using Eq. (4.84), whereupon a trajectory calculation is performed. This may either lead to the desired reaction, in which case $P_R = 1$, or not, in which case $P_R = 0$. If we therefore run N trajectories and N_R trajectories lead to the desired result, then the reaction probability is simply the ratio between these numbers as given in the equation.

Analysis of the trajectory results

There are usually several possible outcomes of a reaction between two reactants:

$$\mathrm{A} + \mathrm{BC}(n, J) \longrightarrow \begin{cases} \mathrm{AB} + \mathrm{C} \\ \mathrm{AC} + \mathrm{B} \\ \mathrm{A} + \mathrm{B} + \mathrm{C} \\ \mathrm{A} + \mathrm{BC} \end{cases} \tag{4.86}$$

For example, in addition to the formation of molecule AB, we may also get a molecule AC, three separated atoms, or no reaction at all. Each possibility is usually referred to as a *channel* for the reaction. At the end of each trajectory, it will therefore be necessary to decide whether the particular trajectory ended up in the relevant channel, here the formation of AB; see Fig. 4.1.17. This will be the case if the distance between C and AB increases monotonically with time and the AB distance is confined and oscillating in time.

The Jacobi coordinates used for the propagation of the system are not convenient for such an analysis, since they were based on reactants A and BC and therefore include the BC distance and the A–BC distance rather than the AB and AB–C distances. It is therefore convenient to replace the coordinates in Eq. (4.66),

$$\begin{pmatrix} Q_1 \\ Q_4 \\ Q_7 \end{pmatrix} = \begin{pmatrix} 0 & -1 & 1 \\ 1 & -\frac{m_B}{m_B+m_C} & -\frac{m_C}{m_B+m_C} \\ \frac{m_A}{m_A+m_B+m_C} & \frac{m_B}{m_A+m_B+m_C} & \frac{m_C}{m_A+m_B+m_C} \end{pmatrix} \begin{pmatrix} r_{A,x} \\ r_{B,x} \\ r_{C,x} \end{pmatrix} \tag{4.87}$$

by the following Jacobi coordinates:

$$\begin{pmatrix} Q'_1 \\ Q'_4 \\ Q'_7 \end{pmatrix} = \begin{pmatrix} 1 & -1 & 0 \\ -\frac{m_A}{m_A+m_B} & -\frac{m_B}{m_A+m_B} & 1 \\ \frac{m_A}{m_A+m_B+m_C} & \frac{m_B}{m_A+m_B+m_C} & \frac{m_C}{m_A+m_B+m_C} \end{pmatrix} \begin{pmatrix} r_{A,x} \\ r_{B,x} \\ r_{C,x} \end{pmatrix} \tag{4.88}$$

To find the relation between the two sets of coordinates Q_i and Q'_i, we invert Eq. (4.87) which gives

$$\begin{pmatrix} r_{A,x} \\ r_{B,x} \\ r_{C,x} \end{pmatrix} = \begin{pmatrix} 0 & \frac{m_B+m_C}{m_A+m_B+m_C} & 1 \\ -\frac{m_C}{m_B+m_C} & -\frac{m_A}{m_A+m_B+m_C} & 1 \\ \frac{m_B}{m_B+m_C} & -\frac{m_A}{m_A+m_B+m_C} & 1 \end{pmatrix} \begin{pmatrix} Q_1 \\ Q_4 \\ Q_7 \end{pmatrix} \tag{4.89}$$

and introduce that result in Eq. (4.88) and find

$$\begin{pmatrix} Q'_1 \\ Q'_4 \\ Q'_7 \end{pmatrix} = \begin{pmatrix} \frac{m_C}{m_B+m_C} & 1 & 0 \\ \frac{m_B(m_A+m_B+m_C)}{(m_A+m_B)(m_B+m_C)} & -\frac{m_A}{m_A+m_B} & 0 \\ 0 & 0 & 1 \end{pmatrix} \begin{pmatrix} Q_1 \\ Q_4 \\ Q_7 \end{pmatrix} \tag{4.90}$$

Similar equations may be obtained for the y-components and z-components of the Jacobi coordinates. Let us denote the (3×3) matrix in Eq. (4.90) by B; then we may write the transformations in shorthand as

$$\begin{aligned} Q' &= BQ \\ P' &= [B^T]^{-1}P \end{aligned} \tag{4.91}$$

The integration of the equations of motion is done in the (P, Q) coordinates and at any time we may apply the transformations in Eq. (4.91) and determine the Jacobi coordinates for the product states.

Now, according to Eq. (4.85) we may determine the reaction probability $\langle P_R \rangle$ simply as the fraction of trajectories leading to the desired reaction, that is,

$$\langle P_R(b; v, n, J) \rangle \rightarrow \frac{N_R(b; v, n, J)}{N(b; v, n, J)} \quad \text{for} \quad N(b; v, n, J) \rightarrow \infty \tag{4.92}$$

where N_R is the number of trajectories leading to the desired reaction, and N is the total number of trajectories. It is usually necessary to run several thousands of trajectories to achieve reliable results with little statistical noise.

In experiments one cannot control the impact parameter b and determine $\langle P_R(b; v, n, J)\rangle$, but rather determine the cross-section which is related to the reaction probability, see Eq. (4.10). Since $\langle P_R(b; v, n, J)\rangle$ is independent of the angle ϕ, defined in Fig. 4.1.1, we obtain

$$\sigma_R(v, n, J) = 2\pi \int_0^\infty \langle P_R(b; v, n, J)\rangle b db \tag{4.93}$$

This implies that we also have to average over the impact parameter b in our simulation. As above, this is done by the Monte Carlo method (Appendix I). We have that

$$f_b = b \tag{4.94}$$

and accordingly

$$db' = \frac{2b}{b_{\text{max}}^2} \tag{4.95}$$

where b_{max} is chosen so large that the reaction probability is zero. Equation (4.95) is integrated and, with the integration constant set to zero, we find

$$b = b_{\text{max}}\sqrt{b'} \tag{4.96}$$

so

$$\begin{aligned}
\sigma_R(v, n, J) &= 2\pi \int_0^\infty \frac{\langle P_R(b, v, n, J)\rangle}{2b/b_{\text{max}}^2} \frac{2b}{b_{\text{max}}^2} b\, db \\
&= 2\pi \int_0^1 \frac{\langle P_R(b(b'), v, n, J)\rangle}{2b/b_{\text{max}}^2} b\, db' \\
&= \pi b_{\text{max}}^2 \int_0^1 \langle P_R(b(b'), v, n, J)\rangle db' \\
&= \pi\, b_{\text{max}}^2 \frac{N_R(v, n, J)}{N(v, n, J)}
\end{aligned} \tag{4.97}$$

where N is the total number of trajectories and N_R the number of trajectories leading to the reaction considered.

Sometimes it will also be of interest to know the distribution over rotational and vibrational states in the products. We therefore need to determine the rotational and vibrational states of the product molecule. It is only meaningful to speak about a specific rotational and vibrational state if the coupling between the two modes is weak, so the rotational and vibrational energies are reasonably constant. The assignment of the quantum state for AB is based on the coordinates Q_i' and momenta P_i', for $i = 1, 2, 3$, and is done in much the same way as when we set up the initial state of the BC molecule. The turning points $q_\pm'$ are identified by monitoring the length of the AB bond, that is, $\sqrt{Q_1'^2 + Q_2'^2 + Q_3'^2}$. At a turning point, all kinetic energy will be rotational energy such that

$$\sum_{i=1}^{3} P_i'^2 = \hbar^2 J(J+1)\frac{1}{q_\pm'^2} \tag{4.98}$$

We solve for J and round the result to an integer to give the rotational quantum number. The vibrational energy is given by

$$E_{\text{vib}} = V_{\text{AB}}(q'_{\pm}) \tag{4.99}$$

and we use the Bohr–Sommerfeld semi-classical prescription for quantization of phase space for a diatomic non-rotating molecule to determine the vibrational quantum number n:

$$\begin{aligned}(n + 1/2)h &= 2\int_{q'_-}^{q'_+} p\,dq \\ &= 2\int_{q'_-}^{q'_+} \sqrt{2\mu_{\text{AB}}\left[E_{\text{vib}} - V_{\text{AB}}(q)\right]}dq\end{aligned} \tag{4.100}$$

where the result is rounded to an integer value of n.

The extension of the trajectory calculations to a system with any number of atoms is straightforward except for the quantization of the vibrational and rotational states of the molecules. For a molecule with three different principal moments of inertia, there does not exist a simple analytical expression for the quantized rotational energy. This is only the case for molecules with some symmetry like a spherical top molecule, where all moments of inertia are identical, and a symmetric top, where two moments of inertia are identical and different from the third. For the vibrational modes, we may use a normal coordinate analysis to determine the normal modes (see Appendix E) and quantize those as for a one-dimensional oscillator.

Finally, when the cross-section (Eq. (4.97)) is known, we can use Eq. (2.29) to obtain the rate constant $k(T)$. Thus, we must average over the relative speed and the quantum numbers of BC, that is, (v, n, J), using the Boltzmann distribution. Alternatively, we could obtain $k(T)$ using a somewhat more direct approach. We can run trajectories in principle for all initial states and determine the final result as a weighted average of the results for given initial states with the Boltzmann factor as the weight factor for the given molecular state. In practice that limits the number of initial states we need to consider since the high energy states occur with a very small probability depending on the temperature.

4.2 Quantum dynamics

After having discussed the approximate quasi-classical dynamics, we return (see Section 1.1) now to exact quantum dynamics.[9] The Schrödinger equation for motion of the atomic nuclei is given by Eq. (1.10):

$$\begin{aligned}i\hbar\frac{\partial\chi(\boldsymbol{R}, t)}{\partial t} &= \hat{H}\chi(\boldsymbol{R}, t) \\ &= (\hat{T}_{\text{nuc}} + E_i(\boldsymbol{R}))\chi(\boldsymbol{R}, t)\end{aligned} \tag{4.101}$$

where $\hat{T}_{\text{nuc}}$ is the operator for the kinetic energy of the nuclei, $E_i(\boldsymbol{R})$ is the electronic energy, i.e., the potential energy surface, and $\chi(\boldsymbol{R}, t)$ the wave function as a function

[9] In order to fully appreciate the content of this section, a good background in quantum mechanics is required; see also Appendix F.

of all the nuclear coordinates $\boldsymbol{R}$. The time-dependent Schrödinger equation is a first-order differential equation in time (like Hamilton's equation of motion), and the time evolution is given once the initial state is specified. Assume that the initial state is given by $\chi(\boldsymbol{R}, t_0)$, then

$$\chi(\boldsymbol{R}, t) = \hat{U}(t - t_0)\chi(\boldsymbol{R}, t_0) \tag{4.102}$$

where the propagator $\hat{U}(t - t_0)$ is given by

$$\begin{aligned} \hat{U}(t - t_0) &= \exp(-i\hat{H}(t - t_0)/\hbar) \\ &\equiv 1 - i\hat{H}(t - t_0)/\hbar - \hat{H}^2(t - t_0)^2/2\hbar^2 + \cdots \end{aligned} \tag{4.103}$$

that is, the exponential of an operator is defined by its (formal) Taylor expansion. Equation (4.102) is seen to be a solution by direct substitution into Eq. (4.101). The propagator propagates the wave function from time t_0 to time t. When the wave function is fairly localized in position (and momentum) space, the designation 'wave packet' is often used for the wave function.

We recall some basic results of quantum dynamics [3]. First, the state of the system and the time evolution can be expressed in a generalized (Dirac) notation, which is often very convenient. The state at time t is specified by $|\chi(t)\rangle$ with the representations $\chi(\boldsymbol{R}, t) = \langle \boldsymbol{R}|\chi(t)\rangle$ and $\chi(\boldsymbol{P}, t) = \langle \boldsymbol{P}|\chi(t)\rangle$ in coordinate and momentum space, respectively. Probability is a concept that is inherent in quantum mechanics. $|\langle \boldsymbol{R}|\chi(t)\rangle|^2$ is the probability density in coordinate space, and $|\langle \boldsymbol{P}|\chi(t)\rangle|^2$ is the same quantity in momentum space. The time evolution (in the Schrödinger picture) can be expressed as

$$|\chi(t)\rangle = \hat{U}(t - t_0)|\chi(t_0)\rangle \tag{4.104}$$

From this equation and the unitarity of the time-evolution operator,

$$\langle \chi(t)|\chi(t)\rangle = \langle \chi(t_0)|\hat{U}(t - t_0)^\dagger \hat{U}(t - t_0)|\chi(t_0)\rangle = \langle \chi(t_0)|\chi(t_0)\rangle \tag{4.105}$$

Since, for example, $\langle \chi(t)|\chi(t)\rangle = \int \langle \chi(t)|\boldsymbol{R}\rangle\langle \boldsymbol{R}|\chi(t)\rangle d\boldsymbol{R}$, the interpretation of this result is that we have a global conservation of probability density.

The Schrödinger equation also leads to a continuity equation that can be interpreted as a local conservation of probability density (here for a single particle with the Hamiltonian $\hat{H} = -\hbar^2/(2m)\nabla^2 + V$):

$$\frac{\partial P}{\partial t} = -\nabla \cdot \boldsymbol{j} \tag{4.106}$$

where $P = \chi(\boldsymbol{R}, t)^\star \chi(\boldsymbol{R}, t) = |\chi(\boldsymbol{R}, t)|^2$ is the probability density, and the probability current density or *flux density*, i.e., the probability current per unit time per unit area, is given by

$$\boldsymbol{j} = \frac{\hbar}{2im}(\chi^\star \nabla \chi - \chi \nabla \chi^\star) = \frac{\hbar}{m}\mathrm{Im}(\chi^\star \nabla \chi) = \mathrm{Re}(\chi^\star \hat{\boldsymbol{P}} \chi)/m \tag{4.107}$$

where the momentum operator is $\hat{\boldsymbol{P}} = -i\hbar\nabla$. An alternative form of this continuity equation can be obtained by integration over a volume V, and application of Gauss' theorem, which implies that the change in probability within the volume is equal to the flux (in or out) of a surface S_V that encloses the entire volume:

$$\frac{d}{dt}\int_V P(\boldsymbol{R},t)\,d\boldsymbol{R} = -\int_V \nabla\cdot\boldsymbol{j}\,d\boldsymbol{R} = -\int_{S_V} \boldsymbol{j}\cdot d\boldsymbol{S} \tag{4.108}$$

Another consequence of the Schrödinger equation is that the time evolution of the expectation value of a physical observable, represented by an operator $\hat{A}$ (with no explicit time dependence), is given by

$$\frac{d}{dt}\langle\hat{A}\rangle = \frac{i}{\hbar}\langle[\hat{H},\hat{A}]\rangle \tag{4.109}$$

where $\langle\hat{A}\rangle = \langle\chi(t)|\hat{A}|\chi(t)\rangle$. Constants of motion commute with the Hamiltonian, i.e., $[\hat{H},\hat{A}] = 0$. Examples of such constants of motion are the total energy and the total angular momentum.

As an application of this equation, consider (a result known as Ehrenfest's theorem) the time evolution of the expectation values of, respectively, position $\boldsymbol{R} = x\hat{\mathbf{i}} + y\hat{\mathbf{j}} + z\hat{\mathbf{k}}$ and momentum $\hat{\boldsymbol{P}} = -i\hbar\nabla = -i\hbar[(\partial/\partial x)\hat{\mathbf{i}} + (\partial/\partial y)\hat{\mathbf{j}} + (\partial/\partial z)\hat{\mathbf{k}}]$:

$$\begin{aligned} \frac{d}{dt}\langle\boldsymbol{R}\rangle &= \frac{\langle\hat{\boldsymbol{P}}\rangle}{m} &&= \left\langle\frac{\partial\hat{H}}{\partial\hat{\boldsymbol{P}}}\right\rangle \\ \frac{d}{dt}\langle\hat{\boldsymbol{P}}\rangle &= -\langle\nabla V\rangle &&= -\langle\nabla\hat{H}\rangle \end{aligned} \tag{4.110}$$

These equations have a close resemblance to Hamilton's equations of classical mechanics, Eq. (4.63). An identity is, however, only obtained for potentials with terms of no more than second order (note, e.g., that $\langle x^2\rangle \neq \langle x\rangle^2$). One simple application is to the dynamics of a free particle (e.g., the motion of the center of mass). The expectation values behave like in classical mechanics—the expectation value of the momentum is constant as is the associated momentum uncertainty. That is, except for the inherent momentum uncertainty of the initial state, the free particle behaves as in classical mechanics.

Unlike classical mechanics, a precise simultaneous specification of position and momentum is not possible. Thus, any valid state will at all times obey the Heisenberg uncertainty relation

$$\Delta x\Delta p \geqslant \hbar/2 \tag{4.111}$$

where $\Delta x = \sqrt{\langle x^2\rangle - \langle x\rangle^2}$ and similar for Δp are the well-known standard deviations for the observables. Note that $\Delta x\Delta p$ will be time dependent.

At this point we will, briefly, describe some of the fundamental qualitative differences between a quantum mechanical and a classical mechanical description. First of all, a trajectory $\boldsymbol{R}(t)$ is replaced by a wave packet, which implies that a deterministic description is replaced by a probabilistic description. $|\chi(\boldsymbol{R},t)|^2$ is a probability density, giving the probability of observing the nuclei at the position $\boldsymbol{R}$ at time t. In connection with the description of a chemical reaction, say (n and m again denote all the required vibrational and rotational quantum numbers)

$$\mathrm{A} + \mathrm{BC}(n) \longrightarrow \begin{cases} \mathrm{AB}(m) + \mathrm{C} \\ \mathrm{AC}(m) + \mathrm{B} \\ \mathrm{A} + \mathrm{B} + \mathrm{C} \\ \mathrm{A} + \mathrm{BC}(m) \end{cases} \tag{4.112}$$

$\chi(\boldsymbol{R}, t_0)$ is chosen such that it represents the reactants at a given total energy. After propagation of this initial state until the scattering (collision) is completed, one will observe the following non-classical features.

- At a given total energy several product channels will be populated with some *probability*, whereas a classical trajectory is deterministic and will always show up in one particular channel, depending on the initial conditions. Consider, e.g., a situation where the energy is above the threshold energy for the first reaction ($\mathrm{AB}(m) + \mathrm{C}$). In this case the wave packet will, typically, split up into two parts; see Fig. 1.1.1. One part will be reflected at the barrier corresponding to no reaction ($\mathrm{A} + \mathrm{BC}(m)$) whereas the other part will pass the barrier and form products.
- Wave functions can penetrate potentials and even '*tunnel*' through barriers. It is well known that wave functions can penetrate into barriers, corresponding to regions of configuration space that are not accessible according to classical mechanics. Consider, e.g., the well-known one-dimensional harmonic oscillator. The eigenstates of the harmonic oscillator show this feature. Thus, the wave functions are non-zero beyond the classical turning points. If barriers on the potential energy surface are not too broad, say the barrier around a saddle point, then the wave function can penetrate through the barrier corresponding to tunneling (see Section 6.3.1).
- *Zero-point* energies are conserved during propagation. Zero-point energies are present in any bound potential; the harmonic oscillator is again a simple example. The quasi-classical approach described in the previous section takes the energy quantization of the reactants (including zero-point energies) into account. However, when quantum dynamics is replaced by classical dynamics proper energy quantization is lost during the collision process. Thus, products can, for example, show up with vibrational energies below the quantum mechanical zero-point energy.

Before we go into the more detailed description of the reaction dynamics of bimolecular reactions, we will describe an important finding that may have practical applications in the future. In Section 3.1, we discussed potential energy surfaces with 'early' and 'late' barriers and the proposition that translational and vibrational energy might not be equally effective in promoting reaction. Thus, at a given total energy, see Eq. (1.16), one will sometimes find that not all combinations (partitionings) of translational and vibrational energy are equally effective. Such effects have been observed for bimolecular gas-phase reactions as well as gas-surface reactions [4,5] and is termed *mode-selective chemistry*. The dynamical reasons for mode-selective chemistry are related to dynamical constraints that do not allow for a complete interchange of the two forms of energy prior to reaction.[10]

Example 4.2: Controlling the chemical nature of products

The vibrational motion of polyatomic molecules encompasses all nuclei in the molecule and, as long as the displacement from the equilibrium configuration is sufficiently small, it can be broken down into the so-called normal-mode vibrations (see Appendix E). In special cases these vibrations take a particular simple form. Consider, e.g., a partially deuterated water molecule HOD. In this molecule, the H–OD and HO–D stretching motions are largely independent and the normal modes are, essentially, equivalent to the local bond-stretching modes. To that end, consider the following reaction that has been studied experimentally [6,7] as well as theoretically [8]:

$$\mathrm{H + HOD} \nearrow \mathrm{H_2 + OD}$$
$$\mathrm{H + HOD} \searrow \mathrm{HD + OH}$$

It is observed that excitation of the stretching vibrations in HOD enhance the rate more than the increase resulting from an equivalent amount of relative translational energy. In addition, the branching ratio between the two product channels can be controlled by appropriate vibrational pre-excitation of HOD. Thus, it is found that after excitation of the H–OD stretch with one or more vibrational quanta, the reaction produces almost exclusively $H_2 + OD$, whereas after excitation of the HO–D stretch, only the HD + OH products are formed.

We mention, in addition, that complete and highly accurate *ab initio*[11] theoretical solutions for a number of simple atom–diatom reactions like $H + HD \rightarrow H_2 + D$ have been obtained. Recently, the quantum dynamics of the four-atom reaction $H_2 + OH \rightarrow H_2O + H$ and its reverse has also been reported [9,10]. The general features of these results at the level of $k(T)$ will typically agree fairly well with the quasi-classical approach, provided the temperature is sufficiently high so that quantum mechanical tunneling is negligible.

4.2.1 Gaussian wave packet dynamics

In this section, we describe wave packet dynamics within a (time-dependent) local harmonic approximation to the potential, since this enables us to write down relatively simple expressions for the time evolution of the wave packet. This provides a valuable insight into quantum dynamics and the approximation may be used, for example, to

[10] In the following chapters, we will consider an approach to the calculation of rate constants—transition-state theory—that do not take into account such details of the reaction dynamics. The theory will be based on the basic axioms of statistical mechanics where all partitionings of the total energy are equally likely, and it is assumed that all these partitionings are equally effective in promoting reaction.

[11] That is, only fundamental constants of nature (Planck constant, elementary charge, and electronic and nuclear masses) have been used as input in the calculation.

study inelastic collisions and direct photodissociation (see Section 7.2.2). We consider only one-dimensional motion but the formalism can be generalized to n-dimensional (non-separable) systems [11].

Consider a (complex-valued) *Gaussian wave packet* of the form

$$G(x,t) = \langle x|G(t)\rangle = \exp\left[i\alpha_t(x-x_t)^2/\hbar + ip_t(x-x_t)/\hbar + i\gamma_t/\hbar\right] \tag{4.113}$$

where x_t and p_t are the expectation values of position and momentum, respectively:

$$\begin{aligned} x_t &= \langle G(t)|x|G(t)\rangle \\ p_t &= \langle G(t)|\hat{p}|G(t)\rangle \end{aligned} \tag{4.114}$$

and α_t and γ_t are complex numbers; α_t is related to the variance in position and momentum:

$$\begin{aligned} (\Delta x)_t^2 &= \hbar/[4\mathrm{Im}(\alpha_t)] \\ (\Delta p)_t^2 &= \hbar|\alpha_t|^2/\mathrm{Im}(\alpha_t) \end{aligned} \tag{4.115}$$

That is,

$$(\Delta x)_t(\Delta p)_t = \hbar/2\sqrt{\mathrm{Re}(\alpha_t)^2 + \mathrm{Im}(\alpha_t)^2}/\mathrm{Im}(\alpha_t) \geqslant \hbar/2 \tag{4.116}$$

in agreement with the general Heisenberg uncertainty relation, Eq. (4.111). From this relation we observe, e.g., that a well-defined momentum implies a large uncertainty in position, that is, a broad wave packet in position space.

We consider now the dynamics of the Gaussian wave packet within the framework of a time-dependent local harmonic approximation (LHA) to the exact potential $V(x)$ around x_t:

$$V_{\mathrm{LHA}}(x,t) = V(x_t) + V^{(1)}(x_t)(x-x_t) + (1/2)V^{(2)}(x_t)(x-x_t)^2 \tag{4.117}$$

where $V^{(n)}(x_t)$ is the nth derivative of $V(x)$ evaluated at $x = x_t$. The Gaussian wave packet is a solution to the Schrödinger equation at all times provided the parameters evolve in time as specified below: x_t and p_t evolve according to Hamilton's equations,

$$\begin{aligned} dx_t/dt &= p_t/m \\ dp_t/dt &= -V^{(1)}(x_t) \end{aligned} \tag{4.118}$$

(in accordance with Ehrenfest's theorem), the time evolution of α_t is given by

$$d\alpha_t/dt = -2\alpha_t^2/m - V^{(2)}(x_t)/2 \tag{4.119}$$

the imaginary part of γ_t accounts for the normalization,

$$i\gamma_t/\hbar = i\mathrm{Re}(\gamma_t)/\hbar + \ln\left(\frac{\pi\hbar}{2\mathrm{Im}(\alpha_t)}\right)^{-1/4} \tag{4.120}$$

and the real part of γ_t implies that the wave packet acquires a phase (the classical action),

$$\mathrm{Re}(\gamma_t) = \int_0^t (p_t^2/(2m) - V(x_t))dt \tag{4.121}$$

Thus, the full quantum dynamics of the Gaussian is described by a small number of parameters with a close correspondence to classical mechanics.

The probability density associated with the Gaussian takes the form

$$|G(x,t)|^2 = (2\pi(\Delta x)_t^2)^{-1/2} \exp\left[-\frac{(x-x_t)^2}{2(\Delta x)_t^2}\right] \tag{4.122}$$

and is shown in Fig. 4.2.1.

The flux density Eq. (4.107) takes the form

$$j_x = [2\mathrm{Re}(\alpha_t)(x-x_t)/m + p_t/m]|G(x,t)|^2 \tag{4.123}$$

It depends on the position; at the center $x = x_t$, it is simply the speed times the probability density, $j_x = [p_t/m](2\pi(\Delta x)_t^2)^{-1/2}$. Note that the dimension of the flux density for one-dimensional motion is speed times an inverse distance, i.e., inverse time (for three-dimensional motion, the dimension of $\boldsymbol{j}$ is speed/volume).

The LHA is exact for potentials that contain, at most, quadratic terms but obviously an approximation for anharmonic potentials. Thus, a single Gaussian wave packet within a local harmonic approximation can, e.g., not tunnel or bifurcate, i.e., there will be no simultaneously reflected and transmitted part in scattering off barriers.

When the LHA is invalid, the Gaussian wave packet will not keep its simple analytical form. One must then solve Eq. (4.101) numerically. To that end various methods have been developed [12,13].

Momentum-space representation

It is instructive to consider the momentum-space representation of the Gaussian wave packet. In this representation, the states are projected onto the eigenstates of the momentum operator, i.e., $\hat{P}|p\rangle = p|p\rangle$, which in the coordinate representation takes the form

$$\langle x|\hat{P}|p\rangle = (\hbar/i)\partial/\partial x\langle x|p\rangle = p\langle x|p\rangle \tag{4.124}$$

with the solution

$$\langle x|p\rangle = (2\pi\hbar)^{-1/2}e^{ipx/\hbar} \tag{4.125}$$

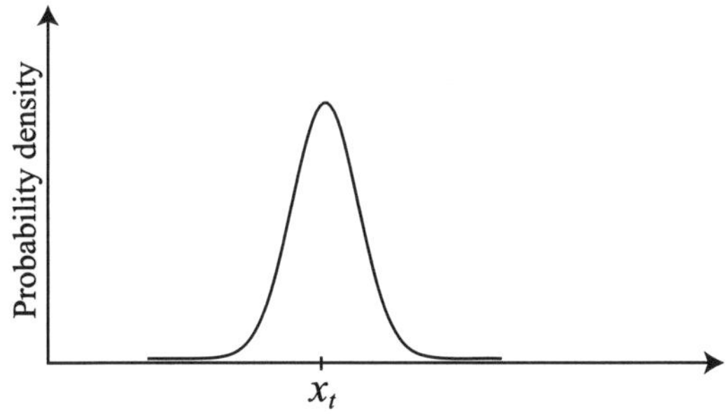

Fig. 4.2.1 The probability density associated with the Gaussian wave packet. The most probable position is at $x = x_t$, which also coincides with the expectation (average) value of the time-dependent position. The width is related to the time-dependent uncertainty $(\Delta x)_t$, i.e., the standard deviation of the position.

These states cannot be normalized in the traditional way, but the normalization constant is fixed using a generalized orthonormality that can be expressed via a well-known representation of the delta function:

$$\langle p'|p\rangle = \int_{-\infty}^{\infty} \langle p'|x\rangle\langle x|p\rangle dx = \frac{1}{2\pi\hbar}\int_{-\infty}^{\infty} e^{i(p-p')x/\hbar}dx = \delta(p-p') \tag{4.126}$$

using the completeness of the position eigenstates $\int_{-\infty}^{\infty} |x\rangle\langle x|dx = 1$. This way of normalizing is referred to as 'delta-function normalization on the momentum scale'. The momentum-space representation of the Gaussian can now be written as

$$\begin{aligned} G(p,t) &= \int_{-\infty}^{\infty} dx\,\langle p|x\rangle\langle x|G\rangle \\ &= (2\pi\hbar)^{-1/2}\int_{-\infty}^{\infty} dx\, e^{-ipx/\hbar}G(x,t) \\ &= (-i2\alpha_t)^{-1/2}\exp\left[-i(p-p_t)^2/(4\hbar\alpha_t) - ix_tp/\hbar + i\gamma_t/\hbar\right] \end{aligned} \tag{4.127}$$

with the associated probability density in momentum space

$$|G(p,t)|^2 = (2\pi(\Delta p)_t^2)^{-1/2}\exp\left[-\frac{(p-p_t)^2}{2(\Delta p)_t^2}\right] \tag{4.128}$$

which is centered around the mean momentum p_t.

4.2.2 Elements of quantum scattering theory

After these general results and remarks concerning quantum dynamics, we turn now to the more detailed description of the scattering of particles, including reactive scattering.

As an introduction, it is useful to recall the well-known simple one-dimensional description of scattering off a potential barrier $V(x)$. Assume, for simplicity, that $\lim[V(x)] = 0$ for $x \to \pm\infty$. A particle, say described by a Gaussian wave packet Eq. (4.113), is moving from the left toward the potential barrier with a constant mean momentum p_0, as long as the particle does not feel the potential. When the wave packet hits the barrier, it breaks into two packets $\psi_R(x,t)$ and $\psi_T(x,t)$ (see also Fig. 1.1.1), corresponding to a reflected and a transmitted part. The *reflection* and *transmission probabilities* are $R = \int |\psi_R(x,t)|^2 dx$ and $T = \int |\psi_T(x,t)|^2 dx$, for $t \to \infty$, respectively, and $R + T = 1$. In general, these probabilities will depend on the exact form of the wave packets, but if the (Gaussian) wave packet was localized sharply in momentum space (corresponding to a broad wave packet in position space) R and T depend only on the mean momentum p_0.

Alternatively, it turns out that these probabilities can be extracted from *stationary scattering states*, i.e., eigenfunctions of the Hamiltonian with energy $E_0 = p_0^2/(2m)$, and the asymptotic behavior

$$\psi_{p_0}(x) \longrightarrow \begin{cases} (e^{ip_0x/\hbar} + Be^{-ip_0x/\hbar})/\sqrt{2\pi\hbar} & x \to -\infty \\ Ce^{ip_0x/\hbar}/\sqrt{2\pi\hbar} & x \to \infty \end{cases} \tag{4.129}$$

Note that these states are eigenfunctions corresponding to a free particle, and strictly speaking not physically acceptable, since they are not square integrable. Again, they have been normalized using delta-function normalization on the momentum scale, see Eq. (4.126). Nevertheless, the reflection and transmission probabilities for a wave packet that is localized sharply in momentum space are related to the different parts of the stationary state in Eq. (4.129), and given by $R = |B|^2$ and $T = |C|^2$.

Even if $E_0 > V_{\max}$, R can differ from 0, i.e., the particle can be reflected. Likewise, if $E_0 < V_{\max}$, T can differ from 0, i.e., the particle can tunnel. If the potential barrier is replaced by a repulsive potential then $R = 1$, corresponding to total reflection.

We continue with the full description in three-dimensional space and consider first, as an introduction to reactive scattering, elastic two-body scattering, and then generalize to reactive (three-body) scattering [14].

Elastic two-body scattering

We consider in this section the collision between two (structureless) particles, i.e., the quantum version of Section 4.1.3. This discussion provides the essence of the quantum mechanical description of scattering.

The aim is to establish the relation between the observable cross-sections and the collision dynamics. We denote the scattering state in the interaction region at $t = 0$ by $|\chi\rangle$ and write the Hamiltonian in the form $\hat{H}_{\text{c.m.}} + \hat{H}_{\text{rel}}$, i.e., the Hamiltonians associated with the center-of-mass motion and the relative motion. The propagator can be written in the form $\hat{U}(t) = \exp(-i\hat{H}_{\text{c.m.}}t/\hbar)\exp(-i\hat{H}_{\text{rel}}t/\hbar)$, and $|\chi(t)\rangle = \hat{U}(t)|\chi\rangle$ describes the time-dependent scattering state at any time, i.e. $\langle \boldsymbol{R}|\chi(t)\rangle$ is the associated wave packet.

Since $\hat{H}_{\text{rel}} = \hat{H}^0_{\text{rel}} + E(R)$, and the interaction potential vanishes at large internuclear separations, it is a reasonable conjecture (as can be proved formally) that

$$\hat{U}(t)|\chi\rangle \overset{t\to-\infty}{\longrightarrow} \hat{U}^0(t)|\chi_{\text{in}}\rangle \tag{4.130}$$

$$\hat{U}(t)|\chi\rangle \overset{t\to+\infty}{\longrightarrow} \hat{U}^0(t)|\chi_{\text{out}}\rangle \tag{4.131}$$

where $\hat{U}^0(t)$ is the propagator associated with the asymptotic free relative motion, $\hat{H}^0_{\text{rel}}$. That is,

$$|\chi\rangle = \hat{\Omega}_+|\chi_{\text{in}}\rangle = \hat{\Omega}_-|\chi_{\text{out}}\rangle \tag{4.132}$$

where the Møller operators are defined by

$$\hat{\Omega}_\pm = \lim_{t\to\mp\infty} \hat{U}^\dagger(t)\hat{U}^0(t) \tag{4.133}$$

Note that the center-of-mass motion drops out in these operators and in the following it is understood that the Hamiltonians refer to the relative motion; the subscript 'rel' will consequently be dropped. Thus, according to Eq. (4.132), the relation between the scattering state at $t = 0$ and the asymptotic scattering states can be illustrated in the following way:

$$|\chi_{\text{in}}\rangle \overset{\hat{\Omega}_+}{\longrightarrow} |\chi\rangle \overset{\hat{\Omega}_-}{\longleftarrow} |\chi_{\text{out}}\rangle \tag{4.134}$$

Equation (4.132) also implies

$$|\chi_{\text{out}}\rangle = \hat{\Omega}_-^\dagger|\chi\rangle = \hat{\Omega}_-^\dagger\hat{\Omega}_+|\chi_{\text{in}}\rangle = \hat{S}|\chi_{\text{in}}\rangle \tag{4.135}$$

where $\hat{S}$ is the *scattering operator*, defined by

$$\hat{S} = \hat{\Omega}_-^\dagger \hat{\Omega}_+ = \lim_{t\to+\infty} \lim_{t'\to-\infty} e^{i\hat{H}^0 t/\hbar} e^{-i\hat{H}(t-t')/\hbar} e^{-i\hat{H}^0 t'/\hbar} \tag{4.136}$$

or for $t' = -t$, $\hat{S} = \lim_{t\to+\infty} e^{i\hat{H}^0 t/\hbar} e^{-i2\hat{H}t/\hbar} e^{i\hat{H}^0 t/\hbar}$. The scattering probability for $|\chi_{\text{in}}\rangle = |\phi\rangle$ and $|\chi_{\text{out}}\rangle = |\chi\rangle$ can be expressed as

$$P(\chi \leftarrow \phi) = |\langle\chi|\hat{\Omega}_-^\dagger \hat{\Omega}_+|\phi\rangle|^2 = |\langle\chi|\hat{S}|\phi\rangle|^2 \tag{4.137}$$

This probability cannot, however, be measured in practice, since the precise form of the $|\phi\rangle$ states cannot be controlled in an experiment.

Before we return to the relevant average over the $|\phi\rangle$ states, we elaborate a little on the dynamics. Since the potential is spherically symmetric it is natural to consider the Hamiltonian in polar coordinates. Thus, the nuclear Hamiltonian for the relative motion takes the well-known form

$$\hat{H} = \hat{H}^0 + E(R) = -\frac{\hbar^2}{2\mu}\left(\frac{\partial^2}{\partial R^2} + \frac{2}{R}\frac{\partial}{\partial R} - \frac{\hat{L}^2}{\hbar^2 R^2}\right) + E(R) \tag{4.138}$$

where μ is the reduced mass, R the internuclear distance, $\hat{L}$ the relative angular momentum operator, and $E(R)$ the electronic energy. Rotational invariance implies $[\hat{H}, \hat{L}] = 0$, and according to Eq. (4.109) the angular momentum associated with the relative motion is a constant of motion. We consider the dynamics of a state with a definite angular momentum, and write the radial part of the wave function in the form $u_l(R,t)/R$. We note that

$$\hat{H}\left(\frac{u_l(R,t)}{R} Y_{lm}(\theta,\phi)\right) = \frac{Y_{lm}(\theta,\phi)}{R}\hat{H}(l)u_l(R,t) \tag{4.139}$$

where Y_{lm} are the spherical harmonics eigenstates associated with the angular momentum and $\hat{H}(l)$ is equivalent to a one-dimensional Hamiltonian given by

$$\hat{H}(l) = -\frac{\hbar^2}{2\mu}\frac{\partial^2}{\partial R^2} + V_l(R) \tag{4.140}$$

with the effective potential

$$V_l(R) = E(R) + \frac{\hbar^2 l(l+1)}{2\mu R^2} \tag{4.141}$$

where the angular momentum quantum number $l = 0, 1, \ldots$. Note the close correspondence to the equivalent classical expression in Eq. (4.32). Using Eq. (4.139), the time evolution of a state that is in a stationary rotational state can then be written in the form

$$e^{-i\hat{H}t/\hbar}\left(\frac{u_l(R,t)}{R} Y_{lm}(\theta,\phi)\right) = \frac{Y_{lm}(\theta,\phi)}{R} e^{-i\hat{H}(l)t/\hbar} u_l(R,t) \tag{4.142}$$

Thus, the radial motion takes place in the effective one-dimensional potential, $V_l(R)$.

We return now to the connection between the S-matrix elements in Eq. (4.137) and the measurable *cross-section*. To that end, the scattering probability must be averaged over the relevant $|\phi\rangle$ states, which all are assumed to be sharply centered around the momentum $\boldsymbol{p} = \boldsymbol{p}_0$.

Thus, the relative motion of the molecules (atoms) is described by a series of wave packets that differ by a *random* lateral displacement vector $\boldsymbol{b}$ (with the polar coordinates b, ϕ) corresponding to different impact points perpendicular to the momentum $\boldsymbol{p}_0$. As shown in Appendix F, such a displacement of $|\phi\rangle$ can be generated by the operator $\exp(-i\boldsymbol{b}\cdot\hat{\boldsymbol{P}}/\hbar)$, where $\hat{\boldsymbol{P}}$ is the momentum operator with eigenstates $|\boldsymbol{p}\rangle$ and associated eigenvalues $\boldsymbol{p}$, i.e.,

$$\hat{\boldsymbol{P}}|\boldsymbol{p}\rangle = \boldsymbol{p}|\boldsymbol{p}\rangle \tag{4.143}$$

$|\boldsymbol{p}\rangle$ is also an eigenstate of the Hamiltonian of the asymptotic free relative motion, i.e.,

$$\hat{H}^0|\boldsymbol{p}\rangle = E_p|\boldsymbol{p}\rangle \tag{4.144}$$

with $E_p = p^2/2\mu$ and $p = |\boldsymbol{p}|$. In the momentum-space representation, the displacement takes the form

$$\chi_{\text{in}}(\boldsymbol{p}) = \phi_{\boldsymbol{b}}(\boldsymbol{p}) = e^{-i\boldsymbol{b}\cdot\boldsymbol{p}/\hbar}\phi(\boldsymbol{p}) \tag{4.145}$$

The scattered state corresponding to $|\chi_{\text{in}}\rangle$ is $|\chi_{\text{out}}\rangle = \hat{S}|\chi_{\text{in}}\rangle$, and in the momentum-space representation

$$\chi_{\text{out}}(\boldsymbol{p}) = \int d\boldsymbol{p}'\langle\boldsymbol{p}|\hat{S}|\boldsymbol{p}'\rangle\chi_{\text{in}}(\boldsymbol{p}') \tag{4.146}$$

Energy conservation implies that the energy of the two states $|\boldsymbol{p}\rangle$ and $|\boldsymbol{p}'\rangle$ must be identical, i.e., $E_p = E_{p'}$. Indeed, more formally, it can be shown that $[\hat{H}^0, \hat{S}] = 0$, which implies that $\langle\boldsymbol{p}|\hat{S}|\boldsymbol{p}'\rangle(E_p - E_{p'}) = 0$ and $\langle\boldsymbol{p}|\hat{S}|\boldsymbol{p}'\rangle$ contains the delta function $\delta(E_p - E_{p'}) = \mu\delta(p - p')/p$.

The scattering probability into the solid angle element $d\Omega$ about the direction of the final momentum $\boldsymbol{p}$ is

$$P(d\Omega \leftarrow \phi_{\boldsymbol{b}}) = d\Omega \int_0^\infty p^2 dp|\chi_{\text{out}}(\boldsymbol{p})|^2 \tag{4.147}$$

Since we have scattering from wave packets uniformly distributed over $\boldsymbol{b}$, the total scattering probability into the solid angle element $d\Omega$ is $\int d\boldsymbol{b}P(d\Omega \leftarrow \phi_{\boldsymbol{b}})$. If this number is multiplied by the (relative) flux density of molecules in the beam, we get the flux of molecules that show up in $d\Omega$. Thus the cross-section is simply

$$\frac{d\sigma}{d\Omega}d\Omega = \int d\boldsymbol{b}P(d\Omega \leftarrow \phi_{\boldsymbol{b}}) \tag{4.148}$$

Note the similarity with the expression derived within the quasi-classical approach, Eq. (4.10).

When Eq. (4.147) is inserted into Eq. (4.148), we have

$$\frac{d\sigma}{d\Omega} = \int d\boldsymbol{b} \int_0^\infty p^2 dp|\chi_{\text{out}}(\boldsymbol{p})|^2 \tag{4.149}$$

where $|\chi_{\text{out}}(\boldsymbol{p})|^2 = \chi_{\text{out}}(\boldsymbol{p})^\star\chi_{\text{out}}(\boldsymbol{p})$; here $\chi_{\text{out}}(\boldsymbol{p})$ is given by another integral as specified by Eqs (4.145) and (4.146), and $\chi_{\text{out}}(\boldsymbol{p})^\star$ is given by a similar integral (with integration variable $\boldsymbol{p}''$). It is possible to simplify and evaluate the integrals, although it looks a bit cumbersome. First, the integral over $d\boldsymbol{b}$ can be evaluated leading to a delta function using the result $\int_{-\infty}^{\infty} \exp[ik(x - x')]dk = 2\pi\delta(x - x')$. Second, the integral over $d\boldsymbol{p}''$ (associated with $\chi_{\text{out}}(\boldsymbol{p})^\star$) can be evaluated using the delta function associated with the S-matrix elements, and finally the integral in Eq.(4.149) over dp is evaluated, giving

$$\frac{d\sigma}{d\Omega} = (2\pi\hbar\mu)^2 \int d\boldsymbol{p}' \frac{p'}{p'_{\|}} |\langle \boldsymbol{p}|\hat{S}|\boldsymbol{p}'\rangle \phi(\boldsymbol{p}')|^2 \tag{4.150}$$

where $p'_{\|}$ is the component of $\boldsymbol{p}'$ along $\boldsymbol{p}_0$. So far, we have not used the assumption that $\phi(\boldsymbol{p}')$ is sharply centered around $\boldsymbol{p}' = \boldsymbol{p}_0$. We can then replace $(p'/p'_{\|})\langle \boldsymbol{p}|\hat{S}|\boldsymbol{p}'\rangle$ by its value at $\boldsymbol{p}' = \boldsymbol{p}_0$, and take it outside the integral, i.e.,

$$\boxed{\left(\frac{d\sigma}{d\Omega}\right)(\boldsymbol{p}_0|\boldsymbol{p}) = (2\pi\hbar\mu)^2 |\langle \boldsymbol{p}|\hat{S}|\boldsymbol{p}_0\rangle|^2} \tag{4.151}$$

This result is often written in a slightly different form. The scattering operator can obviously be written as $\hat{S} = \hat{I} + (\hat{S} - \hat{I})$, where the identity operator $\hat{I}$ corresponds to scattering in the absence of all interactions, i.e., $\boldsymbol{p} = \boldsymbol{p}_0$. The second term $(\hat{S} - \hat{I})$ corresponds to actual scattering. For $\boldsymbol{p} \neq \boldsymbol{p}_0$, the *scattering amplitude*, f, is introduced by the relation $|\langle \boldsymbol{p}|(\hat{S} - \hat{I})|\boldsymbol{p}_0\rangle|^2 = |f(\boldsymbol{p}_0|\boldsymbol{p})|^2/(2\pi\hbar\mu)^2$. Thus, for $\boldsymbol{p} \neq \boldsymbol{p}_0$ the right-hand side in Eq. (4.151) is simply the squared scattering amplitude, $|f(\boldsymbol{p}_0|\boldsymbol{p})|^2$.

In order to evaluate the matrix element in Eq. (4.151), $\langle \boldsymbol{p}|\hat{S}|\boldsymbol{p}_0\rangle$, we must calculate three-dimensional integrals. In the following we show, however, that the matrix element can be reduced to a sum over one-dimensional S-matrix elements. This is obtained via an expansion of the momentum eigenstates $\langle \boldsymbol{R}|\boldsymbol{p}\rangle$ in a basis where we can use that the angular momentum of the relative motion is conserved.

The eigenstates associated with the asymptotic free relative motion, the so-called plane waves, $\langle \boldsymbol{R}|\boldsymbol{p}\rangle = (2\pi\hbar)^{-3/2}\exp(i\boldsymbol{p}\cdot\boldsymbol{R}/\hbar)$ (with delta-function normalization on the momentum scale), can be expanded in terms of the common eigenstates of $\hat{H}^0(l)$, $\hat{L}^2$, and $\hat{L}_z$, i.e.,

$$\langle \boldsymbol{R}|Elm\rangle = i^l \sqrt{\frac{2\mu}{\pi k}} \frac{j_l(kR)}{R} Y_{lm}(\theta, \phi) \tag{4.152}$$

where i^l is a phase factor and the constant $\sqrt{2\mu/(\pi k)}$ comes from the delta-function normalization of the function $j_l(kR)$, to be specified below. The polar coordinates of $\boldsymbol{R}$ are (R, θ, ϕ), $E = p^2/2\mu$, $k = p/\hbar$, and $j_l(kR)$ is a Riccati–Bessel function that satisfies the free radial Schrödinger equation, i.e., an eigenfunction of Eq. (4.140) with $E(R) = 0$,

$$\hat{H}^0(l) = -\frac{\hbar^2}{2\mu}\frac{\partial^2}{\partial R^2} + \frac{\hbar^2 l(l+1)}{2\mu R^2} \tag{4.153}$$

The asymptotic form of the Riccati–Bessel function is

$$\begin{aligned} j_l(kR) &\overset{R\to\infty}{\longrightarrow} \sin(kR - l\pi/2) \\ &= \frac{1}{2i}\left(e^{i(kR - l\pi/2)} - e^{-i(kR - l\pi/2)}\right) \end{aligned} \tag{4.154}$$

The well-known expansion is

$$e^{i\boldsymbol{p}\cdot\boldsymbol{R}/\hbar} = (2\pi\hbar)^{3/2}\langle\boldsymbol{R}|\boldsymbol{p}\rangle = \frac{4\pi}{k}\sum_{l=0}^{\infty}\sum_{m=-l}^{l} i^l \frac{j_l(kR)}{R} Y_{lm}^{\star}(\hat{\boldsymbol{k}})Y_{lm}(\theta,\phi) \tag{4.155}$$

where $k = p/\hbar$ and the unit vector $\hat{\boldsymbol{k}}$ specify, respectively, the magnitude and direction of the relative momentum $\boldsymbol{p} = k\hat{\boldsymbol{k}}\hbar$. The expansion in Eq. (4.155) shows that a state with a well-defined momentum $\boldsymbol{p}$ can be decomposed into a superposition of states, each having a well-defined angular momentum. The expansion can be formally written in the form

$$\langle\boldsymbol{R}|\boldsymbol{p}\rangle = \sum_{l}\sum_{m}\langle\boldsymbol{R}|Elm\rangle\langle Elm|\boldsymbol{p}\rangle \tag{4.156}$$

and by comparison with Eqs (4.152) and (4.155), we get

$$\langle Elm|\boldsymbol{p}\rangle = Y_{lm}^{\star}(\hat{\boldsymbol{k}})/\sqrt{\mu k} \tag{4.157}$$

Note that we have omitted an integral over the energy, E, since Eq. (4.157) must be diagonal in energy.

The desired S-matrix element in Eq. (4.151) can now be written in the form

$$\begin{aligned}\langle\boldsymbol{p}|\hat{S}|\boldsymbol{p}_0\rangle &= \sum_{ll'}\sum_{mm'}\langle\boldsymbol{p}|El'm'\rangle\langle El'm'|\hat{S}|Elm\rangle\langle Elm|\boldsymbol{p}_0\rangle \\ &= \frac{1}{\mu k}\sum_{ll'}\sum_{mm'} Y_{l'm'}(\hat{\boldsymbol{k}})\langle El'm'|\hat{S}|Elm\rangle Y_{lm}^{\star}(\hat{\boldsymbol{k}}_0)\end{aligned} \tag{4.158}$$

Again, we have omitted an integral over the energy, E, since the S-matrix is diagonal in energy. The scattering operator $\hat{S}$ is given by Eq. (4.136) and for a spherically symmetric potential $E(R)$, using Eq. (4.142) and the orthonormality of the spherical harmonics $\langle Y_{lm}|Y_{l'm'}\rangle = \delta_{ll'}\delta_{mm'}$, we get

$$\langle El'm'|\hat{S}|Elm\rangle = \delta_{ll'}\delta_{mm'}s_l(k) \tag{4.159}$$

where the S-matrix element, $s_l(k)$, in the one-dimensional motion of the radial coordinate is given by

$$s_l(k) = \frac{2\mu}{\pi k}\lim_{t\to+\infty}\lim_{t'\to-\infty}\int dR\, j_l(kR)e^{-i\hat{H}(l)[t-t']/\hbar}j_l(kR) \tag{4.160}$$

Now

$$\sum_{m=-l}^{l} Y_{lm}(\hat{\boldsymbol{k}})Y_{lm}^{\star}(\hat{\boldsymbol{k}}_0) = \frac{2l+1}{4\pi}P_l(\cos\chi) \tag{4.161}$$

where $P_l(\cos\chi)$ is the Legendre polynomial and χ is the angle between $\hat{\boldsymbol{k}}_0$ and $\hat{\boldsymbol{k}}$. That is, Eq. (4.151) takes the form

$$\boxed{\left(\frac{d\sigma}{d\Omega}\right)(E,\chi) = \left|\frac{1}{2k}\sum_{l=0}^{\infty}(2l+1)P_l(\cos\chi)s_l(k)\right|^2} \tag{4.162}$$

where $E = (\hbar k)^2/2\mu$. This is the quantum version of the classical Eq. (4.41).

Equation (4.162) displays clearly how the cross-section is determined from the scattering dynamics in the radial coordinate via the time evolution of the initial state and a subsequent projection onto the final state. The angular momentum $L = \sqrt{l(l+1)}\hbar \sim l\hbar$ is according to Eq. (4.30) in the classical description related to the impact parameter, i.e., $L = \mu v_0 b$. Thus, the sum can be interpreted as the contribution of all impact parameters. In the classical description only one impact parameter contributed to the differential cross-section. For a hard-sphere potential, it can be shown that $d\sigma/d\Omega = d^2$ at low energies, which is four times the classical result in Eq. (4.44).

The *differential cross-section* refers, as in the classical description, to the scattering angle in the center-of-mass coordinate system. In order to relate to experimentally observed differential cross-sections, one has to transform to the appropriate scattering angle. This transformation takes the same form as discussed previously, essentially, because the expectation value of the center-of-mass velocity $\boldsymbol{V}$ is conserved just as in classical mechanics.

The *total cross-section* is obtained by integration over $d\Omega$:

$$\boxed{\sigma(E) = \int \frac{d\sigma}{d\Omega} d\Omega = \frac{\pi}{k^2} \sum_{l=0}^{\infty} (2l+1) \left| s_l(k) \right|^2} \tag{4.163}$$

using

$$\int d\Omega P_l(\cos\chi) P_{l'}(\cos\chi) = \frac{4\pi}{2l+1} \delta_{ll'} \tag{4.164}$$

Reactive scattering

We consider now, briefly, the generalization to three-body scattering that is required in order to describe chemical reactions, i.e.,

$$\mathrm{A} + \mathrm{BC}(n) \longrightarrow \begin{cases} \mathrm{AB}(m) + \mathrm{C} \\ \mathrm{AC}(m) + \mathrm{B} \\ \mathrm{A} + \mathrm{B} + \mathrm{C} \\ \mathrm{A} + \mathrm{BC}(m) \end{cases} \tag{4.165}$$

The different groupings of the atoms are referred to as arrangement channels (or simply channels); we label the reactant channel by α, whereas the different product channels are labeled by $\beta = 1, 2, \ldots$. The part of the Hamiltonian that is effective long before the scattering is the channel Hamiltonian $\hat{H}^{\alpha}$, which is the part of the total Hamiltonian that is left when A and BC are so far apart that the interaction between A and BC vanishes. For a scattering state $\hat{U}(t)|\chi\rangle$ that originates in channel α,

$$\hat{U}(t)|\chi\rangle \overset{t \to -\infty}{\longrightarrow} \hat{U}^{\alpha}(t)|\chi_{\mathrm{in}}\rangle \tag{4.166}$$

where $\hat{U}^{\alpha}(t) = \exp(-i\hat{H}^{\alpha}t/\hbar)$ is the propagator associated with the channel Hamiltonian. This result can be rewritten as

$$|\chi\rangle = \hat{\Omega}_{+}^{\alpha}|\chi_{\mathrm{in}}\rangle = \lim_{t \to -\infty} e^{i\hat{H}t/\hbar} e^{-i\hat{H}^{\alpha}t/\hbar} |\chi_{\mathrm{in}}\rangle \tag{4.167}$$

where $\hat{\Omega}_{+}^{\alpha}$ is the channel Møller operator. The wave function associated with the $\mathrm{A} + \mathrm{BC}(n)$ channel can be written in the form

$$\langle \boldsymbol{R}, \boldsymbol{r} | \chi_{\text{in}} \rangle = \chi(\boldsymbol{R}) \phi^n_{\text{BC}}(\boldsymbol{r}) \tag{4.168}$$

where χ is associated with the motion of A relative to BC and ϕ^n_{BC} is the nth (bound) state of BC. A natural choice of coordinates is again, as in the classical description, Jacobi coordinates. The eigenstates of the channel Hamiltonian $\hat{H}^\alpha$ are

$$\langle \boldsymbol{R}, \boldsymbol{r} | n, \boldsymbol{p}, \alpha \rangle = (2\pi\hbar)^{-3/2} \exp(i\boldsymbol{p} \cdot \boldsymbol{R}/\hbar) \phi^n_{\text{BC}}(\boldsymbol{r}) \tag{4.169}$$

with the energy $E = E_{\text{tr}} + E_n$, i.e., the sum of the relative translational energy and the quantized internal (vibrational/rotational) energy of BC. Note that we did consider such states in Eq. (1.15).

Long after the scattering several arrangement channels can be populated, with channel Hamiltonians $\hat{H}^\beta$ ($\beta = 1, 2, \ldots$), and

$$\hat{U}(t)|\chi\rangle \overset{t \to +\infty}{\longrightarrow} \hat{U}^1(t)|\chi^1_{\text{out}}\rangle + \hat{U}^2(t)|\chi^2_{\text{out}}\rangle + \cdots \tag{4.170}$$

or

$$|\chi\rangle = \hat{\Omega}^1_-|\chi^1_{\text{out}}\rangle + \hat{\Omega}^2_-|\chi^2_{\text{out}}\rangle + \cdots \tag{4.171}$$

The scattering probability for $|\chi_{\text{in}}\rangle = |n, \alpha\rangle$ and $|\chi_{\text{out}}\rangle = |m, \beta\rangle$ can be expressed as

$$P(m, \beta \leftarrow n, \alpha) = |\langle m, \beta | \hat{\Omega}^{\beta\dagger}_- \hat{\Omega}^\alpha_+ | n, \alpha \rangle|^2 = |\langle m, \beta | \hat{S}_{\alpha\beta} | n, \alpha \rangle|^2 \tag{4.172}$$

The connection between these S-matrix elements and the measurable cross-sections is worked out by arguments that are similar to the two-body case (see Eqs (4.148) and (4.151)). To that end, the scattering probability must be averaged over the relevant initial states (corresponding to different impact points) and these states will be assumed to be sharply centered around the momentum $\boldsymbol{p} = \boldsymbol{p}_0$.

The expression for the *total reaction cross-section*, say for A + BC(n) $\longrightarrow$ AB(m) + C, is similar to Eq. (4.163), and takes the form

$$\boxed{\sigma_R(n, E|m) = \frac{\pi}{k_n^2} \sum_J (2J+1) |S^J_{nm}(E)|^2} \tag{4.173}$$

where $k_n = \sqrt{2\mu(E - E_n)}/\hbar$ and the state-to-state *reaction probability*, at the total energy E and total angular momentum J, is given by

$$|S^J_{nm}(E)|^2 = |\langle \phi_m | \hat{U}(t - t_0) | \phi_n \rangle|^2 \tag{4.174}$$

Here ϕ_n and ϕ_m are the initial and final states, i.e., eigenstates of the channel Hamiltonians ($\hat{H}^\alpha$ and $\hat{H}^\beta$, respectively) as specified in Eq. (4.169), and t_0 and t are the initial and final times, chosen such that they correspond to pre- and post-collision times, i.e., $t_0 \to -\infty$ and $t \to \infty$.

Equation (4.173) displays clearly how the cross-section is determined from the scattering dynamics via the time evolution of the initial channel state $\hat{U}(t - t_0)|\phi_n\rangle$ and a subsequent projection onto the final channel state. In practice, the plane wave of the initial state in Eq. (4.169) can be replaced by a Gaussian wave packet, as illustrated in Fig. 1.1.1. When this wave packet is sufficiently broad, it will be localized sharply in momentum space.

4.2.3 Non-adiabatic dynamics

So far, we have considered the dynamics of chemical reactions within the adiabatic approximation. The motion of the atomic nuclei is, however, not always confined to a single electronic state as assumed in Eq. (1.5). This situation can, e.g., occur when two potential energy surfaces come close together for some nuclear geometry. The dynamics of such processes are referred to as *non-adiabatic*. This is a purely non-classical phenomenon [15]. Examples of reactions that involve non-adiabatic transitions are charge-transfer reactions, i.e., reactions in which charge is transferred between reactants.

When several electronic states are in play, Eq. (1.10) is replaced by a matrix equation with a dimension given by the number of electronic states. For example, in the space of two electronic states, Eq. (1.5) is replaced by

$$\Psi(\boldsymbol{r}, \boldsymbol{R}, t) = \chi_1(\boldsymbol{R}, t)\psi_1(\boldsymbol{r}; \boldsymbol{R}) + \chi_2(\boldsymbol{R}, t)\psi_2(\boldsymbol{r}; \boldsymbol{R}) \tag{4.175}$$

where $\boldsymbol{r}, \boldsymbol{R}$ denote all electron and nuclear coordinates, respectively. Substitution into the Schrödinger equation, Eq. (1.1), (using orthonormality of electronic states) gives then

$$i\hbar\frac{\partial}{\partial t}\begin{bmatrix}\chi_1(\boldsymbol{R}, t)\\ \chi_2(\boldsymbol{R}, t)\end{bmatrix} = \begin{bmatrix}\hat{H}_1 & \hat{C}_{12}\\ \hat{C}_{21} & \hat{H}_2\end{bmatrix}\begin{bmatrix}\chi_1(\boldsymbol{R}, t)\\ \chi_2(\boldsymbol{R}, t)\end{bmatrix} \tag{4.176}$$

where

$$\hat{H}_i = \hat{T}_{\text{nuc}} + E_i(\boldsymbol{R}) + \langle\psi_i|\hat{T}_{\text{nuc}}|\psi_i\rangle \tag{4.177}$$

and the coupling terms between the electronic states $\hat{C}_{ij}$ are given by

$$\hat{C}_{ij} = \langle\psi_i|\hat{T}_{\text{nuc}}|\psi_j\rangle - \sum_{s=1}^{N}\frac{\hbar^2}{M_s}\langle\psi_i|\nabla_s|\psi_j\rangle\cdot\nabla_s \tag{4.178}$$

where integration in the matrix elements is over electronic coordinates. The coupling terms imply that the nuclear motion in all the electronic states is coupled. Note that these coupling terms between the electronic states take the form of differential operators.

Equation (4.176) would take a simpler form if the coupling terms were simple (scalar) potentials. It turns out that such a form of the equation can be obtained if one changes to an alternative representation of the electronic states, the so-called *diabatic representation*. In this representation, the electronic wave functions are written in the form $\psi_i(\boldsymbol{r}; \boldsymbol{R}_0)$, i.e., the internuclear coordinates are now fixed at some point. Thus, transformation from the adiabatic representation, where coupling terms between the electronic states are differential operators, to the diabatic representation leads to coupling terms that are (scalar) potentials. The diagonal elements take the form $\hat{T}_{\text{nuc}} + U_i(\boldsymbol{R})$, where $U_i(\boldsymbol{R})$ are the diabatic potentials. A closer analysis of the transformation between the adiabatic and the diabatic representations shows that the adiabatic potentials (electronic energies) E_1 and E_2 are obtained by diagonalization of the diabatic potential energy matrix, with the diagonal elements U_1, U_2 and off-diagonal elements V_{12}, $V_{21} = V_{12}$, that is,

$$E_{1,2} = (U_1 + U_2)/2 \pm (1/2)\sqrt{(U_1 - U_2)^2 + 4V_{12}^2} \tag{4.179}$$

Various numerical (exact) methods have been developed in order to solve Eq. (4.176) or the equivalent equation in the diabatic representation.

We consider next the so-called *Landau–Zener model* that provides insight into non-adiabatic dynamics. The Landau–Zener model concerns the transition probability between two one-dimensional linear intersecting diabatic potentials

$$\begin{aligned} U_1(R) &= \beta_1(R - R_c) + U \\ U_2(R) &= \beta_2(R - R_c) + U \end{aligned} \tag{4.180}$$

where, say, $\beta_1 < 0$ and $\beta_2 > 0$. A constant coupling V_{12} is assumed between these states. Then after substitution of Eq. (4.180) into Eq. (4.179),

$$E_{1,2} = U + (\beta_1 + \beta_2)(R - R_c)/2 \pm V_{12}\sqrt{1 + \epsilon^2} \tag{4.181}$$

where $\epsilon^2 = (\beta_1 - \beta_2)^2(R - R_c)^2/(4V_{12}^2)$. For $R = R_c$, i.e., at the crossing of the diabatic curves, $E_1 - E_2 = 2V_{12}$. For R close to R_c, i.e., ϵ small, we have that $\sqrt{1+\epsilon^2} = 1 + \epsilon^2/2 + \cdots$. The adiabatic potentials are then quadratic expressions in R in the neighborhood of the crossing $R = R_c$. The potentials and the associated adiabatic potentials corresponding to this so-called *avoided crossing* are illustrated in Fig. 4.2.2.

In the Landau–Zener model, dynamics is described by a single trajectory which due to the constant force undergoes an accelerated motion in the crossing region. The probability of a transition from diabatic state 1 to state 2 is denoted by P_{12}, which is also the probability of remaining in the lower adiabatic state, and the transition probability from the lower to the upper adiabatic state is then $P_{\text{nonadia.}} = 1 - P_{12}$, which is given by [16,17]

$$\boxed{P_{\text{nonadia.}}(\tilde{p}_0/m) = \exp(-2\pi\gamma/\hbar)} \tag{4.182}$$

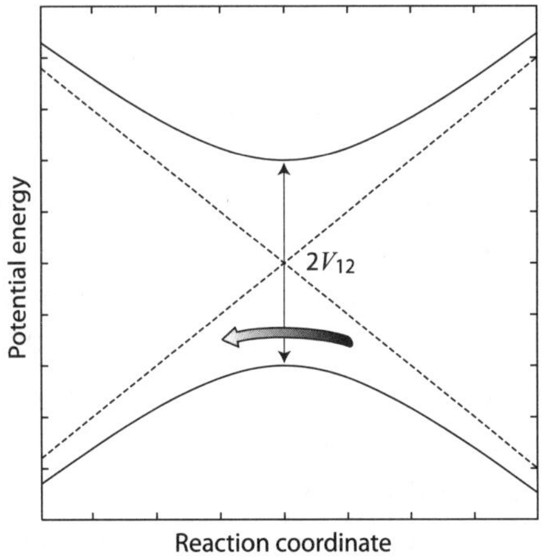

Fig. 4.2.2 Potential energy diagram for the Landau–Zener model. Adiabatic potentials (solid lines) and diabatic potentials (dashed lines), with $\beta_1 < 0$ and $\beta_2 > 0$. The arrow illustrates the dynamics on the lower adiabatic (ground-state) potential.

where

$$\gamma = \frac{|V_{12}|^2}{|\beta_2 - \beta_1|\tilde{p}_0/m} \tag{4.183}$$

β_1, β_2 are the derivatives of the two diabatic potentials, and $\tilde{p}_0/m$ is the speed at the crossing. Note that the probability of staying in the lower adiabatic potential is high when the motion is slow and/or the distance between the two adiabatic potentials is large.

Strictly speaking, the Landau–Zener model is only applicable to the dynamics of the single degree of freedom of a diatomic molecule. In two dimensions, the crossing of the diabatic states will not take place in a single point but along a line, a so-called *crossing seam*. In this case one can, e.g., find corresponding adiabatic potentials that intersect in a single point; the form of the potentials in the vicinity of this point is referred to as a *conical intersection*.

Finally, it is also possible to extend the quasi-classical trajectory approach (Section 4.1) to non-adiabatic dynamics. This is the so-called *trajectory surface hopping approach* [18]. In this approach, nuclei are assumed to move classically on a single adiabatic potential energy surface until a point of large non-adiabatic coupling is reached. At such a point, the trajectory is split into two branches, with appropriate probability factors attached to each branch. The Landau–Zener formula, Eq. (4.182), can be used to estimate the probability. The velocity is, however, now a multidimensional vector but it is a reasonable assumption that only the velocity component perpendicular to the crossing seam of the diabatic surfaces is effective in coupling the states. The trajectory continues on the new surface, and if the potential energies of the two adiabatic surfaces differ at the crossing of the diabatic surfaces, a velocity correction is introduced in order to conserve energy.

Further reading/references

[1] H. Goldstein, *Classical mechanics*, second edition (Addison, Wesley, 1980)

[2] M. Zhao, D.G. Truhlar, N.C. Blais, D.W. Schwenke, and D.J. Kouri, *J. Phys. Chem.* **94**, 6696 (1990)

[3] E. Merzbacher, *Quantum mechanics*, second edition (Addison, Wesley, 1980)

[4] R.N. Zare, *Science* **279**, 1875 (1998)

[5] R.R. Smith, D.R. Killelea, D.F. DelSesto, and A.L. Utz, *Science* **304**, 992 (2004)

[6] A. Sinha, M.C. Hsiao, and F.F. Crim, *J. Chem. Phys.* **92**, 6333 (1990)

[7] M.J. Bronikowski, W.R. Simpson, and R.N. Zare, *J. Phys. Chem.* **97**, 2194 (1993)

[8] X. Zhang, K.L Han, and J.Z.H. Zhang, *J. Chem. Phys.* **116**, 10197 (2002)

[9] D.C. Clary, *Science* **279**, 1879 (1998)

[10] D.H. Zhang, M. Yang, and S.-Y. Lee, *J. Chem. Phys.* **114**, 8733 (2001)

[11] E.J. Heller, *J. Chem. Phys.* **62**, 1544 (1975)

[12] R. Kosloff, *J. Phys. Chem.* **92**, 2087 (1988)

[13] M.H. Beck, A. Jäckle, G.A. Worth, H.-D. Meyer, *Phys. Rep.* **324**, 1 (2000)

[14] J.R. Taylor, *Scattering theory* (Wiley, 1972)

[15] G.A. Worth and L.S. Cederbaum, *Annu. Rev. Phys. Chem.* **55**, 127 (2004)

[16] V. May and O. Kühn, *Charge and energy transfer dynamics in molecular systems* (Wiley–VCH, 2000)

[17] N.E. Henriksen, *Chem. Phys. Lett.* **197**, 620 (1992)

[18] J.C. Tully and R.K. Preston, *J. Chem. Phys.* **55**, 562 (1971)

Problems

4.1 A simple expression for the rate constant is given in Eq. (4.23).

(a) Evaluate $\exp(-E^*/k_B T)$ and $k(T)$ for $T = 1000$ K, 1500 K, and 2000 K, for a system where $\mu = 10$ g/mol, $d = 0.1$ nm, and $E^* = 100$ kJ/mol.

(b) Plot the three values of $k(T)$ on an Arrhenius plot. Is this plot linear?

4.2 Consider the two-dimensional motion of a particle in a central force field. The Lagrange function in Cartesian coordinates is

$$\begin{aligned} L &= T - U(r) \\ &= m(\dot{x}^2 + \dot{y}^2)/2 - U(r) \end{aligned}$$

where $r = \sqrt{x^2 + y^2}$.

(a) Introduce polar coordinates (r, θ) and show that the kinetic energy can be written in the form

$$T = \frac{m}{2}\left(\frac{dr}{dt}\right)^2 + \frac{(mr^2 d\theta/dt)^2}{2mr^2}$$

(b) Replace the velocities by the generalized momenta, p_r and p_θ, conjugate to r and θ, respectively, and write down the Hamilton function.

(c) Write down Hamilton's equations of motion and show that the generalized momentum conjugate to θ is conserved, i.e., $dp_\theta/dt = 0$.

4.3 In connection with Fig. 4.1.8, it was stated that the intersection between the total energy and $V_{\text{eff}}(r)$ gives the distance of closest approach r_c. We consider the hard-sphere potential in Eq. (4.42).

(a) Sketch the effective potential $V_{\text{eff}}(r)$.

(b) Show that $r_c = b$ for $b > d$, and $r_c = d$ for $b \leqslant d$.

4.4 For a hard-sphere potential, the deflection angle χ is given by Eq. (4.43). Evaluate the differential cross-section using Eq. (4.41).

4.5 For the case of the classical scattering of two particles with a repulsive Coulomb potential given by $U(r) = +B/r$, the deflection angle is given by

$$\chi(E, b) = 2\,\text{cosec}^{-1}\left[1 + \left(\frac{2bE}{B}\right)^2\right]^{1/2}$$

Show that the differential cross-section is given by

$$\left(\frac{d\sigma}{d\Omega}\right)(\chi, E) = \left(\frac{B}{4E}\right)^2 \operatorname{cosec}^4(\chi/2)$$

This result is that used by Rutherford in his original α-scattering experiments. Note that $\operatorname{cosec}(x) = 1/\sin(x)$, $\operatorname{cosec}^{-1}(x)$ is the inverse function, and

$$d\operatorname{cosec}^{-1}[u(x)]/dx = -(|u|\sqrt{u^2 - 1})^{-1} du/dx$$

4.6 We consider the so-called Langevin model for ion–molecule reactions. An ion and a neutral non-polar molecule interact—at large distances—through an ion-induced-dipole potential

$$U(r) = -\frac{1}{2}\frac{\alpha e^2}{4\pi\epsilon_0 r^4}$$

where α is the polarizability volume of the neutral non-polar molecule.
The collision is described as a two-body elastic scattering in the spherically symmetric potential $U(r)$ given above. The potential barrier for the reaction is taken as the barrier in the *effective* potential, which is $U(r)$ plus the centrifugal energy $L^2/(2\mu r^2)$ (see Problem 4.2 above), where L is the angular momentum and μ is the reduced mass. Note that $L^2/(2\mu r^2) = E(b/r)^2$, since the total energy is $E = \mu v_0^2/2$ and $L = \mu v_0 b$, where b is the impact parameter.

(a) Sketch the effective potential and find the position r^* of the barrier.

(b) Find the value of the barrier height.

It is assumed that the reaction probability $P = 1$ for kinetic energies larger than or equal to the barrier in the effective potential and $P = 0$ for energies below the barrier.

(c) Find, at the energy E, the maximum value of the impact parameter $b_{\max}$ that can lead to reaction, express the reaction probability in terms of the impact parameter, and find an expression for the reaction cross-section σ_R.

4.7 We consider a chemical reaction that is dominated by the long-range part of the interaction potential between the reactants. The effective potential for the R-coordinate, the distance between the reactants, is

$$V_{\text{eff}}(R) = -\frac{C}{R^n} + \frac{L^2}{2\mu R^2}$$

where L is the orbital angular momentum and μ the reduced mass, and $n > 2$. Assume that the reaction probability $P_r = 1$ for kinetic energies larger than or equal to the barrier in the effective potential.

(a) Find an expression for the reaction cross-section σ_R.

(b) From the cross-section, determine the rate constant $k(T)$. In the integration use the definition of the Gamma function given in Problem 2.3.

(c) Discuss the temperature dependence of the rate constant as a function of n in the range from $n = 3$ to $n = 8$.

(d) For the ion–molecule reaction

$$N^+ + H_2 \rightarrow NH^+ + H$$

the long-range potential can be expressed as

$$U(R) = -\frac{1}{2}\frac{\alpha e^2}{4\pi\epsilon_0 R^4}$$

where α is the polarizability of the H_2 molecule, and e the charge of an electron.
Calculate the cross-sections for the reaction when $\alpha = 0.79\,\text{Å}^3$ for H_2 and at kinetic energies of 30 kJ/mol, 100 kJ/mol, and 300 kJ/mol. Experimental determinations of the cross-sections at those impact energies have given the results 30 Å^2, 15 Å^2, and 10 Å^2, respectively. Compare these values with the calculations.

4.8 We consider a collision between two atoms. A chemical bond between the atoms can be formed, during the collision, via an electromagnetically induced transition to a bound electronic state. This process is called photo-association.
The collision is described here by classical dynamics, and we assume that the motion takes place in a spherically symmetric potential $U(r)$. It is well known that the relative motion of the atoms is equivalent to the motion of a particle with the reduced mass μ, in an effective one-dimensional potential given by

$$V_{\text{eff}}(r) = U(r) + L^2/(2\mu r^2) = U(r) + E(b/r)^2$$

where E is the total energy (corresponding to the relative kinetic energy prior to the collision) and b is the impact parameter of the collision.
The transition to the bound state is assumed to take place at a certain distance $r = r^*$ with the associated photon energy $h\nu^*$. The probability of the process is assumed to take the form

$$P_{h\nu^*}(E) = \begin{cases} 0 & \text{for} \quad E < V_{\text{eff}}(r^*) \\ 1 & \text{for} \quad E \geqslant V_{\text{eff}}(r^*) \end{cases}$$

(a) Find $b_{\text{max}}(E)$, that is, the highest value of b where the process can take place. Determine the reaction cross-section $\sigma_R(E)$.

(b) Determine an expression for the rate constant $k(T)$.

4.9 Show that Hamilton's equations of motion

$$dq_i/dt = \partial H/\partial p_i$$
$$dp_i/dt = -\partial H/\partial q_i$$

with $H(q_i, p_i) = \sum_i (p_i^2/2\mu_i) + V(q_1, \ldots, q_N)$ are equivalent to Newton's law

$$\mu_i d^2 q_i/dt^2 = -\partial V/\partial q_i$$

Show also that the total energy is conserved, i.e., $dH/dt = 0$.

4.10 Consider a bimolecular collision

$$A + BC \longrightarrow$$

between an atom A and a diatomic molecule BC in a collinear model, i.e., all atoms are assumed to move along the same line. Assume that the potential energy surface can be expressed by

$$E(r, R) = D \exp[-\beta(R - \epsilon r)] + (1/2)k(r - r_0)^2$$

where r and R are Jacobi coordinates (the bond length of BC and the distance from A to the center of mass of BC, respectively) and D, β, ϵ, k, and r_0 are constants. The kinetic energy can be written as $T = P_R^2/(2\mu_R) + P_r^2/(2\mu_r)$, where P_R and P_r are the momenta conjugate to R and r, and μ_R and μ_r are reduced masses.

(a) Write down Hamilton's equation of motion.

(b) Describe the motion in the two coordinates for $R \to \infty$.

5

Rate constants, reactive flux

Key ideas and results

In this chapter we consider a direct approach to the calculation of $k(T)$ that bypasses the detailed state-to-state reaction cross-sections. The method is based on the calculation of the *reactive flux* across a dividing surface on the potential energy surface. The results are:

- an approximate result based on the calculation of the classical one-way flux from reactants to products, neglecting possible recrossings of the dividing surface;
- an exact classical expression including recrossings of the dividing surface;
- an exact quantum mechanical expression.

In the previous chapter, we have discussed the reaction dynamics of bimolecular collisions and its relation to the most detailed experimental quantities, the cross-sections obtained in molecular-beam experiments, as well as the relation to the well-known rate constants, measured in traditional bulk experiments. Indeed, in most chemical applications one needs only the rate constant—which represents a tremendous reduction in the detailed state-to-state information.

An important recent theoretical development is the 'direct' approaches for calculating rate constants. These approaches express the rate constant in terms of a so-called flux operator and bypass the necessity for calculating the complete state-to-state reaction probabilities or cross-sections prior to the evaluation of the rate constant [1–3]. This is the theme of this chapter.

First, let us combine some key results from the previous chapters. For a reaction written in the form $\mathrm{A} + \mathrm{BC}(n) \rightarrow \mathrm{AB}(m) + \mathrm{C}$, we have, according to Eq. (2.29),

$$k(T) = \frac{1}{k_B T}\left(\frac{8}{\pi \mu k_B T}\right)^{1/2} \sum_n \sum_m \int_0^\infty p_{\mathrm{BC(n)}} \sigma_R(n, E|m) E_{\mathrm{tr}} e^{-E_{\mathrm{tr}}/k_B T}\, dE_{\mathrm{tr}} \tag{5.1}$$

and from Eq. (4.173),

$$\begin{aligned} \sigma_R(n, E|m) &= \frac{\pi}{k_n^2} \sum_J (2J+1)|S_{nm}^J(E)|^2 \\ &= \frac{\pi \hbar^2}{2\mu} \frac{1}{E_{\mathrm{tr}}} \sum_J (2J+1)|S_{nm}^J(E)|^2 \end{aligned} \tag{5.2}$$

where $k_n = \sqrt{2\mu(E - E_n)}/\hbar = \sqrt{2\mu E_{\rm tr}}/\hbar$, since the total energy $E = E_{\rm tr} + E_n$, i.e., the sum of the relative translational energy and the internal (vibrational/rotational) energy of BC.

We insert Eq. (5.2) into Eq. (5.1), and note that the factors in front of the integral are related to the partition function associated with the relative translational motion of the reactants, see Eq. (A.14):

$$\frac{\pi\hbar^2}{2\mu}\frac{1}{k_BT}\left(\frac{8}{\pi\mu k_BT}\right)^{1/2} = \frac{1}{h}\frac{1}{Q_{\rm trans}/V} \tag{5.3}$$

and use $p_{\rm BC(n)} = \exp(-E_n/k_BT)/Q_{\rm BC}$ to get

$$k(T) = \frac{1}{h}\frac{1}{(Q_{\rm trans}/V)Q_{\rm BC}}\sum_n \int_0^\infty e^{-E_n/k_BT}\sum_m\sum_J(2J+1)|S^J_{nm}(E)|^2 e^{-E_{\rm tr}/k_BT}\,dE_{\rm tr} \tag{5.4}$$

The two exponentials in the integrand are combined and we change the integration variable from the translational energy $E_{\rm tr}$ to the total energy E:

$$k(T) = \frac{1}{h}\frac{1}{(Q_{\rm trans}/V)Q_{\rm BC}}\sum_n \int_{E_n}^\infty \sum_m\sum_J(2J+1)|S^J_{nm}(E)|^2 e^{-E/k_BT}\,dE \tag{5.5}$$

The transition probabilities $|S^J_{nm}(E)|^2 = |S^J_{nm}(E_{\rm tr} + E_n)|^2 = 0$ for $E = E_n$, i.e., when there is no translational energy. We introduce the Heaviside (unit) step function $\theta(\xi)$ by the definition

$$\theta(\xi) = \begin{cases} 1 & \text{for} \quad \xi > 0 \\ 0 & \text{for} \quad \xi < 0 \end{cases} \tag{5.6}$$

and when $|S^J_{nm}(E)|^2$ is replaced by $\theta(E - E_n)|S^J_{nm}(E)|^2$ the lower integration limit E_n can be replaced by 0.

Thus, we have the following exact expression for the rate constant:[1]

$$\boxed{k(T) = \frac{k_BT}{h}\frac{1}{(Q_{\rm trans}/V)Q_{\rm BC}}\int_0^\infty P_{\rm cum}(E)e^{-E/k_BT}\,d(E/k_BT)} \tag{5.7}$$

where

$$P_{\rm cum}(E) = \sum_J(2J+1)\sum_n\sum_m \theta(E - E_n)|S^J_{nm}(E)|^2 \tag{5.8}$$

is the so-called *cumulative reaction probability*. This expression shows how the rate constant depends on the state-to-state reaction probabilities. Note that, at a given total energy E, the summation over n contains a finite number of terms, since the internal energy of the reactant E_n must be smaller than the total energy. Similarly, the summation over m contains a finite number of terms, since $E_m \leqslant E$ where E_m is the internal energy of the product.

[1] Approximate expressions for the rate constant can also be derived (as in Section 4.1.2). Thus, the so-called transition-state theory, to be described in Chapter 6, can be derived based on an approximation to the total reaction probability $\sum_m |S^J_{nm}(E)|^2$; see Section 6.3.

There is another more direct way of calculating the rate constant $k(T)$, i.e., it is possible to bypass the calculation of the complete state-to-state reaction probabilities, $|S^J_{nm}(E)|^2$, or cross-sections prior to the evaluation of the rate constant. The formulation is based on the concept of reactive flux. We start with a version of this formulation based on classical dynamics and, in a subsequent section, we continue with the quantum mechanical version. It will become apparent in the next section that the classical version is valid not only in the gas phase, but in fact in any phase, that is, the foundation for condensed-phase applications will also be provided.

5.1 Classical dynamics

The configuration space may be divided into regions corresponding to the reactants and products, as sketched in Fig. 5.1.1. Consider, for example, the reaction

$$\mathrm{A} + \mathrm{BC} \longrightarrow \begin{cases} \mathrm{AB} + \mathrm{C} & (p1) \\ \mathrm{AC} + \mathrm{B} & (p2) \end{cases} \tag{5.9}$$

It illustrates that the reaction may lead to several products, pj. In the region representing the reactants, region r, the B–C distance will stay close to the equilibrium distance between the two atoms as the system evolves in time, while the distance to A may have all values as well as the momenta. In the region of product $p1$, the distance between A and B will stay close to the equilibrium distance between the two atoms, while there are no constraints on the other variables. For product 2, $p2$, the A–C distance will stay close to the equilibrium distance, and so on.

The various regions in configuration space may be separated by surfaces as shown in the figure. We consider surfaces that completely separate reactants from products, that is, surfaces that are located such that it is impossible to move from the reactant region to the product region, on the potential energy surface, without crossing the

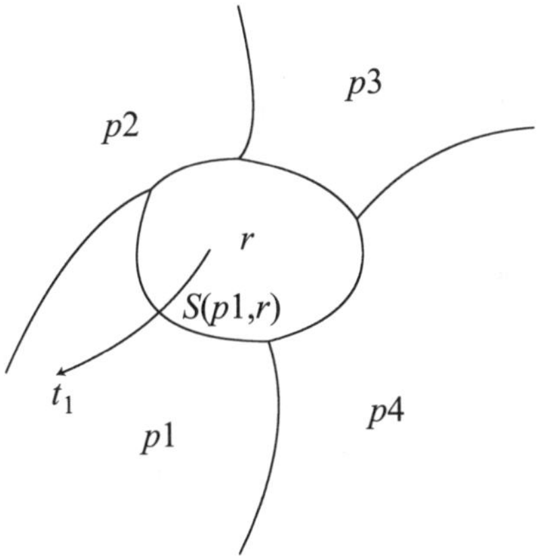

Fig. 5.1.1 Configuration space showing the regions for reactants (r) and products (pj). t_1 is an example of a trajectory that leads to the formation of products $p1$. $S(p1, r)$ is part of the dividing surface between r and $p1$, where the system points have an outward velocity from the reactant region.

surface. Such a surface is referred to as a *dividing surface.* The location of such surfaces is not uniquely given, as we will discuss in the following.

Suppose now a system follows the trajectory t_1 in Fig. 5.1.1; it takes the system across the surface between r and $p1$, which implies that a chemical reaction has taken place in the system. The reaction rate can then be calculated as the number of such trajectories passing the dividing surface.

In Fig. 5.1.2 we have made a sketch with a few examples of trajectories to illustrate how the position of the dividing surface may influence the results.

With the surface placed at D1, trajectory 1 crosses the dividing surface once and leads to reaction. Trajectory 2 crosses one time but does not lead to a reaction since it recrosses once and moves back into the region of the reactants. Trajectory 3 crosses two times from the reactant to the product side and leads to reaction. Trajectory 4 crosses again two times from the reactant to the product side but it does not lead to reaction since it recrosses twice. Trajectory 5 does not cross the dividing surface and therefore does not lead to a reaction.

That is, if we denote by $F(T)$ the number of crossings at temperature T per second and per unit volume from r to p, that is, from left to right in the figure (i.e., the flux), and the *reaction rate* at the same temperature is $R(T)$, then we will always have that

$$R(T) \leqslant F(T) \tag{5.10}$$

which is known as *Wigner's* variational theorem. It states that the number of crossings will always be equal to or larger than the reaction rate because some trajectories will cross and recross the dividing surface a multiple number of times. Some of these trajectories that are *recrossing* the dividing surface will lead to reaction whereas others will be non-reactive. Thus, when a trajectory crosses an even number of times it leads to no reaction, and when a trajectory crosses the dividing surface an odd number of times it leads to reaction. Therefore, by varying the position of the surface separating

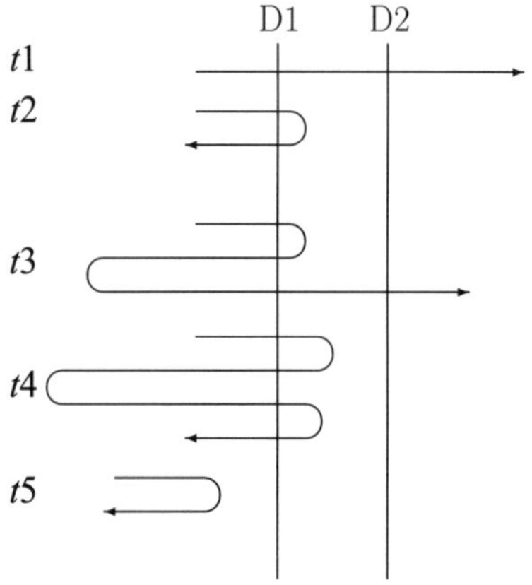

Fig. 5.1.2 Examples of trajectories tj in configuration space. There are six crossings from the left to the right of surface D1, whereas there only are two crossings to the right of surface D2. The position of the latter surface will therefore give a better estimate of the reaction rate according to the variational theorem of Wigner.

reactants and products, we should search for a position that leads to the smallest number of crossings from r to p. That number would then be the best estimate of the reaction rate. With the surface placed at D1, we see that $F(T) = 6$; when placed at D2 it is equal to 2, the 'true value' for the rate. In order words, if we can find a dividing surface with no recrossings, then the flux will coincide with the reaction rate.

5.1.1 Flow of system points in phase space

We have seen that a classical system is represented in phase space by a point, and that an ensemble of identical systems will therefore be represented by a cloud of points. The distribution of system points is determined by the constraints we have imposed on the system, e.g., constant number of particles, constant volume, and constant energy (NVE–ensemble) or constant number of particles, constant volume, and constant temperature (NVT–ensemble).

This cloud of system points is very dense, since we consider a large number of systems, and we can therefore define a number density $\rho(\boldsymbol{p}, \boldsymbol{q}, t)$ such that the number of systems in the ensemble whose phase-space points are in the volume element $d\boldsymbol{p}\, d\boldsymbol{q}$ about $(\boldsymbol{p}, \boldsymbol{q})$ at time t is $\rho(\boldsymbol{p}, \boldsymbol{q}, t) d\boldsymbol{p}\, d\boldsymbol{q}$. Clearly, we must have that

$$\int \cdots \int \rho(\boldsymbol{p}, \boldsymbol{q}, t)\, d\boldsymbol{p}\, d\boldsymbol{q} = A \tag{5.11}$$

where the integration is over all phase space and A is the total number of systems in the ensemble.

We consider an ensemble of systems each containing n atoms. Thus, $\boldsymbol{q} = (q_1, \ldots, q_{3n})$, $\boldsymbol{p} = (p_1, \ldots, p_{3n})$, and $d\boldsymbol{p}\, d\boldsymbol{q} = \Pi_{i=1}^{3n} (dp_i\, dq_i)$. We assume that all interactions are known. As time evolves, each point will trace out a trajectory that will be independent of the trajectories of the other systems, since they represent isolated systems with no coupling between them. Since the Hamilton equations of motion, Eq. (4.63), determine the trajectory of each system point in phase space, they must also determine the density $\rho(\boldsymbol{p}, \boldsymbol{q}, t)$ at any time t if the dependence of ρ on $\boldsymbol{p}$ and $\boldsymbol{q}$ is known at some initial time t_0. This trajectory is given by the *Liouville equation* of motion that is derived below.

Let us determine the change in ρ with time at a given position $(\boldsymbol{p}, \boldsymbol{q})$ in phase space. We surround the point with a small volume element Ω, sufficiently small to make the value of $\rho(\boldsymbol{p}, \boldsymbol{q}, t)$ the same at all points in the volume element, and sufficiently large to contain enough system points so that ρ is well defined and not dominated by large fluctuations. Then the change in the number of system points per second in Ω is equal to the net flow of system points across the bounding surface $S(\Omega)$ of Ω, that is,

$$\frac{\partial}{\partial t} \int_{\Omega} \rho(\boldsymbol{p}, \boldsymbol{q}, t)\, d\boldsymbol{p}\, d\boldsymbol{q} = -\int_{S(\Omega)} \rho(\boldsymbol{p}, \boldsymbol{q}, t) \boldsymbol{V} \cdot \boldsymbol{n}\, ds \tag{5.12}$$

The left-hand side expresses the local rate of change in the number of system points in Ω and the right-hand side expresses the flow of system points with velocity $\boldsymbol{V}$ across the bounding surface $S(\Omega)$. $\boldsymbol{n}$ is by convention the outward unit normal vector to the surface element ds, pointing away from the volume element, and $\boldsymbol{V}$ is the velocity of system points at the surface and given by

$$\boldsymbol{V} = \{\dot{q}_1, \ldots, \dot{q}_{3n}, \dot{p}_1, \ldots, \dot{p}_{3n}\} \tag{5.13}$$

The minus sign is included since, with $\boldsymbol{n}$ pointing outward, $\boldsymbol{V} \cdot \boldsymbol{n} > 0$ when system points flow out of the volume element and therefore cause a decrease in the number of system points in the volume element. It is inconvenient to have a volume and a surface integral in the same equation, so we convert the surface integral to a volume integral using Gauss's theorem:

$$\int_{S(\Omega)} \rho \boldsymbol{V} \cdot \boldsymbol{n}\, ds = \int_{\Omega} \boldsymbol{\nabla} \cdot (\rho \boldsymbol{V})\, d\boldsymbol{p}\, d\boldsymbol{q} \tag{5.14}$$

where the divergence operator is defined as

$$\boldsymbol{\nabla} = \left\{ \frac{\partial}{\partial q_1}, \ldots, \frac{\partial}{\partial q_{3n}}, \frac{\partial}{\partial p_1}, \ldots, \frac{\partial}{\partial p_{3n}} \right\} \tag{5.15}$$

Introduction of Eq. (5.14) into Eq. (5.12) then gives

$$\int_{\Omega} \left[\frac{\partial \rho}{\partial t} + \boldsymbol{\nabla} \cdot (\rho \boldsymbol{V}) \right] d\boldsymbol{p}\, d\boldsymbol{q} = 0 \tag{5.16}$$

and since this expression should be valid at any point of phase space, we must have that the integrand is zero, that is,

$$\frac{\partial \rho}{\partial t} + \boldsymbol{\nabla} \cdot (\rho \boldsymbol{V}) = 0 \tag{5.17}$$

This is a *continuity equation* in phase space.

When this equation is written out,

$$\frac{\partial \rho}{\partial t} + \sum_{i=1}^{3n} \left[\frac{\partial (\rho \dot{q}_i)}{\partial q_i} + \frac{\partial (\rho \dot{p}_i)}{\partial p_i} \right] = 0 \tag{5.18}$$

and the equations for $\dot{q}_i$ and $\dot{p}_i$ from the Hamilton equations of motion are introduced, we get

$$\frac{\partial \rho}{\partial t} + \sum_{i=1}^{3n} \left[\frac{\partial \rho}{\partial q_i} \frac{\partial H}{\partial p_i} - \frac{\partial \rho}{\partial p_i} \frac{\partial H}{\partial q_i} \right] = 0 \tag{5.19}$$

which is *Liouville's equation.* The terms in the square bracket are referred to as the *Poisson bracket.* This is a fundamental equation in classical statistical mechanics.

Since $\rho(\boldsymbol{p}, \boldsymbol{q}, t)$ is a function of $\boldsymbol{p}$, $\boldsymbol{q}$, and t we may write the total differential of ρ as

$$\begin{aligned} \frac{d\rho}{dt} &= \frac{\partial \rho}{\partial t} + \sum_{i=1}^{3n} \left[\frac{\partial \rho}{\partial q_i} \dot{q}_i + \frac{\partial \rho}{\partial p_i} \dot{p}_i \right] \\ &= \frac{\partial \rho}{\partial t} + \sum_{i=1}^{3n} \left[\frac{\partial \rho}{\partial q_i} \frac{\partial H}{\partial p_i} - \frac{\partial \rho}{\partial p_i} \frac{\partial H}{\partial q_i} \right] = 0 \end{aligned} \tag{5.20}$$

according to Hamilton's equations of motion and Eq. (5.19). It shows that there is no change in the density of a volume element that follows the flow of system points, just like the flow of an incompressible fluid. That is, the flow of system points in phase space is analogous to the flow of molecules in an incompressible fluid. Since the Hamilton equations of motion are first-order differential equations, two trajectories will be identical at all times if they have a common point at just one time. That is, two different trajectories will never cross.

Rates and rate constants

The flow of system points is given by the Liouville equation in Eq. (5.17). We consider a volume $\Omega(r)$ in phase space corresponding to the reactants r. Integration of Eq. (5.17) over the entire volume gives the expression

$$\frac{\partial N_r(t)}{\partial t} = -\int_{\Omega(r)} \boldsymbol{\nabla}\cdot(\rho\,\boldsymbol{V})\,d\boldsymbol{p}\,d\boldsymbol{q} \tag{5.21}$$

where

$$N_r(t) = \int_{\Omega(r)} \rho(\boldsymbol{p},\boldsymbol{q},t)\,d\boldsymbol{p}\,d\boldsymbol{q} \tag{5.22}$$

is the number of system points in $\Omega(r)$.

The right-hand side may be converted to a surface integral using Gauss's theorem, Eq. (5.14), and we find

$$\frac{\partial N_r(t)}{\partial t} = -\int_{S(\Omega(r))} \rho\,\boldsymbol{V}\cdot\boldsymbol{n}\,ds \tag{5.23}$$

where $\boldsymbol{n}$ is the outward unit vector normal to ds and $S(\Omega(r))$ is the surface bounding $\Omega(r)$. This equation merely states that the rate of change of the number of system points in $\Omega(r)$ is equal to the negative of the rate at which system points flow out across the surface.

We may rewrite the equation in the form of a simple master equation where the right-hand side is decomposed into rates in and out of r to the different product states pj:

$$\frac{\partial N_r(t)}{\partial t} = \sum_j [R_{r,pj}(t) - R_{pj,r}(t)] \tag{5.24}$$

with

$$\boxed{R_{pj,r}(t) = \int_{S(pj,r)} \rho\,\boldsymbol{V}\cdot\boldsymbol{n}\,ds, \quad \boldsymbol{V}\cdot\boldsymbol{n} > 0} \tag{5.25}$$

and $S(pj,r)$ is that part of the surface between r and pj on which $\boldsymbol{V}\cdot\boldsymbol{n} > 0$. Thus, $R_{pj,r}(t)$ is the rate by which system points pass the surface in the outward direction [4]. Similarly, $R_{pj,r}(t)$ is the rate of system points passing the surface in the inward direction, where $\boldsymbol{V}\cdot\boldsymbol{n} < 0$. The quantity $\rho\,\boldsymbol{V}\cdot\boldsymbol{n}$ is referred to as a *flux density*, i.e., it is the number of system points crossing a unit surface element per unit time. Likewise, the integral in Eq.(5.25) is referred to as the (one-way) *flux*, i.e., the number of system points crossing the surface per unit time.

It is convenient to have an expression for the rates $R_{pj,r}(t)$ in terms of the coordinates $(\boldsymbol{p}, \boldsymbol{q})$ of the system. For that, let the partial surface $S(pj, r)$ be defined by the equations

$$S(\boldsymbol{q}) = 0, \quad \boldsymbol{V} \cdot \boldsymbol{n} > 0 \tag{5.26}$$

The first relation is the equation for the surface separating regions r and pj and the second relation specifies which part of that surface we are considering, namely the one with an outward flow of system points.

Before continuing with an evaluation of the surface integral in Eq. (5.25), let us briefly consider the evaluation of such an integral in ordinary three-dimensional configuration space.

Figure 5.1.3 illustrates a surface F in ordinary three-dimensional Cartesian space. Let it be given by the equation

$$F(x, y, z) = 0 \tag{5.27}$$

where F is a known function of x, y, z, and let us determine the integral

$$I = \int_F \boldsymbol{A} \cdot \boldsymbol{n}\, df \tag{5.28}$$

where $\boldsymbol{A}(x, y, z)$ is some vector quantity at the surface and $\boldsymbol{n}$ the unit vector normal to the differential area element df. Let the projection of F on the x–y plane be R. The projection of the differential area element df on the x–y plane is then $dx\,dy$. The angle between the plane of the differential surface area df and the x–y plane is the same as the angle between the normal $\boldsymbol{n}$ to df and the z-axis (with unit vector $\mathbf{k}$). Then the projected area may also be written as $df\,\boldsymbol{n} \cdot \mathbf{k}$, so

$$df = \frac{dx\,dy}{|\boldsymbol{n} \cdot \mathbf{k}|} \tag{5.29}$$

where we have taken the absolute value of the dot product that may be positive or negative and because dx, dy, and df are all positive. This expression for df may now be

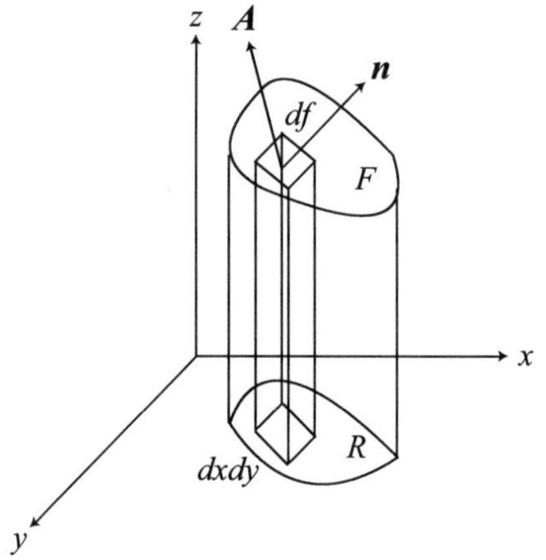

Fig. 5.1.3 Surface F in three-dimensional space.

introduced into Eq. (5.28) and the integral evaluated as a double integral over x and y with $\boldsymbol{n}$ given as $\boldsymbol{\nabla}F/|\boldsymbol{\nabla}F|$, that is, $df = dxdy|\boldsymbol{\nabla}F|/(\partial F/\partial z)$. We use the equation defining the surface to eliminate z from the expressions, to make them functions of x, y only.

This result is easily transferred to the multidimensional phase space. The unit vector, normal to the differential surface element ds, is given by

$$\boldsymbol{n} = \left(\frac{\boldsymbol{\nabla}S}{|\boldsymbol{\nabla}S|}\right)_{S=0} \tag{5.30}$$

This is introduced into Eq. (5.25) and we find

$$R_{pj,r}(t) = \int_{S(pj,r)} \rho\,(\boldsymbol{V}\cdot\boldsymbol{\nabla}S)\,|\boldsymbol{\nabla}S|^{-1}\,ds \tag{5.31}$$

We then project the surface element ds on the $(p_1, \ldots, p_{3n}, q_1, \ldots, q_{3n-1})$ coordinate plane. Then, by analogy to Eq. (5.29), we have

$$ds = \Pi_{i=1}^{3n}\, dp_i\, \Pi_{j=1}^{3n-1} dq_j \frac{|\boldsymbol{\nabla}S|}{|\partial S/\partial q_{3n}|} \tag{5.32}$$

and

$$R_{pj,r}(t) = \int_{S(pj,r)} \rho\, \frac{\boldsymbol{V}\cdot\boldsymbol{\nabla}S}{|\partial S|}\, \Pi_{i=1}^{3n}\,(dp_i\, dq_i) \tag{5.33}$$

where $\partial q_{3n} = dq_{3n}$ is set equal to a small positive increment like the other differentials in the expression, so the absolute value is taken for ∂S only. The dot product in the numerator may be written, using Eqs (5.13) and (5.15) and Hamilton's equation of motion, as follows:

$$\boldsymbol{V}\cdot\boldsymbol{\nabla}S = \sum_{i=1}^{3n}\left[\frac{\partial H}{\partial p_i}\frac{\partial S}{\partial q_i} - \frac{\partial H}{\partial q_i}\frac{\partial S}{\partial p_i}\right] = \sum_{i=1}^{3n}\frac{\partial H}{\partial p_i}\frac{\partial S}{\partial q_i} \tag{5.34}$$

where the last equality holds because the surface in Eq. (5.26) is only specified in terms of position coordinates and not momenta. Then, with the short-hand notation

$$J_i = \frac{\partial H}{\partial p_i}\frac{\partial S}{\partial q_i} \tag{5.35}$$

the expression for the rate may be written

$$\begin{aligned} R_{pj,r}(t) &= \int_{S(pj,r)} \rho \sum_{i=1}^{3n} J_i \frac{\Pi_{j=1}^{3n}\,(dp_j\, dq_j)}{|\partial S|} \\ &= \int_{S(pj,r)} \rho \sum_{i=1}^{3n} \frac{J_i}{|J_1|}\left|\frac{\partial H}{\partial p_1}\frac{\partial S}{\partial q_1}\right| \frac{\Pi_{j=1}^{3n}\,(dp_j\, dq_j)}{|\partial S|} \\ &= \int_{S(pj,r)} \rho \sum_{i=1}^{3n} \frac{J_i}{|J_1|}\, dH\, \Pi_{j=2}^{3n}\,(dp_j\, dq_j) \end{aligned} \tag{5.36}$$

In the second line of the equation, we have just multiplied and divided by $|J_1|$, which is assumed to be non-zero. This makes it possible to eliminate the $|\partial S|$ from the integrand, as seen in the third line of the equation, where we have set $\partial p_1 = dp_1$, $\partial q_1 = dq_1$, and $\partial H = dH$ to small differential increments, just like the other differential increments in the equation. This leaves the absolute value on ∂S alone and cancels with the same term in the denominator.

To make the physical content of Eq. (5.36) more transparent, we note from Eqs (5.34) and (5.30) that

$$\sum_{i=1}^{3n} J_i = \boldsymbol{V} \cdot \boldsymbol{\nabla} S = V_{n,S} \, |\boldsymbol{\nabla} S| \tag{5.37}$$

where $V_{n,S} \equiv \boldsymbol{V} \cdot \boldsymbol{n}$ is the component of the system point velocity that is perpendicular to the dividing surface S. Also, let us write $|J_1|$ as

$$|J_1| = \left| \frac{\partial H}{\partial p_1} \frac{\partial S}{\partial q_1} \right| = \frac{dH}{dp_1} \left| \frac{\partial S}{\partial q_1} \right| \tag{5.38}$$

where we have replaced ∂H and ∂p_1 by small positive differentials.

If these equations are introduced into Eq. (5.36), we may finally write Eq. (5.25) in the form

$$\boxed{\begin{aligned} R_{pj,r}(t) &= \int_{S(pj,r)} \rho \, V_{n,S} \frac{|\boldsymbol{\nabla} S|}{|\partial S/\partial q_1|} \, \Pi_{j=1}^{3n} \, dp_j \, \Pi_{k=2}^{3n} dq_k \\ &= \int_{S(pj,r)} \rho \, V_{n,S} \frac{|\boldsymbol{\nabla} S|}{|\partial S/\partial q_1|} \, \Pi_{j=1}^{3n-3} dp_j \, \Pi_{k=2}^{3n-3} dq_k \end{aligned}} \tag{5.39}$$

where in the last line of the equation we have changed the upper limit from $3n$ to $3n - 3$, since the calculations are always done in a center-of-mass coordinate system.

The expression shows that the rate is determined by the component of the system point velocity that is perpendicular to the chosen surface S, an intuitively reasonable result. The velocity is multiplied by a probability density function ρ and a geometric factor $|\boldsymbol{\nabla} S|/|\partial S/\partial q_1|$, and the integrand is integrated over all momenta and coordinates $q_2, \ldots, q_{3n-3}$, where q_1 is chosen such that $|\partial S/\partial q_1| \neq 0$.

The rate constant is defined as the reaction rate per unit volume of the reaction chamber divided by the product of the reactant concentrations. Thus, for a bimolecular reaction

$$k_{pj,r} = \frac{R_{pj,r}(t)/V}{[\mathrm{A}][\mathrm{B}]} \tag{5.40}$$

where V is the volume of the reaction chamber and $[\mathrm{A}] = N_{\mathrm{A}}/V$ and $[\mathrm{B}] = N_{\mathrm{B}}/V$ are the concentrations (i.e., number densities).

The rate depends on the number of reactant systems $N_r(t)$, Eq. (5.22), as we have seen above through the dependence of ρ, so when we are going to determine the rate constant we need a relation between the number of system points in phase space and the number of reactant molecules in the reaction. Typically, the reactant molecules behave independently of each other, when the concentrations are small, and this is assumed in the following. That is, in the reactant phase space, the total Hamiltonian is a sum of Hamiltonians associated with the A and B molecules, namely $H = H_{\mathrm{A}} + H_{\mathrm{B}}$. Then $\rho = \rho_{\mathrm{A}}\rho_{\mathrm{B}}$ and

$$N_r(t) = \int_{\Omega(r)} \rho(\boldsymbol{p}, \boldsymbol{q}, t)\, d\boldsymbol{p}\, d\boldsymbol{q} = N_{\mathrm{A}} N_{\mathrm{B}} = V^2[\mathrm{A}][\mathrm{B}] \tag{5.41}$$

Combination of Eqs (5.40) and (5.41) gives the following expression for the rate constant:

$$k_{pj,r} = \frac{R_{pj,r}(t) V^{\nu-1}}{N_r(t)} \tag{5.42}$$

where $\nu = 2$ for a bimolecular reaction (it is easily seen that a unimolecular reaction would lead to the same equation with $\nu = 1$).

Before we can determine a rate constant, we must know the density ρ on the surface $S(pj, r)$. In principle, we may choose any distribution function, but usually it is assumed that all degrees of freedom are equilibrated and that the density is given by the (stationary) Boltzmann distribution, that is,

$$\rho = \rho_0 \exp(-H(\boldsymbol{p}, \boldsymbol{q})/k_B T) \tag{5.43}$$

where ρ_0 is given such that Eq. (5.22) is satisfied. This leads to an equilibrium rate constant, sometimes referred to as the canonical rate constant. Combining Eqs (5.43), (5.42), (5.39), and (5.22), we obtain the following expression for the rate constant in the center-of-mass coordinate system:

$$k(T) = Z^{-1}(T) \int_{S(pj,r)} V_{n,S} \exp(-H(\boldsymbol{p}, \boldsymbol{q})/k_B T) \frac{|\boldsymbol{\nabla} S|}{|\partial S/\partial q_1|}\, \Pi_{j=1}^{3n-3} dp_j\, \Pi_{k=2}^{3n-3} dq_k \tag{5.44}$$

where

$$Z(T) = V^{1-\nu} \int_{\Omega(r)} \exp(-H(\boldsymbol{p}, \boldsymbol{q})/k_B T) \Pi_{i=1}^{3n-3} dp_i dq_i \tag{5.45}$$

and the integral is the classical partition function for the reactant system when divided by h^{3n-3}, i.e., $Q_r(T) = Z(T)V^{\nu-1}/h^{3n-3}$, where h is Planck's constant (see Appendix A.2). Again, $\nu = 2$ for a bimolecular reaction.

The optimal choice of the dividing surface $S(pj, r)$ is, according to the Wigner theorem, the surface that gives the smallest rate constant $k(T)$. In principle, it can be determined by a variational calculation of $k(T)$ with respect to the surface such that $\delta k(T) = 0$.

A specific dividing surface

The dividing surface between reactants and products is specified in configuration space. Let us, as an example, choose an orthogonal set of coordinates and a surface $S(pj, r)$ perpendicular to q_1, that is,

$$S(\boldsymbol{q}) = q_1 - a = 0 \tag{5.46}$$

We then pull out the kinetic energy for motion along q_1 from the total energy; it is given by $p_1^2/(2\mu_1)$, where μ_1 is the reduced mass associated with motion in coordinate

q_1 and p_1 is the momentum conjugated to q_1. Then the total energy E may be written as

$$H(\boldsymbol{p}, \boldsymbol{q}) = E = \frac{p_1^2}{2\mu_1} + E_S(p_2, \ldots, p_{3n-3}, q_1, \ldots, q_{3n-3}) \tag{5.47}$$

where, for $q_1 = a$, E_S is the energy associated with motion on the surface.

The geometric factor in Eq. (5.44) is seen to be equal to one with this choice of dividing surface, and the velocity normal to the surface is

$$V_{n,S} \equiv \boldsymbol{V} \cdot \boldsymbol{n} = \dot{q}_1 = p_1/\mu_1 \tag{5.48}$$

This is introduced into Eq. (5.44) and we may write

$$k(T) = Z^{-1}(T) \iint (p_1/\mu_1)\theta(p_1/\mu_1)\delta(q_1 - a) \exp(-H(\boldsymbol{p}, \boldsymbol{q})/k_B T)\, \Pi_{j=1}^{3n-3} dp_j\, dq_j \tag{5.49}$$

where $\delta(x)$ is a delta function that ensures that the integration is over the dividing surface only, and the Heaviside (unit) step function $\theta(\xi)$ defined by

$$\theta(\xi) = \begin{cases} 1 & \text{for} \quad \xi > 0 \\ 0 & \text{for} \quad \xi < 0 \end{cases} \tag{5.50}$$

ensures that only contributions from reactive trajectories that start on the reactant side and end on the product side are included. Note that integration in this expression is over all phase-space coordinates from $-\infty$ to $+\infty$. After Eq. (5.47) is introduced, integration over p_1 gives

$$\int_{-\infty}^{\infty} (p_1/\mu_1)\theta(p_1/\mu_1) \exp[-p_1^2/(2\mu_1 k_B T)] dp_1 = k_B T \tag{5.51}$$

and we find

$$k(T) = \frac{k_\mathrm{B} T}{Z(T)} \int_{S(pj,r)} \exp(-E_S/k_B T)\, \Pi_{j=2}^{3n-3}\, dp_j\, dq_j \tag{5.52}$$

The integral divided by h^{3n-4} is seen to be identical to a partition function $\mathcal{Q}_S$ restricted to the surface $S(pj, r)$, as opposed to an ordinary partition function where the integration extends over all space. Likewise, the partition function $\mathcal{Q}_r$ for the reactants is $\mathcal{Q}_r(T) = Z(T)V^{\nu-1}/h^{3n-3}$. The expression for the bimolecular rate constant may therefore be written in terms of the partition functions as

$$k(T) = \frac{k_\mathrm{B} T}{h} \frac{\mathcal{Q}_S}{(\mathcal{Q}_r/V)} \tag{5.53}$$

It is important to notice that these partition functions can describe a collection of interacting molecules, that is, they need not be partition functions associated with isolated molecules, i.e., molecular partition functions. Since the partition functions are dimensionless, it is easily seen that the expression has the proper unit for a bimolecular rate constant.

Thus, under the assumption that motion from reactants to products can be associated with a single coordinate (orthogonal to the dividing surface) and evaluated as a one-way flux toward the product side, the expressions in Eqs (5.49) and (5.53) give the (classical) rate constant. This result is equivalent to a fully classical version of what is known as conventional *transition-state theory*, to be discussed in detail in subsequent chapters.

Let us be more specific and consider a simple reaction between an atom C and a diatomic molecule AB, that is, a reaction of the type

$$\mathrm{AB} + \mathrm{C} \rightarrow \mathrm{AC} + \mathrm{B} \tag{5.54}$$

and see how the formulation developed above may be used in that case. It will be straightforward to extend the derivations to more complicated reactions between molecules.

The rate constant for the reaction is given by the expression in Eq. (5.44), which has the form

$$k(T) = Z^{-1}(T) \int_{S(pj,r)} V_{n,S} \exp(-H(\boldsymbol{p},\boldsymbol{q})/k_B T) \frac{|\boldsymbol{\nabla} S|}{|\partial S/\partial q_1|} \Pi_{j=1}^{6} dp_j \, \Pi_{k=2}^{6} dq_k \tag{5.55}$$

where $Z(T)$ from Eq. (5.45) is given by

$$Z(T) = V^{-1} \int_{\Omega(r)} \exp(-H(\boldsymbol{p},\boldsymbol{q})/k_B T) \Pi_{i=1}^{6} dp_i dq_i \tag{5.56}$$

q_i and p_i are the conjugated position and momenta coordinates of the atoms, $V_{n,S}$ is the velocity in phase space normal to the dividing surface between reactants and products, and the factor $|\boldsymbol{\nabla} S|/|\partial S/\partial q_1|$ is a geometric factor that relates the surface element dS to the coordinates q_i and p_i; coordinate q_1 is chosen such that the denominator in the geometric factor is different from zero. $H(\boldsymbol{p},\boldsymbol{q})$ is the energy of the atoms in the center-of-mass coordinate system, that is, for the relative motion of the atoms.

We have already seen that a convenient set of coordinates for the relative motion of the atoms is given by the so-called *Jacobi coordinates* in which the kinetic energy is diagonal with no cross terms between different momenta. A systematic way of deriving these coordinates is given in Appendix C, and applied to this system we get

$$\begin{aligned} \boldsymbol{r}_{\mathrm{AB}} &= \boldsymbol{r}_{\mathrm{B}} - \boldsymbol{r}_{\mathrm{A}} \\ \boldsymbol{r}_{\mathrm{C,AB}} &= \boldsymbol{r}_{\mathrm{C}} - \left(\frac{m_{\mathrm{A}}}{m_{\mathrm{A}} + m_{\mathrm{B}}} \boldsymbol{r}_{\mathrm{A}} + \frac{m_{\mathrm{B}}}{m_{\mathrm{A}} + m_{\mathrm{B}}} \boldsymbol{r}_{\mathrm{B}} \right) \end{aligned} \tag{5.57}$$

where the reduced masses associated with the motion in these coordinates are $\mu_{\mathrm{AB}} = m_{\mathrm{A}} m_{\mathrm{B}}/(m_{\mathrm{A}} + m_{\mathrm{B}})$ and $\mu_{\mathrm{C,AB}} = m_{\mathrm{C}}(m_{\mathrm{A}} + m_{\mathrm{B}})/(m_{\mathrm{A}} + m_{\mathrm{B}} + m_{\mathrm{C}})$, respectively.

The total energy in the center-of-mass coordinate system is then (Appendix C)

$$\begin{aligned} H &= \frac{\boldsymbol{p}_{\mathrm{AB}}^2}{2\mu_{\mathrm{AB}}} + \frac{\boldsymbol{p}_{\mathrm{C,AB}}^2}{2\mu_{\mathrm{C,AB}}} + E_{\mathrm{pot}}(\boldsymbol{r}_{\mathrm{AB}}, \boldsymbol{r}_{\mathrm{C,AB}}) \\ &= \frac{\boldsymbol{p}_{\mathrm{AB}}^2}{2\mu_{\mathrm{AB}}} + \frac{\boldsymbol{p}_{\mathrm{C,AB}}^2}{2\mu_{\mathrm{C,AB}}} + E_{\mathrm{pot}}(r_{\mathrm{AB}}, r_{\mathrm{C,AB}}, \xi) \end{aligned} \tag{5.58}$$

where $r_{\mathrm{AB}} = |\boldsymbol{r}_{\mathrm{AB}}|$, and similar for $r_{\mathrm{C,AB}}$. In the last equation we have expressed the potential energy in terms of the magnitude of the distances and the angle ξ between the distance vectors. For a given angle, the potential energy surface may look like that sketched in Fig. 5.1.4.

The next step is to make a reasonable choice of the surface separating reactants from products. It is clear that it should be near the saddle point, if such a point exists. To make the geometrical factor in the expression for the rate constant simple, we choose the dividing surface to be perpendicular to the $r_{\mathrm{C,AB}}$-coordinate spanning the entrance valley from some minimum distance $r_{\mathrm{AB,min}}$ to some maximum distance $r_{\mathrm{AB,max}}$. That is, we identify coordinate q_1 with $r_{\mathrm{C,AB}}$, and the surface is given by the equation

$$S(\boldsymbol{q}) = q_1 - a = r_{\mathrm{C,AB}} - a = 0 \tag{5.59}$$

where a is the value of the $r_{\mathrm{C,AB}}$-coordinate on the surface. With this choice, we immediately see that the geometrical factor is equal to one. Since $q_1 = r_{\mathrm{C,AB}}$ is constant during the integration over the surface S and the potential energy depends on the distances, it will be convenient to switch to spherical coordinates. The spherical coordinates associated with $\boldsymbol{r}_{\mathrm{AB}}$ are r_{AB}, θ_{AB}, and ϕ_{AB}, and similar for $\boldsymbol{r}_{\mathrm{C,AB}}$, i.e., $r_{\mathrm{C,AB}}$, $\theta_{\mathrm{C,AB}}$, and $\phi_{\mathrm{C,AB}}$. The kinetic energy of a particle of mass m is, in spherical coordinates, given by the standard expression

$$E_{\mathrm{kin}} = \frac{1}{2}m\dot{r}^2 + \frac{1}{2}mr\dot{\theta}^2 + \frac{1}{2}mr\sin^2(\theta)\dot{\phi}^2 \tag{5.60}$$

From this and the definition of the conjugated momenta

$$p_i = \left(\frac{\partial L}{\partial \dot{q}_i}\right) = \left(\frac{\partial E_{\mathrm{kin}}}{\partial \dot{q}_i}\right) \tag{5.61}$$

the energy of the system in the center-of-mass coordinate system may be written

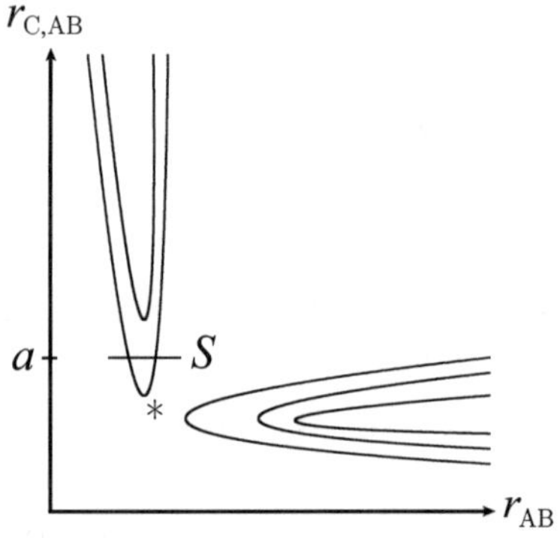

Fig. 5.1.4 Potential energy surface for the reaction in Eq. (5.54) as a function of the distance $r_{\mathrm{C,AB}}$ and r_{AB} for a fixed angle between the two distance vectors. The $*$ marks the position of the saddle point of the potential energy surface and S the surface spanning the reaction channel.

$$E = \frac{p_{r,\mathrm{AB}}^2}{2\mu_{\mathrm{AB}}} + \frac{p_{\theta,\mathrm{AB}}^2}{2\mu_{\mathrm{AB}} r_{\mathrm{AB}}^2} + \frac{p_{\phi,\mathrm{AB}}^2}{2\mu_{\mathrm{AB}} r_{\mathrm{AB}}^2 \sin^2(\theta_{\mathrm{AB}})} + \frac{p_{r,\mathrm{C,AB}}^2}{2\mu_{\mathrm{C,AB}}} + \frac{p_{\theta,\mathrm{C,AB}}^2}{2\mu_{\mathrm{C,AB}} r_{\mathrm{C,AB}}^2} + \frac{p_{\phi,\mathrm{C,AB}}^2}{2\mu_{\mathrm{C,AB}} r_{\mathrm{C,AB}}^2 \sin^2(\theta_{\mathrm{C,AB}})} + E_{\mathrm{pot}}(r_{\mathrm{AB}}, r_{\mathrm{C,AB}}, \xi) \tag{5.62}$$

With the spherical coordinates, the surface and volume elements in the integrals may be written

$$\Pi_{j=1}^6 dp_j \Pi_{k=2}^6 dq_k = dp_{r\mathrm{AB}}\, dp_{\theta\mathrm{AB}}\, dp_{\phi\mathrm{AB}}\, dp_{r\mathrm{C,AB}}\, dp_{\theta\mathrm{C,AB}}\, dp_{\phi\mathrm{C,AB}} \times dr_{\mathrm{AB}}\, d\theta_{\mathrm{AB}}\, d\phi_{\mathrm{AB}}\, d\theta_{\mathrm{C,AB}}\, d\phi_{\mathrm{C,AB}} \tag{5.63}$$

and for the volume element

$$\Pi_{j=1}^6 dp_j dq_j = dp_{r\mathrm{AB}}\, dp_{\theta\mathrm{AB}}\, dp_{\phi\mathrm{AB}}\, dp_{r\mathrm{C,AB}}\, dp_{\theta\mathrm{C,AB}}\, dp_{\phi\mathrm{C,AB}} \times dr_{\mathrm{AB}}\, d\theta_{\mathrm{AB}}\, d\phi_{\mathrm{AB}}\, dr_{\mathrm{C,AB}}\, d\theta_{\mathrm{C,AB}}\, d\phi_{\mathrm{C,AB}} \tag{5.64}$$

since we have used conjugated momenta and coordinates (see Appendix E.3). From Eqs (5.60) and (5.61) we see that

$$V_{n,S} = -\dot{q}_1 = -\dot{r}_{\mathrm{C,AB}} = -\frac{p_{r\mathrm{C,AB}}}{\mu_{\mathrm{C,AB}}} \tag{5.65}$$

The minus sign in Eq. (5.65) indicates that $V_{n,S}$ is the 'outward component of the velocity', that is, from the reactant side at large $r_{\mathrm{C,AB}}$ to the product side at small $r_{\mathrm{C,AB}}$. The angle ξ is the angle between the $\boldsymbol{r}_{\mathrm{C,AB}}$ and $\boldsymbol{r}_{\mathrm{AB}}$ distance vectors. The potential energy is therefore independent of the azimuthal angles ϕ and if we, for example, orient the vector $r_{\mathrm{C,AB}}$ along the z-axis, then we may identify the angle ξ with the polar angle θ_{AB}.

We may now integrate over all variables that do not enter in the potential energy, and the results are compiled in Table 5.1 for both the 'surface' and 'volume' integrals. When combined with what is left in the integrals, we find the following result for the surface integral:

$$I_S(T) = 8\pi^2 k_B T (2\pi\mu_{\mathrm{AB}} k_B T)^{3/2} (2\pi\mu_{\mathrm{C,AB}} k_B T)\, r_{\mathrm{C,AB}}^2 \times \int_0^\pi d\theta_{\mathrm{AB}} \int_{r_{\mathrm{AB,min}}}^{r_{\mathrm{AB,max}}} dr_{\mathrm{AB}}\, r_{\mathrm{AB}}^2 \sin(\theta_{\mathrm{AB}}) \exp(-E_{\mathrm{pot}}(r_{\mathrm{AB}}, r_{\mathrm{C,AB}}, \theta_{\mathrm{AB}})/k_B T) \tag{5.66}$$

where $r_{\mathrm{C,AB}} = a$, and

$$Z(T) = 4\pi (2\pi\mu_{\mathrm{AB}} k_B\, T)^{3/2} (2\pi\mu_{\mathrm{C,AB}} k_B T)^{3/2} \times \int_0^{r_{\mathrm{AB,diss}}} dr_{\mathrm{AB}}\, r_{\mathrm{AB}}^2 \exp(-E_{\mathrm{pot}}(r_{\mathrm{AB}})/k_B T) \tag{5.67}$$

where $r_{\mathrm{AB,diss}}$ is the maximum distance that we have to consider for the reactant AB. We note that the volume integral extends over the reactant space in phase space, where the potential energy only depends on r_{AB} as shown in Eq. (5.67), since atom C

Table 5.1 The integrals over variables that do not enter in the potential energy.

Integration over	Surface integral	Volume integral
$p_{r\mathrm{AB}}$ Range:	$\sqrt{2\pi\mu_{\mathrm{AB}}k_BT}$ $(-\infty,\infty)$	$\sqrt{2\pi\mu_{\mathrm{AB}}k_BT}$ $(-\infty,\infty)$
$p_{r\mathrm{C,AB}}$ Range:	$k_{\mathrm{B}}T$ $(0,-\infty)$	$\sqrt{2\pi\mu_{\mathrm{C,AB}}k_{\mathrm{B}}T}$ $(-\infty,\infty)$
$p_{\theta\mathrm{AB}}$ Range:	$\sqrt{2\pi\mu_{\mathrm{AB}}k_{\mathrm{B}}T}\, r_{\mathrm{AB}}$ $(-\infty,\infty)$	$\sqrt{2\pi\mu_{\mathrm{AB}}k_{\mathrm{B}}T}\, r_{\mathrm{AB}}$ $(-\infty,\infty)$
$p_{\theta\mathrm{C,AB}}$ Range:	$\sqrt{2\pi\mu_{\mathrm{C,AB}}k_{\mathrm{B}}T}\, r_{\mathrm{C,AB}}$ $(-\infty,\infty)$	$\sqrt{2\pi\mu_{\mathrm{C,AB}}k_{\mathrm{B}}T}\, r_{\mathrm{C,AB}}$ $(-\infty,\infty)$
$p_{\phi\mathrm{AB}}$ Range:	$\sqrt{2\pi\mu_{\mathrm{AB}}k_{\mathrm{B}}T}\, r_{\mathrm{AB}}\sin(\theta_{\mathrm{AB}})$ $(-\infty,\infty)$	$\sqrt{2\pi\mu_{\mathrm{AB}}k_{\mathrm{B}}T}\, r_{\mathrm{AB}}\sin(\theta_{\mathrm{AB}})$ $(-\infty,\infty)$
$p_{\phi\mathrm{C,AB}}$ Range:	$\sqrt{2\pi\mu_{\mathrm{C,AB}}k_{\mathrm{B}}T}\, r_{\mathrm{C,AB}}\sin(\theta_{\mathrm{C,AB}})$ $(-\infty,\infty)$	$\sqrt{2\pi\mu_{\mathrm{C,AB}}k_{\mathrm{B}}T}\, r_{\mathrm{C,AB}}\sin(\theta_{\mathrm{C,AB}})$ $(-\infty,\infty)$
ϕ_{AB} Range:	2π $(0,2\pi)$	2π $(0,2\pi)$
$\phi_{\mathrm{C,AB}}$ Range:	2π $(0,2\pi)$	2π $(0,2\pi)$
$\theta_{\mathrm{C,AB}}$ Range:	2 $(0,\pi)$	2 $(0,\pi)$
$r_{\mathrm{C,AB}}$ Range:	Potential energy integration –	$r^3_{\mathrm{C,AB}}/3 = V/(4\pi)$ V
θ_{AB} Range:	Potential energy integration –	2 $(0,\pi)$

is supposed to be far away from molecule AB. Combination of the two expressions for the integrals finally gives the rate constant

$$k(T) = \frac{I_S(T)}{Z(T)} = a^2\sqrt{\frac{2\pi k_BT}{\mu_{\mathrm{C,AB}}}} \times \frac{\int_0^\pi d\theta_{\mathrm{AB}} \int_{r_{\mathrm{AB,min}}}^{r_{\mathrm{AB,max}}} dr_{\mathrm{AB}}\, r_{\mathrm{AB}}^2 \sin(\theta_{\mathrm{AB}}) \exp(-E_{\mathrm{pot}}(r_{\mathrm{AB}}, a, \theta_{\mathrm{AB}})/k_BT)}{\int_0^{r_{\mathrm{AB,diss}}} dr_{\mathrm{AB}}\, r_{\mathrm{AB}}^2 \exp(-E_{\mathrm{pot}}(r_{\mathrm{AB}})/k_BT)} \tag{5.68}$$

From the expression in Eq. (5.68) we may now calculate the rate constant by evaluating the relatively simple integrals in the expression by, for example, a Monte Carlo sampling (see Appendix I).

5.1.2 An exact classical expression for the rate constant

The approach described above will, in general, not give the exact rate constant, since it is based on a quite arbitrary choice of the dividing surface; we do not know if the choice is valid according to the *Wigner theorem*, namely that the rate constant is at a minimum with respect to variations in the choice of dividing surface. A variational determination of the rate constant with respect to the position of the dividing surface is usually not done directly.

Instead, classical trajectory simulations are performed to determine the fraction of trajectories crossing the dividing surface that actually contribute to the formation of product molecules [5,6]. If the surface is the optimal one corresponding to a minimum value for the rate constant, all trajectories crossing the dividing surface from the reactant side to the product side will lead to the formation of products. If not, a certain fraction of the trajectories crossing the dividing surface will turn around, recross the surface, and therefore not make a contribution to the formation of products.

It would be inefficient to start trajectories at some place in the reactant part of phase space, since many of the trajectories would never approach the region of phase space where the dividing surface is located, but would move around in the reactant part of phase space. Instead, all trajectories are *started at the dividing surface*. The initial position of each trajectory is chosen from a Boltzmann distribution proportional to $\exp(-E_{\text{pot}}/k_B T)$ at the given temperature T, see Eq. (5.68). The initial momentum for p_1 is chosen from a distribution proportional to $\exp[-p^2_{\text{C,AB}}/(2\mu_{\text{C,AB}} k_B T)]$ for $-\infty < p < 0$, while the other initial momenta are chosen from equivalent distributions but with the momenta in the range $-\infty < p < \infty$. A small range around the dividing surface is defined arbitrarily as the 'transition region' beyond which we will be in the region of reactants and products, respectively.

If a trajectory makes it to the product side, as sketched in Fig. 5.1.2, then it is also propagated backwards in time from the initial position to check if it originated on the reactant side. If so, the trajectory is marked as successful. In all other cases, that is, the trajectories do not make it to the product side, or if so, do not originate on the reactant side, then they are registered as unsuccessful. The fraction κ of the total number of trajectories that are marked successful is now used to rectify the rate constant for not being calculated with the optimal choice of the dividing surface, and the final result is reported as

$$\boxed{k_{\text{exact}}(T) = \kappa\, k(T)} \tag{5.69}$$

κ is often referred to as a *transmission coefficient*.

The determination of the exact rate constant as given above consists of two steps: (i) a determination of the surface integral in Eq. (5.44), and (ii) a determination of the factor κ in Eq. (5.69) that makes up for the fact that the chosen surface may not be the one leading to a minimum value of the rate constant as required by the *Wigner theorem*. In addition, a determination of the partition function for the reactants is, of course, required.

An alternative formulation [3] of the expression for the rate constant that combines steps (i) and (ii) above is possible, and therefore results in an expression for the rate constant that is independent of the chosen surface, as long as it does not exclude significant parts of the reactant phase space. The expression also forms a convenient basis for developing a quantum version of the theory. We begin with the reformulation of the classical expression and continue with the quantum expression in the following section.

From the basic expression for the rate, Eq. (5.25), using Eqs (5.42), (5.43), and (5.22), we see that the rate constant $k(T)$ in Eq. (5.44) may also be written in the form

$$k(T) = Z^{-1}(T) \int_{S(pj,r)} V_{n,S} \exp(-H(\boldsymbol{p}, \boldsymbol{q})/k_B T)\, ds \tag{5.70}$$

with $V_{n,S} \equiv \boldsymbol{V} \cdot \boldsymbol{n}$, i.e., a canonically averaged flux of system points across some dividing surface S that separates reactants from products in the direction from the reactant side of the surface to the product side. In the previous section we showed one way of determining the surface integral. Here we consider an alternative way that is based on the properties of the Heaviside step function and allows us to combine steps (i) and (ii) above in one expression. The Heaviside (unit) step function is given by Eq. (5.50) and the derivative of the step function is the delta function, that is,

$$\frac{d\theta(\xi)}{d\xi} = \delta(\xi) \tag{5.71}$$

Now suppose, like before, that the dividing surface is given by the equation

$$S(\boldsymbol{q}) = 0 \tag{5.72}$$

in the atomic coordinates $\boldsymbol{q}$ of the system.

The surface integral in Eq. (5.70) may be expressed as a volume integral if we include the delta function $\delta(S(\boldsymbol{q}))$ in the integrand such that we only get contributions when the coordinates satisfy Eq. (5.72). We also need the flux of system points across the surface. Both of these may be introduced by considering the *'flux' function* $F(\boldsymbol{p}, \boldsymbol{q})$ given in terms of the Heaviside step function as

$$\begin{aligned} F(\boldsymbol{p}, \boldsymbol{q}) &\equiv \frac{d}{dt}\theta(S[\boldsymbol{q}(t)]) = \delta(S[\boldsymbol{q}(t)])\frac{d\,S(\boldsymbol{q}(t))}{dt} = \delta(S[\boldsymbol{q}(t)]) \sum_{i=1}^{3n-3} \left(\frac{\partial S(\boldsymbol{q}(t))}{\partial q_i}\right) \dot{q}_i \\ &= \delta(S[\boldsymbol{q}(t)]) \boldsymbol{\nabla} S \cdot \dot{\boldsymbol{q}} = \delta(S[\boldsymbol{q}(t)])\, V_{n,S} \end{aligned} \tag{5.73}$$

In the last line of the equation we have used bold phase notation to emphasize that the sum in the first line of the equation is equivalent to a dot product of the gradient vector perpendicular to the dividing surface and the velocity of system points across the surface, that is, the velocity component perpendicular to the dividing surface, see also Eqs (5.34) and (5.37). Thus, $F(\boldsymbol{p}, \boldsymbol{q})$ describes the flux across the dividing surface of system points starting at $(\boldsymbol{p}, \boldsymbol{q})$. If the coordinates $\boldsymbol{q}$ do not satisfy Eq. (5.72), the flux will be zero, otherwise it will be given by $V_{n,S}$.

We also need a function that shows whether a given trajectory, starting on the reactant side of the dividing surface, ends up at the product side at time $t \to \infty$ and thereby contributes to the formation of products. The Heaviside step function in Eq. (5.50) may also be used to specify whether the phase-space point of a system is at the dividing surface ($S(\boldsymbol{q}(t)) = 0$), on the product side (say $S(\boldsymbol{q}(t)) > 0$), or on the reactant side ($S(\boldsymbol{q}(t)) < 0$). We then define the function $P(\boldsymbol{p}, \boldsymbol{q})$ according to

$$P(\boldsymbol{p}, \boldsymbol{q}) \equiv \lim_{t \to \infty} \theta(S[\boldsymbol{q}(t)]) \tag{5.74}$$

This function will be equal to one if products have been formed because the coordinates will then be on the 'positive side' of the dividing surface, otherwise it will be zero. So, let us consider a set of initial coordinates $(\boldsymbol{p}, \boldsymbol{q})$ somewhere in the reactant space; then the contribution of this trajectory to the formation of products may be written as

$$dk \propto F(\boldsymbol{p}, \boldsymbol{q})\, P(\boldsymbol{p}, \boldsymbol{q}) \tag{5.75}$$

We see that if the coordinates $\boldsymbol{q}(t)$ never have values such that $S(\boldsymbol{q}(t)) = 0$, then $F(\boldsymbol{p}, \boldsymbol{q}) = 0$ and the trajectory will not contribute to the formation of product. If the trajectory has passed the dividing surface then $F(\boldsymbol{p}, \boldsymbol{q}) = V_{n,S} \neq 0$, and if the trajectory is still on the product side at $t \to \infty$ then $P(\boldsymbol{p}, \boldsymbol{q}) = 1$, and the trajectory makes a contribution to the product formation. If the trajectory has recrossed the dividing surface and ends up on the reactant side, then $P(\boldsymbol{p}, \boldsymbol{q}) = 0$ and there will be no contribution to the product formation from that trajectory although it originally crossed the dividing surface from the reactant to the product side.

We then need to consider a canonical weighted average of the contributions in Eq. (5.75) of all trajectories starting on the reactant side, and crossing the dividing surface in the direction from the reactant side to the product side. Thus, after introduction of $P(\boldsymbol{p}, \boldsymbol{q})$, Eq. (5.70) takes the form

$$\boxed{k_{\text{exact}}(T) = \frac{1}{Q_r\, h^{3n-3} V^{1-\nu}} \int_{\Omega(r)} d\boldsymbol{p} d\boldsymbol{q}\, F(\boldsymbol{p}, \boldsymbol{q}) P(\boldsymbol{p}, \boldsymbol{q}) \exp(-H(\boldsymbol{p}, \boldsymbol{q})/k_B T)} \tag{5.76}$$

where $\nu = 2$ for a bimolecular reaction (and $\nu = 1$ for a unimolecular reaction). Note that the integration over the momentum variable perpendicular to the surface is only from 0 to ∞, that is, in the direction from the reactant side to the product side of the dividing surface consistent with the $\boldsymbol{V} \cdot \boldsymbol{n} > 0$ condition in Eq. (5.25).

From the relations in Eqs (5.73) and (5.74) we see that the $P(\boldsymbol{p}, \boldsymbol{q})$ function may also be expressed in terms of the flux function $F(\boldsymbol{p}, \boldsymbol{q})$, that is, we may write $P(\boldsymbol{p}, \boldsymbol{q})$ as the following time integral:

$$\begin{aligned} P(\boldsymbol{p}, \boldsymbol{q}) &= \int_0^\infty dt \frac{d}{dt} \theta(S[\boldsymbol{q}(t)]) \\ &= \int_0^\infty dt\, F(\boldsymbol{p}(t), \boldsymbol{q}(t)) \end{aligned} \tag{5.77}$$

In the first line we have used that $\theta(S[\boldsymbol{q}(0)])$ is always zero since all trajectories start on the reactant side. Inserting into Eq. (5.76) and interchanging the order of the phase space and time integrals gives

$$\boxed{\begin{aligned} k_{\text{exact}}(T) &= \frac{1}{h^{3n-3} V^{1-\nu}} \int_0^\infty dt \langle [F(\boldsymbol{p}(0), \boldsymbol{q}(0))][F(\boldsymbol{p}(t), \boldsymbol{q}(t))] \rangle \\ &\equiv \frac{1}{h^{3n-3} V^{1-\nu}} \int_0^\infty dt\, C_F(t) \end{aligned}} \tag{5.78}$$

where $C_F(t)$ is the canonical average of the flux time-correlation function, except for the fact that we only integrate the momentum perpendicular to the dividing surface from zero to ∞, that is,

$$C_F(t) = \langle [F(\boldsymbol{p}(0), \boldsymbol{q}(0))][F(\boldsymbol{p}(t), \boldsymbol{q}(t))] \rangle$$
$$= \frac{1}{\mathcal{Q}_r} \int_{\Omega(r)} d\boldsymbol{p}(0) d\boldsymbol{q}(0) F(\boldsymbol{p}(0), \boldsymbol{q}(0)) F(\boldsymbol{p}(t), \boldsymbol{q}(t)) \exp(-H(\boldsymbol{p}(0), \boldsymbol{q}(0))/k_B T) \tag{5.79}$$

We see that the rate constant may be determined as the time integral of the canonical averaged *flux autocorrelation function* for the flux across the dividing surface between reactants and products. It is also clear that we only need to calculate the flux correlation function for trajectories starting on the dividing surface, for otherwise $F(\boldsymbol{p}(0), \boldsymbol{q}(0)) = 0$ and there will be no contributions to the product formation.

The observation that the rate constant may be expressed in terms of an auto-time-correlation function of the flux, averaged over an equilibrium ensemble, has a parallel in statistical mechanics. There it is shown, within the frame of linear response theory, that any transport coefficients, like diffusion constants, viscosities, conductivities, etc., may also be expressed in terms of auto-time-correlation functions of proper chosen quantities, averaged over an equilibrium ensemble.

Finally, it is instructive to consider the exact expression under some simplifying assumptions. Let the dividing surface be perpendicular to the reaction coordinate, say q_1, so

$$S(\boldsymbol{q}) = q_1 = 0 \tag{5.80}$$

All trajectories start at the dividing surface, so

$$F(\boldsymbol{p}(0), \boldsymbol{q}(0)) = \delta(S[\boldsymbol{q}(0)])\, \dot{q}_1 = \delta(q_1)\, p_1/\mu_1 \tag{5.81}$$

where we have assumed a Cartesian coordinate system for simplicity. Let us apply the exact expressions to the special situation where no trajectories recross the dividing surface; then

$$F(\boldsymbol{p}(t), \boldsymbol{q}(t)) = \delta(S[\boldsymbol{q}(t)])\, dS/dt$$
$$= \frac{\delta(t)}{|dS/dt|} \frac{dS}{dt} = \delta(t) \tag{5.82}$$

where we have used the relation

$$\delta(g(x)) = \sum_n \frac{\delta(x - x_n)}{|dg(x)/dx|_{x=x_n}} \tag{5.83}$$

x_n are the roots of $g(x)$ (this is easily proven by a Taylor expansion of $g(x)$ around x_n). Note that the assumption of no recrossings implies that there is only one root for $t = 0$. Inserting these relations into Eq. (5.79) gives

$$C_F(t) = \frac{1}{\mathcal{Q}_r} \int d\boldsymbol{p}\, d\boldsymbol{q}\, \delta(q_1) \frac{p_1}{\mu_1} \delta(t) \exp(-H(\boldsymbol{p}, \boldsymbol{q})/k_B T)$$
$$= \frac{k_B T}{\mathcal{Q}_r} \delta(t) \int \exp(-E_S(\boldsymbol{p}, \boldsymbol{q}, q_1 = 0)/k_B T) \Pi_{i=2}^{3n-3}\, dp_i\, dq_i$$
$$= \frac{k_B T}{\mathcal{Q}_r} \mathcal{Q}_S h^{3n-4}\, \delta(t) \tag{5.84}$$

where the integration over p_1 is as in Eq. (5.51), E_S is defined in Eq. (5.47), and $\mathcal{Q}_S$ is the partition function restricted to the surface S. So the flux time-correlation function decays to zero immediately as opposed to a situation where recrossing takes place. Then, according to Eq. (5.78), the rate constant is given by

$$k(T) = \frac{k_B T}{h} \frac{\mathcal{Q}_S}{\mathcal{Q}_r / V^{\nu-1}} \tag{5.85}$$

This result is equivalent to Eq. (5.53), the classical version of what is known as conventional *transition-state theory*, to be discussed in detail in subsequent chapters.

One final remark is that in the derivations given above we did not assume that the chemical reaction took place in the gas phase. Thus, the foundation for condensed-phase applications is also provided. Reaction dynamics in condensed phases will be discussed in the second part of this book.

5.2 Quantum dynamics

After having described the expression for the rate constant within the framework of classical mechanics, we turn now to the quantum mechanical version. We consider first the definition of a flux operator in quantum mechanics.[2] To that end, the flux density operator (for a single particle of mass m) is defined by

$$\hat{\boldsymbol{j}} = (\hat{\boldsymbol{P}}/m|\boldsymbol{R}\rangle\langle\boldsymbol{R}| + |\boldsymbol{R}\rangle\langle\boldsymbol{R}|\hat{\boldsymbol{P}}/m)/2 \tag{5.86}$$

The expectation value of this (Hermitian) operator is

$$\begin{aligned}
\langle\chi|\hat{\boldsymbol{j}}|\chi\rangle &= [\langle\chi|\hat{\boldsymbol{P}}|\boldsymbol{R}\rangle\langle\boldsymbol{R}|\chi\rangle + \langle\chi|\boldsymbol{R}\rangle\langle\boldsymbol{R}|\hat{\boldsymbol{P}}|\chi\rangle]/(2m) \\
&= \left[\left(\frac{\hbar}{i}\nabla\chi(\boldsymbol{R})\right)^{\star}\chi(\boldsymbol{R}) + \chi^{\star}(\boldsymbol{R})\frac{\hbar}{i}\nabla\chi(\boldsymbol{R})\right]/(2m) \\
&= \frac{\hbar}{2im}[\chi^{\star}(\boldsymbol{R})\nabla\chi(\boldsymbol{R}) - \chi(\boldsymbol{R})\nabla\chi^{\star}(\boldsymbol{R})]
\end{aligned} \tag{5.87}$$

where we have used $\langle\boldsymbol{R}|\hat{\boldsymbol{P}}|\chi\rangle = (\hbar/i)\nabla\langle\boldsymbol{R}|\chi\rangle$. Note that in the second line we observe that the second term is just the complex conjugate of the first term. Equation (5.87) is recognized as the probability current density or flux density of Eq. (4.107). The *flux operator* corresponding to the probability flux through a surface separating reactants and products is defined by

$$\hat{F} = \int_S dS\, \mathbf{n}\cdot\hat{\boldsymbol{j}} \tag{5.88}$$

where $\mathbf{n}$ is the unit normal to the surface elements. The surface can be described by $S = \{\boldsymbol{R} \in R^n | s(\boldsymbol{R}) = 0\}$, and we can rewrite the flux integral as a volume integral rather than a surface integral:

$$\begin{aligned}
\hat{F} &= \int_S dS\, \mathbf{n}\cdot\hat{\boldsymbol{j}} \\
&= \int d\boldsymbol{R}\, \delta(s(\boldsymbol{R}))\, \mathbf{n}\cdot\hat{\boldsymbol{j}}
\end{aligned} \tag{5.89}$$

[2]In order to fully appreciate the content of this section, a good background in quantum mechanics is required; see also Appendix F.

where $\mathbf{n} = \nabla s(\boldsymbol{R})/|\nabla s(\boldsymbol{R})|$. Note that in one dimension the 'surface' is a point, whereas in two dimensions the 'surface' is a line, etc. Inserting the expression for $\hat{\boldsymbol{j}}$,

$$\begin{aligned}\hat{F} &= \frac{1}{2m}\int d\boldsymbol{R}\, \delta(s(\boldsymbol{R}))\, \mathbf{n}\cdot\{|\boldsymbol{R}\rangle\langle\boldsymbol{R}|\hat{\boldsymbol{P}} + \hat{\boldsymbol{P}}|\boldsymbol{R}\rangle\langle\boldsymbol{R}|\}\\ &= \frac{1}{2m}\int d\boldsymbol{R}\, \delta(s(\boldsymbol{R}))\, \{|\boldsymbol{R}\rangle\langle\boldsymbol{R}|\,\mathbf{n}\cdot\hat{\boldsymbol{P}} + \mathbf{n}\cdot\hat{\boldsymbol{P}}\,|\boldsymbol{R}\rangle\langle\boldsymbol{R}|\}\end{aligned} \tag{5.90}$$

As an example, specializing to a surface of constant R_1, $s(\boldsymbol{R}) = R_1 - R_0$, implies $\partial s/\partial R_1 = 1$ and $\partial s/\partial R_i = 0$ for $i \neq 1$, that is, $\mathbf{n}\cdot\hat{\boldsymbol{P}} = \hat{P}_{R_1}$ and

$$\begin{aligned}\hat{F} &= \frac{1}{2m}\int d\boldsymbol{R}\, \delta(R_1 - R_0)\, \{|\boldsymbol{R}\rangle\langle\boldsymbol{R}|\, \hat{P}_{R_1} + \hat{P}_{R_1}|\boldsymbol{R}\rangle\langle\boldsymbol{R}|\}\\ &= \frac{1}{2m}\int dR_1\, \delta(R_1 - R_0)\, \{|R_1\rangle\langle R_1|\, \hat{P}_{R_1} + \hat{P}_{R_1}|R_1\rangle\langle R_1|\}\end{aligned} \tag{5.91}$$

where we have used the completeness relation $\int d\tilde{\boldsymbol{R}}|\tilde{\boldsymbol{R}}\rangle\langle\tilde{\boldsymbol{R}}| = 1$, where $\tilde{\boldsymbol{R}}$ refers to all coordinates with the exception of R_1.

5.2.1 One degree of freedom

Let us for simplicity consider $k(T)$ for a one-dimensional 'reaction', where all degrees of freedom are neglected except for one degree of freedom describing the progress of the reaction. Within this framework, reaction corresponds to the crossing of a one-dimensional barrier. Starting from the exact Eq. (5.7) where $Q_{\mathrm{BC}} = 1$, Q_{trans} is the partition function of the relative one-dimensional translation motion, and $P_{\mathrm{cum}} = |S(E)|^2$, the expression for the rate constant takes the form

$$\begin{aligned}k(T) &= \frac{k_BT}{h}\frac{1}{(Q_{\mathrm{trans}}/L)}\int_0^\infty |S(E)|^2 e^{-E/k_BT}\, d(E/k_BT)\\ &= \frac{1}{h}\frac{1}{(Q_{\mathrm{trans}}/L)}\int_0^\infty (p/\mu)T(p)e^{-E/k_BT}\, dp\\ &= \int_0^\infty (p/\mu)T(p)P(p)\, dp\end{aligned} \tag{5.92}$$

where $E = p^2/(2\mu)$, $|S(E)|^2 = T(p)$ is the transmission probability, and $P(p)$ is the Boltzmann distribution,

$$\begin{aligned}P(p)dp &= \frac{1}{h}\frac{1}{(Q_{\mathrm{trans}}/L)}e^{-E/k_BT}dp\\ &= \frac{\exp[-p^2/(2\mu k_BT)]}{\sqrt{2\pi\mu k_BT}}\, dp\end{aligned} \tag{5.93}$$

since in one dimension $Q_{\mathrm{trans}} = \sqrt{2\pi\mu k_BT}L/h$. Thus, the interpretation of Eq. (5.92) for the rate constant is quite simple: it is a thermal average of the speed times the transmission probability. The unit of this rate constant is length/time.

As a simple illustration, assume that $T(p) = 0$ for $p < p_0$, and $T(p) = 1$ for $p \geqslant p_0$; then $k(T) = \sqrt{k_B T/(2\pi\mu)}\exp[-E_0/k_B T]$, where $E_0 = p_0^2/(2\mu)$. The pre-exponential factor is just the average velocity for motion from left to right (see Eq. (6.6)).

Now, in order to introduce the flux operator of Eq. (5.91), we evaluate the expectation value of the flux operator at a point $R = R_0$ in the asymptotic region, using the stationary scattering states in Eq. (4.129), which we now denote by $\langle R|p+\rangle$, with the asymptotic form $\lim_{R\to\infty}\langle R|p+\rangle = \langle R|p\rangle = C\exp(ipR/\hbar)/\sqrt{2\pi\hbar}$:

$$\begin{aligned}\langle p|\hat{F}|p\rangle &= \frac{1}{2\mu}\int dR\ \delta(R-R_0)\ \{\langle p|R\rangle\langle R|\hat{P}|p\rangle + \langle p|\hat{P}|R\rangle\langle R|p\rangle\} \\ &= \frac{1}{2\mu}\left[\int dR\,\delta(R-R_0)C^\star e^{-ipR/\hbar}pCe^{ipR/\hbar} + cc\right]/(2\pi\hbar) \\ &= (p/\mu)T(p)/(2\pi\hbar)\end{aligned} \tag{5.94}$$

since $\hat{P}|p\rangle = p|p\rangle$, and where cc denotes the complex conjugate term and $T(p) = |C|^2$ is the transmission probability, see Section 4.2.2. Thus, this number is independent of R_0; in fact, using Eq. (4.106) the flux associated with any stationary (scattering) state is independent of the position. That is, $\langle p|\hat{F}|p\rangle = \langle p+|\hat{F}|p+\rangle$. Furthermore, since $|p+\rangle$ is an eigenstate of $\hat{H}$ with an eigenvalue that is the same as when $\hat{H}^0$ acts on $|p\rangle$, we have

$$e^{-\hat{H}/k_B T}|p+\rangle = e^{-p^2/2\mu k_B T}|p+\rangle \tag{5.95}$$

Equation (5.92) can now be written in the form

$$k(T) = \frac{1}{(Q_{\text{trans}}/L)}\int_0^\infty \langle p+|\hat{F}e^{-\hat{H}/k_B T}|p+\rangle dp \tag{5.96}$$

and we introduce a projection operator $\hat{P}_\text{r}^+$:

$$\hat{P}_\text{r}^+|p+\rangle = \begin{cases}|p+\rangle & \text{for } p \geqslant 0 \\ 0 & \text{for } p < 0\end{cases} \tag{5.97}$$

then the integral can be extended to $-\infty$, and

$$\boxed{k(T) = \frac{1}{(Q_{\text{trans}}/L)}\text{Tr}[e^{-\hat{H}/k_B T}\hat{F}\hat{P}_\text{r}^+]} \tag{5.98}$$

where Tr is the quantum mechanical trace, i.e., a sum over diagonal matrix elements, and we have used that the trace can be rewritten in various ways:

$$\text{Tr}[\hat{F}e^{-\hat{H}/k_B T}\hat{P}_\text{r}^+] = \text{Tr}[\hat{F}\hat{P}_\text{r}^+e^{-\hat{H}/k_B T}] = \text{Tr}[e^{-\hat{H}/k_B T}\hat{F}\hat{P}_\text{r}^+] \tag{5.99}$$

where we have used that $\hat{H}$ and $\hat{P}_\text{r}^+$ commute, $[\hat{H}, \hat{P}_\text{r}^+] = 0$, which follows directly from $\hat{H}|p+\rangle = E|p+\rangle$ and the definition in Eq. (5.97), and that the trace is invariant to cyclic permutations. The resulting expression for the rate constant takes a form that is well known from statistical mechanics, i.e., a Boltzmann average of an operator, i.e., $\langle\hat{F}\hat{P}_\text{r}^+\rangle$.

Stationary scattering states were used in the derivation of Eq. (5.98). Quantum mechanical traces are, however, independent of the representation in which they are carried out, so that there is no longer any explicit reference to these states, and any other orthonormal set of functions can be used in the trace. The quantum mechanical traces can then, e.g., be evaluated in a coordinate basis.

The trace in Eq. (5.98) can be rewritten in various alternative and more convenient forms [2]. Time evolution can be introduced in the expression using that Eq. (4.132) is also valid for stationary scattering states [7]. Thus,

$$|p+\rangle = \hat{\Omega}_+|p\rangle = \lim_{t\to-\infty} e^{i\hat{H}t/\hbar}e^{-i\hat{H}^0 t/\hbar}|p\rangle \tag{5.100}$$

where $|p\rangle$ is an eigenstate of $\hat{H}^0$. Note that the Møller operator is defined as the limit $t \to -\infty$ of a product of time-evolution operators. This equation relates the scattering states to the asymptotic form. Alternatively, the derivation leading to Eq. (5.98) could have been carried out using the stationary scattering states $|p-\rangle$, given by the analogous equation

$$|p-\rangle = \hat{\Omega}_-|p\rangle = \lim_{t\to\infty} e^{i\hat{H}t/\hbar}e^{-i\hat{H}^0 t/\hbar}|p\rangle \tag{5.101}$$

The projection operator $\hat{P}_{\rm r}^+$ in Eq. (5.97), as well as the equivalent projection operator $\hat{P}_{\rm r}^-$ associated with the $|p-\rangle$ states, can be written in a more explicit form, i.e.,

$$\begin{aligned}
\hat{P}_{\rm r}^{\pm} &= \int_0^{\infty} dp|p\pm\rangle\langle p\pm| \\
&= \int_{-\infty}^{\infty} dp\,\theta(p)|p\pm\rangle\langle p\pm| \\
&= \int_{-\infty}^{\infty} dp\,\theta(p)\hat{\Omega}_{\pm}|p\rangle\langle p|\hat{\Omega}_{\pm}^{\dagger} \\
&= \lim_{t\to\mp\infty} e^{i\hat{H}t/\hbar}\hat{P}_{\rm r}^0 e^{-i\hat{H}t/\hbar}
\end{aligned} \tag{5.102}$$

where $\theta(p)$ is Heaviside's unit step function, defined in Eq. (5.50). Equations (5.100) and (5.101) were used in the third line, and

$$\begin{aligned}
\hat{P}_{\rm r}^0 &= \int_{-\infty}^{\infty} dp\,\theta(p)|p\rangle\langle p| \\
&= \int_{-\infty}^{\infty} dp\,\theta(\hat{P})|p\rangle\langle p| \\
&= \theta(\hat{P})\int_{-\infty}^{\infty} dp|p\rangle\langle p| \\
&= \theta(\hat{P})
\end{aligned} \tag{5.103}$$

since for any function f of an operator defined through its power series (see Eq. (4.103) for an example)

$$f(\hat{P})|p\rangle = f(p)|p\rangle \tag{5.104}$$

Thus, from Eqs (5.102) and (5.103),

$$\hat{P}_{\mathrm{r}}^{\pm} = \lim_{t\to\mp\infty} \hat{P}_{\mathrm{r}}^{0}(t) = \lim_{t\to\mp\infty} e^{i\hat{H}t/\hbar}\theta(\hat{P})e^{-i\hat{H}t/\hbar} = \lim_{t\to\mp\infty} \theta[\hat{P}(t)] \tag{5.105}$$

using the notation

$$\hat{A}(t) = e^{i\hat{H}t/\hbar}\hat{A}e^{-i\hat{H}t/\hbar} \tag{5.106}$$

which is recognized as the time dependence of operators in the *Heisenberg picture* of quantum dynamics [8]. In the Heisenberg picture all the time dependence is carried by the operators, whereas in the Schrödinger picture that we have used so far the operators are fixed in time and all the time dependence is carried by the states. Note that in the Heisenberg picture the observables carry a time dependence exactly as in classical mechanics and if $[\hat{H}, \hat{A}] = 0$, i.e., $\hat{A}$ is a constant of motion, then $\hat{A}(t) = \hat{A}$ is independent of time. In the last step in Eq. (5.105), we used

$$f[\hat{A}(t)] = e^{i\hat{H}t/\hbar}f(\hat{A})e^{-i\hat{H}t/\hbar} \tag{5.107}$$

where $\hat{A}(t)$ is given by Eq. (5.106) and f is again any function of an operator defined through its (Taylor) power series. The projection operators in Eq. (5.105) project onto states that in the infinite past (future) had (have) positive translational momenta.

The trace in Eq. (5.98) can now be rewritten in various alternative forms. We note that $\mathrm{Tr}[e^{-\hat{H}/k_BT}\hat{F}\hat{P}_{\mathrm{r}}^{-}] = \mathrm{Tr}[\hat{F}\hat{P}_{\mathrm{r}}^{-}e^{-\hat{H}/k_BT}] = \mathrm{Tr}[\hat{F}e^{-\hat{H}/2k_BT}\hat{P}_{\mathrm{r}}^{-}e^{-\hat{H}/2k_BT}]$, where $\hat{P}_{\mathrm{r}}^{+}$ was replaced by $\hat{P}_{\mathrm{r}}^{-}$, and in the last step it was used that the Hamiltonian and the projection operator commute. Then, using Eq. (5.105), Eq. (5.98) takes the form

$$k(T) = \frac{1}{(Q_{\mathrm{trans}}/L)} \lim_{t\to\infty} \mathrm{Tr}[\hat{F}e^{i\hat{H}t_c^{\star}/\hbar}\theta(\hat{P})e^{-i\hat{H}t_c/\hbar}] \tag{5.108}$$

where $t_c = t - i\hbar/(2k_BT)$ is a complex time.

We consider the trace at $t = 0$, $\mathrm{Tr}[\hat{F}e^{-\hat{H}/2k_BT}\theta(\hat{P})e^{-\hat{H}/2k_BT}] \equiv \mathrm{Tr}[\hat{F}\hat{A}]$, which defines the operator $\hat{A}$. It is noted that both the flux operator $\hat{F}$ and the $\hat{A}$ operator are Hermitian operators, which implies that $\mathrm{Tr}[\hat{F}\hat{A}]^{\star} = \mathrm{Tr}[(\hat{F}\hat{A})^{\dagger}] = \mathrm{Tr}[\hat{A}^{\dagger}\hat{F}^{\dagger}] = \mathrm{Tr}[\hat{A}\hat{F}] = \mathrm{Tr}[\hat{F}\hat{A}]$, that is, the trace must always be real valued. Furthermore, if the trace is evaluated in a basis of real-valued functions, then the trace must be equal to zero, since the operators contain the momentum operator, which also contains the imaginary unit $i = \sqrt{-1}$. Using this result, Eq. (5.108) can be written in the form

$$\begin{aligned} k(T) &= \frac{1}{(Q_{\mathrm{trans}}/L)} \mathrm{Tr}[\hat{F}e^{i\hat{H}t_c^{\star}/\hbar}\theta(\hat{P})e^{-i\hat{H}t_c/\hbar}]|_{t=0}^{t\to\infty} \\ &= \frac{1}{(Q_{\mathrm{trans}}/L)} \int_0^{\infty} dt\, C_F(t) \end{aligned} \tag{5.109}$$

where

$$\begin{aligned} C_F(t) &= \frac{d}{dt}\mathrm{Tr}[\hat{F}e^{i\hat{H}t_c^\star/\hbar}\theta(\hat{P})e^{-i\hat{H}t_c/\hbar}] \\ &= \mathrm{Tr}\left[\hat{F}\frac{d}{dt}\left\{e^{i\hat{H}t_c^\star/\hbar}\theta(\hat{P})e^{-i\hat{H}t_c/\hbar}\right\}\right] \\ &= \frac{i}{\hbar}\mathrm{Tr}[\hat{F}e^{i\hat{H}t_c^\star/\hbar}[\hat{H},\theta(\hat{P})]e^{-i\hat{H}t_c/\hbar}] \end{aligned} \tag{5.110}$$

and in the last line it was used that the time derivative of an operator in the Heisenberg picture, see Eq. (5.106), is given by

$$\begin{aligned} \frac{d\hat{A}(t)}{dt} &= \frac{i}{\hbar}[\hat{H}(t),\hat{A}(t)] \\ &= \frac{i}{\hbar}\exp(i\hat{H}t/\hbar)[\hat{H},\hat{A}]\exp(-i\hat{H}t/\hbar) \end{aligned} \tag{5.111}$$

(note the similarity with Eq. (4.109) in the Schrödinger picture). Thus, we must now evaluate the commutator $[\hat{H},\theta(\hat{P})]$. It turns out to be useful to replace $\theta(\hat{P})$ by $\theta(R-R_0)$ and it can, indeed, be shown that the projection operator $\hat{P}_\mathrm{r}^-$ in Eq. (5.105) is equivalent to the projection operator $\lim_{t\to\infty}\exp(i\hat{H}t/\hbar)\theta(R-R_0)\exp(-i\hat{H}t/\hbar)$ [2] (see Appendix F); that is, the commutator can be replaced by $[\hat{H},\theta(R-R_0)]$, and

$$\begin{aligned} [\hat{H},\theta(R-R_0)] &= \left[\frac{\hat{P}^2}{2\mu},\theta(R-R_0)\right] \\ &= \frac{1}{2\mu}\left\{\hat{P}\,[\hat{P},\theta(R-R_0)]+[\hat{P},\theta(R-R_0)]\,\hat{P}\right\} \end{aligned} \tag{5.112}$$

where the commutator $[\hat{P},\theta(R-R_0)]=\hbar/i[\partial/\partial R,\theta(R-R_0)]=(\hbar/i)\delta(R-R_0)$, using $d[\theta(x-a)]/dx=\delta(x-a)$. Then,

$$[\hat{H},\theta(R-R_0)] = \frac{\hbar}{2i\mu}\left[\hat{P}\delta(R-R_0)+\delta(R-R_0)\hat{P}\right] = \frac{\hbar}{i}\hat{F} \tag{5.113}$$

where the coordinate representation of Eq. (5.91), $\langle R|\hat{F}|R\rangle$, was used in order to identify the flux operator.

Finally, combining Eqs (5.110) and (5.113), the expression for the thermal rate constant can be written in the form

$$\boxed{k(T) = \frac{1}{(Q_{\mathrm{trans}}/L)}\int_0^\infty dt\, C_F(t)} \tag{5.114}$$

where

$$C_F(t) = \mathrm{Tr}[\hat{F}e^{i\hat{H}t_c^\star/\hbar}\hat{F}e^{-i\hat{H}t_c/\hbar}] \tag{5.115}$$

$t_c = t - i\hbar/(2k_BT)$ and $C_F(t)$ is referred to as the *flux autocorrelation function* or simply the flux correlation function. The expression involves a time evolution of quantum operators over infinite time. Often only short-time evolution near the dividing surface is required in order to get a converged result; this is a key point that can make this expression convenient from a computational point of view.

The form of the expressions in Eqs (5.98) and (5.114) is closely related to the classical expressions for the rate constant given in Section 5.1. The quantum mechanical trace becomes in classical statistical mechanics an integral over phase space [9] and the Heisenberg operators become the corresponding classical (time-dependent) functions of coordinates and momenta [8]. Thus, Eq. (5.78) is the classical version of Eq. (5.114). Furthermore, note that Eq. (5.98) is related to Eq. (5.49), i.e., the relevant classical (one-way) flux through R_0, at a given time, becomes $\delta(R - R_0)(p/\mu)\theta(p/\mu)$, exactly as in Eq. (5.49).

Example 5.1: The flux correlation function of a free particle

To give an idea of the form of the flux autocorrelation function, we consider the dynamics of a free particle with a constant potential energy of E_0, $\hat{H} = \hat{P}^2/(2m) + E_0$, which to a first approximation can describe the dynamics along a relevant reaction coordinate in the barrier region of the potential surface. The flux correlation function (5.115) can, in the coordinate representation, be written in the form [2] (see Appendix F)

$$C_F(t) = \left(\frac{\hbar}{2m}\right)^2 \left(\frac{\partial^2}{\partial x \partial x'}|\langle x'|\hat{U}(t_c)|x\rangle|^2 - 4\left|\frac{\partial}{\partial x'}\langle x'|\hat{U}(t_c)|x\rangle\right|^2\right)_{x=x'=0} \tag{5.116}$$

where $\hat{U}(t_c) = \exp(-i\hat{H}t_c/\hbar)$. Then

$$\begin{aligned}
\langle x'|\hat{U}(t_c)|x\rangle &= \int dp \langle x'|\hat{U}(t_c)|p\rangle\langle p|x\rangle \\
&= \int dp\, e^{-ip^2 t_c/2m\hbar}\langle x'|p\rangle\langle p|x\rangle e^{-iE_0 t_c/\hbar} \\
&= \frac{1}{2\pi\hbar}\int dp\, e^{-ip^2 t_c/2m\hbar} e^{ip(x'-x)/\hbar} e^{-iE_0 t_c/\hbar} \\
&= \left(\frac{m}{2\pi\hbar i t_c}\right)^{1/2} e^{i(x-x')^2 m/2\hbar t_c} e^{-iE_0 t_c/\hbar}
\end{aligned} \tag{5.117}$$

where $t_c = t - i\hbar/(2k_BT)$. The relevant derivatives become

$$\begin{aligned}
\frac{\partial^2}{\partial x \partial x'}|\langle x'|\hat{U}(t_c)|x\rangle|^2_{x=x'=0} &= \frac{m^2}{2\pi\hbar k_B T}\frac{1}{(t^2 + [\hbar/(2k_BT)]^2)^{3/2}} e^{-E_0/k_BT} \\
\left|\frac{\partial}{\partial x'}\langle x'|\hat{U}(t_c)|x\rangle\right|^2_{x=x'=0} &= 0
\end{aligned} \tag{5.118}$$

Then

$$C_F(t) = \left(\frac{\hbar}{2m}\right)^2 \left(\frac{\partial^2}{\partial x \partial x'}|\langle x'|\hat{U}(t_c)|x\rangle|^2 - 4\left|\frac{\partial}{\partial x'}\langle x'|\hat{U}(t_c)|x\rangle\right|^2\right)_{x=x'=0}$$

$$= \frac{k_B T}{h} \frac{[\hbar/(2k_B T)]^2}{(t^2 + [\hbar/(2k_B T)]^2)^{3/2}} e^{-E_0/k_B T} \tag{5.119}$$

Fig. 5.2.1 shows $C_F(t)$ at $T = 300\,\text{K}$; we note that $\hbar/(k_B T) \sim 25\,\text{fs}$ and the correlation function has basically decayed to zero within this time.

The rate constant corresponding to Eq. (5.119) can be evaluated analytically. Thus,

$$\begin{aligned} k(T) &= \frac{1}{Q_r} \int_0^\infty dt\, C_F(t) \\ &= \frac{k_B T}{h} e^{-E_0/k_B T} \frac{1}{Q_r} \int_0^\infty dt\, \frac{[\hbar/(2k_B T)]^2}{(t^2 + [\hbar/(2k_B T)]^2)^{3/2}} \\ &= \frac{k_B T}{h} e^{-E_0/k_B T}/Q_r \end{aligned} \tag{5.120}$$

We note that in one dimension $Q_r = Q_{\text{trans}}/L = \sqrt{2\pi\mu k_B T}/h$ and $(k_B T/h)/Q_r = \sqrt{k_B T/2\pi\mu}$, i.e., $k(T)$ takes exactly the same form as in the result discussed in the model below Eq. (5.92).

5.2.2 The general case

The derivation in the previous subsection was based on the exact quantum mechanical expression for the rate constant given in Eq. (5.7), and it can be generalized to any number of degrees of freedom [1–3]. Consider the reaction $\text{A} + \text{BC}(n) \to \text{AB}(m) + \text{C}$. As in Eq. (5.94) we first evaluate the flux in the product region. We use the asymptotic form of the stationary scattering state (originating from $\text{A}+\text{BC}(n)$) in the product region where the distance between $\text{AB}(m)$ and C is large ($R \to \infty$). When, for simplicity, the reaction is constrained to collinear geometry, we have

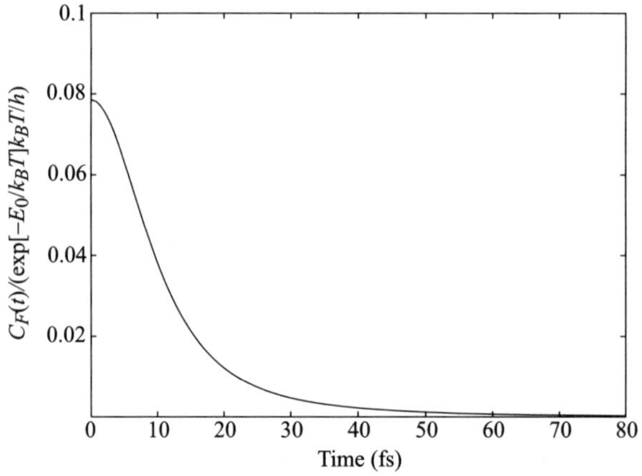

Fig. 5.2.1 The flux autocorrelation function for a free particle at $T = 300\,\text{K}$.

$$\lim_{R\to\infty}\langle R,r|E,n+\rangle = \sum_m \frac{e^{ik_mR}}{\sqrt{2\pi\hbar}}\phi^m_{\mathrm{AB}}(r)\sqrt{\frac{k_n\mu_{\mathrm{AB,C}}}{k_m\mu_{\mathrm{A,BC}}}}S_{nm}(E) \tag{5.121}$$

where $k_n = \sqrt{2\mu_{\mathrm{A,BC}}(E-E_n)}/\hbar$ and $k_m = \sqrt{2\mu_{\mathrm{AB,C}}(E-E_m)}/\hbar$. Using Eq. (5.91), the expectation value of the flux operator becomes

$$\begin{aligned}
&\langle E,n+|\hat{F}|E,n+\rangle \\
&= \frac{1}{2\mu_{\mathrm{AB,C}}}\iint dRdr\,\delta(R-R_0)\Big\{\langle E,n+|R,r\rangle\langle R,r|\hat{P}_R|E,n+\rangle + cc\Big\} \\
&= \frac{1}{2\mu_{\mathrm{AB,C}}}\iint dRdr\,\delta(R-R_0)\left\{\sum_{m'}\frac{e^{-ik_{m'}R}}{\sqrt{2\pi\hbar}}\phi^{m'}_{\mathrm{AB}}(r)\sqrt{\frac{k_n\mu_{\mathrm{AB,C}}}{k_{m'}\mu_{\mathrm{A,BC}}}}S^*_{nm'}(E)\right. \\
&\qquad\left.\times\sum_m \hbar k_m\frac{e^{ik_mR}}{\sqrt{2\pi\hbar}}\phi^m_{\mathrm{AB}}(r)\sqrt{\frac{k_n\mu_{\mathrm{AB,C}}}{k_m\mu_{\mathrm{A,BC}}}}S_{nm}(E) + cc\right\} \\
&= (\hbar k_n/\mu_{\mathrm{A,BC}})\sum_m |S_{nm}(E)|^2/(2\pi\hbar)
\end{aligned} \tag{5.122}$$

where we used the orthonormality of the vibrational wave functions, i.e., $\langle\phi^m_{\mathrm{AB}}|\phi^{m'}_{\mathrm{AB}}\rangle = \delta_{mm'}$, and cc denotes the complex conjugate term. This result generalizes Eq. (5.94), and we can continue rewriting the expressions exactly as in the previous section. In this section we simply summarize the final results.

Again, the quantum mechanical expressions can be written in a form that is analogous to the classical expressions for the rate constant given in Section 5.1, remembering that a classical phase-space integral is equivalent to a quantum mechanical trace [9], and classical functions of coordinates and momenta are equivalent to the corresponding quantum mechanical operators.

Equation (5.98) is valid in general, and analogous to the classical expression in Eq. (5.76),

$$k(T) = \frac{1}{Q_r\,V^{1-\nu}}\,\mathrm{Tr}[e^{-\hat{H}/k_BT}\hat{F}\hat{P}_{\mathrm{r}}] \tag{5.123}$$

where Q_r is the partition function of the reactants, and $\nu = 2$ and 1 for a bimolecular and a unimolecular reaction, respectively.

The flux operator $\hat{F}$ is given by

$$\hat{F} = \frac{1}{2}\left[\delta(S[\boldsymbol{q}])\sum_i\frac{\partial S(\boldsymbol{q})}{\partial q_i}\frac{\hat{p}_i}{m_i} + \sum_i\frac{\hat{p}_i}{m_i}\frac{\partial S(\boldsymbol{q})}{\partial q_i}\delta(S[\boldsymbol{q}])\right] \tag{5.124}$$

a form which is analogous to Eq. (5.90). Again, there is a close analogy to the corresponding classical expression given in Eq. (5.73). The quantum operator corresponds to a symmetrized transcription of the classical expression. Thus, the Hermitian quantum operator corresponding to a product of two classical functions $AB = (AB+BA)/2$ is $(\hat{A}\hat{B}+\hat{B}\hat{A})/2$, where $\hat{A}$ and $\hat{B}$ are Hermitian operators. Hence the factor of 1/2. The projection operator $\hat{P}_{\mathrm{r}} = \lim_{t\to\infty}\exp(i\hat{H}t/\hbar)\theta(S[\boldsymbol{q}])\exp(-i\hat{H}t/\hbar)$ is analogous to the expression discussed in the previous section and, using Eq. (5.107), to the classical expression in Eq. (5.74).

The general result analogous to Eq. (5.114) and to the classical results in Eqs (5.78) and (5.79) is

$$\boxed{k(T) = \frac{1}{Q_r V^{1-\nu}} \int_0^{\infty} dt\, C_F(t)} \tag{5.125}$$

where the time-correlation function for the flux operator is written in the form

$$\begin{aligned} C_F(t) &= \mathrm{Tr}[e^{-\hat{H}/k_B T} \hat{F} e^{i\hat{H}t/\hbar} \hat{F} e^{-i\hat{H}t/\hbar}] \\ &= \mathrm{Tr}[e^{-\hat{H}/k_B T} \hat{F}\, \hat{F}(t)] \end{aligned} \tag{5.126}$$

Q_r is the partition function of the reactants, and $\nu = 2$ and 1 for a bimolecular and a unimolecular reaction, respectively. Note that in the flux autocorrelation function, the position of the Boltzmann operator differs from the form given in Eq. (5.115).

Full numerical evaluations of $C_F(t)$ for simple 'direct' reactions give correlation functions that decay to zero within ~ 30 fs at $T = 300$ K [3]. For the reaction $Cl + H_2 \rightarrow HCl + H$, the correlation function is even quantitatively quite similar to the free-particle result in Eq. (5.119) and Fig. 5.2.1, whereas in the reaction $O + HCl \rightarrow OH + Cl$ the function $C_F(t)$ takes negative values in the time interval $t \sim$ 10–30 fs, which is a signature of *recrossings* of the dividing surface [3].

Further reading/references

[1] W.H. Miller, *J. Chem. Phys.* **61**, 1823 (1974)

[2] W.H. Miller, S.D. Schwartz, and J.W. Tromp, *J. Chem. Phys.* **79**, 4889 (1983)

[3] W.H. Miller, *J. Phys. Chem.* **102**, 793 (1998)

[4] J.C. Keck, *Adv. Chem. Phys.* **13**, 85 (1967)

[5] R.L. Jaffe, J.M. Henry, and J.B. Anderson, *J. Chem. Phys.* **59**, 1128 (1973)

[6] J.B. Anderson, *Adv. Chem. Phys.* **91**, 381 (1995)

[7] J.R. Taylor, *Scattering theory* (Wiley, 1972)

[8] E. Merzbacher, *Quantum mechanics*, second edition (Addison, Wesley, 1980)

[9] D.A. McQuarrie, *Statistical mechanics* (University Science Books, 2000)

6

Bimolecular reactions, transition-state theory

Key ideas and results

In this chapter we consider an approximate approach to the calculation of rate constants for bimolecular reactions. It is assumed that the progress of the reaction corresponds to a direct reaction; the ideal case is reactions with a single saddle point and no wells along the reaction path. In the so-called transition-state theory, one only considers the saddle-point region of the potential energy surface and defines a reaction coordinate that describes the progress of the reaction. The assumptions are now that motion along this coordinate can be treated by classical mechanics, and that this motion always leads to products without 'recrossings' of the saddle point from the product side to the reactant side. The results are as follows.

- The expression for the thermal rate constant $k(T)$ is given as a product of two functions: an exponential function and a prefactor. The prefactor contains the partition function for the reaction complex, the 'supermolecule', at the saddle point (with the reaction coordinate omitted) and partition functions for the reactants. The second factor is an exponential with an argument that contains the energy difference between the zero-point energy level of the supermolecule at the saddle point and of the reactants.
- Corrections to transition-state theory due to quantum tunneling along the reaction coordinate give a thermal rate constant that is larger than the prediction obtained from classical transition-state theory.

In Chapter 5, the direct evaluation of $k(T)$ via the reactive flux through a dividing surface on the potential energy surface was described. As a continuation of that approach, we consider in this chapter an—approximate—approach, the so-called *transition-state theory* (TST).[1] We have already briefly touched upon this approximation, based on an evaluation of a stationary one-way flux, which implies that the rate constant can be obtained without any explicit consideration of the reaction dynamics. In this chapter, we elaborate on this important approach, in a form that takes some quantum effects into account.

[1]Also referred to as 'activated-complex theory'.

As a first approximation, we can assume that a chemical reaction proceeds along the minimum-energy path, i.e., along configurations where the potential energy locally is at a minimum. For potential energy surfaces with a saddle point, the minimum-energy path goes through this point, and the saddle point, separating reactants from products, is the energy 'bottleneck' for the reaction. In transition-state theory, the computation of the thermal rate constant is reduced to the computation of partition functions, and it requires only a knowledge of the potential energy surface in the saddle-point region; the saddle-point energy E_0 and the energies of the internal states at the transition-state configuration are needed.

The intermediate nuclear configurations between reactants and products are all referred to as *transition states* for the reaction. The collection of atoms at the saddle point form a 'supermolecule', referred to as an *activated complex*, and their state is equivalent to *one* particular transition state for the reaction. This transition state obviously has a special status among all the transition states, and when one just refers to the transition state of a chemical reaction, it is tacitly assumed that one refers to the activated complex. The symbol ‡ is used to represent activated complexes.[2]

The *saddle point* is a stationary point on a multidimensional potential energy surface. It is a stable point in all dimensions except one, where the second-order derivative of the potential is negative; see Fig. 6.0.1.

This degree of freedom is the *reaction coordinate* (note that this definition coincides with the definition in Chapter 3). In Appendix E, we show that a multidimensional system close to a stationary point can be described as a set of *uncoupled* harmonic oscillators, expressed in terms of so-called normal-mode coordinates. The 'oscillator' associated with the reaction coordinate has an imaginary frequency, which implies that the motion in the reaction coordinate is unbound.

Example 6.1: Normal-mode frequencies at a saddle point

We consider a linear triatomic molecule XYZ. As shown in Appendix E, close to a stationary point, we can write the potential energy surface in the form

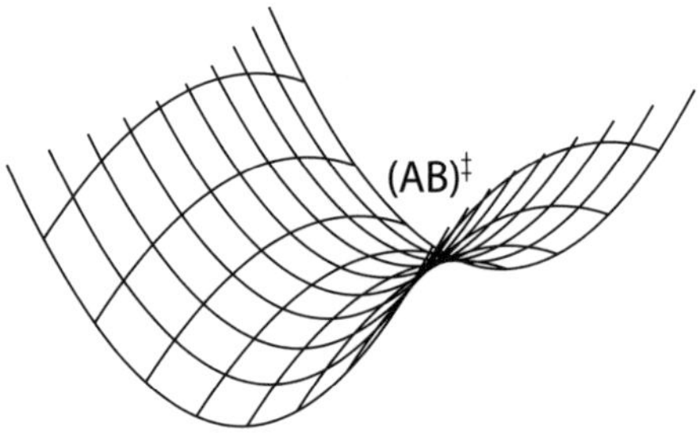

Fig. 6.0.1 The saddle-point region of the potential energy surface.

[2]Note that in this context the word 'complex' does *not* imply an entity that has a chemically significant lifetime. The basic approximations in the theory are only valid for direct reactions.

$$E(\Delta R_{XY}, \Delta R_{YZ}) = (1/2)f_1(\Delta R_{XY})^2 + (1/2)f_2(\Delta R_{YZ})^2 + f_{12}\Delta R_{XY}\Delta R_{YZ}$$

where $\Delta R_{XY} = q_Y - q_X - b_1$, $\Delta R_{YZ} = q_Z - q_Y - b_2$, q_X, q_Y, q_Z are the positions of the three atoms, and b_1 and b_2 are the (stable or unstable) equilibrium distances. The mass-weighted force constant matrix, Eq. (E.4), takes the form

$$\boldsymbol{F} = \begin{bmatrix} f_1/m_X & (f_{12}-f_1)/\sqrt{m_X m_Y} & -f_{12}/\sqrt{m_X m_Z} \\ (f_{12}-f_1)/\sqrt{m_X m_Y} & (f_1+f_2-2f_{12})/m_Y & (f_{12}-f_2)/\sqrt{m_Y m_Z} \\ -f_{12}/\sqrt{m_X m_Z} & (f_{12}-f_2)/\sqrt{m_Y m_Z} & f_2/m_Z \end{bmatrix}$$

The eigenvalues of this matrix are the squared normal-mode frequencies, and they are: $\omega^2 = 0$ (corresponding to free translational motion), and

$$\omega^2 = (B \pm \sqrt{B^2 + 4C})/2$$

where

$$B = f_1/m_X + f_2/m_Z + (f_1 + f_2 - 2f_{12})/m_Y$$
$$C = (f_{12}^2 - f_1 f_2)\frac{M}{m_X m_Y m_Z}$$

and $M = m_X + m_Y + m_Z$ is the total mass. From this expression for the frequencies we observe the following.

(i) For $B > 0$ and $C \leqslant 0$ (and $B^2 \geqslant -4C$), that is, $f_{12}^2 \leqslant f_1 f_2$, then $\sqrt{B^2+4C} \leqslant B$. In this case we obtain two real frequencies. This corresponds to a stationary point associated with a minimum corresponding to a stable molecule.

(ii) For $B > 0$ and $C > 0$, that is, $f_{12}^2 > f_1 f_2$, then $\sqrt{B^2+4C} > B$. In this case we obtain one real and one imaginary frequency, since $\omega^2 = (B - \sqrt{B^2+4C})/2 < 0$ implies that ω is *imaginary*, that is, $\omega = 2\pi i\nu^*$. This corresponds to a stationary point associated with a saddle point, as illustrated in Fig. 6.0.1.

The equation for the frequencies takes a particular simple form when $m_X = m_Y = m_Z \equiv m$ and $f_1 = f_2 \equiv f$, say corresponding to the activated complex $(HHH)^\ddagger$. Then the two frequencies become $\omega^2 = (f + f_{12})/m$, and $\omega^2 = 3(f - f_{12})/m < 0$ for $f_{12}^2 > f^2$.

In conventional transition-state theory, we place the *dividing surface* between reactants and products at the saddle point, perpendicular to the minimum-energy path, and focus our attention on the activated complex. That is, we write the reaction scheme in the form

$$A + B \longrightarrow (AB)^\ddagger \longrightarrow \text{products} \tag{6.1}$$

The basic approximations in the theory are as follows.

- In the activated complex, motion along the reaction coordinate can be separated from the other degrees of freedom (exact within the framework of a normal-mode description) and treated classically as a free translation.

- The activated complex is a configuration of *no return*: when the system has reached this configuration, it will necessarily proceed to the product side, that is, recrossings are neglected.

The reactants *as well as* the activated complex are assumed to be distributed among their states in accordance with the Boltzmann distribution. This also implies (see Appendix A.1.2) that the concentration of the activated complex molecules is related to concentrations of the reactants by a thermal equilibrium constant.

6.1 Standard derivation

Transition-state theory was developed in the 1930s. The derivation presented in this section closely follows the original derivation given by H. Eyring [1]. From Eq. (6.1), the reaction rate may be given by the rate of disappearance of A or, equivalently, by the rate at which activated complexes $(AB)^{\ddagger}$ pass over the barrier, i.e., the flow through the saddle-point region in the direction of the product side.

We use the assumption of equilibrium between the reactants and activated complex in Eq. (6.1) to relate the concentrations according to

$$\begin{aligned} K_c(T) &= \frac{[(\mathrm{AB})^{\ddagger}]}{[\mathrm{A}][\mathrm{B}]} \\ &= \frac{(Q_{(\mathrm{AB})^{\ddagger}}/V)}{(Q_{\mathrm{A}}/V)(Q_{\mathrm{B}}/V)} e^{-E_0/k_B T} \end{aligned} \tag{6.2}$$

where we have used the statistical mechanical expression for the equilibrium constant in the gas phase (Eq. (A.31) of Appendix A.1.2), E_0 is the difference in zero-point energies between the activated complex $(AB)^{\ddagger}$ and the reactants A + B, and the associated partition functions are evaluated with respect to these zero-point energies. Note that this expression for the equilibrium constant is independent of the volume V, as it should be, since the molecular partition functions Q are of the form $f(T)V$.

The reaction coordinate is found by a normal-mode analysis at the saddle point and is therefore separable from the other degrees of freedom in the activated complex; the motion in this coordinate is treated as that of a free particle. Then, according to Eq. (A.13), at sufficiently high temperatures, we have

$$Q_{(\mathrm{AB})^{\ddagger}} = \frac{(2\pi m k_B T)^{1/2} l}{h} Q^{\ddagger}_{(\mathrm{AB})^{\ddagger}} \tag{6.3}$$

where m is the mass associated with the reaction coordinate, l is the length of a one-dimensional box that comprises the activated complex (the actual value of l will later be seen to be immaterial), and $Q^{\ddagger}_{(\mathrm{AB})^{\ddagger}}$ is the partition function for the activated complex with all degrees of freedom except for the reaction coordinate.[3] This is used in Eq. (6.2), and we find

[3]Note that the partition function associated with the reaction coordinate is identical to the expression obtained from classical statistical mechanics, $Q_{\mathrm{trans}} = (1/h)\int_{-l/2}^{l/2}\int_{-\infty}^{\infty} \exp(-H/k_B T) dp_x dx$, where $H = p_x^2/(2m)$. Furthermore, with an atomic-scale value of l, Eq. (6.3) might not give an accurate value for the partition function at relevant temperatures.

$$[(AB)^{\ddagger}] = \frac{(2\pi m k_B T)^{1/2} l}{h} \frac{(Q^{\ddagger}_{(AB)^{\ddagger}}/V)}{(Q_A/V)(Q_B/V)} e^{-E_0/k_BT}[A][B] \tag{6.4}$$

Once an activated complex has been formed, it is assumed that the dynamics in the reaction coordinate is always in the direction of the product side (say to the 'right'), since recrossing, that is, motion toward the reactant side, is neglected in the theory.

The average velocity for the motion from the left to the right over the barrier is then evaluated. From the one-dimensional Maxwell–Boltzmann distribution of velocities, Eq. (2.26),

$$f_1(v_x)dv_x = \left(\frac{m}{2\pi k_B T}\right)^{1/2} \exp\left\{-\frac{mv_x^2}{2k_BT}\right\} dv_x \tag{6.5}$$

we get[4]

$$\begin{aligned} \langle v_x \rangle_+ &= \int_0^{\infty} v_x f_1(v_x) dv_x \Big/ \int_{-\infty}^{\infty} f_1(v_x) dv_x \\ &= \left(\frac{k_BT}{2\pi m}\right)^{1/2} \end{aligned} \tag{6.6}$$

Note that the average velocity for motion in both directions is zero, since the probability distribution is symmetrical with respect to the direction of the motion.

The transition region around the saddle point has the width l and the time for the passage of one complex will then be $l/\langle v_x \rangle_+$ (i.e., the 'lifetime' of the complex, the time it takes to change from reactants to products). So the number of passages per second will therefore be the reciprocal of that time, that is, the frequency with which the complexes pass over the barrier is $\langle v_x \rangle_+/l$. In order to get the rate of the reaction which is defined as a change in concentration per unit time, $[(AB)^{\ddagger}]$ is multiplied by this frequency or, equivalently, divided by the lifetime. Thus,

$$\text{rate} = [(AB)^{\ddagger}]\langle v_x \rangle_+/l \tag{6.7}$$

and the rate constant is identified, after substitution of Eq. (6.4), to be

$$\boxed{\begin{aligned} k_{TST}(T) &= \frac{\langle v_x \rangle_+}{l} \frac{(2\pi m k_B T)^{1/2} l}{h} \frac{(Q^{\ddagger}_{(AB)^{\ddagger}}/V)}{(Q_A/V)(Q_B/V)} e^{-E_0/k_BT} \\ &= \frac{k_BT}{h} \frac{(Q^{\ddagger}_{(AB)^{\ddagger}}/V)}{(Q_A/V)(Q_B/V)} e^{-E_0/k_BT} \end{aligned}} \tag{6.8}$$

Notice again that $Q^{\ddagger}_{(AB)^{\ddagger}}$ is the partition function for the activated complex *except for the one degree of freedom* corresponding to the reaction coordinate. That is, for a non-linear activated complex consisting of N atoms, the partition function incorporates

[4]We could have restricted our discussion to positive velocities. The average velocity would then be twice as big; however, the translational partition function would then take only half of the value given in Eq. (6.3). That is, the final result is unchanged.

$3N - 7$ vibrational degrees of freedom. The expression for the rate constant[5] has the same form as the Arrhenius equation; note, however, that in Eq. (6.8) the pre-exponential factor is temperature dependent. The first factor $(k_B T/h)$ has the units of frequency, and since the partition functions are dimensionless the rate constant has the proper units for a bimolecular rate constant, that is, volume/time.

The partition functions can be factorized into contributions corresponding to the various forms of motion when they are uncoupled (see Appendix A.1), and it is advantageous to rewrite the expression for the rate constant in terms of partition function ratios for the translational, rotational, vibrational, and electronic motion:

$$\begin{aligned} k_{\mathrm{TST}}(T) = \frac{k_B T}{h} &\left(\frac{(Q^{\ddagger}/V)}{(Q_{\mathrm{A}}/V)(Q_{\mathrm{B}}/V)}\right)_{\mathrm{trans}} \\ &\times \left(\frac{Q^{\ddagger}}{Q_{\mathrm{A}}Q_{\mathrm{B}}}\right)_{\mathrm{rot}} \left(\frac{Q^{\ddagger}}{Q_{\mathrm{A}}Q_{\mathrm{B}}}\right)_{\mathrm{vib}} \left(\frac{Q^{\ddagger}}{Q_{\mathrm{A}}Q_{\mathrm{B}}}\right)_{\mathrm{elec}} e^{-E_0/k_B T} \end{aligned} \tag{6.9}$$

The ratio involving the translational partition functions is easily evaluated:

$$\left(\frac{(Q^{\ddagger}/V)}{(Q_{\mathrm{A}}/V)(Q_{\mathrm{B}}/V)}\right)_{\mathrm{trans}} = \frac{h^3}{(2\pi\mu k_B T)^{3/2}} \tag{6.10}$$

where $\mu = m_{\mathrm{A}} m_{\mathrm{B}}/(m_{\mathrm{A}} + m_{\mathrm{B}})$ is the reduced mass of the reactants. This result is identical to the translation partition function $[(Q_{\mathrm{trans}}/V)^{-1}]$ of the relative motion of the reactants. The rotational partition functions are easily determined once the geometry, that is, the internuclear distances, of the activated complex is known. In practice, the vibrational part of the partition function for the activated complex is evaluated following a normal-mode analysis of the activated complex. The vibrational part of the partition functions for polyatomic reactants is, likewise, obtained by a normal-mode description of the reactants. Some applications of transition-state theory are presented in Section 6.5.

When the temperature dependence of the pre-exponential factor in Eq. (6.9) is analyzed (see Problem 6.6), one finds

$$\begin{aligned} k_{\mathrm{TST}}(T) &= Z(T) e^{-E_0/k_B T} \\ &\propto T^{\beta} e^{-E_0/k_B T}, \quad \beta \in \{\text{rational numbers}\} \end{aligned} \tag{6.11}$$

in the limit where the vibrational energy spacings of the reactants and the activated complex are either $\gg k_B T$ or $\ll k_B T$. Equation (6.11) gives a representation of the temperature dependence that is superior to the Arrhenius equation where the temperature dependence of the pre-exponential factor is neglected.

It is important to notice that the energy threshold that enters the expression in Eq. (6.8) is E_0 and *not* the classical threshold E_{cl}; see Fig. 6.1.1. Within the normal-mode description, that is, the local harmonic approximations to the potential energy

[5] We will encounter alternative derivations of this expression in the following sections.

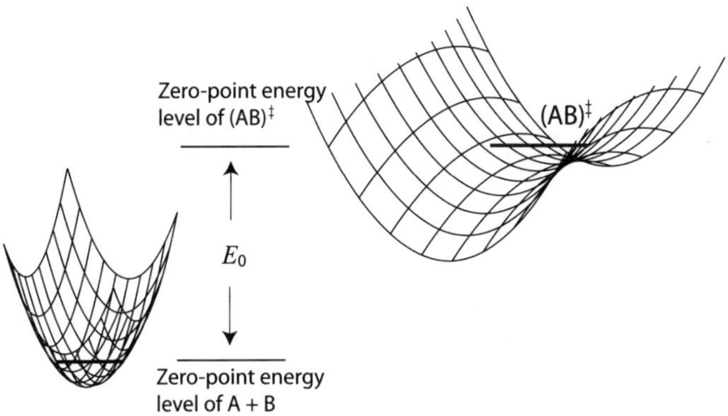

Fig. 6.1.1 An illustration of the barrier height E_0. The zero-point energy levels of the activated complex and the reactants are indicated by solid lines. Note that the zero-point energy in the activated complex comes from vibrational degrees of freedom orthogonal to the reaction coordinate. In the classical barrier height E_{cl}, vibrational zero-point energies are not included.

surface around the saddle point and around the potential well(s) of the reactants, respectively, we have

$$E_0 = E_{\text{cl}} + \sum_{(\text{AB})^\ddagger} \hbar\omega^\ddagger/2 - \sum_{\text{A,B}} \hbar\omega/2 \tag{6.12}$$

where $\omega^\ddagger$ and ω are vibrational frequencies associated with the activated complex and the reactants, respectively, and the summations run over all vibrational degrees of freedom.

The classical threshold energy E_{cl}, which we also refer to as the classical barrier height, is the energy that can be inferred directly from the potential energy surface.

Let us end this section by a *very* important observation about the expression for the rate constant in Eq. (6.8): the exponential dependence on E_0 in Eq. (6.8) implies that the rate constant is very sensitive to small changes in E_0. For example, assume that an energy δE is added to E_0 (say, corresponding to a numerical error in the determination of the electronic energy). The rate constant will then be multiplied by the factor $\exp(-\delta E/(k_B T))$. If $|\delta E| = k_B T$, an error corresponding to the factor 2.7 will be introduced. At $T = 298$ K, $k_B T = 4.1 \times 10^{-21}$ J $= 0.026$ eV ~ 2.5 kJ/mol. Thus, the potential energy surface must be highly accurate in order to provide the basis for an accurate determination of a macroscopic rate constant $k(T)$.

6.2 A dynamical correction factor

Various alternative formulations of transition-state theory have been presented [2–4]. The treatment given below [5] is reminiscent of previous derivations (see especially [3,6]) but differs in some details from those derivations.

In this section, we present a derivation of the conventional transition-state theory expression for the rate constant, Eq. (6.8), that avoids the artificial constructs of the standard derivation, in particular the one-dimensional box of length l. Furthermore, this derivation will also show how we can modify the basic assumption concerning the classical motion in the reaction coordinate.

The rate constant predicted by conventional transition-state theory can turn out to be too small, compared to experimental data, when quantum tunneling plays a role. We would like to correct for this deviation, in a simple fashion. That is, to keep the basic theoretical framework of conventional transition-state theory, and only modify the assumption concerning the motion in the reaction coordinate. A key assumption in conventional transition-state theory is that motion in the reaction coordinate can be described by classical mechanics, and that a point of no return exists along the reaction path.

First, as before, the potential is expanded to second order in the atomic displacements around the saddle point. From a normal-mode analysis, it follows that, in the vicinity of the saddle point, motion in the reaction coordinate is decoupled from the other degrees of freedom of the activated complex. Furthermore, it is assumed that the motion in the reaction coordinate (r.c.), in this region of the potential energy surface, can be described as classical free (translational) motion. Thus, the Hamiltonian takes the form

$$\begin{aligned} H &= H^{\ddagger}_{\text{r.c.}} + H^{\ddagger} \\ &= \frac{(p^{\ddagger}_1)^2}{2m} + E_{\text{cl}} + H^{\ddagger} \end{aligned} \tag{6.13}$$

where m is the effective mass associated with the reaction coordinate, E_{cl} is the saddle-point energy, and $H^{\ddagger}$ is the Hamiltonian of the activated complex, with the reaction coordinate omitted. The motion in the reaction coordinate, in the saddle-point region, is then given by Hamilton's equation of motion:

$$\begin{aligned} \frac{dq^{\ddagger}_1}{dt} &= \frac{\partial H^{\ddagger}_{\text{r.c.}}}{\partial p^{\ddagger}_1} = \frac{p^{\ddagger}_1}{m} \\ \frac{dp^{\ddagger}_1}{dt} &= -\frac{\partial H^{\ddagger}_{\text{r.c.}}}{\partial q^{\ddagger}_1} = 0 \end{aligned} \tag{6.14}$$

The second major assumption related to the dynamics is that recrossings of the saddle point are absent, i.e., the classical transmission probability is given by

$$P_{\text{cl}}(E) = \begin{cases} 0 & \text{for} \quad E < E_{\text{cl}} \\ 1 & \text{for} \quad E \geqslant E_{\text{cl}} \end{cases} \tag{6.15}$$

where E is the total kinetic and potential energy in the reaction coordinate measured from the energy minimum of the reactants.

The third assumption is that the energy states of the reactants as well as the (short-lived) activated complex are populated according to the Boltzmann distribution. Now, we focus on activated complexes where the reaction coordinate $q^{\ddagger}_1$ is fixed at the saddle-point value and where the associated momentum is $p^{\ddagger}_1$, i.e., with a position and momentum in the range $(q^{\ddagger}_1, q^{\ddagger}_1 + dq^{\ddagger}_1)$ and $(p^{\ddagger}_1, p^{\ddagger}_1 + dp^{\ddagger}_1)$. The probability of finding such a complex is denoted by $p^*(q_1^\ddagger, p_1^\ddagger)$, and the probabilities of finding the reactants are p_A and p_B, respectively. Using the Boltzmann distribution, we have (Appendix A.1.2; see also [6]) $p^*(q_1^\ddagger, p_1^\ddagger)/(p_A p_B) = Q^*(q_1^\ddagger, p_1^\ddagger)/(Q_A Q_B)$, where Q_A and Q_B are the molecular partition functions of the reactants and $Q^*(q_1^\ddagger, p_1^\ddagger)$ is the 'partial' molecular partition function of the activated complex with the reaction coordinate $q_1^\ddagger$ and the conjugate momentum $p_1^\ddagger$ fixed. Since the motion in the reaction coordinate is decoupled from other degrees of freedom of the activated complex, $Q^*(q_1^\ddagger, p_1^\ddagger)$ can be written in the form

$$Q^*(q_1^\ddagger, p_1^\ddagger) = \frac{1}{h} dq_1^\ddagger dp_1^\ddagger e^{-E/k_BT} \times Q^\ddagger \tag{6.16}$$

where $E = (p_1^\ddagger)^2/(2m) + E_{cl}$ is the energy in the reaction coordinate, which is described within classical statistical mechanics (Appendix A.2), and $Q^\ddagger$ is the partition function for the activated complex with the reaction coordinate omitted. Note that, at this point of the derivation, all energies are measured from the energy minimum of the reactants.

The probabilities are proportional to the number of activated complexes $dN^*(q_1^\ddagger, p_1^\ddagger)$ and the number of reactants N_A and N_B, respectively. Thus, we obtain the standard result

$$\frac{dN^*(q_1^\ddagger, p_1^\ddagger)/V}{(N_A/V)(N_B/V)} = \frac{Q^*(q_1^\ddagger, p_1^\ddagger)/V}{(Q_A/V)(Q_B/V)} e^{-E_0/k_BT} \tag{6.17}$$

where the zero of energy for the partition functions is now chosen as the zero-point energy levels of $(AB)^\ddagger$, A, and B, respectively, and E_0 is the difference between the zero-point energy levels of the activated complex and the reactants, that is, E_{cl} plus the zero-point energy of $(AB)^\ddagger$ minus the zero-point energy of the reactants A + B.

Using Eqs (6.16) and (6.17), we get

$$dc^*(q_1^\ddagger, p_1^\ddagger) = \frac{1}{h} dq_1^\ddagger dp_1^\ddagger e^{-E_1/k_BT} \frac{Q^\ddagger/V}{(Q_A/V)(Q_B/V)} e^{-E_0/k_BT} c_A c_B \tag{6.18}$$

where $E_1 = (p_1^\ddagger)^2/(2m)$ is the kinetic energy in the reaction coordinate and c_A, c_B, and dc^* are number densities of the reactants and the activated complex. From Hamilton's equation of motion, Eq. (6.14), we have $dt = dq_1^\ddagger/(p_1^\ddagger/m)$ is the time it takes to cross the transition-state region defined by the length $dq_1^\ddagger$. The number of passages per second will therefore be the reciprocal of this 'lifetime'. The rate is then equal to $dc^*(q_1^\ddagger, p_1^\ddagger)$ divided by dt:

$$\frac{dc^*(E_1)}{dt} = \frac{1}{h} dE_1 e^{-E_1/k_BT} \frac{Q^\ddagger/V}{(Q_A/V)(Q_B/V)} e^{-E_0/k_BT} c_A c_B \tag{6.19}$$

since $dE_1 = (p_1^\ddagger/m)dp_1^\ddagger$.

We now have an expression for the reaction rate at a fixed kinetic energy in the reaction coordinate. The total reaction rate is obtained after multiplication of Eq. (6.19) with $P_{cl}(E_1)$, followed by integration over E_1:

$$\frac{dc^\ddagger}{dt} = R_{cl}(T) \frac{Q^\ddagger/V}{(Q_A/V)(Q_B/V)} e^{-E_0/k_BT} c_A c_B \tag{6.20}$$

where the classical thermally-averaged transmission rate (frequency) is

$$\begin{aligned} R_{\rm cl}(T) &= \frac{1}{h}\int_0^\infty P_{\rm cl}(E_1)e^{-E_1/k_BT}dE_1 \\ &= \frac{1}{h}\int_0^\infty e^{-E_1/k_BT}dE_1 \\ &= \frac{k_BT}{h} \end{aligned} \tag{6.21}$$

and $E_1 = E - E_{\rm cl} \geqslant 0$, that is, the available translational energy at the saddle point.

The well-known expression for the rate constant of a bimolecular gas-phase reaction then takes the form

$$\begin{aligned} k_{\rm TST}(T) &= R_{\rm cl}(T)\frac{Q^\ddagger/V}{(Q_{\rm A}/V)(Q_{\rm B}/V)}e^{-E_0/k_BT} \\ &= \frac{k_BT}{h}\frac{Q^\ddagger/V}{(Q_{\rm A}/V)(Q_{\rm B}/V)}e^{-E_0/k_BT} \end{aligned} \tag{6.22}$$

The objective is now to modify this equation such that quantum dynamical corrections to the classical transmission probability, Eq. (6.15), are introduced.

In the present context, it is relevant to consider the barrier penetration that is associated with the traditional (one-dimensional) picture of tunneling. When we consider the time-dependent description of tunneling where a (broad) wave packet hits, e.g., a rectangular barrier, one finds that the center of the wave packet moves as a classical particle. The part of the packet that penetrates the barrier and tunnels through is not slowed down, i.e., it has exactly the same position and velocity as a wave packet that did not experience a barrier (see also [7] for a general discussion of the time-dependent picture of tunneling). The classical transmission probability of Eq. (6.15) is now replaced by the quantum mechanical transmission probability $P_{\rm qm}(E)$ (see Fig. 6.4.2). Thus, as a natural extension of the conventional formulation based on classical mechanics, in the derivation above we replace $P_{\rm cl}$ by $P_{\rm qm}$. That is, we can replace Eq. (6.21) by

$$\begin{aligned} R_{\rm tunnel}(T) &= \frac{1}{h}\int_0^\infty P_{\rm qm}(E = E_1 + E_{\rm cl})e^{-E_1/k_BT}dE_1 \\ &= \frac{1}{h}e^{E_{\rm cl}/k_BT}\int_0^\infty P_{\rm qm}(E)e^{-E/k_BT}dE \end{aligned} \tag{6.23}$$

where we have introduced the total energy $E = E_1 + E_{\rm cl}$ in the reaction coordinate and, in the second integral, extended the lower limit of integration below $E_{\rm cl}$, where $P_{\rm qm}(E) > 0$. The correction factor to the conventional transition-state theory expression in Eq. (6.22) is then

$$\boxed{\begin{aligned} \kappa_{\rm tunnel}(T) &= \frac{R_{\rm tunnel}(T)}{R_{\rm cl}(T)} \\ &= (k_BT)^{-1}e^{E_{\rm cl}/k_BT}\int_0^\infty P_{\rm qm}(E)e^{-E/k_BT}dE \end{aligned}} \tag{6.24}$$

This *tunneling correction factor* appeared as a natural consequence of the fundamental formulation of transition-state theory. This clarifies the situation, since the precise definition of this factor is occasionally discussed in the literature [8]. We will return to a discussion of tunneling and the tunneling correction factor in Section 6.4.

The rate constant predicted by conventional transition-state theory is also frequently too large, e.g., when the saddle point is not a true point of no return along the path to a particular product. It is tacitly assumed that the relevant dynamics can be described within the framework of the separability of the reaction coordinate from all other degrees of freedom. A violation of this assumption might lead to recrossings, and a correction factor $\kappa < 1$. In this context κ is often called the *transmission coefficient*. Another dynamical origin of a $\kappa < 1$ is related to non-adiabatic ('surface hopping') dynamics along the reaction coordinate in the saddle-point region. In the latter case, it is again possible to make a quantitative estimate of κ [5].

6.3 Systematic derivation

In the preceding sections we derived an approximate expression for the thermal rate constant $k(T)$. These derivations were not based on the general expressions for the rate constant that were derived in the first chapters. We consider here a derivation of the TST result that is based on an exact expression for a bimolecular rate constant.

Thus, according to Eq. (5.7), for a reaction written in the form $A + BC(n) \rightarrow AB(m) + C$, we have the following expression for the rate constant:

$$k(T) = \frac{k_B T}{h} \frac{1}{(Q_{\text{trans}}/V)Q_{\text{BC}}} \int_0^\infty P_{\text{cum}}(E) e^{-E/k_B T} \, d(E/k_B T) \tag{6.25}$$

where

$$P_{\text{cum}}(E) = \sum_J (2J+1) \sum_n \sum_m |S^J_{nm}(E)|^2 \tag{6.26}$$

is the so-called *cumulative reaction probability*. As explained in connection with Eq. (5.7), the summations contain a finite number of terms because the internal energies of reactants and products must be smaller than the total energy E.

When we compare Eqs (6.8) and (6.25), we note that the TST expression for the rate constant would be obtained provided that the integral appearing in Eq. (6.25) was identical to the partition function of the activated complex. To that end, we introduce the *approximation*

$$P_{\text{cum}}(E) = 0 \quad \text{for} \quad E \leqslant E_0 \tag{6.27}$$

where E_0 is the barrier height introduced in Section 6.1. The lower limit in the integral can now be replaced by E_0, we write the total energy in the form $E = E^\ddagger + E_0$, and rewrite the integral using partial integration. We get

$$
\begin{aligned}
Q &= \int_{E_0}^{\infty} P_{\text{cum}}(E) e^{-E/k_BT}\, d(E/k_BT) \\
&= e^{-E_0/k_BT} \int_0^{\infty} P_{\text{cum}}(E^\ddagger + E_0) e^{-E^\ddagger/k_BT} d(E^\ddagger/k_BT) \\
&= -e^{-E_0/k_BT} \int_0^{\infty} P_{\text{cum}}(E^\ddagger + E_0) \left[\frac{d}{dE^\ddagger} e^{-E^\ddagger/k_BT}\right] dE^\ddagger \\
&= -e^{-E_0/k_BT} \left\{ \left[P_{\text{cum}}(E^\ddagger + E_0) e^{-E^\ddagger/k_BT}\right]_0^{\infty} - \int_0^{\infty} \frac{dP_{\text{cum}}(E^\ddagger + E_0)}{dE^\ddagger} e^{-E^\ddagger/k_BT} dE^\ddagger \right\} \\
&= e^{-E_0/k_BT} \int_0^{\infty} \frac{dP_{\text{cum}}(E^\ddagger + E_0)}{dE^\ddagger} e^{-E^\ddagger/k_BT} dE^\ddagger
\end{aligned}
\tag{6.28}
$$

where $P_{\text{cum}}(E_0) = 0$ was used in the last line.

In order to proceed, we need to know the precise form of the cumulative reaction probability, and introduce the following *approximation*:

$$
P_{\text{cum}}(E^\ddagger + E_0) = \sum_J (2J+1) \sum_i \Theta(E^\ddagger - E^\ddagger_{i,J})
\tag{6.29}
$$

where $\Theta(x)$ is the unit step function, defined by

$$
\Theta(x) = \begin{cases} 0 & x \leqslant 0 \\ 1 & x > 0 \end{cases}
\tag{6.30}
$$

and $E^\ddagger_{i,J} = E^\ddagger_i + E^\ddagger_J$ are the vibrational/rotational energy levels of the activated complex. Note that this approximation implies that the cumulative reaction probability is zero when the total energy $E^\ddagger$ (i.e., the energy measured relative to the zero-point level of the activated complex) is below the zero-point level of the activated complex, and when the energy exceeds one of the quantized energy levels the cumulative reaction probability is, for $J = 0$, increased by one.

Thus, according to this description the reaction probability increases in a stepwise manner with increasing energy, as the quantized states of the activated complex become energetically open. Since the derivative of the step function is a delta function, we get

$$
\frac{dP_{\text{cum}}(E^\ddagger + E_0)}{dE^\ddagger} = \sum_J (2J+1) \sum_i \delta(E^\ddagger - E^\ddagger_{i,J})
\tag{6.31}
$$

and

$$
\begin{aligned}
Q &= e^{-E_0/k_BT} \int_0^{\infty} \frac{dP_{\text{cum}}(E^\ddagger + E_0)}{dE^\ddagger} e^{-E^\ddagger/k_BT} dE^\ddagger \\
&= e^{-E_0/k_BT} \sum_J (2J+1) \sum_i e^{-E^\ddagger_{i,J}/k_BT} \\
&= e^{-E_0/k_BT} \sum_J (2J+1) e^{-E^\ddagger_J/k_BT} \sum_i e^{-E^\ddagger_i/k_BT} \\
&= Q^\ddagger e^{-E_0/k_BT}
\end{aligned}
\tag{6.32}
$$

where $Q^{\ddagger}$ is the partition function of the activated complex. When this approximate result is used in Eq. (6.25), the expression for the rate constant becomes identical to the TST expression in Eq. (6.8).

6.4 Quantum mechanical corrections

The derivations given in the previous sections show that the assumptions related to the crossing of the saddle point are based on classical mechanics. The approximation of classical mechanics was introduced in the standard approach of Section 6.1, and is also implicitly assumed in the approach of Section 6.3, say in Eq. (6.27).

With a dividing surface placed at a saddle point, the potential has a concave-down shape along the reaction coordinate. It is incorrect to treat the motion in the reaction coordinate as being classical, since *quantum mechanical tunneling* can occur through the potential.

To correct for tunneling, we consider in the next section the problem of crossing the barrier quantum mechanically rather than classically.

6.4.1 Tunneling through an Eckart barrier

As discussed in Section 4.2, wave functions can penetrate into potential energy barriers, corresponding to regions of configuration space that are not accessible according to classical mechanics. Thus, the wave function is non-zero beyond a classical turning point, i.e., a purely quantum mechanical effect without any counterpart in classical mechanics. If the barrier around the saddle point is not too broad, then the wave function can penetrate through the barrier corresponding to tunneling.

An analytical solution to the Schrödinger equation can be obtained for the so-called *Eckart barrier* given by [9,10],

$$\begin{aligned} V(x) &= \frac{E_{\text{cl}}}{\cosh^2\{x\pi/l\}} \\ &= \frac{E_{\text{cl}}}{\cosh^2\{x(F^*/2E_{\text{cl}})^{1/2}\}} \end{aligned} \tag{6.33}$$

where E_{cl} is the maximum of the potential at $x = 0$, i.e., the (classical) barrier height. This potential is shown in Fig. 6.4.1.

The second derivative at $x = 0$ is $F = -F^*$, where $F^* = 2E_{\text{cl}}(\pi/l)^2$ is the magnitude of the second derivative.

If we send in a particle (say, from the left) with kinetic energy E, then it can be shown that the probability of crossing the barrier (i.e., the transmission probability) is

$$\boxed{P_{\text{qm}}(\zeta) = \frac{\cosh(2\alpha\zeta^{1/2}) - 1}{\cosh(2\alpha\zeta^{1/2}) + \cosh(\sqrt{4\alpha^2 - \pi^2})}} \tag{6.34}$$

where

$$\begin{aligned} \zeta &= E/E_{\text{cl}} \\ \alpha &= 2\pi E_{\text{cl}}/h\nu^* \\ \nu^* &= \frac{1}{2\pi}\left(\frac{F^*}{m}\right)^{1/2} \end{aligned} \tag{6.35}$$

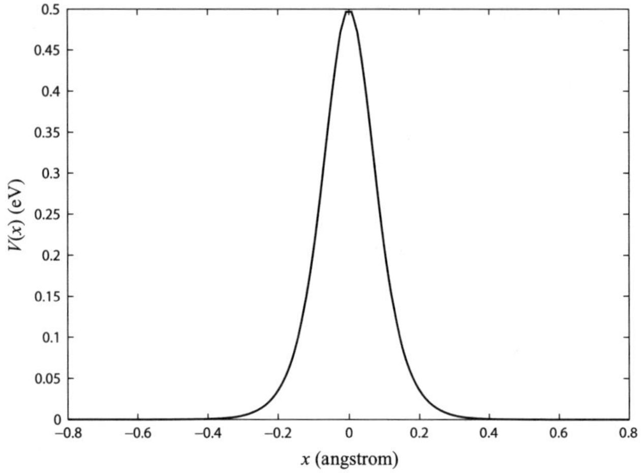

Fig. 6.4.1 An Eckart barrier. $E_{\text{cl}} = 0.5$ eV and $F^* = 110$ eV(angstrom)$^{-2}$.

Note that the frequency of the motion on top of the barrier is

$$\begin{aligned}\nu &= \frac{1}{2\pi}\left(\frac{F}{m}\right)^{1/2} \\ &= \frac{1}{2\pi}\left(\frac{-F^*}{m}\right)^{1/2} \\ &= i\nu^* \end{aligned} \tag{6.36}$$

that is, an imaginary frequency corresponding to passage over the barrier.

The probability in Eq. (6.34) is shown in Fig. 6.4.2 as a function of E/E_{cl}. It shows that there is a finite non-zero probability of 'crossing' the barrier even for kinetic energies smaller than the height of the barrier! This phenomenon is called *tunneling*. Note also that when the kinetic energy exceeds the height of the barrier the probability does not immediately take a value of one. Thus, above-barrier *reflection* takes place. Both phenomena are quantum effects.

The probability in Eq. (6.34) depends on the kinetic energy relative to the barrier height as well as the oscillation frequency associated with the inverted barrier. This frequency depends on the width of the barrier, that is, F^*, as well as the mass of the particle. Thus, tunneling and reflection are enhanced when:

- the barrier height (E_{cl}) is low;
- the width of the barrier is small (F^* large);
- the particle mass (m) is small.

Tunneling plays a role for atomic particles. If one considers heavy particles, then the results obtained from Eq. (6.34) will be indistinguishable from the step function predicted by classical mechanics,

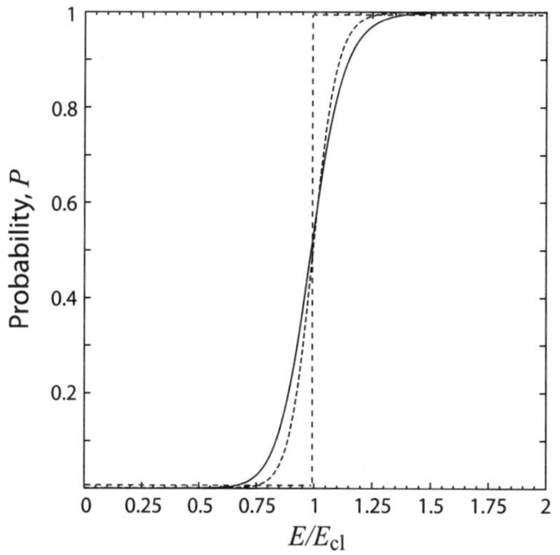

Fig. 6.4.2 Tunneling probabilities for an Eckart barrier. The barrier height is 40 kJ/mol and the magnitudes of the imaginary frequency associated with the reaction coordinate are 1511 cm^{-1} (solid line), corresponding to the reaction $H + H_2$, and $1511/\sqrt{2}$ cm^{-1} (dashed line), corresponding to the reaction $D + D_2$. The step function is the transmission 'probability' according to classical mechanics.

$$P_{\mathrm{cl}}(\zeta) = \begin{cases} 0 & \text{for} \quad E/E_{\mathrm{cl}} \leqslant 1 \\ 1 & \text{for} \quad E/E_{\mathrm{cl}} > 1 \end{cases} \tag{6.37}$$

In a classical calculation, the transmission probability will therefore be underestimated at kinetic energies smaller than the barrier energy and overestimated at kinetic energies above the barrier energy.

Finally, before closing this section, we note that only in the neighborhood of the saddle point is the motion in the reaction coordinate strictly separable from the other degrees of freedom. In the one-dimensional description of tunneling given in this section, we have however considered the full 'straightened-out' reaction path. This implies that we have assumed separability along the full reaction path, which is not strictly valid. Various more advanced treatments of tunneling have been discussed in the literature [14]. For the sake of completeness we also note that tunneling in *multidimensional* systems is more subtle. Thus, in such systems we can, e.g., have a non-zero reaction probability at total energies below the height of the lowest barrier, without having the barrier penetration that is the essence of tunneling in one-dimensional barrier crossing.

6.4.2 Tunneling correction factor

In Section 6.2, we derived a tunneling correction factor for transition-state theory, see Eq. (6.24). The derivation showed that $k_{\mathrm{TST}}(T)$ of Eq. (6.8) is replaced by $\kappa_{\mathrm{tunnel}}(T) \times k_{\mathrm{TST}}(T)$ when tunneling is included.

The correction factor due to quantum tunneling, $\kappa_{\mathrm{tunnel}}(T)$, is sometimes introduced in transition-state theory [8,11] as

$$\kappa_{\text{tunnel}}(T) = \frac{\int_0^\infty dE P_{\text{qm}}(E) e^{-E/k_B T}/(k_B T)}{\int_0^\infty dE P_{\text{cl}}(E) e^{-E/k_B T}/(k_B T)}$$
$$= \frac{\int_0^\infty dp (p/\mu) P_{\text{qm}}(p) e^{-p^2/2\mu k_B T}}{\int_0^\infty dp (p/\mu) P_{\text{cl}}(p) e^{-p^2/2\mu k_B T}} \tag{6.38}$$

where $E = p^2/(2\mu)$ and $dE = (p/\mu)dp$. At temperature T, this factor is the ratio of the number of transmitted systems in the quantum case to the number expected from classical mechanics. In the second line, we observe that the numerator and the denominator are proportional to a thermal average of a one-dimensional flux (see Eq. (5.92) and Section 5.2.1) with, respectively, quantum and classical transmission probabilities associated with the one-dimensional reaction coordinate.[6] Now

$$\int_0^\infty dE P_{\text{cl}}(E) e^{-E/k_B T}/(k_B T) = \int_{E_{\text{cl}}}^\infty dE e^{-E/k_B T}/(k_B T)$$
$$= e^{-E_{\text{cl}}/k_B T} \tag{6.39}$$

which is the fraction of systems with an energy above E_{cl}, and

$$\kappa_{\text{tunnel}}(T) = (k_B T)^{-1} e^{E_{\text{cl}}/k_B T} \int_0^\infty dE P_{\text{qm}}(E) e^{-E/k_B T} \tag{6.40}$$

Note that this result is identical to Eq. (6.24). Although the quantum mechanical transmission probability at kinetic energies above the barrier energy is less than one, that is, particles are reflected above the barrier, the transmission in the tunneling region dominates in the integral due to the Boltzmann factor $\exp(-E/k_B T)$.

An important general conclusion can be made at this point: *when the temperature T is lowered, $\kappa_{\text{tunnel}}(T)$ will increase*, since (due to the Boltzmann factor) the fraction of systems with an energy above E_{cl} will decrease (the denominator in Eq.(6.38)) and the number of transmitted systems in the tunneling region with an energy below E_{cl} will increase (the numerator in Eq.(6.38)) at low temperatures.

The correction factor due to quantum tunneling for an *Eckart barrier* cannot be evaluated analytically. An approximate expression can, however, be derived.

First, we note that Eq. (6.34) has the asymptotic form

$$P_{\text{qm}}(E) \overset{\alpha \gg 1}{\longrightarrow} \frac{1}{1 + \exp[2\alpha(1 - \sqrt{E/E_{\text{cl}}})]} \tag{6.41}$$

Since $\alpha = 2\pi E_{\text{cl}}/(h\nu^*)$, this limit corresponds to a high and/or broad barrier, i.e., the tunneling probability is small. After the introduction of the variable $x = \alpha(E - E_{\text{cl}})/E_{\text{cl}}$ in Eq. (6.40), we obtain

$$\kappa_{\text{tunnel}}(T) = \beta \int_{-\alpha}^\infty dx \frac{\exp[-\beta x]}{1 + \exp[2\alpha(1 - \sqrt{1 + x/\alpha})]} \tag{6.42}$$

[6] The correction factor is expected to be best suited as the tunneling correction to the crossing of an early barrier that occurs in the entrance channel while the reactants are approaching each other; see Fig. 3.1.4. In this case, the reaction coordinate is separable from the other degrees of freedom along the full one-dimensional reaction path associated with barrier crossing.

Table 6.1 Tunneling corrections as a function of temperature according to Eq. (6.40). κ_H is the correction factor for $H + H_2$, where the magnitude of the imaginary frequency associated with the reaction coordinate is 1511 cm^{-1}. κ_D is the correction factor for $D + D_2$, where the magnitude of the imaginary frequency is $1511/\sqrt{2}$ cm^{-1}. The tunneling probabilities are calculated for an Eckart barrier. The barrier height is $E_{cl} = 40$ kJ/mol.

Temperature (K)	κ_H	κ_D
200	864	30.5
225	169	12.3
250	54.1	6.92
300	13.5	3.55
400	3.95	1.98
500	2.38	1.55
1000	1.26	1.12

where $\beta = h\nu^*/(2\pi k_B T)$. Now $\sqrt{1+y} \sim 1 + y/2$, for y small, and for $\alpha \gg 1$ as above we have[7] $\kappa_{\text{tunnel}}(T) = \beta \int_{-\alpha}^{\infty} dx \exp[-\beta x]/(1 + \exp[-x])$. An analytical evaluation of this integral is only possible when the lower limit $-\alpha$ is replaced by $-\infty$, and the integral takes the form $\int_{-\infty}^{\infty} dx \exp[-\beta x]/(1 + \exp[-x]) = \pi/\sin(\pi\beta)$ for $\beta < 1$. The error introduced by this change of lower limit is large unless $\beta \ll 1$. Thus,

$$\kappa_{\text{tunnel}}(T) = \frac{h\nu^*/(2k_B T)}{\sin[h\nu^*/(2k_B T)]} \qquad \begin{array}{ll} \text{for} & \alpha = 2\pi E_{cl}/(h\nu^*) \gg 1 \\ \text{and} & \beta = h\nu^*/(2\pi k_B T) \ll 1 \end{array} \tag{6.43}$$

Since $\beta \ll 1$, corresponding to a high temperature, we can use $y/\sin(y) \sim 1 + y^2/6$, for $y = \pi\beta$ small, and obtain

$$\boxed{\kappa_{\text{tunnel}}(T) \sim 1 + \frac{1}{24}\left(\frac{h\nu^*}{k_B T}\right)^2} \qquad \begin{array}{ll} \text{for} & \alpha = 2\pi E_{cl}/(h\nu^*) \gg 1 \\ \text{and} & \beta = h\nu^*/(2\pi k_B T) \ll 1 \end{array} \tag{6.44}$$

This expression is called the *Wigner tunneling correction factor*, and is valid for a small degree of tunneling (high and/or broad barrier, $h\nu^* \ll E_{cl}$) and, in addition, high temperatures ($h\nu^* \ll k_B T$) are required. Note that the correction factor is larger than one.

When Eq. (6.44) is compared to the result obtained by a numerical evaluation of the integral in Eq. (6.40) (see Table 6.1), one will find good agreement at high temperatures. At low temperatures, the analytical formula can, however, deviate substantially from the exact result.

6.5 Applications of transition-state theory

It is fairly simple to apply transition-state theory once the barrier height E_0, the vibrational energy levels, and the geometry of the activated complex are known.

In special cases, say when we want to account for *isotope effects* in chemical reactions, we just need to consider the changes in the above-mentioned quantities (see, e.g., Problem 6.7). As an example, consider the two exchange reactions

[7]This expression is identical to the tunneling correction factor for a parabolic barrier.

$$\begin{aligned} XH + Y &\xrightarrow{k_H} X + HY \\ XD + Y &\xrightarrow{k_D} X + DY \end{aligned} \tag{6.45}$$

The potential energy surface is unchanged under isotope substitution, and the major contribution to the isotope effect is, typically, related to changes in vibrational zero-point energies, and hence the barrier height E_0. Thus, the zero-point energy in XH is larger than the zero-point energy in XD, and when the X–H/X–D bond breaking is directly related to motion in the reaction coordinate, this difference in zero-point energies is absent in the activated complex. The first reaction in Eq. (6.45) has accordingly the smallest barrier height E_0, and one finds that $k_H > k_D$. The two rate constants can differ by as much as a factor of 10 at $T = 300$ K (depending on the exact identity of the groups X and Y), and even more when quantum tunneling is taken into account.

Transition-state theory has been applied to numerous reactions. In order to evaluate the accuracy of the theory in direct comparison to experimental results, highly accurate data for the activated complex is required as basic input to the theory. To that end, we can, e.g., consider the simple reaction D + H_2 → HD + H; the rate constant at 1000 K (a temperature where tunneling plays a minor role) is 1.78×10^{-12} cm^3/(molecule s) (see Problem 6.8) which is in excellent agreement with the experimental value 2.13×10^{-12} cm^3/(molecule s) (with an experimental uncertainty of about 25%).

Before we consider in more detail some specific applications of transition-state theory, we consider briefly two general points. In applications of transition-state theory, it is sometimes discussed how the theory is properly applied to two special reaction types: symmetric reactions (H + H_2 → H_2 + H is the standard example) and reactions involving optically active species. Thus, for symmetric reactions it is often argued that a factor of 2 is missing in the expression for the rate constant. Transition-state theory gives the rate constant for a reaction in a given direction. Defining the rate of a symmetric reaction is a little subtle. If we define the rate from the lifetime of an H_2 against H atom exchange, we must include the rate of passage in both directions. Since the rate constants are identical in either direction, the total rate is twice the rate obtained for each direction [12]. The rate constant at 1000 K (a temperature where tunneling plays a minor role) is, according to transition-state theory, 8.44×10^{-13} cm^3/(molecule s) (see Problem 6.4). The experimentally measured rate constant is about twice as big, which again is in excellent agreement (given the experimental uncertainty and the problems of defining the rate, as described above). Another situation with some pitfalls is reactions involving optically active species. For example, if the activated complex is optically active, there are in fact two elementary reactions going on, where the two activated complexes are mirror images of each other. These two reactions have identical rate constants, and the overall rate is therefore twice the rate of each reaction.

In connection with transition-state theory, one will also occasionally meet the concept of a *statistical factor* [13]. This factor is defined as the number of different activated complexes that can be formed if all identical atoms in the reactants are labeled. The statistical factor is used instead of the *symmetry numbers* that are associated with each rotational partition function (see Appendix A.1) and, properly applied, the two approaches give the same result. The statistical factor approach has, however, some pitfalls [12], and in the following we use the standard expression for the rate constant involving symmetry numbers.

Example 6.2: The rate constant for the $F + H_2$ reaction

The reaction $F + H_2 \rightarrow HF + H$ is of special theoretical interest because it is one of the simplest examples of an exothermic chemical reaction. Furthermore, the reaction is characterized by an early and very small barrier.

The relevant parameters for the reactants and the activated complex are given in Table 6.2. The activated complex $F \cdots H \cdots H$ is bent with an angle of 119°. The resulting three principal moments of inertia of this asymmetric rotor are given in the table. Note that the interatomic distances in the activated complex are characteristic of an early barrier; $H \cdots H$ is close to its initial equilibrium bond length and $F \cdots H$ is much longer than the final equilibrium bond length (which is 0.9168 Å). The vibrational frequencies of the activated complex are obtained from a normal-mode analysis at the saddle point (Appendix E). There are two real frequencies and one imaginary frequency. The imaginary frequency at $723i$ corresponds to the reaction coordinate and passage over the barrier. The frequency at 3772 corresponds, approximately, to the H–H stretch in the complex, and the frequency at 296 is the bending frequency of the non-linear complex.

In order to evaluate the electronic partition function of $F(1s^2 2s^2 2p^5)$, we need to know that in its electronic ground state (with the term $^2P_{3/2}$ when the spin–orbit coupling is taken into account) the total angular momentum (orbital + spin) is $J = 3/2$, and the degeneracy is therefore $2J + 1 = 4$. Furthermore, for the fluorine atom it is necessary to include the doubly-degenerate first electronically excited state ($^2P_{1/2}$) because its energy relative to the ground state is just $\Delta = 0.05$ eV. For $H_2(1\sigma_g^2)$ the total electronic angular momentum is 0, and the degeneracy is therefore 1. The electronic degeneracy of the activated complex (with an electronic state denoted by $^2A'$, and in the special case of a linear configuration by $^2\Sigma$) is 2.

We will now calculate the rate constant at $T = 300$ K using Eq. (6.9) and the data in Table 6.2. The relevant ratio between the translational partition functions is

$$\left(\frac{(Q^\ddagger/V)}{(Q_F/V)(Q_{H_2}/V)}\right)_{\text{trans}} = \frac{h^3}{(2\pi\mu k_B T)^{3/2}} = 4.16 \times 10^{-31}\ \text{m}^3 \tag{6.46}$$

where μ is the reduced mass of F and H_2. The ratio between the rotational partition functions is

$$\left(\frac{Q^\ddagger}{Q_{H_2}}\right)_{\text{rot}} = \sqrt{\pi}\frac{\sigma_{H_2}}{\sigma_\ddagger}\frac{\sqrt{I_a I_b}}{I_{H_2}}\sqrt{8\pi^2 I_c k_B T/h^2} = 110.30 \tag{6.47}$$

where the two symmetry numbers are $\sigma_\ddagger = 1$ and $\sigma_{H_2} = 2$. The ratio between the vibrational partition functions is

Table 6.2 Properties of the reactants and the activated complex for the F + H_2 reaction. Data from an *ab initio* potential energy surface [*J. Chem. Phys.* **104**, 6515 (1996) and *Chem. Phys. Lett.* **286**, 35 (1998)].

Parameters	$F\cdots H\cdots H$	F	H_2
r(F–H), Å	1.546		
r(H–H), Å	0.771		0.7417
m, amu	21.014	18.9984	2.016
I, amu Å^2			0.277
I_a, amu Å^2	0.173		
I_b, amu Å^2	5.807		
I_c, amu Å^2	5.981		
$\tilde{\nu}_1$, cm^{-1}	3772		4395.2
$\tilde{\nu}_2$, cm^{-1}	296		
$\tilde{\nu}_3$, cm^{-1}	$723i$		
E_0, kJ/mol	4.56		
ω_{elec} (ground state)	2	4	1

$$\left(\frac{Q^\ddagger}{Q_{\mathrm{H}_2}}\right)_{\text{vib}} = \frac{1-\exp(-h\nu_{\mathrm{H}_2}/k_BT)}{\Pi_{i=1}^2[1-\exp(-h\nu_i^\ddagger/k_BT)]} = 1.32 \tag{6.48}$$

where it is remembered that the wavenumber $\tilde{\nu}$ and frequency ν are related by the relation $\tilde{\nu} = \nu/c$. The ratio between the electronic partition functions is

$$\left(\frac{Q^\ddagger}{Q_{\mathrm{F}}Q_{\mathrm{H}_2}}\right)_{\text{elec}} = 2/(4+2\exp[-\Delta/k_BT]) = 0.466 \tag{6.49}$$

After multiplication of all the factors, we obtain

$$k_{\mathrm{TST}}(T = 300\ \mathrm{K}) = 2.83\times10^{-11}\ \frac{\mathrm{cm}^3}{\text{molecule s}} = 1.71\times10^{10}\ \frac{\text{liter}}{\text{mol s}} \tag{6.50}$$

The experimental data can, in the temperature range $T \in [190, 373]$ K, be represented by the equation $k(T) = 1.0\times10^{-10}\exp(-(432.8\pm50.32)/T)\ \mathrm{cm}^3/(\text{molecule s})$ [*J. Chem. Phys.* **92**, 4811 (1980)], and in a larger temperature range by $k(T) = 2.7\times10^9T^{0.5}\exp(-319/T)$ liter/(mol s), for $T \in [190, 800]$ K [*J. Phys. Chem. Ref. Data* **12**, No. 3, (1983)], where the uncertainty in the experimental data is estimated as $\log_{10}[k(T)] = \pm0.15$. Thus, at $T = 300$ K, the experimental value is $k(T = 300\ \mathrm{K}) = (2.36\pm0.4)\times10^{-11}\ \mathrm{cm}^3/(\text{molecule s})$, and the transition-state theory rate constant agrees with the experimental data to within the experimental uncertainty.

As shown in the previous example, for the reaction $\mathrm{F} + \mathrm{H}_2 \to \mathrm{HF} + \mathrm{H}$, transition-state theory is sufficiently accurate to reproduce the experimental result for the thermal rate constant at $T = 300$ K.

Table 6.3 A comparison of different theoretical approaches to the evaluation of the thermal rate constant for the $F + H_2 \rightarrow HF + H$ reaction at $T = 300$ K. TST is transition-state theory (Example 6.2), QCT is the quasi-classical trajectory method [*Chem. Phys. Lett.* **254**, 341 (1996)], and QM is (exact) quantum mechanics [*J. Phys. Chem.* **102**, 341 (1998)].

Method	TST	QCT	QM
$k/10^{-11}$ cm^3/(molecule s)	2.83	2.07	2.26

Clearly, the successful reproduction of the experimental result is, in part, related to the high quality of the potential energy surface. A more direct evaluation of the accuracy of transition-state theory can be obtained via a comparison to other (more exact) theoretical approaches to the calculation of the rate constant, all using the same potential energy surface. Table 6.3 shows such a comparison. We observe that transition-state theory does overestimate the rate constant but the agreement is quite reasonable, especially when the simplicity of the calculation is taken into account.

The reaction rate depends critically on the barrier height and the shape of the potential energy surface in the vicinity of the saddle point. As an illustration of this, we consider again the $F + H_2 \rightarrow HF + H$ reaction, where a second saddle point has been identified (Section 3.1) which corresponds to a linear activated complex with a classical barrier height that is slightly higher than for the bent complex. An evaluation of the rate constant based on the linear activated complex would have resulted in a rate constant that is about ten times smaller than the experimental result.

Example 6.3: Collision between two atoms

The simplest application of transition-state theory is to the collision between two structureless molecules, i.e., two atoms A and B:

$$A + B \longrightarrow (AB)^{\ddagger} \longrightarrow \text{product} \tag{6.51}$$

The reaction coordinate is the interatomic distance, and the energy profile along the reaction coordinate is identical to a one-dimensional interatomic potential energy curve; that is, there is no saddle point along the reaction path. Furthermore, the total energy is conserved along the reaction coordinate, and any intermediate AB configuration will always have the same energy as the reactants; it will consequently dissociate after the turning point on the potential energy curve has been reached. If, however, there is a mechanism for releasing the energy, a stable molecule may be formed and the rate constant can be related to the formation of this molecule. An example is the so-called *radiative recombination*, where an electronically excited molecule may emit a photon and form a stable AB molecule in the electronic ground state. Thus, $(AB)^{\ddagger} \rightarrow AB + h\nu$; see Fig. 6.5.1.

We place the dividing surface between reactants and products at a critical separation r_c and calculate the rate constant according to Eq. (6.8) (assuming, as in Section 4.1.2, that the atoms 'react' with a probability of one when $r = r_c$). In the relevant partition function for the activated complex, there are both translational degrees of freedom for the center-of-mass motion and rotational degrees of freedom

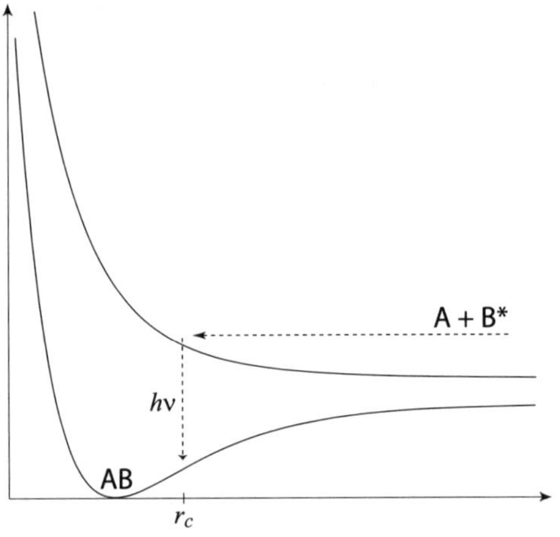

Fig. 6.5.1 Radiative recombination. B* is an electronically excited atom.

(with the electronic degrees of freedom omitted). Thus (see Appendix A.1),

$$(Q^{\ddagger}_{(\mathrm{AB})^{\ddagger}}/V) = \frac{(2\pi[m_{\mathrm{A}} + m_{\mathrm{B}}]k_B T)^{3/2}}{h^3} \frac{8\pi^2 I^{\ddagger} k_B T}{h^2} \tag{6.52}$$

with the moment of inertia for the diatomic complex, $I^{\ddagger} = \mu r_c^2$, and μ is the reduced mass. The reactants have only translational degrees of freedom. So,

$$(Q_{\mathrm{A}}/V) = (2\pi m_{\mathrm{A}} k_B T)^{3/2}/h^3 \tag{6.53}$$

with an equivalent expression for the B atom.

The rate constant, Eq. (6.8), then takes the form

$$\begin{aligned} k_{\mathrm{TST}}(T) &= \frac{k_B T}{h} \frac{(Q^{\ddagger}_{(\mathrm{AB})^{\ddagger}}/V)}{(Q_{\mathrm{A}}/V)(Q_{\mathrm{B}}/V)} e^{-E_0/k_B T} \\ &= \frac{k_B T}{h} \frac{h^3}{(2\pi\mu k_B T)^{3/2}} \frac{8\pi^2 \mu r_c^2 k_B T}{h^2} e^{-E_0/k_B T} \\ &= \pi r_c^2 \left(\frac{8 k_B T}{\pi\mu}\right)^{1/2} e^{-E_0/k_B T} \end{aligned} \tag{6.54}$$

We observe that the result is the same as in Eq. (4.23).

The expression for the rate constant in Example 6.3 can also be considered as an application of transition-state theory to a bimolecular reaction where the molecules are considered as structureless 'atoms' with the relevant masses. Thus, when we compare this with the proper expression for a bimolecular reaction, we can gain insight into how the internal molecular degrees of freedom affect the rate constant. In Section 4.1.2, we noticed that the pre-exponential factor predicted by Eq. (6.54), when compared to experimental data for a number of bimolecular reactions, typically, is much too large. Now when we compare Eq. (6.54) to the proper TST expression for a bimolecular reaction, Eq. (6.9), we observe that the pre-exponential factor contains vibrational and electronic partition functions, as well as rotational partition functions associated with the reactants. The vibrational and electronic partition functions are typically of the order of one, whereas the value of the rotational partition functions is well above one (remember that the partition function gives an indication of the average number of states that are populated at a given temperature). Since the rotational partition functions of the reactants are in the denominator of the expression, transition-state theory will typically, in agreement with experimental observations, predict smaller pre-exponential factors than the simple collision model where molecules are treated as structureless spheres.

Finally, we must stress that transition-state theory is not an exact theory. Thus, complete agreement with experimental data cannot be expected in all cases. In particular, recrossings of the saddle-point region were neglected. The validity of this assumption will depend on the exact topology of the potential energy surface; that is, the barrier height, the position of the barrier along the reaction coordinate, etc.[8]

6.6 Thermodynamic formulation

For some reactions, especially those involving large molecules, it might be difficult to determine the precise structure and energy levels of the activated complex. In such cases, it can be useful to phrase the transition-state theory result for the rate constant in thermodynamic terms. It does not bring any new information but an alternative way of interpreting the result. This formulation leads to an expression where the pre-exponential factor is related to an *entropy of activation* that, at least qualitatively, can be related to the structure of the activated complex. We will encounter the thermodynamic formulation again in Chapter 10, in connection with chemical reactions in solution, where this formulation is particularly useful.

According to Eq. (6.8),

$$\begin{aligned} k_{\mathrm{TST}}(T) &= \frac{k_B T}{h} \frac{Q^{\ddagger}/V}{(Q_{\mathrm{A}}/V)(Q_{\mathrm{B}}/V)} e^{-E_0/k_B T} \\ &= \frac{k_B T}{h} V \frac{Q^{\ddagger}}{Q_{\mathrm{A}} Q_{\mathrm{B}}} \end{aligned} \tag{6.55}$$

where, in the second line, the zero point of energy for all partition functions has been chosen to coincide with the zero-point energy of A + B. Note that the rate constant depends on T but is independent of the volume V (or pressure) since the molecular partition functions are of the form $f(T)V$.

According to statistical mechanics, the partition function for a system of volume V, at the temperature T, containing N independent (non-interacting) indistinguishable

[8]Note that for two reactions that only differ by the position of the saddle point along the reaction path, i.e., 'early' and 'late' barriers (Section 3.1), respectively, transition-state theory will predict exactly the same rate.

molecules is $Q^N/N!$. The entropy, S, of the system is related to the molecular partition function Q according to[9]

$$S = Nk_BT\left(\frac{\partial \ln Q}{\partial T}\right) + Nk_B \ln\left(Q\frac{e}{N}\right)$$
$$= E/T + Nk_B \ln\left(Q\frac{e}{N}\right) \tag{6.56}$$

where $E = \langle E \rangle$ according to Eq. (A.22) is the temperature-dependent internal energy. Now, the difference in entropy between the activated complex $(\mathrm{AB})^\ddagger$ (where the reaction coordinate is pulled out) and the reactants A + B, that is, the *entropy of activation*, is (per mole, $N = nN_A$, where $n = 1$ mole)

$$\begin{aligned}\Delta S_\ddagger^\ominus &\equiv S_\ddagger - (S_\mathrm{A} + S_\mathrm{B})\\ &= E_\ddagger/T + R\ln[Q^\ddagger e/N] - (E_\mathrm{A}/T + R\ln[Q_\mathrm{A} e/N] + E_\mathrm{B}/T + R\ln[Q_\mathrm{B} e/N])\\ &= \{E_\ddagger - (E_\mathrm{A} + E_\mathrm{B})\}/T + R\ln\left[\frac{Q^\ddagger}{Q_\mathrm{A}Q_\mathrm{B}}\frac{N}{e}\right]\\ &= \Delta E_\ddagger^\ominus/T + R\ln\left[\frac{Q^\ddagger}{Q_\mathrm{A}Q_\mathrm{B}}\frac{N}{e}\right]\end{aligned} \tag{6.57}$$

where the energy of activation was defined as $\Delta E_\ddagger^\ominus \equiv E_\ddagger - (E_\mathrm{A} + E_\mathrm{B})$. Note that the change in entropy refers to the entropy of pure (unmixed) activated complexes minus the entropy of pure (unmixed) reactants. Furthermore, the entropies refer to the same values of $N, V,$ and T (and the pressure, since $p = (N/V)k_BT$). These conditions are identical to the *standard changes* known from chemical thermodynamics, and they are indicated by the symbol $^\ominus$. Thus,

$$\frac{Q^\ddagger}{Q_\mathrm{A}Q_\mathrm{B}} = \frac{e}{N}\exp(\Delta S_\ddagger^\ominus/R)\exp(-\Delta E_\ddagger^\ominus/RT) \tag{6.58}$$

The enthalpy $H = E + pV = E + RT$ for one mole of an ideal gas, and the *enthalpy of activation* $\Delta H_\ddagger^\ominus$ is

$$\begin{aligned}\Delta H_\ddagger^\ominus &\equiv E_\ddagger + RT - (E_\mathrm{A} + RT + E_\mathrm{B} + RT)\\ &= \Delta E_\ddagger^\ominus - RT\end{aligned} \tag{6.59}$$

Note that $\Delta E_\ddagger^\ominus = \Delta E_\ddagger$ and $\Delta H_\ddagger^\ominus = \Delta H_\ddagger$, since there is no interaction between the molecules. Thus, for these energy changes the symbol $^\ominus$ is not needed, since the standard changes and the changes in a mixture of ideal gases are identical. We have now $\Delta E_\ddagger^\ominus = \Delta H_\ddagger^\ominus + RT$, and

$$k_\mathrm{TST} = \frac{k_BT}{h}\frac{1}{c^\ominus}\exp(\Delta S_\ddagger^\ominus/R)\exp(-\Delta H_\ddagger^\ominus/RT) \tag{6.60}$$

where $c^\ominus = N/V$ is a concentration, typically chosen as 1 mole/liter. The rate constant is now related to the activation entropy $\Delta S_\ddagger^\ominus$ and the enthalpy of activation $\Delta H_\ddagger^\ominus = \Delta H_\ddagger$.

[9]Using Sterling's approximation, i.e., $\ln N! \sim N(\ln N - 1) = N\ln(N/e)$.

Alternatively, the *Gibbs energy of activation* can be introduced, i.e., $\Delta G_{\ddagger}^{\ominus} = \Delta H_{\ddagger}^{\ominus} - T\Delta S_{\ddagger}^{\ominus}$, and

$$k_{\mathrm{TST}} = \frac{k_B T}{h} \frac{1}{c^{\ominus}} \exp(-\Delta G_{\ddagger}^{\ominus}/RT) \tag{6.61}$$

It might be argued that in a (N, V, T) system, the Helmholtz energy is a more natural choice than the Gibbs energy. Since $A = G - pV$, the *Helmholtz energy of activation* is $\Delta A_{\ddagger}^{\ominus} = \Delta G_{\ddagger}^{\ominus} + RT$, and

$$k_{\mathrm{TST}} = \frac{k_B T}{h} \frac{e}{c^{\ominus}} \exp(-\Delta A_{\ddagger}^{\ominus}/RT) \tag{6.62}$$

Finally, these expressions can be related to the Arrhenius equation, where the activation energy E_a is identified from $RT^2 d\ln k/dT$. Although the rate constant is independent of volume (or pressure), we note that the exponential factor, as well as the pre-exponential factor, in Eq. (6.61) depends on volume (or pressure). Thus, we choose to perform the differentiation under constant pressure. Using that $1/c^{\ominus} = V/N = k_B T/p$, we get

$$\begin{aligned} \frac{d\ln k_{\mathrm{TST}}}{dT} &= \frac{2}{T} - \frac{1}{R}\left(\frac{\partial(\Delta G_{\ddagger}^{\ominus}/T)}{\partial T}\right)_p \\ &= \frac{2}{T} + \frac{\Delta H_{\ddagger}^{\ominus}}{RT^2} \\ &= \frac{2RT + \Delta H_{\ddagger}^{\ominus}}{RT^2} \\ &\equiv \frac{E_a}{RT^2} \end{aligned} \tag{6.63}$$

where the Gibbs–Helmholtz equation was used in the second line. We observe that $E_a = 2RT + \Delta H_{\ddagger}^{\ominus} = RT + \Delta E_{\ddagger}^{\ominus} = RT + E_{\ddagger} - (E_{\mathrm{A}} + E_{\mathrm{B}})$, that is, we have a simple interpretation of E_a in terms of the enthalpy of activation $\Delta H_{\ddagger}^{\ominus}$ and the energy of activation $\Delta E_{\ddagger}^{\ominus}$, and from Eq. (6.60) we get

$$\boxed{k_{\mathrm{TST}} = \frac{k_B T}{h} \frac{e^2}{c^{\ominus}} \exp(\Delta S_{\ddagger}^{\ominus}/R) \exp(-E_a/RT)} \tag{6.64}$$

From this expression originates the statement that 'entropy effects' are determining the pre-exponential factor in the Arrhenius expression. Thus we have the following.

- A positive entropy of activation implies that there has been an increase in entropy in forming the activated complex, and the activated complex must be more 'disordered' or 'floppy' than the reactants.
- A negative entropy of activation implies that the activated complex must be more 'ordered' than the reactants.

Bimolecular reactions correspond to a negative entropy of activation. Note that $\Delta E_{\ddagger}^{\ominus}$, $\Delta H_{\ddagger}^{\ominus}$, $\Delta S_{\ddagger}^{\ominus}$, $\Delta G_{\ddagger}^{\ominus}$, and $\Delta A_{\ddagger}^{\ominus}$ all refer to an activated complex where the reaction coordinate is pulled out.

Further reading/references

[1] H. Eyring, *J. Chem. Phys.* **3**, 107 (1935)
[2] D.G. Truhlar, W.L. Hase, and J.T. Hynes, *J. Phys. Chem.* **87**, 2664 (1983)
[3] B.H. Mahan, *J. Chem. Educ.* **51**, 709 (1974)
[4] W.H. Miller, *Acc. Chem. Res.* **9**, 306 (1976)
[5] N.E. Henriksen and F.Y. Hansen, *Phys. Chem. Chem. Phys.* **4**, 5995 (2002)
[6] J.H. Knox, *Molecular thermodynamics* (John Wiley, New York, 1978)
[7] M.S. Marinov and B. Segev, *Phys. Rev. A* **55**, 3580 (1997)
[8] R.P. Bell, *The tunnel effect in chemistry* (Chapman and Hall, London, 1980)
[9] C. Eckart, *Phys. Rev.* **35**, 1303 (1930)
[10] H. Eyring, J. Walter, and G.E. Kimball, *Quantum chemistry* (John Wiley, New York, 1944)
[11] I. Shavitt, *J. Chem. Phys.* **31**, 1359 (1959)
[12] E. Pollak and P. Pechukas, *J. Am. Chem. Soc.* **100**, 2984 (1978)
[13] K.J. Laidler, *Chemical kinetics* (Harper Collins, New York, 1987)
[14] D.G. Truhlar, B.C. Garrett, and S.J. Klippenstein, *J. Phys. Chem.* **100**, 12771 (1996)

Problems

6.1 The 'lifetime' of the activated complex was defined in connection with the standard derivation of transition-state theory. Estimate the numerical value of the lifetime (in femtoseconds) using the following numbers: $l = 0.1$ Å, $m = 1$ amu, and $T = 300$ K.

6.2 Consider a collinear reaction of the form A + BC $\rightarrow$ AB + C, i.e., all atoms are assumed to move along the same line. Imagine that a calculation of the (real-valued) vibrational frequency of the activated complex, at two different levels of accuracy, gives $\tilde{\nu}_1$ cm^{-1} and $\tilde{\nu}_2$ cm^{-1}, respectively, and $\tilde{\nu}_1 > \tilde{\nu}_2$.
Using transition-state theory, which of the two frequencies will give the larger pre-exponential factor? (Remember that the partition function for a harmonic oscillator is given by $Q_{\text{vib}} = (1 - \exp[-\hbar\omega/(k_B T)])^{-1}$.)

6.3 Consider a linear triatomic AAA molecule where the atoms have mass m. Assume that the potential energy for the linear internal bond-stretching motion is given by

$$V = (1/2)k(q_2 - q_1 - b)^2 + (1/2)k(q_3 - q_2 - b)^2$$

where q_1, q_2, and q_3 are the positions of atoms 1, 2, and 3, respectively, and b is the bond length at equilibrium.

(a) Find expressions for the normal-mode frequencies (note that one of the frequencies, that for translation, will be zero).

(b) Find expressions for the normal-mode coordinates expressed as linear combinations of the atomic displacement coordinates.

(c) Show the result from (b) in graphical form, that is, for each normal mode, using 'arrows' to represent (the magnitude and sign of) the coefficients in the linear combinations of the atomic displacement coordinates.

6.4 Using transition-state theory, calculate the rate constant for the exchange reaction

$$\mathrm{H} + \mathrm{H}_2 \rightarrow (\mathrm{H}\cdots\mathrm{H}\cdots\mathrm{H})^{\ddagger} \rightarrow \mathrm{H}_2 + \mathrm{H}$$

Assume that $(\mathrm{H}\cdots\mathrm{H}\cdots\mathrm{H})^{\ddagger}$ has a linear symmetric configuration with a ground-state electronic degeneracy of 2 and a symmetry number of 2. The electronic degeneracies for ground-state H_2 and H are 1 and 2, respectively. For the $(\mathrm{H}\cdots\mathrm{H}\cdots\mathrm{H})^{\ddagger}$ transition state (from accurate *ab initio* electronic structure calculations; *J. Chem. Phys.* **68**, 2457 (1978)) the $\mathrm{H}\cdots\mathrm{H}$ distance is 0.93 Å and the fundamental frequencies are 2058 cm^{-1}, 909 cm^{-1}, and 909 cm^{-1}. The *classical* barrier height is 41.0 kJ/mol. The H–H distance in H_2 is 0.74 Å, and its fundamental vibrational frequency is 4395 cm^{-1}.

(a) Calculate numerical values for the transition-state theory rate constants for the reaction at 300 K and 1000 K (the experimentally measured rate coefficient is $[2.1 \pm 0.6] \times 10^{-12}$ cm^3 molecule^{-1} s^{-1} at 1000 K).

(b) Estimate the tunneling correction for the reaction at 400 K, 500 K, and 1000 K. Use the Wigner tunneling correction factor for a barrier with $\nu^* =$ 1511 cm^{-1}.

6.5 We consider a forward and its reverse bimolecular elementary reaction at thermal equilibrium:

$$\mathrm{A} + \mathrm{B} \underset{k_r}{\overset{k_f}{\rightleftarrows}} \mathrm{C} + \mathrm{D}$$

(a) Write down the expressions for the rate constants $k_f(T)$ and $k_r(T)$ according to transition-state theory.

(b) Show that $k_f(T)/k_r(T)$, calculated according to transition-state theory, fulfills the principle of detailed balance (for a potential energy surface with one saddle point).

6.6 Determine, using transition-state theory, the temperature dependence of the pre-exponential factor for each of the reactions:

(a) $\mathrm{H}_2 + \mathrm{F}_2 \rightarrow 2\mathrm{HF}$

(b) $\mathrm{C}_2\mathrm{H}_4 + \mathrm{HCl} \rightarrow \mathrm{C}_2\mathrm{H}_5\mathrm{Cl}$

Assume that the activated complexes are non-linear. Determine the temperature dependence for $hv \ll k_BT$, corresponding to classical partition functions for the harmonic vibrational degrees of freedom, as well as for $hv \gg k_BT$.

6.7 In order to illustrate an isotope effect for a bimolecular reaction, we consider the following reactions:

(i) $\mathrm{H} + \mathrm{H}_2 \xrightarrow{k_1} \mathrm{H} + \mathrm{H}_2$

(ii) $D + H_2 \xrightarrow{k_2} DH + H$

where 'D' is deuterium. The potential energy surfaces and electronic states for the two systems (i) and (ii) are, according to the adiabatic (Born–Oppenheimer) approximation, identical. The activated complexes, ABC, of the three atoms are linear with interatomic distances $R_{AB} = R_{BC} = R = 0.93$ Å, and the vibrational frequencies for the complex are, respectively, 2058 cm^{-1}, 909 cm^{-1}, and 909 cm^{-1} for reaction (i) and 1764 cm^{-1}, 870 cm^{-1}, and 870 cm^{-1} for reaction (ii).

In the following we use transition-state theory in order to calculate the ratio between the rate constants k_2/k_1 (tunneling is neglected).

(a) Calculate $E_0^2 - E_0^1$ (in kJ/mol), that is, the difference between the barrier heights for the two reactions.

(b) Show that k_2/k_1 can be written in the following form:

$$\frac{k_2}{k_1} = \left(\frac{\mu_1}{\mu_2}\right)^{3/2} \left(\frac{Q_2^\ddagger}{Q_1^\ddagger}\right)_{\text{rot}} \left(\frac{Q_2^\ddagger}{Q_1^\ddagger}\right)_{\text{vib}} e^{-(E_0^2 - E_0^1)/k_B T}$$

where μ_i is the reduced mass of the reactants in reaction 'i', and the second and third factors are the ratios between the rotational and vibrational partition functions of the activated complexes of the 'i'th reaction.

(c) Calculate $(Q_2^\ddagger/Q_1^\ddagger)_{\text{rot}}$, using that the moment of inertia for a linear 3-atomic molecule ABC is $I = (m_A + m_C)R^2 - (m_A - m_C)^2 R^2/(m_A + m_B + m_C)$, with $R_{AB} = R_{BC} = R$, where R is the bond distance.

(d) Calculate k_2/k_1 at $T = 450$ K, and compare with the experimental values $k_1 = 3.85 \times 10^9$ cm^3/(mol s) and $k_2 = 9 \times 10^9$ cm^3/(mol s).

(e) Calculate k_2/k_1 at $T = 450$ K, when tunneling is included in the transition-state calculation, using the Wigner tunneling correction factor. The ratio of the imaginary frequencies associated with the activated complexes $(HHH)^\ddagger$ and $(DHH)^\ddagger$ is $\nu_H/\nu_D = 1.05$, and the imaginary frequency (in wave number) for the $(HHH)^\ddagger$ complex is $\tilde{\nu}_H = i\tilde{\nu}_H^\star = i\,1511$ cm^{-1}.

The atomic masses are 1.007825 amu for ^1H and 2.0140 amu for $D \equiv {}^2H$.

6.8 Calculate the rate constant for the reaction $D + H_2 \rightarrow DH + H$ at $T = 1000$ K. Use transition-state theory and the data and results of Problems 6.4 and 6.7. The experimentally measured rate coefficient is $[2.13 \pm 28\%] \times 10^{-12}$ cm^3 molecule^{-1} s^{-1} at 1000 K [*J. Phys. Chem.* **94**, 3318 (1990)].

6.9 A molecule AB can be formed from the two molecules A and B in an association reaction via the following mechanism:

(i) $A + B \rightleftharpoons AB^*$

(ii) $AB^* + M \rightarrow AB + M$

where AB^* is an energy-rich (unstable) state of AB (with an energy that exceeds the bond dissociation energy of AB) and M is a molecule that in an inelastic collision can absorb the excess energy in AB^*. At low pressures of M one finds a third-order rate law, $v = k(T)[A][B][M]$, $k(T) = K_1(T)k_2(T)$, where $K_1(T)$ is the equilibrium constant for (i) and $k_2(T)$ is the rate constant for (ii).

An important example of an association reaction is the formation of ozone in the stratosphere from atomic and molecular oxygen: $O + O_2 + M \rightarrow O_3 + M$. At low pressures, a third-order rate law is found with rate constants given in the table below for $M = N_2$ [*J. Phys. Chem. Ref. Data* **26**, 1329 (1997)].

T/K	100	200	300
$k(T)$/m^6 molecule^{-2} s^{-1}	1.19×10^{-44}	1.71×10^{-45}	5.49×10^{-46}

(a) Express the equilibrium constant $K_1(T)$ for (i) in terms of molecular partition functions for a reaction in an ideal gas.

(b) Write down the expression for the rate constant $k_2(T)$ according to transition-state theory, where it can be assumed that (ii) proceeds via a particular activated complex $(AB^*M)^{\ddagger}$. It can be assumed that $E_0 = 0$, where E_0 is the difference between the zero-point energy level of the activated complex and the zero-point energy level of the reactants A + B + M. Also, write down an expression for $k(T)$.

(c) Determine the temperature dependence of $k(T)$ for $O + O_2 + M \rightarrow O_3 + M$, where it can be assumed that the activated complex is non-linear, that $E_0 = 0$, and the temperature dependence of the vibrational and electronic partition functions can be neglected.

(d) The experimental values in the table can be represented by an expression of the form $k(T) = AT^a$, where A and a are constants. Compare the theoretically determined value of a to the value that can be determined from the experimental values in the table. The experimental uncertainty on a is ± 0.5.

(e) The pre-exponential factor for ('trimolecular') association reactions is very small compared to typical values for bimolecular reactions. Calculate how much bigger the rate constant would be at $T = 300$ K, if $M = N_2$ did not participate in the reaction. It can be assumed that the difference is solely due to the translational contributions from $M = N_2$.

6.10 Consider the reaction

$$F + H_2 \underset{k_r}{\overset{k_f}{\rightleftarrows}} HF + H$$

(which we also considered in Section 6.5). The rate constant of the forward reaction is $k_f(T) = 2.0 \times 10^{11} \exp\{-800/T(\mathrm{K})\}$ liter/(mol s). Calculate $k_r(T)$ at $T = 300$ K and $T = 1000$ K.
Use the following information:

- $\Delta E_0 = -1.391$ eV, that is, the difference between the zero-point energies of the products and reactants;
- vibrational frequencies: $\tilde{\nu} = 4400$ cm^{-1} for H_2 and $\tilde{\nu} = 4138$ cm^{-1} for HF;
- equilibrium bond lengths: $r_e = 0.7414$ Å for H_2 and $r_e = 0.9168$ Å for HF;
- electronic ground-state degeneracies: $\omega_{\text{elec}} = 4$ for F, $\omega_{\text{elec}} = 1$ for H_2, $\omega_{\text{elec}} = 1$ for HF, and $\omega_{\text{elec}} = 2$ for H.

6.11 We consider, in the gas phase, the S_N2 reaction

$$Cl^- + CH_3Cl \longrightarrow CH_3Cl + Cl^-$$

The data for the activated complex are given below—essentially corresponding to a calculation at the Hartree–Fock (HF) level using a 6–31 G* basis set.

- The barrier height (corrected for vibrational zero-point energies) is $E_0 = 16.6$ kJ/mol.
- The principal moments of inertia for CH_3Cl are: $I_a = I_b = 38.1$ amu Å^2 and $I_c = 3.08$ amu Å^2, and the symmetry number is 3.
- The principal moments of inertia for the activated complex, $(Cl \cdots CH_3 \cdots Cl)^{\ddagger}$, are: $I_a = I_b = 405.0$ amu Å^2 and $I_c = 3.46$ amu Å^2, and the symmetry number is 6.
- The degeneracies of the electronic states can be set to one.
- The vibrational frequencies (in wave numbers), for the reactants and activated complex, are given below.

Molecule	Wave numbers/cm^{-1}
CH_3Cl	3042(2) 2933(1) 1463(1) 1403(2) 1013(2) 736(1)
$(Cl \cdots CH_3 \cdots Cl)^{\ddagger}$	3430(1) 3274(2) 1379(2) 1015(2) 1004(1) 195(1) 188(2)

The degeneracies of the vibrational states are given in parentheses after the frequencies. The imaginary frequency of the activated complex is not included. The vibrational frequencies of CH_3Cl correspond to the data in G. Herzberg, *Electronic spectra of polyatomic molecules.*

Calculate the rate constant, at $T = 300$ K, according to transition-state theory. The vibrational partition functions for vibrations with wave numbers larger than 1000 cm^{-1} can be set to 1.

The atomic masses are H: 1.008 amu, C: 12.01 amu, and Cl: 35.45 amu.

7 Unimolecular reactions

Key ideas and results

When a molecule is supplied with an amount of energy that exceeds some threshold energy, a unimolecular reaction can take place, that is, a dissociation or an isomerization. We distinguish between a true unimolecular reaction that can be initiated by absorption of electromagnetic radiation (photo-activation) and an apparent unimolecular reaction initiated by bimolecular collisions (thermal activation). For the apparent unimolecular reaction, the time scales for the activation and the subsequent reaction are well separated. When such a separation is possible, for true or apparent unimolecular reactions, the reaction is also referred to as an indirect reaction. We will discuss the following.

- Elements of classical dynamics of unimolecular reactions; in particular, the Slater theory for indirect reactions, where the molecule is modeled as a set of uncoupled harmonic oscillators. Reaction is defined to occur when a particular bond length attains a critical value, and the rate constant is given as the frequency with which this occurs.
- Elements of quantum dynamics of unimolecular reactions; in particular, photo-activated reactions.
- RRKM theory, an approach to the calculation of the rate constant of indirect reactions that, essentially, is equivalent to transition-state theory. The reaction coordinate is identified as being the coordinate associated with the decay of an activated complex. It is a statistical theory based on the assumption that every state, within a narrow energy range of the activated complex, is populated with the same probability prior to the unimolecular reaction. The microcanonical rate constant $k(E)$ is given by an expression that contains the ratio of the sum of states for the activated complex (with the reaction coordinate omitted) and the total density of states of the reactant. The canonical $k(T)$ unimolecular rate constant is given by an expression that is similar to the transition-state theory expression of bimolecular reactions.
- How thermal activation can take place following the Lindemann and the Lindemann–Hinshelwood mechanisms. An effective rate constant is found that shows the interplay between collision activation and unimolecular reaction. In the high-pressure limit, the effective rate constant approaches the microcanonical rate

constant of a unimolecular reaction multiplied by the probability of finding the molecule at a given energy.

- Basic concepts of femtochemistry, i.e., the real-time detection and control of chemical dynamics.

The relation between the key quantities (the rate constant $k(T)$, the microcanonical rate constant $k(E)$, and the reaction probability P) and various approaches to the description of the nuclear dynamics is illustrated below.

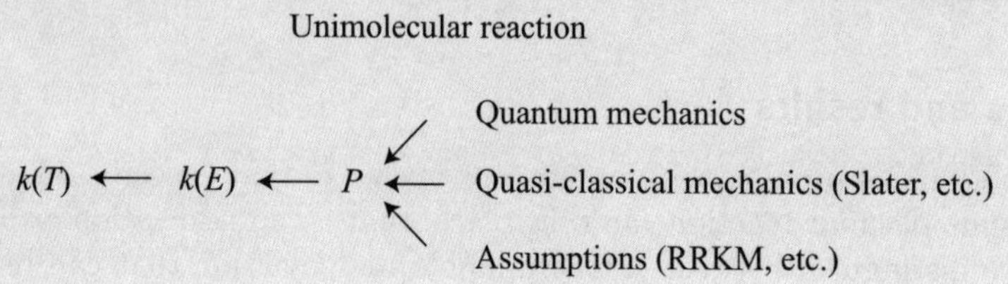

So far, we have only considered bimolecular reactions. Elementary reactions can, however, also be unimolecular; that is, just one molecule takes part in the chemical reaction. Such reactions correspond to either *isomerization* (rearrangement) or *dissociation* (fragmentation).

In chemical kinetics, we learn that an elementary unimolecular reaction

$$\mathrm{A} \longrightarrow \text{products} \tag{7.1}$$

obeys a first-order rate law, given by

$$-\frac{d[\mathrm{A}]}{dt} = k[\mathrm{A}] \tag{7.2}$$

where $k \equiv k(T)$ is the temperature-dependent unimolecular rate constant. The purpose of this chapter is to obtain an in-depth understanding of this relation.

Before we go on, we note that it is easy to interpret the physical significance of the unimolecular rate constant. The integrated rate law takes the form

$$\begin{aligned}[\mathrm{A}] &= [\mathrm{A}]_{t_0} \exp[-k(t - t_0)] \\ &= [\mathrm{A}]_{t_0}[1 - k(t - t_0) + \cdots]\end{aligned} \tag{7.3}$$

That is, for $k(t-t_0) \ll 1$, $k(t-t_0)$ can be interpreted as the fraction of molecules that has disappeared, and k is *the fraction of molecules that disappear per unit time*. For example, $k(T)$ may have the value 0.001 s^{-1}, which means that during each second 1/1000 of the molecules would react. Note that this fraction can also be interpreted as the *probability* that a single molecule undergoes reaction per unit time.

7.1 True and apparent unimolecular reactions

In unimolecular reactions only one molecule is involved in the reaction. Since, however, the molecules prior to the reaction are stable, it is necessary to 'activate' them in order to initiate the reaction. Based on the different methods of activation, we distinguish between the following two types of unimolecular reaction:

- a *true* unimolecular reaction;
- an *apparent* unimolecular reaction.

In a true unimolecular reaction, the activation is done by exposing the molecules to electromagnetic radiation, whereas activation is accomplished by inelastic collisions with other molecules in an apparent unimolecular reaction. The condition for the latter process to be unimolecular is that the time scales of the activation process and the chemical reaction are very different, so that the chemical reaction is much slower than the activation process.

7.1.1 True unimolecular reactions

A *true* unimolecular reaction is induced by electromagnetic radiation. That is, only one molecule takes part in the reaction and the energy is provided by the electromagnetic field. In fact, chemical reactions induced by electromagnetic radiation form such an important subfield of chemistry that it has its own designation: *photochemistry*.

A well-known example is the photodissociation of ozone:

$$O_3 + h\nu \longrightarrow O_2 + O$$

There are also many interesting photochemical rearrangements of organic molecules. One example is the cis-trans isomerization of retinal induced by visible light, a basic step in the chemistry of vision.

The electromagnetic radiation will normally induce a transition from the electronic ground state to an excited electronic state, where the reaction takes place.

It is also possible that the unimolecular reaction takes place with the molecule in the electronic ground state, but it requires very intense fields to generate so-called multiphoton or direct overtone transitions; that is, transitions from the vibrational ground state of the type $0 \rightarrow n$, where $n > 1$. The opening of the cyclo-butene ring to form butadiene is an example of a unimolecular reaction induced by direct overtone excitation:

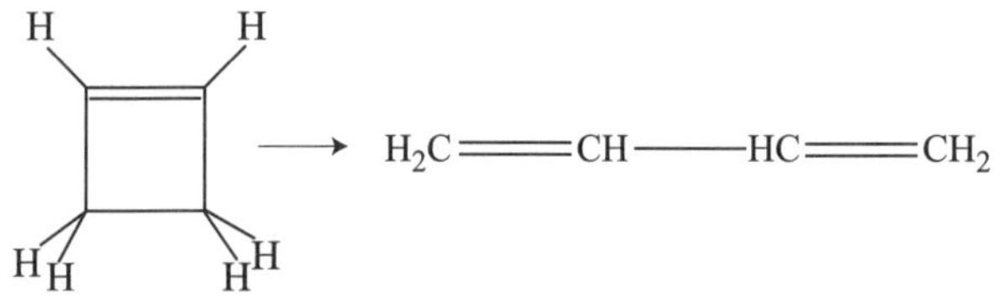

In a normal-mode picture, the C–H stretching of cyclo-butene is, essentially, a pure normal mode. There are two types of CH bonds, and the normal modes involve either olefinic CH stretch or methylenic CH stretch. Experimentally, it has been demonstrated that direct overtone excitation of these modes, e.g., to the fourth excited C–H stretch mode, provides enough energy for the ring-opening to 1,3-butadiene.

Unimolecular reactions induced by electromagnetic radiation are often further divided into *direct* and *indirect* reactions. In a direct reaction, the unimolecular reaction is over within the order of a vibrational period after the initial excitation. In an indirect reaction, the unimolecular reaction starts long after the initial excitation, since a long-lived intermediate complex is formed. Evidently, the cyclo-butene reaction is not a direct reaction, since the energy first has to flow from the C–H stretch to the C–C bond connecting the two $-CH_2-$ units in the ring. These definitions correspond to two extremes; real unimolecular reactions can fall in between these limiting cases and, in general, we have to consider the interplay between activation and reaction.

It is quite easy to derive the rate law for a unimolecular reaction. Consider the situation where molecules, all in the quantum state n, are irradiated by radiation of frequency ν, which induces an electronic transition:

$$\mathrm{A}(n) + h\nu \xrightarrow{P_n(h\nu)} \text{products} \tag{7.4}$$

e.g., as illustrated in Fig. 7.1.1. *Assume* a constant reaction probability per unit time, at the energy $E = h\nu$, denoted by $P_n(h\nu)$. Here we limit the discussion to direct reactions where products are formed directly without delay following electronic excitation. Indirect reactions can be handled as in Section 7.1.2.

The number of molecules that disappear per unit volume and per unit time is then $P_n(h\nu) \times [\mathrm{A}(n)]$, where $[\mathrm{A}(n)]$ is the concentration (number/volume) at time t. The rate of reaction must then be given by

$$d[\mathrm{A}(n)]/dt = -P_n(h\nu)[\mathrm{A}(n)] \tag{7.5}$$

that is, a first-order rate law. When we make the identification

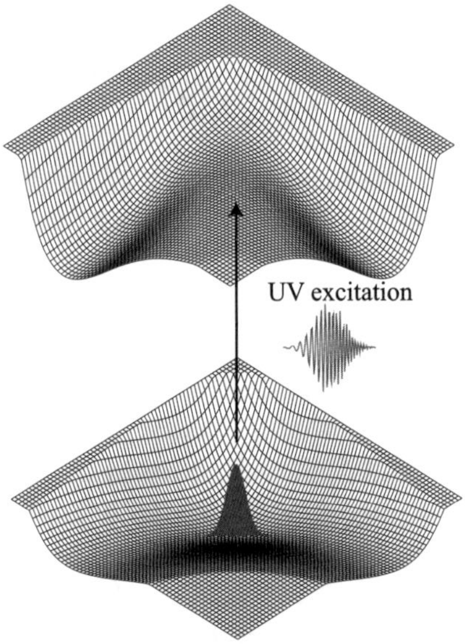

Fig. 7.1.1 Photodissociation (at a fixed bending angle) for a symmetric triatomic molecule like ozone. The vibrational ground state is superimposed on the potential energy surface of the electronic ground state; an illustration of a true (direct) unimolecular reaction. (Note that in this figure all potential energies above a fixed cut-off value $E_{\max}$ have been replaced by $E_{\max}$, in the electronic ground state as well as in the excited electronic state.)

$$k_n(h\nu) \equiv P_n(h\nu) \tag{7.6}$$

and denote by $k_n(h\nu)$ the rate constant of the unimolecular reaction, the rate law is clearly equivalent to a standard first-order rate law. The *prerequisite* used in this derivation, a rate constant that is equivalent to a constant reaction probability per unit time, is obtained from a microscopic quantum mechanical treatment of the interaction between the molecule and electromagnetic radiation; see Section 7.2.2.

Except for high vacuum systems, where isolated reactions occur, the energy of the molecules is not fixed. We must then, as in Chapter 2, consider the transition from the microscopic to the macroscopic description. Again, it is quite easy to derive the rate law of macroscopic reaction kinetics for the unimolecular reaction. We now write the number density of A(n) in a form that is equivalent to Eq. (2.16), that is,

$$[\mathrm{A}(n)] = [\mathrm{A}]p_{\mathrm{A}(n)} \tag{7.7}$$

where [A] is the total concentration of A molecules and $p_{\mathrm{A}(n)}$ is the probability of finding A in the quantum state denoted by the quantum number n.

The macroscopic rate of reaction, $d[\mathrm{A}]/dt$, is obtained from Eq. (7.5) by summation over the quantum states. We get

$$d[\mathrm{A}]/dt = -k(T)[\mathrm{A}] \tag{7.8}$$

with the rate constant

$$\boxed{k(T) = \sum_n k_n(h\nu)p_{\mathrm{A}(n)}} \tag{7.9}$$

where we have used that $\sum_n p_{\mathrm{A}(n)} = 1$. Since $p_{\mathrm{A}(n)}$ is typically the Boltzmann distribution, for the vibrational and rotational quantum states, we have indicated that the macroscopic rate constant will be a function of the temperature T.

In Chapter 2 where bimolecular reactions were discussed, we considered the formation of products in particular quantum states and at particular orientations in space. The description above for unimolecular reactions can be generalized in a similar way. Thus, experiments similar to the *molecular-beam* set-up in Fig. 2.1.1 are carried out for unimolecular reactions, however, with one of the molecular beams replaced by a beam of photons from a laser.

7.1.2 Apparent unimolecular reactions

In an *apparent* unimolecular reaction, the molecule is activated in a bimolecular collision process. In addition, it is assumed that the preparation of the initial metastable state of the molecule can be separated from its unimolecular decay. Thus, the apparent unimolecular reaction can also be classified as an *indirect* reaction. One writes such an apparent unimolecular reaction in the form

$$\mathrm{A}^* \longrightarrow \text{products} \tag{7.10}$$

where A^* is a highly vibrationally (and rotationally) excited A molecule with a total vibrational and rotational energy[1] $E > E_0$, where E_0 is the threshold energy for the

[1] The center-of-mass translational motion can be separated out (as described in Section 1.1) and it plays no role in unimolecular or bimolecular reactions. Thus, the translational energy of the molecule is not included in the internal energy that is relevant to unimolecular reactions.

unimolecular reaction. A* is created in an activation step, which can be: (i) chemical activation, where a long-lived intermediate is formed in a bimolecular complex mode reaction, and (ii) thermal activation, due to inelastic collisions where translational energy is converted to vibrational energy; this type of activation takes place in the electronic ground state, see Fig. 7.1.2.

Thus, it is clear what is meant by the words 'apparent' unimolecular reaction, since the activation process can be bimolecular. A key point is the separation in time scales between the activation step and the subsequent unimolecular reaction dynamics. This was also a distinctive feature of the *indirect* photo-activated reactions considered in the previous section. Thus, the theoretical description of these reaction types is very similar; photo-activation makes it, in particular, possible to deposit a precise amount of energy according to $E = h\nu$.

After (and during) the creation of A*, intramolecular vibrational energy redistribution—denoted *IVR*—will take place. Consider again the ring-opening of cyclo-butene to 1,3-butadiene. After the excitation of the C–H stretch, IVR will occur. The normal-mode description is not accurate for highly excited states due to anharmonic coupling terms in the potential. Thus, the vibrational energy will redistribute and spread to all vibrational degrees of freedom in the molecule.

Assume a constant reaction probability per unit time at the internal energy E. In anticipation of the role played by this probability in the final result, the probability is denoted by $k(E)$. The unimolecular reaction for an A* molecule at the fixed internal energy E is

$$A^*(E) \xrightarrow{k(E)} \text{products} \tag{7.11}$$

The molecule is highly vibrationally (and rotationally) excited with an energy above the threshold for unimolecular reaction. The density of vibrational eigenstates will be very high when the energy exceeds the energy threshold for reaction (recall, e.g., that the energy spacing in, say, a Morse potential is decreasing as the dissociation limit is

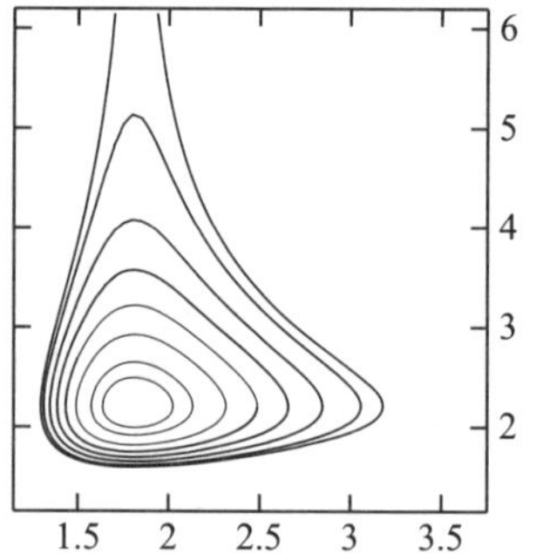

Fig. 7.1.2 Contour plot (at a fixed bending angle) of the electronic ground-state potential energy surface for a (asymmetric) triatomic molecule. The last contour exceeds the dissociation limit of one of the bonds.

approached). Thus, the internal energy can be considered as continuous.[2]

The number of molecules that disappear per unit volume and per unit time is then $k(E) \times [\mathrm{A}^*(E)]dE$, where $[\mathrm{A}^*(E)]dE$ is the concentration (number/volume) of $\mathrm{A}^*(E)$ molecules at time t, in the internal energy range $[E, E+dE]$. $[\mathrm{A}^*(E)]$ denotes, consequently, a concentration per unit energy at the energy E. The rate of reaction must then be given by

$$d\{[\mathrm{A}^*(E)]dE\}/dt = -k(E)\{[\mathrm{A}^*(E)]dE\} \tag{7.12}$$

that is, a first-order rate law resulting in an exponential decay of the population. The *prerequisite* used in this derivation, a rate constant that is equivalent to a constant reaction probability per unit time, is, e.g., obtained in RRKM theory, to be described below.

Except for high vacuum systems, where isolated reactions may occur, the internal energy is not fixed at a given value. We must therefore, as in Chapter 2, consider a transition from the microscopic to a macroscopic description. We write the number density of $\mathrm{A}^*(E)$ $(E > E_0)$ in a form that is equivalent to Eq. (2.16), that is,

$$[\mathrm{A}^*(E)]dE = [\mathrm{A}^*]P(E)dE \Big/ \int_{E_0}^{\infty} P(E)dE \tag{7.13}$$

where $[\mathrm{A}^*] = \int_{E_0}^{\infty}[\mathrm{A}^*(E)]dE$ is the total concentration of A^* molecules and $P(E)dE$ is the probability of finding A in the internal energy range $[E, E + dE]$.

The macroscopic rate of reaction, $d[\mathrm{A}^*]/dt$, is obtained from Eqs (7.12) and (7.13) by integration over the energy, E, from E_0 to infinity. We get

$$d[\mathrm{A}^*]/dt = -k(T)[\mathrm{A}^*] \Big/ \int_{E_0}^{\infty} P(E)dE \tag{7.14}$$

with the rate constant

$$\boxed{k(T) = \int_{E_0}^{\infty} k(E)P(E)dE} \tag{7.15}$$

Since $P(E)$ is normally the Boltzmann distribution, we have indicated that the macroscopic rate constant will be a function of the temperature T. Note that typically the magnitude of $k(T)$ and $k(E)$ will differ by *many* orders of magnitude. For example, $k(E)$ may have the value 0.01 fs^{-1} (corresponding to a typical vibrational frequency of 10^{13} s^{-1}), which means that during each femtosecond 1/100 of the molecules, at an energy $E > E_0$, will react. However $k(T)$ will be much smaller; it may have, say, the value 0.001 s^{-1}, which means that during each second 1/1000 of the molecules would react at the temperature T. This difference is, of course, due to the Boltzmann distribution $P(E)$ in Eq. (7.15). Thus, the fraction of reactants with sufficient internal energy to react is very small at typical temperatures.

[2]There is most likely a high degeneracy at the energy E. These states will, in general, not decay with the same rate constant. Thus, to be more precise, one should introduce additional quantum numbers, n, to distinguish between the degenerate states, and replace $k(E)$ by $k_n(E)$ in the discussion.

Now using $d[\mathrm{A}^*]/dt = d[\mathrm{A}]/dt$, and that $[\mathrm{A}]/[\mathrm{A}^*] = \int_0^\infty P(E)dE/\int_{E_0}^\infty P(E)dE$, where $\int_0^\infty P(E)dE = 1$, the well-known first-order rate law is obtained,[3] i.e.,

$$d[\mathrm{A}]/dt = -k(T)[\mathrm{A}] \tag{7.16}$$

In the following we discuss how to *calculate* the rate constants for a unimolecular decay.

7.2 Dynamical theories

In dynamical theories, one solves the equation of motion for the individual nuclei, subject to the potential energy surface. This is the exact approach, provided one starts with the Schrödinger equation. The aim is to calculate $k(E)$ and $k_n(h\nu)$, the microcanonical rate constants associated with, respectively, indirect (apparent or true) unimolecular reactions and true (photo-activated) unimolecular reactions.

7.2.1 Classical mechanics, Slater theory

This is an approach for the calculation of the microcanonical rate constant $k(E)$ for *indirect* unimolecular reactions that is based on several approximations. The molecule is represented by a collection of s uncoupled harmonic oscillators. According to Appendix E, such a representation is exact close to a stationary point on the potential energy surface. Furthermore, the dynamics is described by classical mechanics.

The Hamiltonian takes, according to Eqs (E.12) and (E.10), the form

$$\begin{aligned} H &= T + V \\ &= \sum_{n=1}^{s} [(1/2)P_n^2 + (1/2)\omega_n^2 Q_n^2] \end{aligned} \tag{7.17}$$

where Q_n are normal-mode coordinates. Hamilton's equations of motion, Eq. (4.63), take the form

$$\begin{aligned} \dot{Q}_i &= \left(\frac{\partial H}{\partial P_i}\right) = P_i \\ \dot{P}_i &= -\left(\frac{\partial H}{\partial Q_i}\right) = -\omega_i^2 Q_i \end{aligned} \tag{7.18}$$

where $i = 1, \ldots, s$. These equations are equivalent to

$$\frac{d^2 Q_i}{dt^2} + \omega_i^2 Q_i = 0 \tag{7.19}$$

that is, a linear second-order differential equation with constant coefficients, which has the complete solution

$$\begin{aligned} Q_i(t) &= Q_i(0)\cos(\omega_i t) + P_i(0)\sin(\omega_i t)/\omega_i \\ &= A_i \cos(\omega_i t - \phi_i) \\ &= \sqrt{2E_i/\omega_i^2}\cos(\omega_i t - \phi_i) \end{aligned} \tag{7.20}$$

[3] The same result is traditionally derived from the high-pressure limit of the Hinshelwood–Lindemann mechanism; see Section 7.4.1.

where $A_i = Q_i(0)/\cos(\phi_i)$ is the *amplitude* of the oscillation, and the *phase* ϕ_i is given by $\tan(\phi_i) = P_i(0)/[\omega_i Q_i(0)]$ (using the relation $\cos(\omega_i t - \phi_i) = \cos(\phi_i)\cos(\omega_i t) + \sin(\phi_i)\sin(\omega_i t)$). The time dependence of the normal-mode coordinate is illustrated in Fig. 7.2.1. The energy in the ith mode is related to the amplitude, i.e., $E_i = 1/2[P_i^2(t) + \omega_i^2 Q_i^2(t)] = (1/2)A_i^2\omega_i^2$. When $Q_i(0)$ and $P_i(0)$ are chosen, the energy and phase are fixed and Eq. (7.20) gives the time evolution of the particular normal-mode coordinate.

Note that the time it takes to complete one full oscillation, i.e., the period of the oscillation, is $P = 2\pi/\omega_i$, since $\cos(\omega_i t - \phi_i) = \cos(\omega_i[t + P] - \phi_i)$. Furthermore, the number of complete oscillations per unit time is $1/P = \omega_i/(2\pi) = \nu_i$, that is, the frequency of the oscillation. For typical molecules these vibrational frequencies are in the range 10^{13} to 10^{14} s^{-1}.

Since there is no coupling between the normal modes, the energy in each mode is constant. The description is rigorous only for small displacements from a stationary point, typically chosen as the molecular equilibrium geometry. Slater suggested the use of this description also when molecules are highly vibrationally excited, corresponding to large amplitude motions. This is obviously a serious approximation.

Since harmonic oscillators are considered in this theory, the bonds will never break, so it is necessary to introduce an *ad hoc criterion* for when a reaction occurs. Reaction is normally defined to occur when a particular bond length attains a critical value. The bond length cannot be extracted directly from a particular normal-mode coordinate, since these coordinates, typically, involve the motion of several atoms in the molecule. The bond length can, however, be calculated quite readily, by noting that the displacement of a coordinate associated with an atom of mass m_r is given by Eq. (E.5):

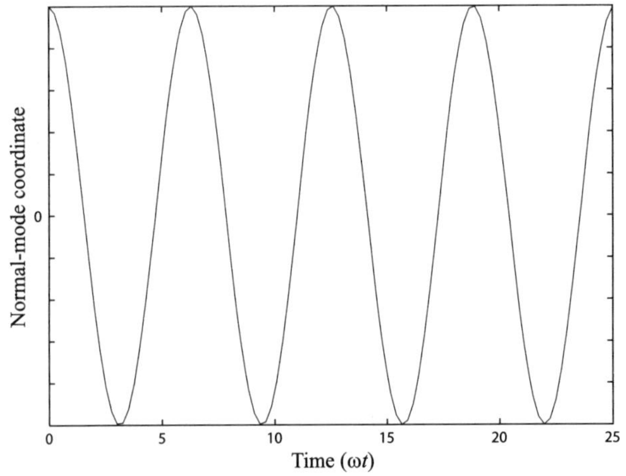

Fig. 7.2.1 Time evolution of a harmonic-oscillator coordinate, according to Eq. (7.20). The phase is chosen as $\phi = 0$, and note that the amplitude of the oscillation is related to the energy.

$$q_r(t) - q_r^0 = \frac{1}{\sqrt{m_r}} \sum_{j=1}^{s} L_{rj} Q_j(t) \tag{7.21}$$

where q_r^0 is the equilibrium position of coordinate q_r; that is, the displacement can be expressed as a linear combination of the normal-mode coordinates. Unimolecular reaction is assumed to occur when the bond length exceeds a critical extension at a particular time when the normal modes will be in phase; see Fig. 7.2.2.

The rate constant $k(E)$ at the total vibrational energy

$$E = \sum_{i=1}^{s} E_i \tag{7.22}$$

is calculated from the number of times per second a superposition of harmonic terms (Eq. (7.21)) exceeds the critical value.

The molecule undergoing reaction can, accordingly, be pictured as an assembly of s harmonic oscillators of particular amplitudes. The time evolution will, however, also depend on the phases of each oscillator (see Eq. (7.20)), and we have to consider a large number of these 'oscillator assemblies' and average over the initial phase of each oscillator.[4]

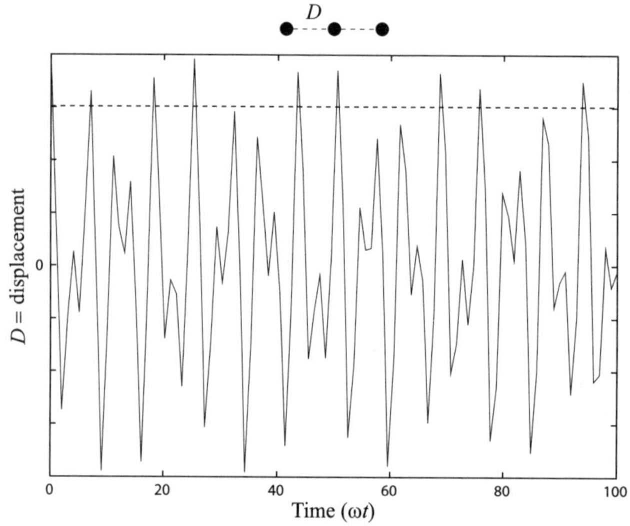

Fig. 7.2.2 Displacement of bond length from its equilibrium value for a linear chain of three identical atoms. The energy and phase of the symmetric and antisymmetric normal modes are assumed to be identical. The displacement is proportional to $\cos(\omega t) + \cos(\sqrt{3}\omega t)$. The dashed line marks a critical value.

[4]Note that this average over phases is equivalent to the approach used in quasi-classical trajectory calculations for bimolecular reactions; see, e.g., Fig. 4.1.2.

When the average behavior over a long time interval (much longer than the longest period associated with the normal modes) is considered, the problem of finding the fraction of molecules with a particular bond that—per unit time—exceeds the critical value can be solved analytically [1]. $k(E)$ will depend on the particular distribution of energy $E_1, E_2, \ldots, E_s$ in the normal modes. Thus, since the theory does not allow for redistribution of energy between the normal modes, i.e., IVR, the rate constant will depend on how the molecule was excited initially. Such mode specificity is, in general, not observed in practice. Overall, the Slater theory is unsuccessful in interpreting experiments; the theory has, however, still some pedagogical value.

The harmonic approximation is unrealistic in a dynamical description of the dissociation dynamics, because anharmonic potential energy terms will play an important role in the large amplitude motion associated with dissociation. An accurate potential energy surface must be used in order to obtain a realistic dynamical description of the dissociation process and, as in the quasi-classical approach for bimolecular collisions, a numerical solution of the classical equations of motion is required [2].

To that end, we note that a formally exact classical expression for a unimolecular rate constant was given in Eq. (5.76).

7.2.2 Quantum mechanical theory

True unimolecular reactions

When an electromagnetic field is present, a term describing the coupling between the molecule and the field must be added to the Hamiltonian. The most important contribution related to absorption of electromagnetic radiation is

$$\hat{H}_I = -\boldsymbol{\mu} \cdot \boldsymbol{E}(t) \tag{7.23}$$

where $\boldsymbol{\mu}$ is the dipole-moment operator of the molecule and $\boldsymbol{E}(t)$ is the electric field usually given by

$$\boldsymbol{E}(t) = \boldsymbol{E}_0 a(t) \cos(\omega t) \tag{7.24}$$

The field is linearly polarized in the direction given by $\boldsymbol{E}_0$. $a(t)$ is an envelope function corresponding to a pulsed field; when $a(t)$ is constant $\boldsymbol{E}(t)$ describes a continuous wave (cw) field oscillating with the frequency $\omega = 2\pi\nu$.

When the field oscillates with a frequency that matches the energy spacing of the electronic states, all the relevant electronic states must be taken into consideration in the equation of motion. When, for example, two electronic states are coupled by the field, the time-dependent Schrödinger equation for the atomic nuclei takes the form

$$i\hbar\frac{\partial}{\partial t}\begin{bmatrix}\chi_1(\boldsymbol{R},t)\\ \chi_2(\boldsymbol{R},t)\end{bmatrix} = \begin{bmatrix}\hat{H}_1 & \hat{C}_{12}\\ \hat{C}_{21} & \hat{H}_2\end{bmatrix}\begin{bmatrix}\chi_1(\boldsymbol{R},t)\\ \chi_2(\boldsymbol{R},t)\end{bmatrix} \tag{7.25}$$

where $\hat{H}_1$ and $\hat{H}_2$ are nuclear Hamiltonians associated with the two electronic states, the coupling term is $\hat{C}_{12} = -\langle\psi_1|\boldsymbol{\mu}|\psi_2\rangle \cdot \boldsymbol{E}(t) \equiv -\boldsymbol{\mu}_{12} \cdot \boldsymbol{E}(t) = -\mu_{12}E(t)\cos\theta$, and

$$\mu_{12} = |\langle\psi_1|\boldsymbol{\mu}|\psi_2\rangle| \tag{7.26}$$

is the magnitude of the *electronic transition-dipole moment*, where ψ_i are electronic wave functions and integration is over all electronic coordinates. θ is the angle between the transition-dipole vector, $\boldsymbol{\mu}_{12}$, and the direction of the (linearly polarized) electric field, $\boldsymbol{E}_0$. Electronic transitions are clearly subject to the condition of a non-vanishing electronic transition-dipole moment. The initial state $\chi_1(\boldsymbol{R}, t_0)$, $\chi_2(\boldsymbol{R}, t_0) = 0$ is one of the nuclear eigenstates of the electronic ground state $i = 1$, for example, the vibrational (and rotational) ground state of the molecule. Equation (7.25) is a set of coupled first-order differential equations in time, where the coupling terms can be considered as an inhomogeneous term in each differential equation. The solution to such first-order differential equations can be written as a complete solution to the homogeneous equation plus a specific solution to the inhomogeneous equation:

$$\chi_1(\boldsymbol{R}, t) = \exp(-i\hat{H}_1 t/\hbar)\chi_1(\boldsymbol{R}, 0) - \frac{i}{\hbar}\int_0^t \exp(-i\hat{H}_1(t-t')/\hbar)\hat{C}_{12}(t')\chi_2(\boldsymbol{R}, t')dt'$$
$$\chi_2(\boldsymbol{R}, t) = -\frac{i}{\hbar}\int_0^t \exp(-i\hat{H}_2(t-t')/\hbar)\hat{C}_{21}(t')\chi_1(\boldsymbol{R}, t')dt' \tag{7.27}$$

i.e., we have replaced the differential equations in time by integral equations. We note that Eq. (7.27) reduces to the right solution when $\hat{C}_{12}(t') = 0$ and for $t = 0$. For arbitrary strengths of the electric field, Eq. (7.27) must be solved numerically.

When the interaction with the field is sufficiently weak, the second term on the right-hand side of the equation for $\chi_1(\boldsymbol{R}, t)$ can be neglected, and the wave packet associated with the excited electronic state is identical to the result obtained within the framework of so-called *first-order perturbation theory*, giving [3,4]

$$\chi_2(\boldsymbol{R}, t) = \frac{i}{\hbar}\int_0^\infty dt' e^{-i\epsilon_0 t'/\hbar} E(t')\phi(\boldsymbol{R}, t-t') \tag{7.28}$$

where $E(t)$ is the laser field, $\phi(\boldsymbol{R}, t-t') = \langle \boldsymbol{R}| \exp(-i\hat{H}_2(t-t')/\hbar)|\phi\rangle$ and

$$|\phi\rangle = \mu'_{12}|\chi_1(0)\rangle \tag{7.29}$$

$|\phi\rangle$ is often referred to as the *Franck–Condon wave packet* with $|\chi_1(0)\rangle$ being the initial ($t = 0$) stationary nuclear state in electronic state '1' with energy ϵ_0, and $\mu'_{12} = \mu_{12}\cos\theta$ is the projection of the electronic transition-dipole moment on the polarization vector of the electric field. The nature of the excited-state wave packet clearly depends on the form of the electric field.

Consider, e.g., $E(t) = E_0\delta(t - t_0)$, that is, an ultrashort pulse. According to Eq. (7.28), $\chi_2(\boldsymbol{R}, t)$ is then proportional to $\phi(\boldsymbol{R}, t - t_0)$. For $t = t_0$, the wave packet is the Franck–Condon wave packet, that is, the ground-state wave function times the transition-dipole moment is transferred to the excited state at $t = t_0$ and, subsequently, this wave packet evolves in time on the excited-state surface. The transition-dipole moment depends on the internuclear coordinates via the electronic wave functions. Neglecting this dependence is referred to as the *Condon approximation*. This is often an excellent approximation since, typically, the transition-dipole moment is essentially constant in the range of internuclear distances where the initial vibrational state is non-vanishing. The electronic transition described above is referred to as a 'vertical transition'—the energy is increased but the distribution of internuclear positions, described by $\chi_1(\boldsymbol{R}, t_0)$, is unchanged. Figure 7.2.3 illustrates the vertical transition

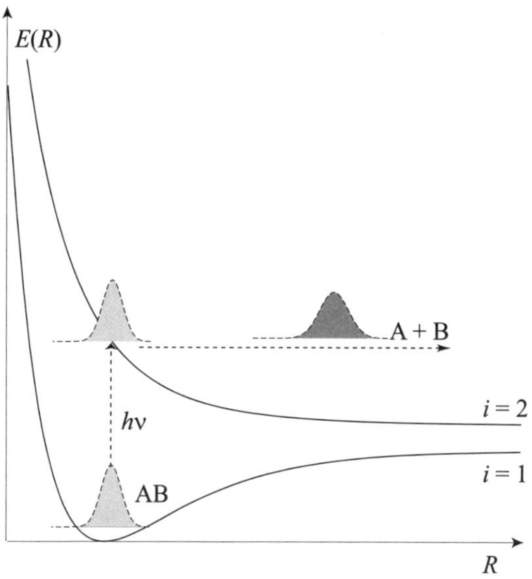

Fig. 7.2.3 A true unimolecular reaction; here photodissociation of a diatomic molecule.

(which is also referred to as the Franck–Condon principle) and the subsequent dynamics. Note that for the purpose of illustration it is assumed that $\mu_{12} = 1$, but typical values are of the order of 1 debye $= 3.3356 \times 10^{-30}$ Cm. The vibrational ground state is normally well described by a Gaussian wave function and the subsequent dynamics can (at least semi-quantitatively) be described by the well-known dynamics of a Gaussian wave packet (Section 4.2.1).

For a general form of $E(t)$, the excited-state wave function can be thought of as a coherent superposition of Franck–Condon wave packets promoted to the upper state at times t' with different weighting factors and phases. At time t each of these wave packets has evolved for a time $t - t'$.

The general outcome of photodissociation, say for a triatomic molecule,

$$\mathrm{ABC}(n) \longrightarrow \begin{cases} \mathrm{AB}(m) + \mathrm{C} \\ \mathrm{AC}(m) + \mathrm{B} \\ \mathrm{A} + \mathrm{B} + \mathrm{C} \\ \mathrm{A} + \mathrm{BC}(m) \end{cases} \tag{7.30}$$

is very similar to the bimolecular reaction described in Eq. (4.112). Photodissociation can be considered as a 'half collision', where the dynamics start in the transition-state region of the excited-state potential energy surface. State-to-state transition probabilities can be calculated from the projection of $\chi_2(\boldsymbol{R}, t \sim \infty)$ onto the various final states.

The coupling to the electromagnetic field depends on the orientation of the molecule, and this will be reflected in the *spatial distribution* of the products. Consider, as an example, a diatomic molecule with $\boldsymbol{\mu}_{12}$ parallel to the molecular axis. For a spherically symmetric initial state, the angular distribution is given by $(\cos\theta)^2$. That is, no products will show up in the direction perpendicular to the field $\boldsymbol{E}_0$, whereas most products are found parallel or anti-parallel to $\boldsymbol{E}_0$.

When $a(t) = 1$, the field $\boldsymbol{E}(t)$ in Eq. (7.24) describes a continuous wave with amplitude E_0. The transition probability to the excited state is given by $\langle \chi_2(t)|\chi_2(t)\rangle$, and in this case a constant transition probability per unit time is found (after a few oscillations of the electromagnetic field). For a *direct reaction*, this is equal to the *rate constant* of Eq. (7.5), $k_n(h\nu)$. Using Eq. (7.28), it is found [3,4] that

$$k_n(h\nu) = \frac{E_0^2}{2\hbar^2} \int_0^\infty \mathrm{Re}\left\{ e^{iE_l t/\hbar} \langle \phi|\hat{U}_2(t)|\phi\rangle \right\} dt \tag{7.31}$$

where $E_l = h\nu + \epsilon_0$, and

$$\hat{U}_2(t) = \exp(-i\hat{H}_2 t/\hbar) \tag{7.32}$$

is the propagator for nuclear motion in the excited electronic state, with the potential energy surface $E_2(R)$, which for a direct reaction does not support any bound states (as in Fig. 7.2.3). Again, the general features of the time evolution of the wave packet are equivalent to the description in Section 4.2. The rate constant is proportional to the strength of the electric field or, more precisely, the squared amplitude E_0^2, which can be shown to be related to the intensity I of the field, or the photon density. Within the Condon approximation, μ_{12} can be taken outside the integral and the rate then also becomes proportional to μ_{12}^2. Furthermore, it is easy to show that the integral is identical to the so-called Franck–Condon factors, i.e., overlaps between the initial stationary state and the stationary states of the excited electronic state.

Examples of photo-activated reactions that have been studied intensively are the direct $\mathrm{ICN} \to \mathrm{I} + \mathrm{CN}$, and the indirect $\mathrm{NaI} \to \mathrm{NaI}^* \to \mathrm{Na} + \mathrm{I}$.

Example 7.1: Photodissociation of NaI: an indirect reaction

NaI dissociates when it is excited from the electronic ground state to the first electronically excited state. In the excited state the molecule, NaI^*, is trapped for several vibrational periods, with a vibrational period close to 1 ps. The finite lifetime of NaI^* is due to non-adiabatic coupling to the electronic ground state. This coupling is largest when the energy spacing between the electronic states is at a minimum, which is at the so-called avoided crossing at an internuclear distance around 7 Å; see Fig. 7.2.4. Each time the wave packet passes through this avoided crossing, a small fraction of the wave packet leaks into $\mathrm{Na} + \mathrm{I}$. The transition can be described by the Landau–Zener formula, Eq. (4.182), with a transition probability that is about 10%.

The decay of NaI^* can be described in an alternative way [K.B. Møller, N.E. Henriksen, and A.H. Zewail, *J. Chem. Phys.* **113**, 10477 (2000)]. In the 'bound' region of the excited-state potential energy surface, one can define a discrete set of quasi-stationary states that are (weakly) coupled to the continuum states in the dissociation channel $\mathrm{Na} + \mathrm{I}$. These quasi-stationary states are also called resonance states and they have a finite lifetime due to the coupling to the continuum. Each quasi-stationary state has a time-dependent amplitude with a time evolution that can be expressed in terms of an effective (complex, non-Hermitian) Hamiltonian.

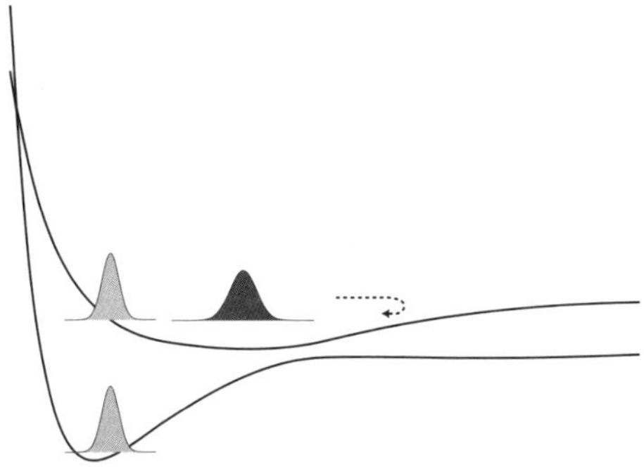

Fig. 7.2.4 Photodissociation of NaI. Potential energy curves for the electronic ground state and the first electronically excited state. The wave packet created by an ultrashort pulse and its subsequent dynamics is indicated.

The decay of the individual quasi-bound (metastable) resonance states follows an exponential law. The wave packet prepared by an ultrashort pulse can be represented as a (coherent) superposition of these states. The decay of the associated norm (i.e., population) follows a multi-exponential law with some superimposed oscillations due to quantum mechanical interference terms. The description given above is confirmed by experimental data.

Apparent unimolecular reactions

The basic equation of motion is again the time-dependent Schrödinger equation for the atomic nuclei, given by Eq. (4.101):

$$i\hbar\frac{\partial\chi(\boldsymbol{R},t)}{\partial t} = \hat{H}_i\chi(\boldsymbol{R},t) \tag{7.33}$$

where $\hat{H}_i = \hat{T}_{\text{nuc}} + E_i(\boldsymbol{R})$. A subscript '$i$' has been added here to the Hamiltonian, in order to emphasize that the equation describes the dynamics in this particular electronic state (the same index is on the electronic energy). In a thermally-activated unimolecular reaction, the dynamics is normally associated with the electronic ground state $i = 1$.

When the reaction probability of the activated molecule $A^*(E)$ is sufficiently small, see Example 7.1 for an illustration, one can introduce the concept of *quasi-stationary states*. These states, $|n\rangle$, associated with the slowly decaying 'bound states' of $A^*(E)$ are also called *resonance states*. The dynamics of the states is, essentially, given by [3] $\hat{U}(t)|n\rangle = e^{-i(E_n^0 - i\Gamma_n/2)t/\hbar}|n\rangle$. Note that, if $|n\rangle$ was a true stationary state, i.e., an eigenstate of the full Hamiltonian $\hat{H}_i$, then $\Gamma_n = 0$. The probability of observing the molecule in a quasi-stationary state is given by $|\hat{U}(t)|n\rangle|^2 = e^{-\Gamma_n t/\hbar}$; that is, a probability that decreases with time according to the anticipated exponential law, and with a rate constant that can be identified as $k_n = \Gamma_n/\hbar$.

The information about the decay of the resonance states can be extracted from a wave packet propagation. The initial state $\chi(\boldsymbol{R}, t_0) = \langle\boldsymbol{R}|\chi(0)\rangle$ is chosen such that it represents a highly excited vibrational state of the reactant at an energy $E > E_0$, and such that it has substantial overlap, $\langle n|\chi(0)\rangle$, with the resonance states and no overlap with the products. The lifetimes $\tau_n = \hbar/\Gamma_n$ can now be extracted from the *autocorrelation function* $\langle\chi(0)|\chi(t)\rangle$. Thus,

$$\begin{aligned}\langle\chi(0)|\chi(t)\rangle &= \langle\chi(0)|\hat{U}(t)|\chi(0)\rangle \\ &= \sum_n \langle\chi(0)|\hat{U}(t)|n\rangle\langle n|\chi(0)\rangle \\ &= \sum_n \langle\chi(0)|e^{-i(E_n^0 - i\Gamma_n/2)t/\hbar}|n\rangle\langle n|\chi(0)\rangle \\ &= \sum_n |\langle n|\chi(0)\rangle|^2 e^{-iE_n^0 t/\hbar} e^{-\Gamma_n t/2\hbar}\end{aligned} \tag{7.34}$$

The Fourier transformation of this autocorrelation function, $\int_{-\infty}^{\infty} e^{iEt/\hbar}\langle\chi(0)|\chi(t)\rangle dt$, gives an energy spectrum with lines centered at E_n^0, and line shapes of a Lorentzian form. The widths of these lines are $\Gamma_n = \hbar/\tau_n$ and the rate constants can then be obtained as $k_n = 1/\tau_n$.

Examples of unimolecular reactions that have been studied intensively are HCO^* $\rightarrow$ H + CO [S.K. Gray, *J. Chem. Phys.* **96**, 6543 (1992)] and $HO_2{}^* \rightarrow H + O_2$ [A.J. Dobbyn *et al.*, *J. Chem. Phys.* **104**, 8357 (1996)]. Larger molecules will have a high density of states and the resonances will most likely be overlapping, i.e., Γ_n can be larger than the energy spacing between adjacent lines, making it impossible to extract the decay rates of the individual resonance states.

7.3 Statistical theories

In *statistical theories* of unimolecular reactions, the rate is determined from an approach that does not involve any explicit consideration of the reaction dynamics.

The basic assumption in statistical theories is that the initially prepared state, in an *indirect* (true or apparent) unimolecular reaction $A^*(E) \rightarrow$ products, prior to reaction has relaxed (via IVR) such that any distribution of the energy E over the internal degrees of freedom occurs with the same probability. This is illustrated in Fig. 7.3.1, where we have shown a constant energy surface in the phase space of a molecule. Note that the assumption is equivalent to the basic 'equal a priori probabilities' postulate of statistical mechanics, for a microcanonical ensemble where every state within a narrow energy range is populated with the same probability. This uniform population of states describes the system regardless of where it is on the potential energy surface associated with the reaction.

An average decay rate $k(E)$ is obtained, since one does not distinguish between the various degenerate states.

7.3.1 The RRK theory

The RRK (after Rice, Ramsperger, and Kassel) theory is, like the Slater theory, a model for a unimolecular reaction rather than a faithful representation. The molecule is again represented by a collection of s uncoupled harmonic oscillators, which is an exact representation close to a stationary point on the potential energy surface. One of these

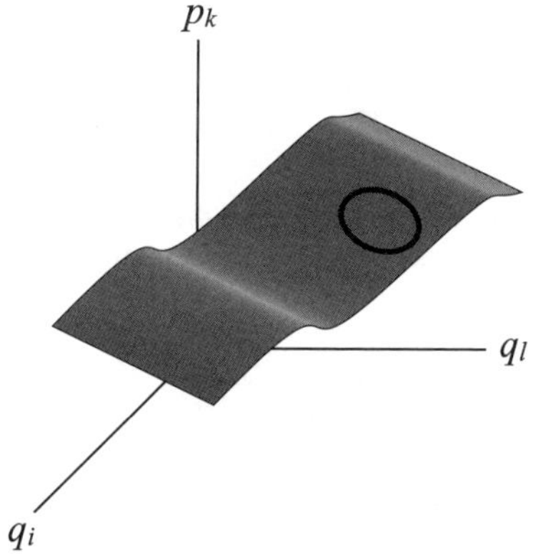

Fig. 7.3.1 A surface in the phase space where all states (corresponding to each point) have the same energy. Furthermore, all states are populated with the same probability, indicated by the uniform shading. A range of different molecular configurations is considered; the states within the closed curve can undergo unimolecular reaction.

oscillators, say oscillator number r, is associated with the reaction coordinate. Reaction occurs if the energy E_r exceeds the threshold energy of unimolecular reaction in the reaction coordinate, and the rate constant is then equal to the associated vibrational frequency, ν_r. As discussed in connection with Eq. (7.20), a frequency is the number of complete oscillations per unit time, and, together with the energy criterion, the rate constant is then equivalent to the number of times (per unit time) the amplitude of an oscillator attains a critical value. Thus, the microcanonical rate constant is, according to this model, given by

$$k(E) = \nu_r P_{E_r > E_0}(E) \tag{7.35}$$

where $P_{E_r>E_0}(E)$ is the probability that $E_r > E_0$ for oscillator number r at the total energy $E \geqslant E_r > E_0$.

The probability is calculated according to classical statistical mechanics (Appendix A.2). According to Eq. (A.49), the density of states (number of states per unit energy) for s uncoupled harmonic oscillators with frequencies ν_i is

$$N(E) = \frac{E^{s-1}}{(s-1)!\Pi_{i=1}^{s} h\nu_i} \tag{7.36}$$

The assumption about a uniform probability for any distribution of the energy between the harmonic oscillators may now be used to determine the probability $P_{E_r>E_0}(E)$. It can be expressed as the ratio between the density of states corresponding to the situation where the energy exceeds the threshold energy in the reaction coordinate and the total density of states at energy E, that is, $N(E)$ of Eq. (7.36).

Assume that the energy in the reaction coordinate is $E_r = E_0 + E'$, where $E' \in [0, E - E_0]$. Then the energy in the remaining $s-1$ vibrational degrees of freedom is $E-(E_0+E')$. The density of states corresponding to this particular partitioning of the energy is $(E-[E_0+E'])^{s-2}/((s-2)!\Pi_{i=1}^{s-1} h\nu_i) \times (E_0+E')^0 dE'/(h\nu_r)$, where $dE'/h\nu_r$

is the number of states in the reaction coordinate in the energy range $[(E_0+E'),(E_0+E')+dE']$, and the first term is the density of states in the remaining $s-1$ vibrational degrees of freedom at the energy $E-(E_0+E')$. Now we invoke the assumption that all partitionings of the total energy between the reaction coordinate and the remaining vibrational degrees of freedom are equally probable, and sum (integrate) over all partitionings of the energy in order to get the total density. Thus,[5]

$$P_{E_r>E_0}(E) = \frac{1}{N(E)}\int_0^{E-E_0} \frac{(E-[E_0+E'])^{s-2}}{(s-2)!\Pi_{i=1}^{s-1}h\nu_i}\frac{(E_0+E')^0}{h\nu_r}dE' \tag{7.37}$$

The probability can be written in the following form, using Eq. (7.36):

$$\begin{aligned} P_{E_r>E_0}(E) &= \frac{(s-1)}{E^{s-1}}\int_0^{E-E_0}(E-E_0-E')^{s-2}dE' \\ &= \frac{(s-1)}{E^{s-1}}\left[-\frac{(E-E_0-E')^{s-1}}{(s-1)}\right]_0^{E-E_0} \\ &= \left(\frac{E-E_0}{E}\right)^{s-1} \end{aligned} \tag{7.38}$$

The microcanonical rate constant takes, accordingly, the form

$$\boxed{k(E) = \nu_r\left(\frac{E-E_0}{E}\right)^{s-1}} \tag{7.39}$$

for $E>E_0$, and $k(E)=0$ otherwise. Note that $k(E)=\nu_r$ for $s=1$, and $k(E)\to\nu_r$ for $E/E_0\to\infty$ when $s>1$. Furthermore, at a fixed energy $E>E_0$, $k(E)$ decreases as a function of s, that is, as a function of the number of degrees of freedom (see Fig. 7.3.2) simply because the probability of having the 'right' distribution of the energy decreases with the number of oscillators.

When the energy levels are populated according to thermal equilibrium, we get according to Eq. (7.15) the rate constant

$$k(T) = \int_{E_0}^{\infty} k(E)P(E)dE \tag{7.40}$$

where $P(E)$ is the Boltzmann distribution for s uncoupled harmonic oscillators at temperature T. This distribution is given in Eq. (A.51), and

[5]Note that the integral is a so-called *convolution* of two densities, i.e.,

$$N_{\text{tot}}(E) = \int_{E_0}^{E} N_1(E-E_2)N_2(E_2)dE_2$$

It can be shown formally that the total density of states for two independent degrees of freedom with the densities $N_1(E)$ and $N_2(E)$ is obtained as a convolution of the densities.

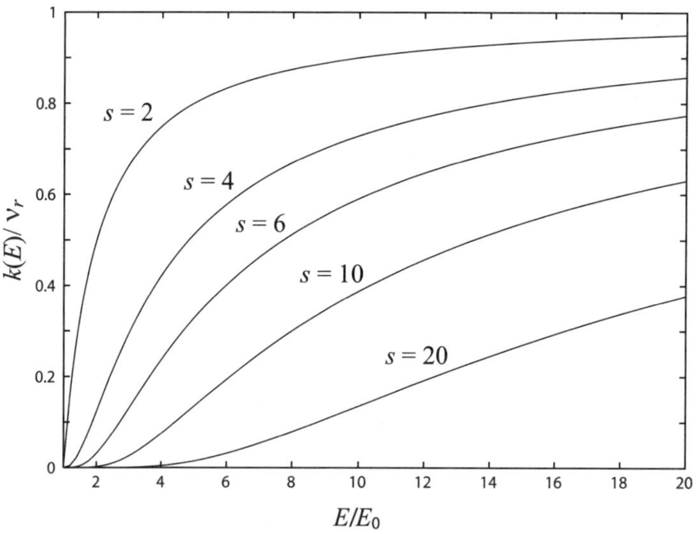

Fig. 7.3.2 The rate constant $k(E)$ according to RRK theory, for a molecule with s vibrational degrees of freedom.

$$
\begin{aligned}
k(T) &= \int_{E_0}^{\infty} k(E)\frac{1}{(s-1)!}\left(\frac{E}{k_BT}\right)^{s-1} \exp(-E/(k_BT))\frac{dE}{k_BT} \\
&= \frac{\nu_r}{(s-1)!}\int_{E_0}^{\infty}\left(\frac{E-E_0}{k_BT}\right)^{s-1} \exp(-E/(k_BT))\frac{dE}{k_BT} \\
&= \frac{\nu_r \exp(-E_0/(k_BT))}{(s-1)!}\int_0^{\infty} x^{s-1}e^{-x}dx
\end{aligned}
\tag{7.41}
$$

where the variable $x = (E - E_0)/(k_BT)$ was introduced in the last line. The integral in the last line defines the Gamma function $\Gamma(s)$, which is equal to $(s-1)!$. Thus,

$$\boxed{k(T) = \nu_r \exp(-E_0/(k_BT))} \tag{7.42}$$

which has the form of the Arrhenius equation with a constant prefactor. Vibrational frequencies (ν_r) are typically in the range 10^{13} to 10^{14} s^{-1}. Experimentally one often finds, however, pre-exponential factors that are larger than 10^{14} s^{-1}. Thus, a more complete theory is needed.

7.3.2 The RRKM theory

The RRKM (after Rice, Ramsperger, Kassel, and Marcus) theory is, basically, *transition-state theory* (see, in particular, the description in Section 6.2) applied to a unimolecular reaction. Thus, one focuses on the activated complex

$$A^* \longrightarrow A^{\ddagger} \longrightarrow \text{products} \tag{7.43}$$

which for a potential energy surface with a barrier is located at a saddle point. Unlike RRK theory, the reaction coordinate is then precisely identified as being the coordinate associated with the formation and decay of the activated complex, and the energy threshold is determined from the properties of the potential energy surface at the position of the complex; see Fig. 7.3.3.

The microcanonical rate constant $k(E)$

First, we want to derive an expression for the *microcanonical rate constant* $k(E)$ when the total internal energy of the reactant is in the range E to $E + dE$. From Eq. (7.43), the rate of reaction is given by the rate of disappearance of A or, equivalently, by the rate at which activated complexes $A^{\ddagger}$ pass over the barrier, i.e., the flow through the saddle-point region. The essential assumptions of RRKM theory are equivalent to the assumptions underlying transition-state theory.

Besides the assumption of separability of the reaction coordinate from the other degrees of freedom in the saddle-point region, and the assumption of classical dynamics along the reaction coordinate, there are two key assumptions in the derivation:

- all possible ways of partitioning a given total energy between the internal degrees of freedom of the activated complex and the translational energy of the reaction coordinate are equally probable; and
- a point of no return (corresponding to the activated complex) exists along the reaction coordinate, i.e., there are no *recrossings* of trajectories at this point.

The first assumption, that phase space is populated statistically prior to reaction, implies that the ratio of activated complexes to reactants is obtained by the evaluation of the ratio between the respective volumes in phase space. If this assumption is *not* fulfilled, then the rate constant $k(E, t)$ may depend on time and it will be different from $k_{\text{RRKM}}(E)$. If, for example, the initial excitation is localized in the reaction coordinate, $k(E, t)$ will be larger than $k_{\text{RRKM}}(E)$. However, when the initially prepared state has relaxed via IVR, the rate constant will coincide with the predictions of RRKM theory (provided the other assumptions of the theory are fulfilled).

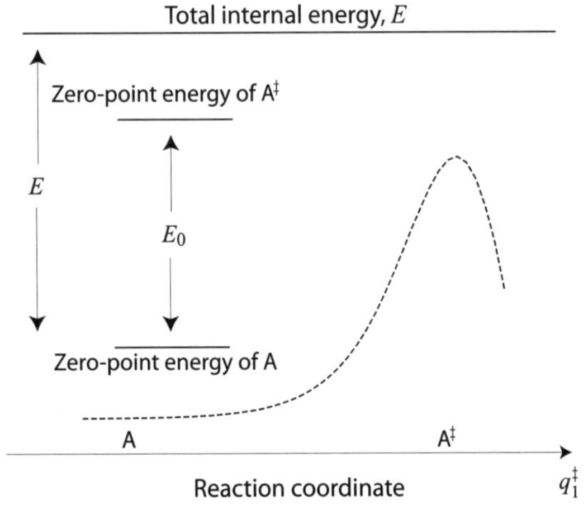

Fig. 7.3.3 Schematic illustration of the energetics in RRKM theory.

The derivation of the expression for $k(E)$ consists of the following four steps.

- In the first step, we define the relevant activated complexes as 'microcanonical transition states' having a total energy $H = E$ and a value for the reaction coordinate q_1 that lies between $q_1^\ddagger$ and $q_1^\ddagger + dq_1^\ddagger$. The separation of the reaction coordinate from the other degrees of freedom in the saddle-point region implies that the Hamiltonian in this region can be written as

$$H = H_{\text{trans}}^\ddagger + H^\ddagger \tag{7.44}$$

where

$$H_{\text{trans}}^\ddagger = \frac{(p_1^\ddagger)^2}{2\mu_1} = E_1^\ddagger \tag{7.45}$$

corresponding to free motion along the reaction coordinate, and $H^\ddagger$ is the Hamiltonian for the internal degrees of freedom of the activated complex, with the reaction coordinate omitted.

- In the second step, we evaluate the fraction of molecules in the transition state, for a particular partitioning of the total energy between the internal degrees of freedom of the activated complex and the translational energy associated with the reaction coordinate. This fraction of molecules in the transition state is evaluated as the number of states in the transition state divided by the total number of states of the reactant[6] and is denoted by $dN^\ddagger(q_1^\ddagger, p_1^\ddagger)/N(E)$. The number of states at the energy E is in quantum mechanics referred to as the *degeneracy*, and when the energy is continuous (i.e., classical mechanics) as the density of states, that is, the number of states per unit energy. We denote in the following $N(E)$ as a density of states.
According to classical statistical mechanics (Appendix A.2) $dq_1^\ddagger dp_1^\ddagger/h$ is the number of states in the reaction coordinate at the position $q_1^\ddagger$ and the momentum $p_1^\ddagger$ (corresponding to the particular energy $E_1^\ddagger$ in the reaction coordinate). The energy, measured relative to the zero-point level at the transition state, for all degrees of freedom except the reaction coordinate is $E' = E - E_1^\ddagger - E_0$, and we get

$$\frac{dN^\ddagger(q_1^\ddagger, p_1^\ddagger)}{N(E)} = \frac{(dq_1^\ddagger dp_1^\ddagger/h)N^\ddagger(E - E_1^\ddagger - E_0)}{N(E)} \tag{7.46}$$

where $N(E)$ is the density of states of the reactant and $N^\ddagger(E)$ is the density of states of the transition state (except for the reaction coordinate). Note that the 'microcanonical transition states' have a *fixed* value for the reaction coordinate $q_1^\ddagger$ (within an infinitesimal region $dq_1^\ddagger$), whereas the remaining coordinates as well as all the momenta can take *any* value as long as the total energy equals $H = E$.

- In the third step, the number of activated complexes that pass over the barrier per unit time is evaluated, assuming the same partitioning of the total energy as above, i.e., at the momentum $p_1^\ddagger$.
From Hamilton's equation of motion, $dq_1^\ddagger/dt = \partial H_{\text{trans}}^\ddagger/\partial p_1^\ddagger = p_1^\ddagger/\mu_1$, we have that $dq_1^\ddagger = (p_1^\ddagger/\mu_1)dt$, or $dt = dq_1^\ddagger/(p_1^\ddagger/\mu_1)$ which is the time it takes the activated

[6]The number of states in the transition state is neglected in the denominator, since this number will be negligible compared to the reactant phase space.

complexes to cross the transition-state region as defined by $dq_1^\ddagger$ (i.e., the 'lifetime' of the activated complex). The rate is then equal to $dN^\ddagger(q_1^\ddagger, p_1^\ddagger)$ divided by dt; that is, from Eq. (7.46),

$$\frac{dN^\ddagger(p_1^\ddagger)}{dt} = \frac{dE_1^\ddagger N^\ddagger(E - E_1^\ddagger - E_0)}{hN(E)} N(E) \tag{7.47}$$

where Eq.(7.45) was used, which implies that $dE_1^\ddagger = (p_1^\ddagger/\mu_1)dp_1^\ddagger$.
By identifying $dN^\ddagger(p_1^\ddagger)/dt$ with the rate of reaction, we used the assumption of a configuration of *no return*, which in the context of dynamics implies that it is not necessary to follow the trajectories all the way from the reactant to the product valley; it is sufficient to count the rate at which molecules pass through the critical configuration. When there is a barrier along the reaction path, the location of the barrier is a good choice for a point of no return. Trajectory calculations show that the assumption of no *recrossings* holds for trajectories with low energies that just make it past the barrier. In general, the assumption of no recrossings will overestimate the true rate.

- In the final step, the total number of activated complexes that pass over the barrier per unit time is evaluated, allowing all possible partitionings of the total energy; that is, we evaluate the total rate of reaction.
  This number is obtained by integration over $E_1^\ddagger$ from 0 to $E - E_0$ (corresponding to the situation where all available energy is in the reaction coordinate). Thus,

$$\frac{dN^\ddagger}{dt} = \frac{\int_0^{E-E_0} dE_1^\ddagger N^\ddagger(E - E_1^\ddagger - E_0)}{hN(E)} N(E) \tag{7.48}$$

In this step, we used the assumption that all possible ways of partitioning a given total energy between the internal degrees of freedom of the activated complex and the translational energy of the reaction coordinate are equally probable. In terms of dynamics, an efficient interchange of vibrational and translational energy along the reaction path is required. That is, it is assumed that there is no bias for the energy to go into particular degrees of freedom, i.e., a total randomization of the energy is required.
The rate constant $k(E)$ is defined by the relation $-dN(E)/dt = k(E)N(E)$, and from $dN^\ddagger/dt = -dN(E)/dt$ we may identify the rate constant in Eq. (7.48) to be

$$k(E) = \frac{\int_0^{E-E_0} dE_1^\ddagger N^\ddagger([E - E_0] - E_1^\ddagger)}{hN(E)} \equiv \frac{G^\ddagger(E - E_0)}{hN(E)} \tag{7.49}$$

where $G^\ddagger(E - E_0)$ is the *sum of states* in the transition state (with the reaction coordinate omitted) at the energy $E - E_0$; that is, the total number of states at or below the energy $E - E_0$. $N(E)$ is the *density of states* of the reactant.

We now introduce a correction/amendment to this expression. The (overall) rotational motion was implicitly neglected in the derivation of $k(E)$. The overall rotational energy is not active (available) in overcoming the energy threshold; therefore we subtract the rotational energy from the total energy in Eq. (7.49). The rotational energy is, however, not always constant when a molecule undergoes unimolecular reaction. Consider, as an example, unimolecular dissociation; the activated complex is usually larger than the stable molecule, since at least one of the bonds extends as the molecule dissociates; the activated complex therefore has larger moments of inertia. Since conservation of angular momentum requires that, as the molecule undergoes unimolecular reaction, it does not change its total angular momentum quantum number J, the total rotational energy of the complex $E_r^\ddagger$ is then less than the rotational energy in the reactant E_r (consider, e.g., the rotational energy of a linear molecule, $E_J = \hbar^2 J(J+1)/2I$). The total internal energy is conserved and therefore the difference in rotational energy is released into the vibrational degrees of freedom of the activated complex. Thus, we subtract the overall rotational energy, for the reactant and the activated complex, respectively:

$$\boxed{k_{\mathrm{RRKM}}(E) = \frac{G^\ddagger(E - E_0 - E_r^\ddagger)}{hN(E - E_r)}} \tag{7.50}$$

where $E = E_v + E_r$ is the sum of vibrational and rotational energy and $E - E_0 = E^\ddagger + E_r^\ddagger$, where $E^\ddagger$ is the available energy of the activated complex excluding rotational energy. Note that Planck's constant h has the unit [energy × time], the sum of states is dimensionless, the density of states has the unit $[\text{energy}]^{-1}$, that is, the rate constant has the proper unit for a unimolecular rate constant, i.e., $[\text{time}]^{-1}$.

Motion along the reaction coordinate was limited to classical mechanics, whereas the sum and density (or, to be precise, the degeneracy) of states should be evaluated according to quantum mechanics. The integral in Eq. (7.49) should really be replaced by a sum; $N^\ddagger(E)$ is not a continuous function of the energy, but due to the quantization of energy, it is only defined at the allowed quantum levels of the activated complex. That is, the sum of states $G^\ddagger(E^\ddagger)$ should be calculated exactly by a direct count of the number of states:

$$G^\ddagger(E^\ddagger) = \sum_i \Theta(E^\ddagger - E_i^\ddagger) \tag{7.51}$$

where the sum runs over all states of energy $E_i^\ddagger$ of the activated complex, and the unit step function is defined in Eq. (6.30). Thus, according to RRKM theory, the rate constant increases in a stepwise manner as the internal energy increases; see Fig. 7.3.4. This prediction has been confirmed experimentally [5,6]. Since the density of states $N(E_v)$ is essentially continuous at high energies ($E > E_0$), a classical evaluation of this quantity suffices in most cases. It will be evaluated in the following.

An additional consequence of the omission of rotation is that symmetry numbers (see Appendix A.1.1) were ignored in calculating the sum and density of states. Thus, the sum of states of the activated complex and the density of states of the reactant should be divided by their symmetry numbers, $\sigma^\ddagger$ and σ, respectively.

The information needed for calculating the sum and density of states in Eq. (7.50) includes the barrier height corrected for zero-point energies E_0 (see Eq. (6.12)), and the geometry and the vibrational frequencies of the activated complex and the reactant. From the geometry one can determine the moments of inertia and the rotational energy for a given rotational quantum state. Obtaining the structure and properties of the activated complex including E_0 is the most difficult part. These properties can

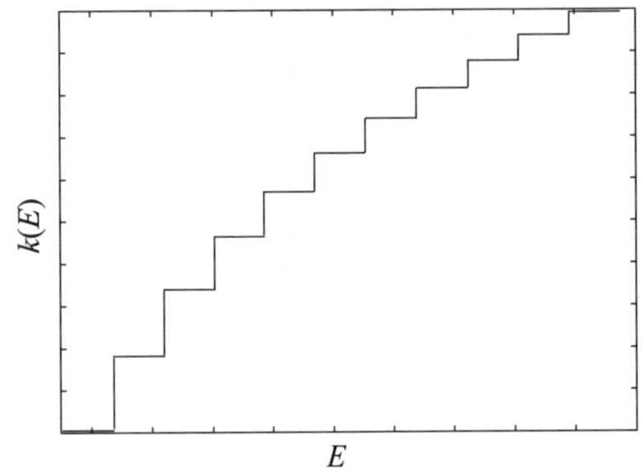

Fig. 7.3.4 Schematic illustration of the rate constant $k(E)$ according to RRKM theory.

all be derived from the potential energy surface, and with modern *ab initio* programs (Chapter 3) this information can be obtained with high accuracy, especially for molecules where the number of electrons is fairly small. There are two general situations: (i) a saddle point exists and the above-mentioned programs can directly determine the properties of the activated complex; (ii) no saddle point exists on the potential energy surface—this happens quite often for unimolecular reactions, see, e.g., Fig. 3.1.6. In the second case where there is no barrier on the surface we have to generalize our definition of the activated complex.

To that end, *variational transition-state theory* has been introduced, which is based on Wigner's variational theorem, Eq. (5.10). When a saddle point exists, it represents a 'bottleneck' between reactant and products. It is the point along the reaction coordinate where we have the smallest rate of transformation from the reactant to products. This can be seen from Eq. (7.50), where it should be noted that only the sum of states $G^{\ddagger}(E^{\ddagger})$ changes as the reaction proceeds along the reaction coordinate. We have the smallest sum of states of the activated complex $G^{\ddagger}(E^{\ddagger})$ on top of the barrier because at this point the available energy $E^{\ddagger}$ is at a minimum; see Fig. 7.3.3.

In the absence of a saddle point, we have to search for the configuration corresponding to a minimum in the sum of states. The potential energy will in this case be constantly rising as the reaction proceeds from the configuration of the stable reactant to the products; see Fig. 7.1.2 for an illustration. The existence of a minimum in the sum of states is due to the interplay between the constantly rising potential energy and the decrease in the vibrational frequencies that are evolving into product rotations and translations. As the reaction proceeds, the reduction in the available energy will reduce the sum of states while the lowering of the vibrational frequencies increases the sum of states. These opposing factors result in a minimum in the sum of states at some point along the reaction coordinate. Thus, in variational transition-state theory the rate constant, that is, $G^{\ddagger}(E^{\ddagger})$, is calculated as a function of the reaction coordinate, and the minimum identifies the activated complex.

More explicit insights into the predictions of RRKM theory are obtained when the following approximations are introduced:

- rotational motion is neglected;
- the sum and density of vibrational states are calculated classically.

From Appendix A.2, we have classical expressions for the sum and density of states of s uncoupled harmonic oscillators. Thus, the sum of states is

$$G(E) = \frac{E^s}{s!\Pi_{i=1}^s h\nu_i} \tag{7.52}$$

and the density of states is

$$\begin{aligned} N(E) &= dG(E)/dE \\ &= \frac{E^{s-1}}{(s-1)!\Pi_{i=1}^s h\nu_i} \end{aligned} \tag{7.53}$$

Now, if there are s degrees of freedom in the reactant with frequencies ν_i, there are $s-1$ degrees of freedom in the activated complex with frequencies $\nu_i^\ddagger$, when the reaction coordinate is excluded. When the above approximations are introduced in RRKM theory, we get

$$\begin{aligned} k(E) &= \frac{G^\ddagger(E-E_0)}{hN(E)} \\ &= \frac{(E-E_0)^{s-1}}{(s-1)!\Pi_{i=1}^{s-1} h\nu_i^\ddagger} \Bigg/ h\frac{E^{s-1}}{(s-1)!\Pi_{i=1}^s h\nu_i} \\ &= \frac{\Pi_{i=1}^s \nu_i}{\Pi_{i=1}^{s-1}\nu_i^\ddagger}\left(\frac{E-E_0}{E}\right)^{s-1} \end{aligned} \tag{7.54}$$

The dimension of the factor $\Pi_{i=1}^s \nu_i/\Pi_{i=1}^{s-1}\nu_i^\ddagger$ is that of a frequency. If the frequencies of the reactant and the activated complex are not too different, this frequency is roughly a typical vibrational frequency ν_r (typically in the range 10^{13} to 10^{14} s^{-1}). Since the energy-dependent factor is less than one, we have that the microcanonical rate constant $k(E) < \nu_r$, i.e., it is less than a typical vibrational frequency. The energy dependence as a function of the number of vibrational degrees of freedom was illustrated in Fig. 7.3.2, and as shown previously in Eq. (7.38) it can be interpreted as the probability that the energy in one out of s vibrational modes exceeds the energy threshold E_0 for the reaction. Note that if we make the identification $\nu_r \sim \Pi_{i=1}^s \nu_i/\Pi_{i=1}^{s-1}\nu_i^\ddagger$, we have recovered RRK theory, Eq. (7.39), from RRKM theory.

The classical evaluation of the sum and density of vibrational states can introduce quite large errors, especially at low energies and for molecules with many vibrational degrees of freedom. Equation (7.52) gives the sum of vibrational states when we have a vibrational energy E, measured relative to the lowest possible vibrational energy, i.e., zero in this classical description. Consider, as a simple example, the evaluation of the sum of states for a single harmonic oscillator. According to Eq. (7.52), the classical result is $G_{\rm cl}(E) = E/(h\nu_1)$, whereas the quantum result according to Eq. (7.51) is an integer given by $G_{\rm qm}(E) = [E/(h\nu_1) + 1]_{\rm int}$; the subscript 'int' means the integer part and E is the energy in excess of the lowest possible vibrational energy $E_z = h\nu_1/2$.

We see that the classical expression in Eq. (7.52) underestimates the true number of states. At low energies, the relative error introduced by the classical expression is quite large, even for this simple example where only a single oscillator is considered. The two expressions are similar when the energy is large compared to $h\nu_1$.

A simple correction to the classical approach to the evaluation of the sum of vibrational states has been suggested. As before, the sum of states is evaluated at a vibrational energy E, measured relative to the lowest possible energy, and we now use that, according to quantum mechanics, this vibrational energy is E_z, i.e., the zero-point energy. We can obtain an improved estimate of the sum of states simply by replacing E by $E + E_z$ in Eq. (7.52), where E_z is the zero-point energy (see Appendix A.2.1 and Problems 7.3 and 7.6). This approach is an improvement but still not highly accurate. For a detailed account on the evaluation of sums and densities of molecular quantum states, the reader should consult the literature [1,7,8].

Example 7.2: The lifetime of CH_3CO

Using femtosecond spectroscopy, the lifetime of the acetyl radical (CH_3CO),

$$CH_3CO \longrightarrow CH_3 + CO$$

has been measured as a function of the internal energy [*J. Chem. Phys.* **103**, 477 (1995)]. We can estimate the lifetime via RRKM theory.

The potential energy surface for the reaction has been calculated and the classical barrier height associated with the activated complex is $E_{cl} = 90.8$ kJ/mol. The relevant vibrational frequencies are given in the table below. (Note that the imaginary frequency of the activated complex is not included in the table.)

CH_3CO $\tilde{\nu}_i/\text{cm}^{-1}$	$(CH_3CO)^\ddagger$ $\tilde{\nu}_i/\text{cm}^{-1}$
3193	3325
3188	3321
3175	3225
1928	2027
1478	1452
1477	1445
1361	901
1062	568
960	566
884	276
468	43
101	

First, we calculate the barrier height (E_0), corrected for vibrational zero-point energies, using Eq. (6.12). Thus (1 cm^{-1} = 0.01196 kJ/mol),

$$\begin{aligned} E_0 &= 90.8 \text{ kJ/mol} + (8574.5 - 9637.5) \times 0.01196 \text{ kJ/mol} \\ &= 78.2 \text{ kJ/mol} \end{aligned}$$

i.e., a significant difference between E_0 and E_{cl}.

Second, we calculate the unimolecular rate constant at the internal energy E via the RRKM theory. We use Eq. (7.54), where the rotational energy is neglected and where the sum and density of vibrational states are evaluated classically. Thus at $E = 184$ kJ/mol we get

$$\begin{aligned} k(E) &= \frac{\Pi_{i=1}^{s}\nu_i}{\Pi_{i=1}^{s-1}\nu_i^{\ddagger}} \left(\frac{E-E_0}{E}\right)^{s-1} \\ &= \frac{\Pi_{i=1}^{12}\nu_i}{\Pi_{i=1}^{11}\nu_i^{\ddagger}} \left(\frac{184-78.2}{184}\right)^{11} \\ &= 15147.8 \times 2.27 \times 10^{-3}\ \text{cm}^{-1} \\ &= 1.03 \times 10^{12}\ \text{s}^{-1} \end{aligned}$$

where in the last line we have used the connection between frequency and wave number, $\nu = c\tilde{\nu}$.

The lifetime expressed as the half-life time, $\tau = \ln 2/k$, is then 671 fs.

Before closing the general discussion of Eq. (7.50), we note that quite often a molecule can dissociate (or isomerize) to *more than one* set of products; for a triatomic molecule, e.g.,

$$\text{ABC}^* \longrightarrow \begin{cases} \text{AB} + \text{C} \\ \text{A} + \text{BC} \end{cases} \tag{7.55}$$

which means that the reaction may proceed via several activated complexes. Then, for each reaction, we have a rate constant given by Eq. (7.50), and the total decay rate of ABC is given by the sum of all the dissociation rate constants. The branching ratio between the channels is given by the ratio of the rate constants (see, e.g., Problem 7.7).

An important question is: how accurate is RRKM theory? One answer comes from detailed dynamical calculations on small molecules. These studies have, e.g., shown that the energy dependence of the rate constant can be much stronger than predicted by RRKM theory. For example, the decay rates of the individual resonance states in $HO_2{}^* \rightarrow H + O_2$ can vary over several orders of magnitude even within a very narrow energy range. However, when the rates are convoluted with a Gaussian with a width corresponding to a few hundred cm^{-1}, the rate constant agrees well with the predictions of RRKM theory. Larger molecules will have a high density of states and the resonances will most likely be overlapping (see Section 7.2.2), making it impossible to extract the decay rates of the individual resonance states. Thus, the question of the accuracy of RRKM theory must be analyzed in comparison to experimental data. Here, a number of studies show that the microcanonical rate constant predicted by RRKM theory agrees with experiments, within the experimental uncertainty.

The canonical rate constant $k(T)$

Finally, when the energy levels are populated according to thermal equilibrium, we get the rate constant $k(T)$ by applying Eq. (7.15):

$$k(T) = \int_{E_0}^{\infty} k(E)P(E)dE \tag{7.56}$$

where $P(E)$ is the Boltzmann distribution, which in the following calculation we use in its classical statistical mechanical form, i.e., $P(E) = N(E)\exp(-E/k_BT)/Q$, according to Eq. (A.6), where $N(E)$ is the density of states.

The total internal energy is written in the form $E = E^\ddagger + E_0$, and with $k(E)$ given by Eq. (7.49) we obtain[7]

$$\begin{aligned}
k(T) &= \frac{1}{hQ}\int_{E_0}^{\infty} G^\ddagger(E - E_0)e^{-E/k_BT}dE \\
&= \frac{e^{-E_0/k_BT}}{hQ}\int_0^{\infty} G^\ddagger(E^\ddagger)e^{-E^\ddagger/k_BT}dE^\ddagger \\
&= -k_BT\frac{e^{-E_0/k_BT}}{hQ}\int_0^{\infty} G^\ddagger(E^\ddagger)\left[\frac{d}{dE^\ddagger}e^{-E^\ddagger/k_BT}\right]dE^\ddagger \\
&= -k_BT\frac{e^{-E_0/k_BT}}{hQ}\left\{\left[G^\ddagger(E^\ddagger)e^{-E^\ddagger/k_BT}\right]_0^{\infty} - \int_0^{\infty}\frac{dG^\ddagger(E^\ddagger)}{dE^\ddagger}e^{-E^\ddagger/k_BT}dE^\ddagger\right\} \\
&= \frac{k_BT}{h}\frac{\int_0^{\infty} N^\ddagger(E^\ddagger)e^{-E^\ddagger/k_BT}dE^\ddagger}{Q}e^{-E_0/k_BT}
\end{aligned} \tag{7.57}$$

where partial integration and $G^\ddagger(E^\ddagger = 0) = 0$ was used in the fourth line and in the last line $N^\ddagger(E^\ddagger) = dG^\ddagger(E^\ddagger)/dE^\ddagger$. The integral is identified as $Q^\ddagger$ and the well-known expression of transition-state theory is obtained:

$$\boxed{k(T) = \frac{k_BT}{h}\frac{Q^\ddagger}{Q}\exp(-E_0/(k_BT))} \tag{7.58}$$

where $Q^\ddagger$ and Q are vibrational–rotational partition functions for the activated complex (with the reaction coordinate omitted) and the reactant. The expression in Eq. (7.58) is often referred to as the unimolecular rate constant in the high-pressure limit. For apparent unimolecular reactions activated by inelastic collisions, it is shown below (Section 7.4.1) that under high-pressure conditions an equilibrium (Boltzmann) distribution of reactants is maintained. This condition is equivalent to the assumption behind Eq. (7.56).

From Eq. (7.58), we see that the pre-exponential factor in $k(T)$, roughly, corresponds to a typical vibrational frequency, $k_BT/h = 6.25 \times 10^{12}$ s^{-1} at $T = 300$ K. We can also understand why the pre-exponential factors can be larger than typical vibrational frequencies (which could not be accounted for in RRK theory). If we, for a moment, assume that the rotational contributions to the partition functions cancel (corresponding to a negligible change in the geometry between the reactant and the activated complex) and focus on the vibrational degrees of freedom in the classical (high-temperature) limit, we get, according to Eq. (A.50),

[7]The same expression for $k(T)$ can be derived from Eq. (7.50).

$$k(T) \sim \frac{k_BT}{h}\frac{(k_BT)^{s-1}/\Pi_{i=1}^{s-1}h\nu_i^{\ddagger}}{(k_BT)^s/\Pi_{i=1}^{s}h\nu_i}\exp(-E_0/(k_BT))$$
$$= \frac{\Pi_{i=1}^{s}\nu_i}{\Pi_{i=1}^{s-1}\nu_i^{\ddagger}}\exp(-E_0/(k_BT)) \tag{7.59}$$

Thus, we see again that the pre-exponential factor in $k(T)$, roughly, corresponds to a typical vibrational frequency, but if $\Pi_{i=1}^{s-1}\nu_i^{\ddagger} < \Pi_{i=1}^{s-1}\nu_i$ the pre-exponential factor will be larger than the vibrational frequency ν_s. This corresponds to a 'loose' transition state, as illustrated by Example 7.2.

It should be noted that the result in Eq. (7.59) is strictly valid only in the classical high-temperature limit which, except for very high temperatures, is not well satisfied for typical vibrational frequencies. Qualitatively, a similar result will also be obtained when the exact vibrational partition functions are used in Eq. (7.58). Rotational contributions were also neglected, but the moments of inertia associated with the activated complex are often larger than the moments of inertia of the reactant. Thus, we have very often that $Q^{\ddagger} > Q$, and large pre-exponential factors may often arise due to a 'loose' transition state as well as due to a substantial change in geometry between the reactant and the activated complex.

7.4 Collisional activation and reaction

In a true unimolecular reaction, the energized molecules are formed by absorption of electromagnetic radiation; see Section 7.2.2. In an apparent unimolecular reaction, the first step is the formation of the energized molecules by collisions with other molecules. We consider in the following subsection the interplay between the formation of energized molecules—by bimolecular collision—and their subsequent reaction.

7.4.1 Inelastic collisions, Lindemann mechanism

The energy needed for activation of a molecule can be provided by collisions with other molecules, as in

$$\mathrm{M} + \mathrm{A}(n) \longrightarrow \mathrm{A}(m) + \mathrm{M} \tag{7.60}$$

where M is an (inert) buffer gas or any molecule that does not react with the A molecule, which could be A itself. We distinguish between an *elastic collision* process, if quantum states n and m are identical, and an *inelastic collision* process when $n \neq m$. Note that inelastic collisions are equivalent to energy transfer between molecules. In the present case, for example, there will be transfer of relative translational energy between M and A to vibrational energy in A, when $m > n$. The dynamics of such an inelastic collision process can, of course, be described in detail by the equation of motion of the atomic nuclei, the time-dependent Schrödinger equation in Eq. (1.10).

Activation by inelastic collisions is also called *thermal activation*. After a large number of collisions, a distribution over internal (rotational and vibrational) states will be established as given by the Boltzmann distribution at the given temperature.

The activation and its interplay with chemical reaction is often described by the so-called (generalized) Hinshelwood–Lindemann mechanism:

$$\begin{aligned}\mathrm{A}+\mathrm{M} &\overset{dk_1(E)}{\longrightarrow} \mathrm{A}^*(E)+\mathrm{M}\\ \mathrm{A}^*(E)+\mathrm{M} &\overset{k_{-1}}{\longrightarrow} \mathrm{A}+\mathrm{M}\\ \mathrm{A}^*(E) &\overset{k(E)}{\longrightarrow} \text{products}\end{aligned} \tag{7.61}$$

where $dk_1(E)$ is the rate constant for activation of the molecule to an energy in the range $[E, E + dE]$. Note that k_{-1} is assumed to be independent of the energy. We introduce the steady-state approximation for the activated molecule:

$$\begin{aligned}\frac{d[\mathrm{A}^*(E)]}{dt} &= dk_1(E)[\mathrm{A}][\mathrm{M}] - k_{-1}[\mathrm{A}^*(E)][\mathrm{M}] - k(E)[\mathrm{A}^*(E)]\\ &= 0\end{aligned} \tag{7.62}$$

which implies

$$[\mathrm{A}^*(E)] = \frac{dk_1(E)[\mathrm{A}][\mathrm{M}]}{k_{-1}[\mathrm{M}] + k(E)} \tag{7.63}$$

The rate of reaction is

$$\begin{aligned}\text{Rate} &= k(E)[\mathrm{A}^*(E)]\\ &= \frac{k(E)dk_1(E)[\mathrm{M}]}{k_{-1}[\mathrm{M}] + k(E)}[\mathrm{A}]\\ &\equiv k_{\text{uni}}(E)[\mathrm{A}]\end{aligned} \tag{7.64}$$

The apparent unimolecular rate constant k_{uni} may be written

$$\begin{aligned}k_{\text{uni}}(E) &= \frac{k(E)dk_1(E)[\mathrm{M}]}{k_{-1}[\mathrm{M}] + k(E)}\\ &= \frac{k(E)dk_1(E)/k_{-1}}{1 + k(E)/(k_{-1}[\mathrm{M}])}\\ &= \frac{k_{\text{uni}}^{\infty}}{1 + k_{\text{uni}}^{\infty}/(dk_1(E)[\mathrm{M}])}\end{aligned} \tag{7.65}$$

where $k_{\text{uni}}^{\infty}(E)$ is the rate constant in the high-pressure limit ($[\mathrm{M}] \to \infty$), which is given by

$$\begin{aligned}k_{\text{uni}}^{\infty}(E) &= \lim_{[\mathrm{M}]\to\infty} k_{\text{uni}}(E)\\ &= k(E)(dk_1(E)/k_{-1})\end{aligned} \tag{7.66}$$

Note that $k_{\text{uni}}(E)$, in general, is a function of [M].

In high pressure ($[\mathrm{M}] \to \infty$), the rate in Eq. (7.64) is given by a first-order expression in [A] with rate constant $k_{\text{uni}}^{\infty}(E)$. In the low-pressure limit ($[\mathrm{M}] \to 0$) we find

$$\begin{aligned}k_{\text{uni}}^{0}(E) &= \lim_{[\mathrm{M}]\to 0} k_{\text{uni}}(E)\\ &= dk_1(E)[\mathrm{M}]\end{aligned} \tag{7.67}$$

and the rate in Eq. (7.64) is given by a second-order expression, when the activation step is the rate-limiting step.

The ratio of rate constants in Eq. (7.66) has a simple interpretation. When the rate of reaction is small compared to the rates of activation/deactivation, we have from Eq. (7.62) $dk_1(E)[\mathrm{A}][\mathrm{M}] = k_{-1}[\mathrm{A}^*(E)][\mathrm{M}]$, which implies

$$\begin{aligned} dk_1(E)/k_{-1} &= [\mathrm{A}^*(E)]/[\mathrm{A}] \\ &= P(E)dE \end{aligned} \qquad (7.68)$$

where $P(E)dE$ is the probability of finding an A molecule in the energy range $[E, E+dE]$.

To obtain the thermal rate constant, we must integrate $k_{\mathrm{uni}}(E)$ over energy. In the high-pressure limit $[\mathrm{M}] \to \infty$, we use Eqs (7.66) and (7.68), and obtain

$$k_{\mathrm{uni}}^{\infty}(T) = \int_{E_0}^{\infty} k(E)P(E)dE \qquad (7.69)$$

and the expression is identical to Eq. (7.15).

In the low-pressure limit $[\mathrm{M}] \to 0$, we use Eqs (7.67) and (7.68), and obtain

$$k_{\mathrm{uni}}^{0}(T) = k_{-1}[\mathrm{M}] \int_{E_0}^{\infty} P(E)dE \qquad (7.70)$$

Thus, the rate is in this limit is proportional to the fraction of molecules with energies above the energy threshold for unimolecular reaction.

7.5 Detection and control of chemical dynamics

Recently, two basic questions of chemical dynamics have attracted much attention: first, is it possible to detect ('film') the nuclear dynamics directly on the femtosecond time scale; and second, is it possible to direct (control) the nuclear dynamics directly as it unfolds? These efforts of real-time detection and control of molecular dynamics are also known as *femtosecond chemistry*. Most of the work on the detection and control of chemical dynamics has focused on unimolecular reactions where the internuclear distances of the initial state are well defined within, of course, the quantum mechanical uncertainty of the initial vibrational state. The discussion in the following builds on Section 7.2.2, and we will in particular focus on the real-time control of chemical dynamics. It should be emphasized that the general concepts discussed in the present section are not limited to reactions in the gas phase.

7.5.1 Detection of chemical dynamics

We know that the key events of chemical reactions take place on a femtosecond time scale. Recent fundamental breakthroughs in experimental methods have made it possible to resolve—in real time—the transformation from reactants to products via transition states [9–11]. This spectacular achievement was made possible by the development of femtosecond lasers, that is, laser pulses with a duration as short as a few femtoseconds.

To monitor fast events in time, firstly, a precise zero of time must be established. To that end, a femtosecond pulse is used to excite the system and initiate the dynamics. This pulse is called a *pump pulse*, and a nuclear wave packet is created as described in Section 7.2.2. Secondly, the different dynamical states must be detected as the dynamics unfolds. Two approaches can, in principle, be taken: (1) the detector operates in a continuous mode and it has to be fast enough to resolve every step of the dynamical process, or (2) the detector operates in a pulsed mode, i.e., a short (femtosecond) pulse of radiation is used to detect ('illuminate') the instantaneous state of the system. In the latter approach, the pulse is called a *probe pulse*. The probe pulse must arrive at the system at a well-defined time delay with respect to the pump pulse (this delay can be introduced by letting the probe pulse travel a slightly longer distance than the pump pulse). The *pump–probe approach* has so far been the only method that can resolve dynamics in the femtosecond regime.

Various detection methods can be used, depending on the (central) frequency of the probe pulse. So far the shortest pulses have been laser pulses generated at wavelengths close to visible light. In this case, the probe pulse can be tuned to be in resonance with an electronic transition, coupling the electronic state where the wave packet was created with another higher-lying electronic state. Note that, since the energy difference between electronic states will depend on the internuclear distances, the resonance condition is associated with a certain set of distances. The radiation of the probe pulse will then be absorbed by the molecular system, at the time where the wave packet created by the pump pulse has reached the internuclear positions corresponding to resonance. An early application of this approach was to the direct photodissociation of ICN into I + CN, where it was possible to detect the different stages of the bond breaking and to witness the 'birth' of CN. Another method of detection is based on ultrafast scattering, e.g., x-ray scattering; that is, an extension of the well-known structural determination for static structures into dynamical non-equilibrium structures. An advantage of the scattering method is that the relation between the internuclear distances and the time-resolved signal is more direct, i.e., a detailed knowledge of various excited electronic states is not required.

In the theoretical analysis of such experiments, the finite duration of the pulses must be taken into account and, consequently, that nuclear motion might occur during probing [12].

7.5.2 Control of chemical dynamics

A central objective of chemistry is how to 'steer' the reactants into a particular desired product. In order to attain this goal, the reaction might be assisted by a catalyst, and perhaps by a proper choice of solvent. The outcome of chemical reactions is, in addition, determined by the energy that is supplied to the reactants.

The way that energy is injected into the molecular system affects the dynamics, i.e., the motion under the influence of the forces, and hence the outcome of the reaction. Traditionally, the necessary energy of activation that is required in order for a chemical reaction to take place is supplied in the form of heat, i.e., by thermal activation. This approach gives rise to a distribution of population over the molecular energy states that follows the Boltzmann distribution. Alternatively, the energy can be supplied in the form of electromagnetic radiation, as it is known from traditional photochemistry where a monochromatic light source can create an energy state in the molecular system with a very well-defined energy. With short laser pulses of so-called coherent electromagnetic radiation it is possible to create dynamical states in molecules that cannot be created by heating or by traditional photochemistry. Experimentally, one can now 'design' tailored laser pulses with a duration that is shorter than typical vibrational periods. This provides, in principle, new possibilities for controlling molecular dynamics and chemical reactions directly on the femtosecond time scale where they take place.

Thus, the key point is [13–15] to circumvent the Boltzmann distribution (i.e., 'the temperature') by replacing the 'heat bath' by a controllable electromagnetic field. This generates distributions over the molecular states that are inaccessible when energy is supplied in the form of heat. In addition, when the electromagnetic field is generated by a laser there is a more fundamental difference: laser excitation is a coherent excitation of states; this is explained in more detail below.

Example 7.3: Controlling the chemical nature of products

Consider the reaction

$$
\begin{array}{lll}
 & & \mathrm{H} + \mathrm{OD} \\
 & \nearrow & \\
\mathrm{HOD} & & \\
 & \searrow & \\
 & & \mathrm{D} + \mathrm{OH}
\end{array}
$$

The potential energy surfaces of the ground as well as the first electronically excited state of HOD are shown in Fig. 7.1.1. When the photodissociation is induced by ultraviolet (UV) light corresponding to an excitation to the first electronically excited state of HOD, the branching ratio between H + OD and D + OH depends on the frequency of the radiation but in such a way that one will always get, at least, about twice as much H + OD as D + OH.

As in Example 4.2, the branching ratio between the two product channels can be controlled by appropriate vibrational pre-excitation of HOD [16]. For example, when the initial state is a vibrationally excited state of HOD corresponding to four quanta in the HO–D stretch, the channel D + OH is exclusively populated in a subsequent unimolecular photodissociation reaction induced by a UV-photon. The energy of the UV-photon must, however, lie within a rather narrow energy range.

Perhaps the simplest approach is obtained by noting that with electromagnetic radiation we can control the supply of energy into specific molecular degrees of freedom *prior* to reaction, e.g., in the form of pre-excitation of vibrational degrees of freedom of the reactants. This might affect the reactivity; we have already, in Section 4.2, discussed the concept of *mode-selective chemistry* for bimolecular reactions, i.e., the notion that not all partitionings of the total energy might be equally effective in promoting a reaction. Example 7.3 illustrates this concept for a unimolecular reaction: the direct photodissociation of HOD.

The approach illustrated in Example 7.3 works only in special cases. It is clear that a more general approach to microscopic control would involve an active intervention *during* the course of the reaction; that is, femtosecond-laser pulses are required [17,18].

To that end, we take a second look at the basic equations of motion in the presence of an electromagnetic field. We have already discussed the equation of motion for the atomic nuclei, in the case where two electronic states are coupled by the field, that is Eq. (7.25). This situation describes, typically, the dynamics induced by fields in the visible or ultraviolet (UV) spectral region. We observe from the equation of motion that a necessary condition for electronic excitation is that the electronic states are coupled by a non-vanishing transition-dipole moment.

When the frequency of the field is lower, the dynamics will be confined to the electronic ground state. This is, typically, dynamics induced by fields in the infrared (IR) spectral region. In this case, the time-dependent Schrödinger equation for the atomic nuclei takes the form

$$i\hbar\frac{\partial}{\partial t}\chi_1(\boldsymbol{R},t) = \left[\hat{H}_1 + \hat{C}_{11}\right]\chi_1(\boldsymbol{R},t) \tag{7.71}$$

where $\hat{H}_1$ is the nuclear Hamiltonian associated with the electronic ground state, and the coupling term is $\hat{C}_{11} = -\langle\psi_1|\boldsymbol{\mu}|\psi_1\rangle \cdot \boldsymbol{E}(t) \equiv -\boldsymbol{\mu}_{11} \cdot \boldsymbol{E}(t) = -\mu E(t)\cos\theta$, and

$$\mu = |\langle\psi_1|\boldsymbol{\mu}|\psi_1\rangle| \tag{7.72}$$

is the magnitude of the *dipole moment*. The dipole moment (as the transition-dipole moment) depends on the internuclear coordinates via the electronic wave functions. In order to vibrationally excite a molecule, a dipole moment that changes as a function of the internuclear coordinates is required. This is easy to demonstrate; if we consider for simplicity one-dimensional motion, then Eq. (4.109) shows that

$$d\langle\hat{H}_1\rangle/dt = -i(E(t)/\hbar)\langle\chi_1(R,t)|[\mu(R),\hat{T}_{\text{nuc}}]|\chi_1(R,t)\rangle \tag{7.73}$$

Clearly, if $\mu(R)$ is constant, the commutator is zero, and there will be no change in energy.

As shown in Example 7.3, vibrational pre-excitation of stationary vibrational states in HOD gives a large degree of control in the subsequent photochemical decomposition. This scheme can be readily extended beyond the case where the normal-mode vibrations coincide with local bond-stretching modes [18]. Consider, as an example, an ozone molecule with isotopic substitution, i.e., $^{16}O^{16}O^{18}O$. The relevant potential energy surfaces are similar to the ones in Fig. 7.1.1. In this case, the isotopic substitution leads only to a minor asymmetry in the wave functions. The normal modes of the electronic ground state are quite close to the symmetric and asymmetric bond-stretching modes of the normal ozone molecule. One way to control the outcome of the fragmentation, i.e., to produce $^{16}O + {}^{16}O^{18}O$ or $^{16}O^{16}O + {}^{18}O$, is to use two laser pulses. The first IR-pulse forces the molecule into a non-stationary vibrational state, corresponding to asymmetric bond stretching in $^{16}O^{16}O^{18}O$. We note that, when $\mu(R)$ is linear in the internuclear coordinate and the potential is harmonic, then Eq. (7.71) can be solved analytically, and the dynamics is described by Gaussian wave packet dynamics (Section 4.2.1). Thus, an oscillating Gaussian wave packet is created by an (intense) IR-pulse with a center frequency that matches the frequency of the asymmetric bond-stretching mode. The second UV-pulse initiates the reaction at an appropriate time delay relative to the first pulse. We have seen that an ultrashort pulse in the ultraviolet region of the spectrum can launch, at a well-defined time, a localized wave packet on a potential energy surface corresponding to an excited electronic state. We note in particular that the short duration of the second (control) pulse implies that intramolecular vibrational relaxation (IVR) of the electronic ground state is bypassed. Note that this scheme has a close formal correspondence to femtosecond pump–probe spectroscopy.

The scheme considered here is based on a limited number of optimization parameters, primarily the center frequencies and durations of the laser pulses, and the time delay between the two pulses. In more complex systems these parameters may not be sufficient and, furthermore, the separation of the laser pulse into two distinct non-overlapping pulses might not be possible.

In principle, one can induce and control unimolecular reactions directly in the electronic ground state via intense IR fields. Note that this resembles traditional thermal unimolecular reactions, in the sense that the dynamics is confined to the electronic ground state. High intensities are typically required in order to 'climb up the vibrational ladder' and induce bond breaking (or isomerization). The dissociation probability is substantially enhanced when the frequency of the field is time dependent, i.e., the frequency must decrease as a function of time in order to accommodate the anharmonicity of the potential. Selective bond breaking in polyatomic molecules is, in addition, complicated by the fact that the dynamics in various bond-stretching coordinates is coupled due to anharmonic terms in the potential.

The problem of laser-controlled chemical reactions is in general how to design a laser pulse that can guide the system into the desired final state. In general, it is not possible to guess what form an optimal pulse takes. The problem can, however, be handled systematically by the optimization technique called optimal control theory [19]. This leads to a procedure where the time-dependent Schrödinger equation, Eq. (7.71), must be solved iteratively. A hint to the 'mechanism' and the resulting optimal pulses of such numerical optimization procedures can be obtained from Eq. (7.73). Suppose the objective is to pump energy into a bond (neglecting, for simplicity, coupling to all other bonds). This goal can be achieved when the right-hand side of the equation takes a positive value at every time step, that is, when $E(t)$ is proportional to $i\langle\chi_1(R,t)|[\mu(R),\hat{T}_{\text{nuc}}]|\chi_1(R,t)\rangle^*$.

Unimolecular reactions can, of course, also be induced by UV-laser pulses. As pointed out above, in order to reach a specific reaction channel, the electric field of the laser pulse must be specifically designed to the molecular system. All features of the system, i.e., the Hamiltonian (including relativistic terms), must be completely known in order to solve this problem. In addition, the full Schrödinger equation for a large molecular system with many electrons and nuclei can at present only be solved in an approximate way. Thus, in practice, the precise form of the laser field cannot always be calculated in advance.

In order to bypass this problem, a clever idea has been introduced: the laboratory feedback control technique [20]. The optimization procedure is based on the feedback from the observed experimental signal (e.g., a branching ratio) and an optimization algorithm that iteratively improves the applied femtosecond-laser pulse. This iterative optimum-seeking process has been termed 'training lasers to be chemists' [21].

Another important development for laser control is due to the latest advances in laser technology. One can now create laser fields with complex pulse shapes. Thus, it is, e.g., possible to modulate the phases of each spectral component of a short pulse [22]. In this way it is possible to experimentally tailor, essentially, any laser pulse shape.

Based on these developments, the experimental implementation of automated feedback optimized laser control has been achieved [23]. This type of control at the molecular level can be much more selective than traditional methods of control where only macroscopic parameters like the temperature can be varied.

We end this section with a comparison of the basic concepts of laser control and traditional temperature control. This discussion includes an elementary explanation and definition of concepts such as incoherent superpositions of stationary states versus coherent superpositions of stationary states and quantum interference.

Incoherent excitation

According to quantum mechanics, the allowed energy eigenstates of an isolated molecule are described by wave functions, $\Psi_n(x,t) = \psi_n(x)e^{-iE_nt/\hbar}$, where E_n is the energy of the state. The wave functions depend on the time, t, and coordinates, x, which specify the configuration of the molecule. The molecule can be found in any of these energy states, where $n = 0, 1, 2, \ldots$. At $T = 0$ K, all molecules will be found in the lowest energy state with the energy E_0. Higher temperatures correspond to molecular motion and via inelastic collisions energy will be transferred between translational and internal (vibrational/rotational) degrees of freedom. This implies that the molecular energy states will be populated in a specific way according to the Boltzmann distribution.

When molecules are produced in chemical reactions, the energy states are normally not populated according to the Boltzmann distribution. However, this distribution will be quickly established. Thus, say at 1 atm, the Boltzmann distribution will typically be established among the vibrational energy levels within the order of a millisecond, and within the order of a nanosecond among the rotational energy levels. When the pressure is reduced, the distribution will, however, be established more slowly (see Example 2.2). In liquids, the Boltzmann distribution will typically be established among the vibrational energy levels within the order of a picosecond.

Consider, as an example, the probability of finding the system at the configuration specified by the coordinate x, given by

$$\begin{aligned} P_{\text{incoherent}}(x) &= \sum_n p_n |\Psi_n(x,t)|^2 \\ &= \sum_n p_n |\psi_n(x)|^2 \end{aligned} \qquad (7.74)$$

where $|\Psi_n(x,t)|^2 = \Psi_n(x,t)\Psi_n(x,t)^*$, that is, the product of the wave function and its complex conjugate, and p_n is the probability (population) according to the Boltzmann distribution. Equation (7.74) defines what is meant by a so-called *incoherent* (classical) sum of quantum states. For a harmonic oscillator (i.e., a normal-mode vibration in a molecule), the probability takes a Gaussian form, $P_{\text{incoherent}}(x) \propto \exp[-f(T)x^2]$, where $f(T) = (m\omega/\hbar)\tanh[\hbar\omega/(2k_BT)]$ (m and ω are the mass and the frequency, respectively, of the oscillator). At low temperatures, $T \to 0$, the probability distribution corresponds to the well-known ground state of a harmonic oscillator, that is, $f(T) = m\omega/\hbar$ independent of T, whereas in the high-temperature limit $f(T) = m\omega^2/(2k_BT)$. Note that the probability distribution broadens when the temperature is raised and, consequently, the probability of finding configurations with high potential energy is increasing.

When we consider a thermally-activated unimolecular reaction, say the fragmentation of a triatomic molecule with two product channels,

$$\begin{array}{ccc} & & \mathrm{A + BC} \\ & \overset{k_1}{\nearrow} & \\ \mathrm{ABC} & & \\ & \overset{k_2}{\searrow} & \\ & & \mathrm{C + AB} \end{array}$$

the branching ratio, b, is at all times determined by the ratio of the rate constants, i.e., $b = [\mathrm{A}]/[\mathrm{C}] = k_1/k_2$. It is easy to estimate the temperature dependence of this ratio. According to transition-state (RRKM) theory, $k_i = (Q_i^\ddagger/Q)e^{-E_0^i/RT}$, where Q and $Q_i^\ddagger$ are partition functions associated with the reactant and the activated complex leading to product channel i, respectively, and E_0^i are the barrier heights associated with the product channels. Thus, when we neglect the weak temperature dependence of the pre-exponential factors, $b = k_1/k_2 \sim e^{-\Delta E_a/RT}$, where $\Delta E_a = E_0^1 - E_0^2$ is the difference between the barrier heights ($\sim$ activation energies) of the two channels.

The branching ratio can be controlled to some extent, by the temperature, provided $\Delta E_a \neq 0$. Thus, when T is small, that is, $\Delta E_a \gg RT$, then $b \ll 1$ when $\Delta E_a > 0$ and $b \gg 1$ when $\Delta E_a < 0$. On the other hand, at higher temperatures both channels will be populated more evenly. Thus in the limit where T is large, that is, $\Delta E_a \ll RT$, then $b \sim 1$.

Coherent excitation

It is possible to create a coherent linear superposition of energy eigenstates via laser excitation. This is a state of the form $\sum_n c_n\Psi_n(x,t)$, where $c_i = |c_i|\exp(i\delta_i)$, with a particular phase relation between the coefficients c_n. The value of these coefficients is determined by the pulsed laser excitation.

We consider again the probability of finding the system at the configuration specified by the coordinate x. Then, according to the rules of quantum mechanics,

$$\begin{aligned} P_{\mathrm{coherent}}(x,t) &= \left|\sum_n c_n\Psi_n(x,t)\right|^2 \\ &= \sum_n |c_n|^2|\Psi_n(x,t)|^2 + \text{interference terms} \\ &= \sum_n |c_n|^2|\psi_n(x)|^2 + \text{interference terms} \end{aligned} \tag{7.75}$$

Equation (7.75) defines what is meant by a so-called *coherent* sum of quantum states. The diagonal terms resemble the incoherent sum in Eq. (7.74); the values of the populations $|c_n|^2$ are, however, determined by the laser pulse. The off-diagonal terms are called interference terms; these terms are the key to quantum control. They are time dependent and we use the term 'coherent dynamics' for the motion associated with the coherent excitation of quantum states. A particular simple form of Eq. (7.75) is obtained in the special case of two states. Then

$$P_{\text{coherent}}(x,t) = |c_1|^2|\psi_1(x)|^2 + |c_2|^2|\psi_2(x)|^2 \\ + 2|c_1||c_2||\psi_1(x)||\psi_2(x)|\cos[\Delta Et/\hbar - \delta] \quad (7.76)$$

where real-valued wave functions were assumed, $\Delta E = E_1 - E_2$, $\delta = \delta_1 - \delta_2$, and $c_i = |c_i|\exp(i\delta_i)$. The coherent superposition of quantum states differs clearly from the incoherent superposition of Eq. (7.74).

In the control scheme [13,17] that we have focused on, the time evolution of the interference terms plays an important role. We have already discussed more explicit forms of Eq. (7.75). One example is the Franck–Condon wave packet considered in Section 7.2.2; another example, which we considered above, is the oscillating Gaussian wave packet created in a harmonic oscillator by an (intense) IR-pulse. Note that the interference term in Eq. (7.76) becomes independent of time when the two states are degenerate, that is, $\Delta E = 0$. The magnitude of the interference term still depends, however, on the phase δ. This observation is used in another important scheme for coherent control [14].

The time evolution in Eq. (7.75) is described by the time-dependent Schrödinger equation, provided the molecule is isolated from the rest of the universe. In practice, there are always perturbations from the environment, say due to inelastic collisions. The coherent sum in Eq. (7.75) will then relax to the incoherent sum of Eq. (7.74), that is, the off-diagonal interference terms will vanish and $|c_n|^2 \to p_n$ corresponding to the Boltzmann distribution. As mentioned earlier, the relaxation time depends on the pressure. In order to take advantage of coherent dynamics it is, of course, crucial that relaxation is avoided within the duration of the relevant chemical dynamics.

Further reading/references

[1] K.A. Holbrook, M.J. Pilling, and S.H. Robertson, *Unimolecular reactions*, second edition (Wiley, 1996)

[2] T. Uzer, J.T. Hynes, and W.P. Reinhardt, *J. Chem. Phys.* **85**, 5791 (1986)

[3] R. Schinke, *Photodissociation dynamics* (Cambridge University Press, 1993)

[4] N.E. Henriksen, *Adv. Chem. Phys.* **91**, 433 (1995)

[5] E.R. Lovejoy, S.K. Kim, and C.B. Moore, *Science* **256**, 1541 (1992)

[6] S.K. Kim, E.R. Lovejoy, and C.B. Moore, *J. Chem. Phys.* **102**, 3202 (1995)

[7] T. Baer and W.L. Hase, *Unimolecular reaction dynamics. Theory and experiments* (Oxford, 1996)

[8] R.G. Gilbert and S.C. Smith, *Theory of unimolecular and recombination reactions* (Blackwell, 1990)

[9] A.H. Zewail, *Femtochemistry, Vols 1, 2* (World Scientific, Singapore, 1994)
[10] *Femtosecond chemistry*, edited by J. Manz and L. Wöste (VCH, Weinheim, 1995)
[11] A.H. Zewail, *J. Phys. Chem.* **104**, 5660 (2000)
[12] N.E. Henriksen and V. Engel, *Int. Rev. Phys. Chem.* **20**, 93 (2001)
[13] S.A. Rice and M. Zhao, *Optical control of molecular dynamics* (Wiley, 2000)
[14] M. Shapiro and P. Brumer, *Principles of quantum control of molecular processes* (Wiley, 2003)
[15] N.E. Henriksen, *Chem. Soc. Rev.* **31**, 37 (2002)
[16] D.G. Imre and J. Zhang, *Chem. Phys.* **139**, 89 (1989)
[17] D.J. Tannor and S.A. Rice, *J. Chem. Phys.* **83**, 5013 (1985)
[18] B. Amstrup and N.E. Henriksen, *J. Chem. Phys.* **105**, 9115 (1996)
[19] A.P. Peirce, M.A. Dahleh, and H. Rabitz, *Phys. Rev. A* **37**, 4950 (1988)
[20] R.S. Judson and H. Rabitz, *Phys. Rev. Lett.* **68**, 1500 (1992)
[21] R.F. Service, *Science* **279**, 1847 (1998)
[22] A.M. Weiner, *Rev. Sci. Instr.* **71**, 1929 (2000)
[23] T. Brixner and G. Gerber, *Chem. Phys. Chem.* **4**, 418 (2003)

Problems

7.1 The (high-pressure) pre-exponential factor for the ring-opening of cyclo-butene into butadiene is $10^{13.4}$ s^{-1}, and the activation energy is 137.6 kJ/mol. Using the RRK theory calculate the unimolecular rate constant for the reaction at an excitation energy of 200 kJ/mol.

7.2 We consider two molecules, X–H and X–D, where X is a heavy group of atoms, which in the following is considered as a point mass. The two molecules have, according to the Born–Oppenheimer approximation, the same potential. The unimolecular bond breakage is described within the framework of the RRKM theory.

(a) Show that $\nu_{XH} = \sqrt{2}\nu_{XD}$, for the vibrational frequencies within the harmonic approximation, when X is considered as infinitely heavy compared to H (1.00 amu) and D (2.00 amu).

(b) Determine the difference in the barrier heights for the breaking of the X–H and X–D bonds, that is, $(E_0^{XD} - E_0^{XH})$ expressed by ν_{XH}.

(c) Derive an expression for the ratio $k_{XH}(T)/k_{XD}(T)$ between the unimolecular rate constants. Use RRKM theory (in the high-pressure limit) and express the result in terms of ν_{XH}.

(d) Assume that the zero-point energy in the X–H bond is 17.2 kJ/mol and calculate the ratio $k_{XH}(T)/k_{XD}(T)$ at $T = 298$ K.

7.3 Consider three harmonic oscillators (normal modes) with frequencies 1600 cm^{-1}, 3650 cm^{-1}, and 3750 cm^{-1}. We want to calculate the sum of vibrational states at the total energy $E = 6000\ \text{cm}^{-1}$ relative to the lowest possible (zero-point) energy.

(a) Calculate the sum of states, $G(E)$, based on the classical expression Eq. (7.52):

(i) use Eq. (7.52);

(ii) use the modification where the zero-point energy, E_z, is added to the argument, i.e., $G(E + E_z)$;

(iii) use the modification $G(E + E_z) - G(E_z)$, where we also subtract the states that are not allowed according to quantum mechanics.

(b) Label the possible states by the quantum numbers associated with each oscillator and, by direct count, show that the exact sum of states at $E = 6000\ \text{cm}^{-1}$ is 8. Compare with the estimates in part (a).

7.4 We consider the isomerization

$$\text{HCN} \longrightarrow \text{HNC}$$

The unimolecular rate constant $k(E)$ is described within the framework of RRKM theory. In the following, we neglect the rotational energy in HCN as well as in the activated complex. The classical barrier height is $E_{\text{cl}} = 1.51$ eV.

(a) Calculate the barrier height E_0 for isomerization.

(b) Calculate $k(E)$ at $E = 1.5$ eV and $E = 2.5$ eV, respectively.

The vibrational frequencies (wave numbers) are:

- HCN: $3300(1)\ \text{cm}^{-1}$, $2100(1)\ \text{cm}^{-1}$, $713(2)\ \text{cm}^{-1}$;

and for the activated complex

- $(\text{HCN})^{\ddagger}$: $3000(1)\ \text{cm}^{-1}$, $2000(1)\ \text{cm}^{-1}$.

The degeneracy of the vibrational states is given in the parentheses after each vibrational frequency.

7.5 Consider the thermal unimolecular rearrangement of hydrogen isocyanide,

$$\text{HNC} \longrightarrow \text{HCN}$$

Assume that the reaction proceeds via a three-center transition state with the atoms at the corners of a triangle.

(a) Evaluate $k_B T/h$ at $T = 1000$ K.

(b) Evaluate the pre-exponential factor according to the RRKM theory at $T = 1000$ K, given the following values:

- HNC: $I = 11.3\ \text{amu}\,\text{Å}^2$; $\tilde{\nu} = 3300\ \text{cm}^{-1}$ (stretch), $713\ \text{cm}^{-1}$ (doubly-degenerate bend), $2100\ \text{cm}^{-1}$ (stretch);
- transition state: $I^{\ddagger} = 1.02\ \text{amu}\,\text{Å}^2$, $10.5\ \text{amu}\,\text{Å}^2$, $12\ \text{amu}\,\text{Å}^2$; $\tilde{\nu} = 3000\ \text{cm}^{-1}$ (stretch), $1000i\ \text{cm}^{-1}$ (bend), $2000\ \text{cm}^{-1}$ (ring-breathing stretch).

7.6 The classical expression in Eq. (7.52) often underestimates the true number of vibrational states. In order to correct the formula, it has been suggested to replace E by $E + E_z$ in the expression, where E_z is the zero-point energy.

(a) Consider an apparent unimolecular reaction $\text{A}^*(E) \rightarrow$, where the molecule A has s vibrational degrees of freedom (and rotation is neglected). Using the classical expression with E replaced by $E + E_z$, write down the rate constant $k(E)$, according to RRKM theory.

(b) Repeat the calculation in Example 7.2, using the formula derived in part (a).

(c) Consider now a system of two uncoupled harmonic oscillators with identical frequencies. An energy level for the system can, as for a single harmonic oscillator, be characterized by a single quantum number n ($n = 0, 1, 2, \ldots$). What is the zero-point energy, E_z, and what is the energy and degeneracy of the energy level E_n with quantum number n?

7.7 We consider a thermally-activated unimolecular reaction with two product channels:

$$CH_2DCH_2Cl \xrightarrow{k_1} CHD=CH_2 + HCl$$

$$CH_2DCH_2Cl \xrightarrow{k_2} CH_2=CH_2 + DCl$$

The reaction takes place at high pressure where it obeys first-order kinetics.

(a) Write down the rate law $-d[\mathrm{A}]/dt$ for the reactant, here denoted by A, and integrate this expression in order to obtain an expression for the concentration of A as a function of time.

(b) Determine the ratio $[\mathrm{HCl}]/[\mathrm{DCl}]$ expressed in terms of the rate constants, when the initial concentrations are $[\mathrm{HCl}]_0 = [\mathrm{DCl}]_0 = 0$.

(c) Using RRKM theory, calculate in the high-pressure limit k_1/k_2 at $T = 1000\,\mathrm{K}$. The vibrational frequencies and the barrier heights (E_0) for the two activated complexes (T.S. 1 and T.S. 2) are given in the table below [data taken from J.I. Steinfeld, J.S. Francisco, and W.L. Hase, *Chemical kinetics and dynamics*, second edition (Prentice Hall, 1999)].

CH_2DCH_2Cl	T.S. 1 (HCl elimination)	T.S. 2 (DCl elimination)
$\tilde{\nu}_i/\mathrm{cm}^{-1}$	$\tilde{\nu}_i/\mathrm{cm}^{-1}$	$\tilde{\nu}_i/\mathrm{cm}^{-1}$
2940(4)	3000(3)	3000(4)
2160(1)	2200(2)	2088(1)
1340(4)	1380(2)	1380(1)
1270(2)	1115(1)	1100(1)
960(3)	960(5)	960(4)
720(2)	850(1)	850(3)
330(1)	820(1)	750(1)
200(1)	645(1)	570(1)
	403(1)	501(1)
Barrier(kJ/mol)	232.7	237.9

The degeneracy of the vibrational states is specified in parentheses after the frequency. The imaginary frequency is not included. In order to simplify the calculations, one can assume that $h\nu \gg k_B T$ when the vibrational frequencies are larger than 1000 cm^{-1}. Assume also that the moments of inertia and the symmetry numbers for the activated complexes are identical.

8

Microscopic interpretation of Arrhenius parameters

Key ideas and results

We return to the microscopic interpretation of the Arrhenius parameters, i.e., the pre-exponential factor (A) and the activation energy (E_a) known from classical chemical kinetics.

- The pre-exponential factor of an apparent unimolecular reaction is, roughly, expected to be of the order of a vibrational frequency, i.e., 10^{13} to 10^{14} s^{-1}. The pre-exponential factor of a bimolecular reaction is, roughly, related to the collision frequency, i.e., the number of collisions per unit time and per unit volume.
- The activation energy of an elementary reaction is related to the classical barrier height of the potential energy surface corrected, however, for zero-point energies and average internal molecular energies. The activation energy is the average energy of the molecules that react minus the average energy of the reactants. Specializing to transition-state theory, the activation energy is equal to the difference in zero-point energy levels of the activated complex and the reactants plus the difference in the average internal energies (relative to the zero-point levels). When tunneling corrections are introduced in conventional transition-state theory, the activation energy is reduced compared to the interpretation given above.

In Chapter 2, the first chapter of the gas-phase part of the book, we began the transition from microscopic to macroscopic descriptions of chemical kinetics. In this last chapter of the gas-phase part, we will assume that the Arrhenius equation forms a useful parameterization of the rate constant, and consider the microscopic interpretation of the Arrhenius parameters, i.e., the pre-exponential factor (A) and the activation energy (E_a) defined by the Arrhenius equation: $k(T) = A\exp(-E_a/k_BT)$.

In the previous chapters, we have seen that the rate constant for unimolecular as well as bimolecular elementary reactions can be written in a form similar to the Arrhenius equation, provided we allow for a (weak) temperature dependence of the pre-exponential factor A. Experimentally, the parameters may be determined from an Arrhenius plot, that is, a plot of $\ln[k(T)]$ versus $1/T$, which according to the Arrhenius equation will be a straight line with slope $-E_a/R$ and an intercept of $\ln A$. The question is: what is the molecular origin of these parameters?

8.1 The pre-exponential factor

In the previous chapters, we have already discussed the calculation, and to some extent the microscopic interpretation, of the pre-exponential factor. In this section, we review the microscopic interpretation. In the next section, we will show that the activation energy is typically (when tunneling is neglected) related to the barrier height E_0 according to $E_a = E_0 + \beta k_B T$, where β is a small rational number. In the context of interpreting the order of magnitude of the pre-exponential factor, we will neglect the small factor $e^{-\beta}$.

8.1.1 Unimolecular reactions

The general relation between the rate constant of (apparent) unimolecular reactions and the microscopic dynamics is illustrated in Fig. 8.1.1.

Based on the theoretical descriptions of Chapter 7, the pre-exponential factor of an apparent unimolecular reaction is, roughly, expected to be of the order of a vibrational frequency, i.e., 10^{13} to 10^{14} s^{-1}.

Thus, according to RRKM theory for an apparent unimolecular reaction, Eq. (7.58) gives the (canonical) rate constant for such an elementary reaction:

$$k(T) = \frac{k_B T}{h} \frac{Q^{\ddagger}}{Q} e^{-E_0/k_B T} \tag{8.1}$$

We see that the pre-exponential factor in $k(T)$, roughly, corresponds to a typical vibrational frequency, since $k_B T/h = 6.25 \times 10^{12}$ s^{-1} at $T = 300$ K. We can also understand why the pre-exponential factors can be somewhat larger than typical vibrational frequencies, because very often $Q^{\ddagger} > Q$. This situation will arise when the (product of the) vibrational frequencies of the activated complex are smaller than the vibrational frequencies of the reactant. Furthermore, the rotational contribution in the ratio of partition functions, which is proportional to $\sqrt{I_a^{\ddagger} I_b^{\ddagger} I_c^{\ddagger}/(I_a I_b I_c)}$, will often also be larger than one since, typically, the principal moments of inertia associated with the activated complex are larger than the moments of inertia of the reactant.

8.1.2 Bimolecular reactions

The general relation between the rate constant of a bimolecular reaction and the microscopic dynamics is illustrated in Fig. 8.1.2.

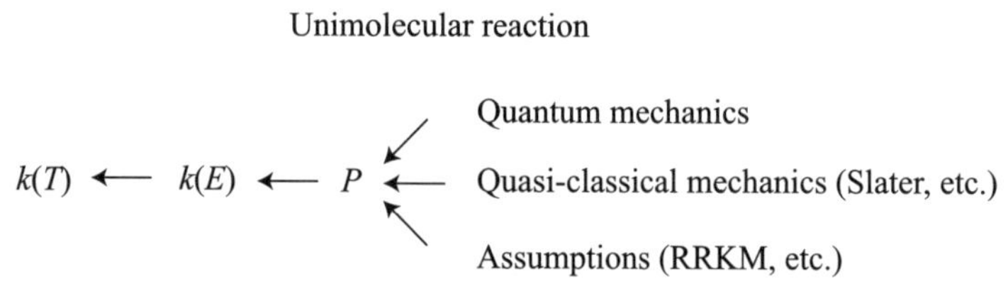

Fig. 8.1.1 An illustration of the relations between the rate constants, $k(T)$ and $k(E)$, and the reaction probability P as obtained from either quantum mechanics, quasi-classical mechanics, or various assumptions (approximations) for the reaction dynamics.

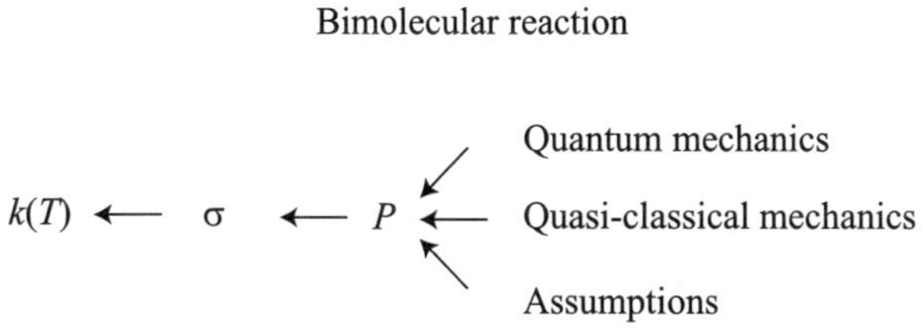

Fig. 8.1.2 An illustration of the relations between the rate constant $k(T)$, the reaction cross-section σ, and the reaction probability P as obtained from either quantum mechanics, quasi-classical mechanics, or various assumptions (approximations) for the reaction dynamics.

Based on the theoretical descriptions of Chapters 2, 4, and 6, the pre-exponential factor of a bimolecular reaction $A + B \to$ products is expected to be related to the collision frequency, i.e., the number of collisions per unit time and per unit volume.

The pre-exponential factor of a bimolecular reaction is related to the reaction cross-section (see Problem 2.3). A relation that is fairly easy to interpret can be obtained within the framework of transition-state theory. Combining Eqs (6.9) and (6.54), we can write the expression for the rate constant in a form that gives the relation to the (hard-sphere) collision frequency:

$$\begin{aligned} k_{\rm TST}(T) &= \frac{k_B T}{h} \frac{h^3}{(2\pi\mu k_B T)^{3/2}} \left(\frac{Q^{\ddagger}_{(\rm AB)^\ddagger}}{Q_{\rm A} Q_{\rm B}} \right)_{\rm int} e^{-E_0/k_B T} \\ &= Z \frac{1}{Q^{\rm AB}_{\rm rot}} \left(\frac{Q^{\ddagger}_{(\rm AB)^\ddagger}}{Q_{\rm A} Q_{\rm B}} \right)_{\rm int} e^{-E_0/k_B T} \end{aligned} \tag{8.2}$$

Here the partition functions refer to internal degrees of freedom (subscript 'int' for internal), $Q^{\rm AB}_{\rm rot} = 8\pi^2(\mu d^2)k_B T/h^2$, that is, a rotational partition function where A and B are considered as point masses separated by the distance d, and $Z = \pi d^2 \langle v \rangle$ is related to the (hard-sphere) collision frequency $Z_{\rm AB}$ defined in Eq. (4.16), that is, $Z_{\rm AB} = Z[{\rm A}][{\rm B}]$.

Within transition-state theory, Eq. (8.2) is an exact expression for the rate constant. We observe that the pre-exponential factor deviates from the simple interpretation, as being related to the collision frequency $Z_{\rm AB}$ via Z, due to the presence of internal degrees of freedom. Typically, the calculated value of Z is of the order of 10^{11} dm^3 mol^{-1} s^{-1} $\sim 10^{-16}$ m^3 molecule^{-1} s^{-1} (see Example 4.1). The magnitude of the partition functions in Eq. (8.2) is typically small compared to this number. Thus, if we neglect the internal degrees of freedom of the reactants and the activated complex, except for rotational degrees of freedom of the activated complex $({\rm AB})^\ddagger$, and assume that the associated partition function can be approximated by $Q^{\rm AB}_{\rm rot}$, we will get a pre-exponential factor given by Z.

8.2 The activation energy

The activation energy E_a is a macroscopic quantity *defined* by the Arrhenius equation $k(T) = A\exp(-E_a/k_B T)$ and given by

$$
\begin{aligned}
E_a &= -k_B \frac{d \ln k(T)}{d(1/T)} \\
&= k_B T^2 \frac{d \ln k(T)}{dT}
\end{aligned}
\tag{8.3}
$$

This definition is used whether or not A is independent of temperature, that is, whether or not the Arrhenius plot is linear. Note that Eq. (8.3) shows that a large value of E_a implies that the rate constant depends strongly on the temperature.

When we have expressed $k(T)$ in terms of microscopic information, we can obtain a corresponding microscopic interpretation of the activation energy. It should be stressed that the following relations are only valid for elementary gas-phase reactions.

8.2.1 Activation energy and detailed balance

From the definition of activation energy in Eq. (8.3), we obtain the following expression for the *difference* in activation energies between the forward and the reverse direction of a reaction:

$$
\begin{aligned}
E_a^f - E_a^r &= k_B T^2 \frac{d}{dT} (\ln k_f - \ln k_r) \\
&= k_B T^2 \frac{d}{dT} \ln \left(\frac{k_f}{k_r} \right)
\end{aligned}
\tag{8.4}
$$

We note, in passing, that this equation is consistent with the well-known equation for the temperature dependence of an equilibrium constant $K = k_f/k_r$, i.e., the van't Hoff equation. From the general principle of detailed balance, one can obtain a microscopic interpretation of the difference in activation energies between the forward and the reverse direction of an elementary reaction. Detailed balance, Eq. (2.34), implies

$$
\frac{k_f}{k_r} = \left(\frac{\mu_{\mathrm{CD}}}{\mu_{\mathrm{AB}}} \right)^{3/2} \frac{Q_{\mathrm{C}} Q_{\mathrm{D}}}{Q_{\mathrm{A}} Q_{\mathrm{B}}} e^{-\Delta E_0 / k_B T}
\tag{8.5}
$$

where the partition functions are associated with the internal degrees of freedom of the molecules. Now,

$$
\begin{aligned}
E_a^f - E_a^r &= k_B T^2 \frac{d}{dT} \left(\ln \left(\frac{Q_{\mathrm{C}} Q_{\mathrm{D}}}{Q_{\mathrm{A}} Q_{\mathrm{B}}} \right) - \frac{\Delta E_0}{k_B T} \right) \\
&= \Delta E_0 + k_B T^2 \left(\frac{d \ln Q_{\mathrm{C}}}{dT} + \frac{d \ln Q_{\mathrm{D}}}{dT} - \frac{d \ln Q_{\mathrm{A}}}{dT} - \frac{d \ln Q_{\mathrm{B}}}{dT} \right) \\
&= \Delta E_0 + (\langle E_{\mathrm{int}}^{\mathrm{C}} \rangle + \langle E_{\mathrm{int}}^{\mathrm{D}} \rangle) - (\langle E_{\mathrm{int}}^{\mathrm{A}} \rangle + \langle E_{\mathrm{int}}^{\mathrm{B}} \rangle)
\end{aligned}
\tag{8.6}
$$

where the expression for the average energy $\langle E \rangle$ (Eq. (A.22)) was used in the last line. Thus, the difference between activation energies of the forward and the reverse reaction is equal to the difference in zero-point energies between products and reactants ΔE_0 plus the difference between the average internal energy of the products and the reactants at a given temperature. If we had chosen to include the zero-point energies in the average energies, the difference in activation energies would simply be equal to the average internal energy of the products minus the average internal energy of the reactants. This relation is illustrated in Fig. 8.2.1.

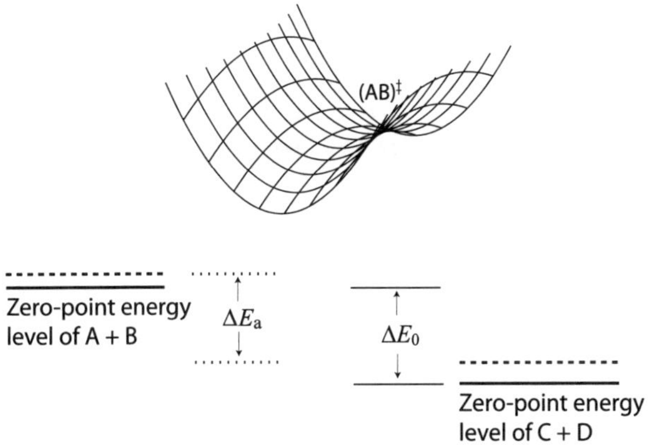

Fig. 8.2.1 An illustration of Eq. (8.6). The zero-point energy levels for an exothermic reaction are indicated by solid lines and the average internal energies of the products and the reactants (relative to the zero-point levels) are given by dashed lines.

The average internal energies depend on the temperature, and can according to Eq. (A.9) be written in the form

$$\langle E_{\text{int}} \rangle = \langle E_{\text{vib}} \rangle + \langle E_{\text{rot}} \rangle \tag{8.7}$$

when the coupling between vibration and rotation is neglected. The average vibrational energy of a harmonic oscillator can be calculated from the partition function given in Eq. (A.15) (remember that the energy is measured relative to the zero-point level); that is,

$$\begin{aligned} \langle E_{\text{vib}} \rangle &= k_B T^2 \frac{d \ln Q_{\text{vib}}}{dT} \\ &= \frac{k_B T^2}{Q_{\text{vib}}} \frac{dQ_{\text{vib}}}{dT} \\ &= \frac{h\nu_s}{\exp(h\nu_s/k_B T) - 1} \end{aligned} \tag{8.8}$$

In the (classical) high-temperature limit, when $h\nu_s \ll k_B T$, the exponential can be expanded to first order and the average vibrational energy is $\sim k_B T$. The approximation $h\nu_s \ll k_B T$ is, however, not well satisfied for typical molecular vibrational frequencies, except at temperatures that exceed several thousand degrees. The average rotational energy of a rigid rotor is

$$\langle E_{\text{rot}} \rangle = \begin{cases} k_B T & \text{for a linear molecule} \\ (3/2) k_B T & \text{for a non-linear molecule} \end{cases} \tag{8.9}$$

where the rotational partition functions in Eqs (A.19) and (A.20) were used.

Example 8.1

The temperature dependence of $(E_a^f - E_a^r)$ is often negligible. Consider, e.g., the reaction

$$F + H_2 \longrightarrow HF + H$$

The difference between the average internal energies of the products and the reactants, $((\langle E_{\rm vib}^{\rm HF}\rangle + k_BT) - ((\langle E_{\rm vib}^{\rm H_2}\rangle + k_BT) = \langle E_{\rm vib}^{\rm HF}\rangle - \langle E_{\rm vib}^{\rm H_2}\rangle \sim 0$, since the vibrational frequencies $\tilde{\nu}_{\rm HF} = 4138\ {\rm cm}^{-1}$ and $\tilde{\nu}_{\rm H_2} = 4395\ {\rm cm}^{-1}$ are quite similar. Thus, in this case the difference between the average energies is zero, essentially, and $(E_a^f - E_a^r) = \Delta E_0$.

8.2.2 Activation energy and transition-state theory

As demonstrated in the previous section, it is quite easy to obtain an exact and transparent interpretation of the *difference* in activation energies between the forward and the reverse direction of an elementary reaction.

It is also possible to obtain an exact interpretation of the individual activation energies, often referred to as 'Tolman's theorem'. Starting from an exact expression for the bimolecular rate constant, Eq. (5.7), we have $k(T) = aI(T)/Q_{\rm react}$, where a is a constant, I is an integral over the Boltzmann factor $\exp(-E/k_BT)$ times the reaction probability $P_{\rm cum}(E)$, and $Q_{\rm react}$ is the partition function of the reactants. Then using Eq. (8.3), $E_a = k_BT^2(1/I)dI/dT - \langle E\rangle$, where $\langle E\rangle$ is the average energy of the reactants. The first term can be identified as an average value of the energy E, averaged over the distribution $P_{\rm cum}(E)\exp(-E/k_BT)$. This leads to the result $E_a = \langle E\rangle^* - \langle E\rangle$, i.e., the activation energy is the average energy of the molecules that react minus the average energy of the reactants.

In the following, we elaborate on this interpretation of the activation energy within the framework of transition-state theory, where a particularly simple and transparent microscopic interpretation of the activation energy can be obtained. This interpretation agrees with Tolman's theorem. The rate constant for a *bimolecular* reaction is, according to transition-state theory (Eq. (6.8)),

$$k_{\rm TST}(T) = \frac{k_BT}{h}\frac{(Q^{\ddagger}/V)}{(Q_{\rm A}/V)(Q_{\rm B}/V)}e^{-E_0/k_BT} \tag{8.10}$$

Now using Eq. (8.3), the activation energy takes the form

$$\begin{aligned} E_a &= k_BT^2\frac{d}{dT}\left(\ln T + \ln(Q^{\ddagger}/V) - \ln(Q_{\rm A}/V) - \ln(Q_{\rm B}/V) - \frac{E_0}{k_BT}\right) \\ &= E_0 + k_BT + k_BT^2\left(\frac{d\ln(Q^{\ddagger}/V)}{dT} - \frac{d\ln(Q_{\rm A}/V)}{dT} - \frac{d\ln(Q_{\rm B}/V)}{dT}\right) \\ &= E_0 + k_BT + \langle E^{\ddagger}\rangle - (\langle E^{\rm A}\rangle + \langle E^{\rm B}\rangle) \\ &= E_0 + \langle E_{\rm int}^{\ddagger}\rangle - (\langle E_{\rm int}^{\rm A}\rangle + \langle E_{\rm int}^{\rm B}\rangle) - (1/2)k_BT \end{aligned} \tag{8.11}$$

where the relation (see Eqs (A.8) and (A.14))

$$\langle E \rangle = k_B T^2 \left(\frac{d \ln(Q_{\text{trans}}/V)}{dT} + \frac{d \ln Q_{\text{int}}}{dT} \right)$$
$$= (3/2) k_B T + \langle E_{\text{int}} \rangle \tag{8.12}$$

was used in the last line. Note that in the third line the term $k_B T$ corresponds to the average energy in the ('low frequency') reaction coordinate pulled out of the activated complex, and this internal degree of freedom is not included in the energy $\langle E^{\ddagger} \rangle$.

Thus, the activation energy is equal to the difference in zero-point energies between the activated complex and the reactants plus the difference between the average energy of the activated complex and the reactants. Alternatively, if we had chosen to include the zero-point energies in the average energies, the activation energy would simply be equal to the average energy of the activated complex minus the average energy of the reactants. This relation is illustrated in Fig. 8.2.2.

Example 8.2

As an example, we consider again the reaction

$$\text{F} + \text{H}_2 \longrightarrow (\text{FHH})^{\ddagger} \longrightarrow \text{HF} + \text{H}$$

Recent calculations (see Section 3.1) show that the activated complex is non-linear, that is, the average rotational energy is $(3/2)k_B T$ and $E_a = E_0 + \langle E^{\ddagger}_{\text{vib}} \rangle - \langle E^{\text{H}_2}_{\text{vib}} \rangle$. $\tilde{\nu}_{\text{H}_2} = 4395$ cm^{-1} and the two vibrational frequencies associated with the activated complex are 3772 cm^{-1} and 296 cm^{-1}, respectively (remember that the third vibrational degree of freedom of the non-linear triatomic molecule is the reaction coordinate which is not included in $Q^{\ddagger}$). The thermal energies associated with the

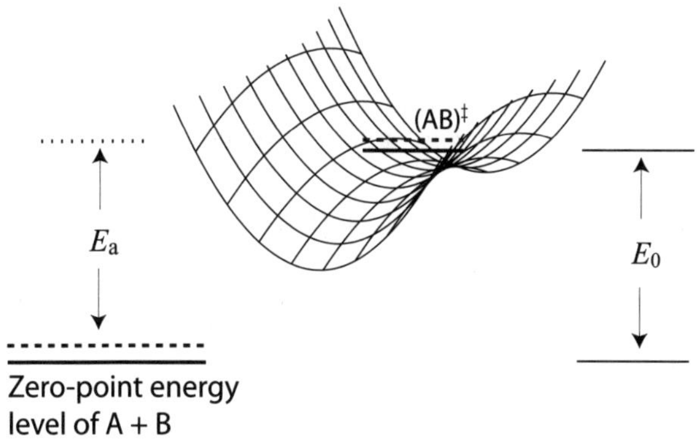

Fig. 8.2.2 An illustration of Eq. (8.11) for the activation energy. The zero-point energy levels of the activated complex and the reactants are indicated by solid lines and the associated average internal energies (relative to the zero-point levels) are given by dashed lines. Note that the zero-point energy in the activated complex comes from vibrational degrees of freedom orthogonal to the reaction coordinate.

two high frequency modes are small and cancel, essentially, whereas the energy associated with the low frequency mode is k_BT, at sufficiently high temperatures. So, according to transition-state theory, at high temperatures $E_a \sim E_0 + k_BT$. Since, $E_0 \sim 4.5$ kJ/mol and $k_BT \sim 8.31$ kJ/mol at $T = 1000$ K, we obtain $E_a \sim$ 12.8 kJ/mol at $T = 1000$ K. Thus, in this example, E_a deviates substantially from E_0.

Example 8.3: On negative activation energy

For the reaction

$$O + OH \longrightarrow H + O_2$$

it is observed [*J. Chem. Phys.* **96**, 1077 (1992)] that the rate constant can decrease when the temperature is increased. Thus, $k(T = 1000\text{ K}) = 22.2 \times 10^{-12}$ cm^3/s and $k(T = 2000\text{ K}) = 16.5 \times 10^{-12}$ cm^3/s. The activation energy becomes

$$\begin{aligned} E_a &= -k_B \frac{d \ln k(T)}{d(1/T)} \\ &= -k_B \frac{\ln[k(T = 1000\text{ K})/k(T = 2000\text{ K})]}{1/1000 - 1/2000} \\ &= -8.19 \times 10^{-21}\text{ J} \\ &= -4.93\text{ kJ/mol} \end{aligned} \tag{8.13}$$

This result clearly shows that the activation energy is not equivalent to a barrier height. Such negative activation energies can occur for reactions without barriers ($E_0 \sim 0$), typically for bimolecular reactions between radicals, as above. Within the framework of transition-state theory, the activation energy is given by Eq. (8.11). From this equation, we observe that E_a can become negative when no barrier is present, provided the average thermal energies of the reactants exceed the energy of the activated complex. Such energies are of the order of RT, which at 1000 K amounts to 8.31 kJ/mol, in good agreement with the result in Eq. (8.13).

These two examples clearly demonstrate that E_a is not equal to E_0. The temperature dependence of E_a will, however, often be negligible since typically $k_BT \ll E_0$. Thus, we have a rough identification of E_a with E_0. It should be remembered that this is not the potential energy barrier height, E_{cl}, but the difference in zero-point energies between the activated complex and the reactants.

Within the framework of transition-state (RRKM) theory for *unimolecular* reactions, one can obtain a microscopic interpretation of the activation energy that is analogous to the one presented above (Problem 8.4).

8.2.3 Activation energy and quantum tunneling

We consider again the interpretation of activation energy within transition-state theory and now the implications of quantum tunneling. A correction factor due to quantum tunneling is given in Eq. (6.44):

$$\kappa_{\text{tunnel}}(T) \sim 1 + \frac{1}{24}\left(\frac{h\nu^*}{k_B T}\right)^2 \tag{8.14}$$

which is valid at high temperatures ($h\nu^* \ll k_B T$) and for high and/or broad barriers ($h\nu^* \ll E_0$).

We let '$\cdots$' denote the four temperature-dependent average energies on the right-hand side in Eq. (8.11), and the activation energy now takes the form

$$\begin{aligned} E_a &= E_0 + \cdots + k_B T^2 \frac{d \ln \kappa_{\text{tunnel}}(T)}{dT} \\ &= E_0 + \cdots + \frac{k_B T^2}{\kappa_{\text{tunnel}}(T)} \frac{d\kappa_{\text{tunnel}}(T)}{dT} \\ &= E_0 + \cdots - \frac{(h\nu^*)^2/12}{\kappa_{\text{tunnel}}(T)} \frac{1}{k_B T} \end{aligned} \tag{8.15}$$

Since the last term is negative, we see that the activation energy is lowered due to quantum tunneling. Now

$$\begin{aligned} \kappa_{\text{tunnel}}(T) k_B T &= k_B T + \frac{(h\nu^*)^2}{24 k_B T} \\ &\sim k_B T \end{aligned} \tag{8.16}$$

where the limit $h\nu^* \ll k_B T$ is taken in the last line. Thus, the activation energy decreases ($\propto -T^{-1}$) when the temperature is lowered and Arrhenius plots show deviations from linearity as a consequence of quantum tunneling; see Fig. 8.2.3.

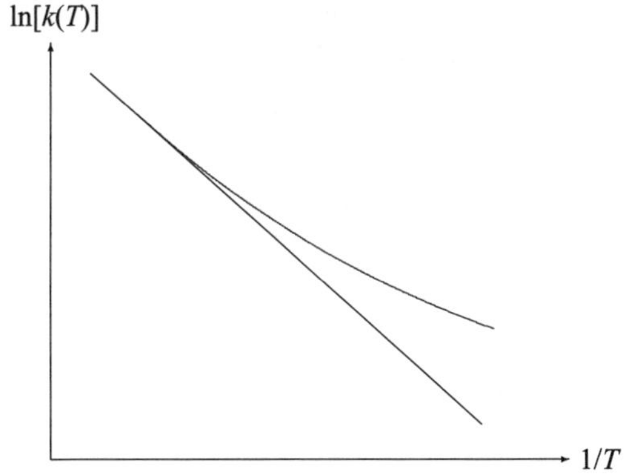

Fig. 8.2.3 An Arrhenius plot, i.e., $\ln[k(T)]$ versus the reciprocal temperature. The straight line represents 'normal' Arrhenius behavior whereas the curved behavior is found in systems with quantum tunneling. It shows that at low temperatures the rate constant is larger than in a classical system.

Problems

8.1 Assume that the temperature dependence of the rate constant can be represented as in Eq. (6.11). Derive an expression for the activation energy.

8.2 Using the exact expression for the average vibrational energy, calculate the activation energy for the reaction of Example 8.2, at $T = 298$ K and $T = 1000$ K.

8.3 We consider the gas-phase reaction

$$\mathrm{F} + \mathrm{H_2} \longrightarrow \mathrm{HF} + \mathrm{H}$$

The rate constant has been determined experimentally in the temperature range $T \in [190, 1000]$ K, and it can be represented by the expression $k(T) = 4.5 \times 10^{-12} T^{0.5} \times \exp(-319/T)$ cm^3 molecule^{-1} s^{-1}.

(a) Calculate the activation energy E_a (in the unit kJ/mol) at $T = 298$ K.

(b) The harmonic vibrational frequency for H_2 is 4395 cm^{-1}, and from the latest calculations of the potential energy surface it is known that the activated complex is non-linear, and the harmonic vibrational frequencies are 3772 cm^{-1}, 296 cm^{-1}, and i723 cm^{-1}.
We assume that the rate constant is calculated according to transition-state theory. Calculate the barrier height E_0 (in the unit kJ/mol), using E_a and the exact average vibrational energies.

(c) The transition-state theory is now corrected for quantum tunneling according to Eq. (6.44). Calculate again the barrier height E_0 (in the unit kJ/mol) when quantum tunneling is included.

8.4 We consider a unimolecular reaction where the rate constant is described by the RRKM theory. Show how the Arrhenius activation energy, E_a, is related to the barrier height, E_0, of the reaction.

Part II

Condensed-phase dynamics

9 Introduction to condensed-phase dynamics

Key ideas and results

In this chapter we consider chemical reactions in solution; first, how solvents modify the potential energy surface of the reacting molecules, and second the role of diffusion. The reactants of bimolecular reactions are brought into contact by diffusion, and there will therefore be an interplay between diffusion and chemical reaction that determines the overall reaction rate. The results are as follows.

- As a first approximation, solvent effects can be described by models where the solvent is represented by a dielectric continuum, e.g., the Onsager reaction-field model.
- The overall (effective) reaction rate of a bimolecular reaction is, in general, determined by both the diffusion rate and the chemical reaction rate. A steady-state limit for the effective reaction rate is approached at long times and the rate constant takes a simple form. When the chemical reaction is very fast, the overall rate is determined by the diffusion rate, which is proportional to the diffusion constant. In the opposite limit where the chemical reaction is very slow, the overall rate is equal to the intrinsic rate of the chemical reaction.

Most reactions of interest to chemists take place in either solution or at the gas–solid interface. At the atomic level, much less is known about the reaction dynamics in such systems than about the dynamics of gas-phase reactions. In the gas phase one may follow the detailed evolution from reactants to products without disturbing collisions with other molecules, at least in the low pressure limit. Contrary to that, in solution, where reactants and products are continually perturbed by collisions with solvent molecules, it is much more complicated to follow a chemical reaction.

The complexity of condensed-phase reaction dynamics implies that both from a practical (computational) as well as a conceptual (i.e., insight) point of view, it is desirable to change the approach somewhat compared to gas-phase reaction dynamics. For example, we will see approximate representations of the interaction potentials and the dynamics—the latter, e.g., in the form of so-called stochastic dynamics.

The *elementary reactions* in a solvent include *unimolecular* and *bimolecular* reactions as in the gas phase. In the gas phase the products always fly apart, but in a solvent we also have elementary reactions of the type A + BC $\rightarrow$ ABC that can take place because ABC may be stabilized due to energy transfer to the solvent; that is, a *bimolecular association/recombination reaction.*

The solvent influences the chemical reaction in several ways.

- Solvents may modify the potential energy surface for the reaction.
- Solvents may create the so-called cage effect, where the separation of products may be impeded, leading to a high probability of association/recombination.
- Solvents may enhance and impede molecular motions. There will be energy exchanges between reactants and solvents that may act as an energy sink or source.

These effects are related to the electric properties of the reacting molecules, like their dipole moments and polarizability, as well as to solvent properties, like their dielectric constants and viscosity.

A solvent can dramatically alter the potential energy surface of a reaction from that in the gas phase, as illustrated schematically in Fig. 9.0.1 for the generic S_N2 reaction

$$X^- + CH_3Y \rightarrow XCH_3 + Y^- \tag{9.1}$$

Representative potential energies as functions of the reaction coordinate are shown. Path (a) is followed in the gas phase and path (b) is followed in a polar solvent; the change in the potential energy in a polar solvent shows that both reactants and products are more stabilized than the transition state, because the electrons are less localized in the transition state than in the ionic state. The solvent is seen to have a strong effect on the reaction barrier, so the rate constant in solution may be many

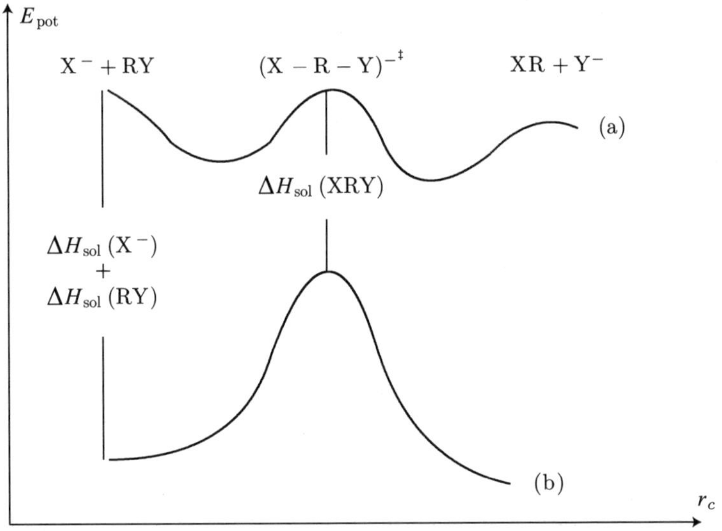

Fig. 9.0.1 Schematic diagram of the potential energy as a function of the reaction coordinate for an S_N2 reaction in (a) the gas phase and in (b) a polar solvent. [Adapted from W.N. Olmstead and J.I. Brauman, *J. Am. Chem. Soc.* **99**, 4219 (1977).]

orders of magnitude smaller than in the gas phase due to the different stabilization of transition state and reactants and products.

It is, however, also possible to find elementary reactions with very similar rates in gas and solution phases. Thus, the thermally-activated unimolecular isomerization of some substituted cyclo-butenes in a solution (of dimethyl phtalate at temperatures around 275°C) has been found to proceed at, essentially, the same rate as in the gas phase.

The well-known Maxwell–Boltzmann distribution for the velocity or momentum associated with the translational motion of a molecule is valid not only for free molecules but also for interacting molecules in a liquid phase (see Appendix A.2.1). The average kinetic energy of a molecule at temperature T is, accordingly, $(3/2)k_BT$. For the molecules to react in a bimolecular reaction they should be brought into contact with each other. This happens by diffusion when the reactants are dispersed in a solution, which is a quite different process from the one in the gas phase. For fast reactions, the diffusion rate of reactant molecules may even be the limiting factor in the rate of reaction.

We begin this part by considering the energetics of solvation, and the combination of diffusion and chemical reactions in order to study the importance of diffusion for the effective rate constant of bimolecular reactions. In the next chapter we study a generalization of transition-state theory to reactions in a solution. The average effect of the solvent on the reaction is studied in a mean-field approximation, where the influence of all dynamical fluctuations has been averaged out. In the last chapter we study a model that includes the dynamical fluctuations. It is a model originally proposed by Kramers, and since it is based on stochastic dynamics, as opposed to deterministic dynamics, we include a discussion of random processes as represented by the Langevin equation and the Fokker–Planck equation to provide a basis for understanding Kramers theory.

9.1 Solvation, the Onsager model

When a molecule is placed in a solvent, the interaction between the molecule (the solute) and the solvent can, roughly, be divided into long-range effects and short-range effects, that is, effects related to the first few 'solvation shells'.

In an exact representation of the interaction between a solute and a solvent, i.e., *solvation*, the solvent molecules must be explicitly taken into account. That is, the solvent is described on a microscopic level, where the individual solvent molecules are considered explicitly. The interaction potential between solvent molecules and between solvent molecules and the solute can, *in principle*, be found by solving the electronic Schrödinger equation for a system consisting of all the involved molecules. Typically, in practice, a more empirical approach is followed where the interaction potential is described by parameterized energy functions. These potential energy functions (often referred to as force fields) are typically parameterized as pairwise atom–atom interactions.

Alternatively, as a first approximation, the solvent can be described on a macroscopic level, where the solvent is considered as a *continuous medium*, i.e., the molecular structure of the solvent molecules is not considered. This description cannot, of course, describe specific short-range effects between the solute and nearby solvent molecules. Within such a model, where the solvent is a continuous medium with *dielectric constant* ϵ_r, the potential energy of two particles with charges q_1 and q_2 separated by a distance R is given by Coulombs law,

$$V(R) = \frac{q_1 q_2}{4\pi\epsilon_r\epsilon_0 R} \tag{9.2}$$

where ϵ_0 is the vacuum permittivity and $\epsilon_r \geqslant 1$, with $\epsilon_r = 1$ in vacuum. The effect of the solvent is accordingly to reduce the interaction between two charges, and for solvents with high dielectric constants (polar solvents) the magnitude of the interaction potential is strongly reduced compared to vacuum.

9.1.1 The Onsager model

When a collection of point charges, a molecule with a permanent dipole moment, or any configuration of atoms corresponding to the transition states in chemical reactions (in the following simply referred to as a 'molecule') is placed in a solvent described as a polarizable continuous medium it will experience: (i) a lowering of its energy, i.e., a net stabilization, and (ii) an increased dipole moment, depending on its polarizability. This is described quantitatively within the so-called Onsager reaction-field model [1,2].

In a simplified molecular model, a net point charge is associated with each atom in a molecule. The permanent *dipole moment* of the isolated molecule is denoted by $\boldsymbol{\mu}$. The dipole moment associated with a set of n point charges is defined by

$$\boldsymbol{\mu} = \sum_{i=1}^{n} q_i \boldsymbol{r}_i \tag{9.3}$$

where q_i and $\boldsymbol{r}_i$ are the charge and the position vector, respectively, of the point charge i. The unit of a dipole is charge (Coulomb) times length, i.e., C $\times$ m.

The molecule is surrounded by the solvent described as a continuous medium with dielectric constant ϵ_r, and it is assumed that the molecule fills a spherical cavity of radius a, which reflects the size of the molecule.

The physical picture is now as illustrated in Fig. 9.1.1. The electric field associated with the molecular dipole ($\boldsymbol{\mu}$) polarizes the dielectric medium (i.e., the solvent). This polarization of the solvent will give rise to an additional electric field ($\boldsymbol{E}_\mathrm{R}$) at the molecular dipole (in addition to the field created by the isolated molecular dipole). The field is called the reaction field of the dipole. The dipole and the field are parallel, i.e.,

$$\boldsymbol{E}_\mathrm{R} = f\boldsymbol{\mu} \tag{9.4}$$

where the factor f will be specified below for two cases, one where the molecule is non-polarizable and one where it is polarizable.

A non-polarizable dipole

It can be shown that the electric (reaction) field at the dipole, due to the solvent, is given by [1,2]

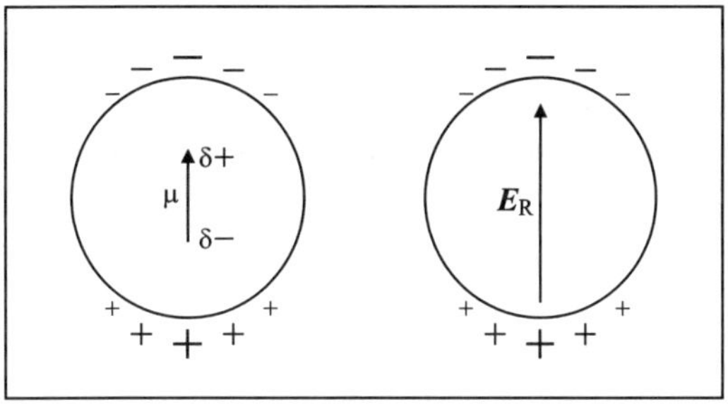

Fig. 9.1.1 The dipole moment $\boldsymbol{\mu}$ associated with two point charges inside a spherical cavity of a dielectric medium. The induced polarization of the medium is illustrated with + and −. The resulting electric field is $\boldsymbol{E}_{\mathrm{R}}$.

$$\boldsymbol{E}_{\mathrm{R}} = f(a, \epsilon_r)\boldsymbol{\mu} = \frac{1}{4\pi\epsilon_0 a^3}\frac{2(\epsilon_r - 1)}{2\epsilon_r + 1}\boldsymbol{\mu} \qquad (9.5)$$

Note that the unit of the electric field is V/m [V/m = J/(C m) = ($\mathrm{J^{-1}\ C^2\ m^{-1}\ m^3}$)$^{-1}$ C m]. The interaction energy between the dipole and the field is given by

$$\boxed{V_{\mathrm{int}} = -\boldsymbol{\mu} \cdot \boldsymbol{E}_{\mathrm{R}} = -\frac{1}{4\pi\epsilon_0 a^3}\frac{2(\epsilon_r - 1)}{2\epsilon_r + 1}\mu^2} \qquad (9.6)$$

From this expression, we observe that the energy stabilization depends on:

(i) the magnitude of the dipole moment (μ^2);

(ii) the dielectric constant of the solvent (ϵ_r);

(iii) the radius of the molecule (cavity radius a).

The dependence on the dielectric constant is given by the factor $2(\epsilon_r - 1)/(2\epsilon_r + 1)$, which can take values in the interval between 0 (for $\epsilon_r = 1$) and 1 (e.g., 0.98 for $\epsilon_r = 78.5$, the dielectric constant of water). Furthermore, a small value of a gives a large stabilization. The magnitude of this radius can, e.g., be estimated from the molecular volume obtained from the ratio between the molecular weight and the molecular density.

Solvent effects for ions can be described by a similar continuum solvent model: the so-called *Born model.* This model predicts a stabilization proportional to the square of the charge, and inversely proportional to the size (radius) of the ion; that is, small and highly charged ions are strongly stabilized in solution.

A polarizable dipole

From Fig. 9.1.1, it is clear that the polarization of the solvent will tend to enhance the electrical asymmetry of the molecule, i.e., enhance the molecular dipole moment. The resulting dipole moment $\boldsymbol{\mu}^*$ is then the permanent dipole plus an induced dipole moment:

$$\boldsymbol{\mu}^* = \boldsymbol{\mu} + \alpha \boldsymbol{E}_{\mathrm{R}}^* \tag{9.7}$$

where α is the *polarizability* of the molecule, and $\boldsymbol{E}_{\mathrm{R}}^*$ is the reaction field of the polarizable dipole. The polarizability measures, roughly speaking, how easy it is to move the electrons of the molecule. Equation (9.4) now takes the form

$$\begin{aligned} \boldsymbol{E}_{\mathrm{R}}^* &= f\boldsymbol{\mu}^* \\ &= f(1-\alpha f)^{-1}\boldsymbol{\mu} \end{aligned} \tag{9.8}$$

where Eq. (9.7) was used in the last line. The factor f is defined in Eq. (9.5). The resulting dipole moment is according to this equation related to the permanent dipole by the relation

$$\boldsymbol{\mu}^* = (1-\alpha f)^{-1}\boldsymbol{\mu} \tag{9.9}$$

The increase compared to the permanent dipole moment can be of the order of 20% to 50%.

9.1.2 Quantum mechanical energy, potential energy surface

The description given above is purely classical, i.e., based on classical electrostatics. In order to transcribe the result into quantum mechanics, we must first replace the dipole moment of Eq. (9.3) with the dipole moment operator, which formally takes the same form. The charges q_i and position vectors $\boldsymbol{r}_i$ are associated with the nuclei and the electrons, respectively, of the molecular system.

We add the operator $-\boldsymbol{\mu}\cdot\boldsymbol{E}_{\mathrm{R}}$ to the total molecular Hamiltonian. According to Eq. (3.1), the electronic Hamiltonian of the molecule in the field due to the solvent is then $\hat{H}_e - \boldsymbol{\mu}\cdot\boldsymbol{E}_{\mathrm{R}}$. The electronic Schrödinger equation is then solved using this modified Hamiltonian. This leads to a self-consistent solution where the electronic wave function and the electronic energy are modified due to the solvent field. Thus, polarization of the molecular electronic density (as described approximately above) is automatically included in this approach.

We can estimate the effect of the solvent within first-order perturbation theory. The corrected energy takes the form

$$\begin{aligned} \tilde{E}_i(\boldsymbol{R}') &= E_i(\boldsymbol{R}') + \langle\psi_i(\boldsymbol{r};\boldsymbol{R}')| - \boldsymbol{\mu}\cdot\boldsymbol{E}_{\mathrm{R}}|\psi_i(\boldsymbol{r};\boldsymbol{R}')\rangle \\ &= E_i(\boldsymbol{R}') - f\langle\psi_i(\boldsymbol{r};\boldsymbol{R}')|\mu^2|\psi_i(\boldsymbol{r};\boldsymbol{R}')\rangle \end{aligned} \tag{9.10}$$

where $E_i(\boldsymbol{R}')$ is the electronic energy, $\psi_i(\boldsymbol{r};\boldsymbol{R}')$ is the stationary electronic wave function of the isolated (gas-phase) molecule, and $f = (4\pi\epsilon_0 a^3)^{-1}2(\epsilon_r - 1)/(2\epsilon_r + 1)$. Note that the integration in the matrix element is over all coordinates ($\boldsymbol{r}$) associated with the electrons. Thus, the quantum mechanical expectation value of the (squared) dipole moment will depend on the internuclear distances. The perturbation from the solvent will accordingly introduce a modified potential energy surface. This will lead to a shift of vibrational frequencies and barrier heights compared to the gas phase [3].

The order of magnitude of the energy correction is easy to estimate. Thus, with a dipole moment equivalent to a unit (electron) charge separated by 1 ångström, we get $\mu \sim 1.602 \times 10^{-19} \times 10^{-10}$ C m, and with $f \sim (4\pi\epsilon_0 a^3)^{-1}$ and $a \sim 3 \times 10^{-10}$ m the energy correction is 8.5×10^{-20} J = 0.5 eV $\sim$ 50 kJ/mol. This number is similar to barrier heights in the gas phase and the energetics of solvation can clearly play an essential role in condensed-phase reaction dynamics.

Since the development of the Onsager model, there have been a number of elaborations on the model [4,5]. For example, the spherical cavity has been replaced by 'molecularly-shaped' cavities. The state of the art within the field of solvent effects described by continuum solvent models is now implemented in, e.g., the Gaussian program package.

9.2 Diffusion and bimolecular reactions

In order for two reactants A and B to react in a bimolecular reaction, they need to be brought in the vicinity of each other. When dispersed in a fluid, this happens by diffusive motion, which is entirely different from the free motion in the gas phase. Once an encounter between two reactants takes place, they will usually stay together much longer than in a gas phase due to a 'cage' effect of the surrounding fluid molecules. This allows for numerous exchanges of energy between reactants and fluid, and thereby for activation and deactivation of the reaction complex. A complicated interplay between diffusion rates and reaction rates may determine the overall reaction rate in a fluid. We shall study an example of how diffusive motion and chemical reactions are combined in a description of chemical reactions in solution.

9.2.1 Introduction, a macroscopic description

A simplified picture of the situation is often given by the reaction scheme

$$\mathrm{A} + \mathrm{B} \underset{k_D'}{\overset{k_D}{\rightleftarrows}} \mathrm{AB} \overset{k_s'}{\longrightarrow} \mathrm{P}$$

k_D is determined by the diffusion of reactants toward each other and k_D' is the rate constant for the reverse process which implies that the reactants in the encounter pair may diffuse away from each other without reaction. k_s' denotes the reaction rate constant for the pair. Assuming *steady state* with respect to AB leads to

$$\frac{dC_{\mathrm{AB}}}{dt} = k_D C_{\mathrm{A}} C_{\mathrm{B}} - k_D' C_{\mathrm{AB}} - k_s' C_{\mathrm{AB}} \equiv 0 \tag{9.11}$$

so

$$C_{\mathrm{AB}} = \frac{k_D C_{\mathrm{A}} C_{\mathrm{B}}}{k_s' + k_D'} \tag{9.12}$$

and

$$\frac{dC_{\mathrm{P}}}{dt} = k_s' C_{\mathrm{AB}} = \frac{k_s' k_D}{k_s' + k_D'} C_{\mathrm{A}} C_{\mathrm{B}} \equiv k C_{\mathrm{A}} C_{\mathrm{B}} \tag{9.13}$$

We note that the effective rate constant k is determined by both the diffusion rate constant k_D and the reaction rate constant k_s'. Let us consider two limits. If $k_D' \ll k_s'$, then the effective rate constant is seen to be $k \sim k_D$, that is, the rate constant is determined by the rate of diffusion. This is often referred to as the *diffusion-controlled limit*. When $k'_s \ll k'_D$, then

$$k \sim \frac{k_D}{k'_D} k'_s = K k'_s \tag{9.14}$$

where K is the equilibrium constant for $\mathrm{A} + \mathrm{B} \rightleftharpoons \mathrm{AB}$. The effective rate constant is given as a product of the equilibrium constant K and the rate constant k'_s. This is usually referred to as the *activation-controlled limit*, since the effective rate constant is determined by the rate constant k'_s.

9.2.2 Fick's second law of diffusion and chemical reaction

After these qualitative considerations we will go on and study chemical reactions in solution in more detail and determine the effective rate constant in terms of fundamental properties such as a diffusion constant D and the intrinsic bimolecular rate constant k_s.

The equation we have to solve may be written as

$$\frac{\partial C_\mathrm{B}}{\partial t} = D \nabla^2 C_\mathrm{B} - k_s C_\mathrm{B} C_\mathrm{A} \tag{9.15}$$

when we neglect convection in the system. The left-hand side of the equation expresses the local rate of change in the concentration of B. The first term on the right-hand side of the equation expresses the contribution to the rate of change from diffusion and the second term the contribution from the chemical reaction. The change in the concentration of A may usually be found from a material balance equation. This is a very complicated partial differential equation that is difficult to solve and that cannot be solved analytically. For a qualitative understanding of the behavior of a system with diffusion and chemical reactions it would, however, be useful if an approximate analytical solution could be found. This is indeed possible, as shown in the following. The idea is that, instead of solving the complete equation above, we solve the equation with the diffusion term included but without the chemical reaction term. That term is then introduced in the boundary condition. This gives equations that may be solved analytically and the solution may help us to understand qualitative features of such a system.

We consider the diffusive motion of a B molecule relative to an A molecule. In order for a reaction to occur, the reactants must be brought close together by the diffusive motion, that is, a B molecule must approach an A molecule. For the sake of solving the differential equation used to describe this problem we need to specify some distance R_c between A and B at which reaction may take place. It is a *necessary* condition for a reaction to occur that the molecules must get close to each other, say at a distance R_c, but not a *sufficient* condition. Whether or not they will react is determined by the reaction rate constant k_s in a simple second-order reaction scheme according to $k_s C_\mathrm{B}(R_c, t)$ (it is second order because the concentration of A is one at R_c and therefore not seen explicitly in the expression). $C_\mathrm{B}(R_c, t)$ is the concentration of B at a distance R_c from the A molecule. The diffusive motion of B is described by *Fick's second law* of diffusion:

$$\frac{\partial C_\mathrm{B}}{\partial t} = D \nabla^2 C_\mathrm{B} \tag{9.16}$$

We shall assume that our system is spherically symmetric; so with the nabla operator in spherical coordinates, the diffusion equation may be written

$$\frac{\partial C_{\mathrm{B}}(r,t)}{\partial t} = D\left[\frac{\partial^2}{\partial r^2} + \frac{2}{r}\frac{\partial}{\partial r}\right] C_{\mathrm{B}}(r,t) \tag{9.17}$$

Here r denotes the distance of B from A. The equation is solved with the following boundary conditions:

(i) for $r < R_c$: $C_{\mathrm{B}}(r,0) = 0$;

(ii) for $r \geqslant R_c$: $C_{\mathrm{B}}(r,0) = C_{\mathrm{B}}$ (the bulk concentration);

(iii) for $t > 0$: $C_{\mathrm{B}}(r \to \infty, t) = C_{\mathrm{B}}$;

(iv) $I_{\mathrm{B}}(R_c,t) = 4\pi R_c^2 |J_{\mathrm{B}}(R_c,t)| = 4\pi R_c^2 D(\partial C_{\mathrm{B}}(r,t)/\partial r)_{r=R_c} = k_s C_{\mathrm{B}}(R_c,t)$.

In boundary condition (iv), $I_{\mathrm{B}}(R_c,t)$ is the flow of B molecules across the surface of a sphere with radius R_c, the distance between the reactants, where a reaction may take place. This flow is expressed by the flux density, $J_{\mathrm{B}}(R_c,t)$ (i.e., the number of molecules per second that are passing through an area of one square meter), as given by *Fick's first law*

$$J_{\mathrm{B}}(R_c,t) = -D\left(\frac{\partial C_{\mathrm{B}}(r,t)}{\partial r}\right)_{r=R_c} \tag{9.18}$$

multiplied by the area of the sphere. The flow of B is set equal to the rate by which B molecules disappear in the chemical reaction. The condition expresses a *steady state* at all times at $r = R_c$ between the diffusive influx of B and the rate by which B molecules disappear by the chemical reaction. The boundary conditions are summarized in the sketch in Fig. 9.2.1.

Equation (9.17) is solved by a Laplace transformation. In chemical kinetics and diffusion, the problems may often be formulated in terms of partial differential equations

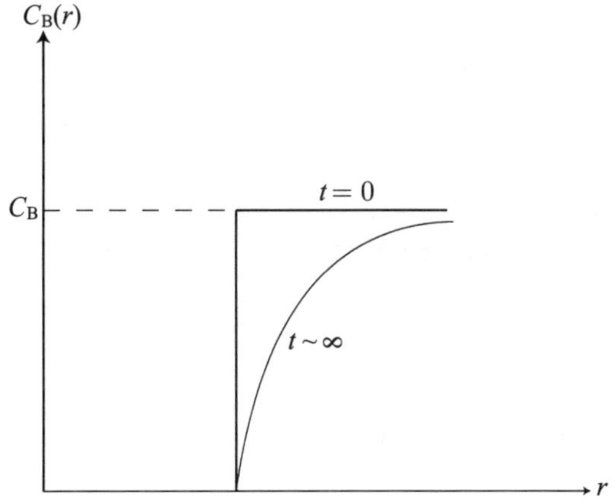

Fig. 9.2.1 Sketch of boundary conditions (i), (ii), and (iii) at infinite time. The concentration profile at time $t \sim \infty$ is a sketch of the solution when every encounter at $r = R_c$ leads to reaction.

that are first order with respect to time and second order with respect to position coordinates. In order to solve the problem we seek a solution to the differential equation for given initial and boundary conditions. The Laplace transformation technique is often used in solving these differential equations. One transforms the original function in time and coordinates $F(r,t)$ to a Laplace transformed function $\tilde{F}(r,s)$ in frequency s and coordinates r, by

$$\tilde{F}(r,s) = \int_0^\infty dt F(r,t) \exp(-st) \tag{9.19}$$

The idea of using this technique is that we may transform the partial differential equation in both time and coordinates into an ordinary differential equation in just the coordinates, which is usually easier to solve. The elimination of the derivative with respect to time dc/dt is easily seen by partial integration:

$$\begin{aligned} &\int_0^\infty dt(dc(t)/dt)\exp(-st) = \\ &\qquad [\exp(-st)c]_0^\infty + s\int_0^\infty dt c(t)\exp(-st) = s\tilde{c}(s) - c(0) \end{aligned} \tag{9.20}$$

When we have found a solution for the Laplace transformed function, then we need to make an inverse transformation to find the solution in terms of time and coordinates. There are elegant techniques for doing this based on the theory of complex functions, but often these are not necessary since there exist extensive tables in mathematical handbooks of functions and their Laplace transformed functions. Only in cases where the relevant functions have not been tabulated will it be necessary to carry out the inverse transformation using these techniques.

Taking the Laplace transform of Eq. (9.17) gives

$$s\tilde{C}_{\mathrm{B}}(r,s) - C_{\mathrm{B}}(r,0) = D\left[\frac{\partial^2}{\partial r^2} + \frac{2}{r}\frac{\partial}{\partial r}\right]\tilde{C}_{\mathrm{B}}(r,s) \tag{9.21}$$

This is a standard second-order differential equation with the solution

$$\frac{\tilde{C}_{\mathrm{B}}(r,s)}{C_{\mathrm{B}}(r,0)} = \frac{1}{s} + \frac{A}{r}\exp(r\sqrt{s/D}) + \frac{B}{r}\exp(-r\sqrt{s/D}) \tag{9.22}$$

which may easily be checked by substitution into the differential equation.

Boundary condition (iii) requires that $A = 0$. The Laplace transform of boundary condition (iv) is

$$k_s\tilde{C}_{\mathrm{B}}(R_c,s) = 4\pi R_c^2 D\left(\frac{\partial\tilde{C}_{\mathrm{B}}(r,s)}{\partial r}\right)_{r=R_c} \tag{9.23}$$

This may be used to determine B. We find after some lengthy but straightforward manipulations that

$$\begin{aligned} B &= -\frac{R_c}{s}\frac{k_s}{k_s + 4\pi R_c D + 4\pi R_c^2\sqrt{sD}}\exp(R_c\sqrt{s/D}) \\ &= -\frac{1}{s}\frac{R_c k_s}{(k_s + 4\pi R_c D)}\frac{(k_s + 4\pi R_c D)/(4\pi R_c^2\sqrt{D})}{(k_s + 4\pi R_c D)/(4\pi R_c^2\sqrt{D}) + \sqrt{s}}\exp(R_c\sqrt{s/D}) \end{aligned} \tag{9.24}$$

so the Laplace transformation of the concentration is given by

$$\frac{\tilde{C}_{\mathrm{B}}(r,s)}{C_{\mathrm{B}}} = \frac{1}{s} - \frac{R_c}{r}\frac{k_s}{(k_s+4\pi R_c D)}\frac{(k_s+4\pi R_c D)/(4\pi R_c^2\sqrt{D})}{s((k_s+4\pi R_c D)/(4\pi R_c^2\sqrt{D})+\sqrt{s})}\exp(-(r-R_c)\sqrt{s/D}) \tag{9.25}$$

where we have used boundary condition (ii).

We shall now determine the inverse Laplace transform of this function to find the concentration of B at time t. In tables of Laplace transforms the following pair has been found:

$$\tilde{F}(s) = \frac{a}{s(a+\sqrt{s})}\exp(-b\sqrt{s}) \tag{9.26}$$

with the inverse given by

$$F(t) = \mathrm{erfc}(b/(2\sqrt{t})) - \exp(ab)\exp(a^2 t)\mathrm{erfc}(a\sqrt{t}+b/(2\sqrt{t})) \tag{9.27}$$

where erfc denotes the error function complement, which is defined by the error function, erf, according to

$$\begin{aligned}\mathrm{erfc}(x) &= 1-\mathrm{erf}(x)\\ &\equiv 1-\frac{2}{\sqrt{\pi}}\int_0^x dz\,\exp(-z^2)\end{aligned} \tag{9.28}$$

The Laplace transform $\tilde{F}(s)$ is seen to have exactly the same form as the second term in Eq. (9.25) with

$$b = (r-R_c)/\sqrt{D} \tag{9.29}$$

and

$$a = (k_s+4\pi R_c D)/(4\pi R_c^2\sqrt{D}) \tag{9.30}$$

so the solution may be written

$$\begin{aligned}C_{\mathrm{B}}(r,t)/C_{\mathrm{B}} = 1 - \frac{R_c}{r}\frac{k_s}{k_s+4\pi R_c D}\Bigg[&\mathrm{erfc}\left(\frac{r-R_c}{\sqrt{4Dt}}\right)\\ &-\exp\left(\frac{(k_s+4\pi R_c D)(r-R_c)}{4\pi R_c^2 D}\right)\exp\left(\frac{(k_s+4\pi R_c D)^2}{(4\pi R_c^2\sqrt{D})^2}t\right)\\ &\times\mathrm{erfc}\left(\frac{r-R_c}{\sqrt{4Dt}}+\frac{k_s+4\pi R_c D}{4\pi R_c^2\sqrt{D}}\sqrt{t}\right)\Bigg]\end{aligned} \tag{9.31}$$

where we have used that the inverse transform of $1/s$ is 1, as easily shown or found in the tables.

The time evolution of the concentration is illustrated in Fig. 9.2.2, using a diffusion constant that is typical for diffusional motion in liquid water. The curve at $t = 25$ ns is very close to being stationary, consistent with the steady-state boundary condition (iv).

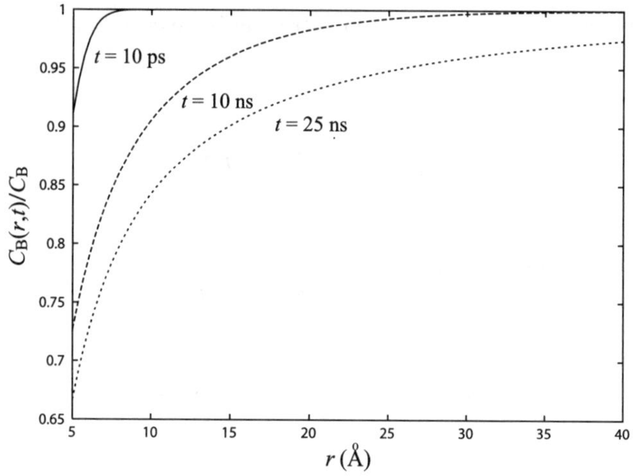

Fig. 9.2.2 Concentration profiles according to Eq. (9.31). The diffusion constant is $D = 10^{-9}$ m^2 s^{-1}, $R_c = 5$ Å (see Example 9.2), and $4\pi R_c D/k_s = 2$.

The rate of reaction may then be expressed in terms of the flow $I_{\mathrm{B}}(R_c, t)$ of B across a sphere of radius R_c around A using boundary condition (iv):

$$-\frac{dC_{\mathrm{A}}}{dt} = I_{\mathrm{B}}(R_c, t)C_{\mathrm{A}} = k_s C_{\mathrm{B}}(R_c, t)C_{\mathrm{A}} \equiv k(t)C_{\mathrm{A}}C_{\mathrm{B}} \tag{9.32}$$

which defines an *effective rate constant* $k(t)$ that may be a function of time. After a little algebra using Eq. (9.31), we find the following expression for the rate constant:

$$k(t) = \frac{4\pi R_c D k_s}{k_s + 4\pi R_c D}\left[1 + \frac{k_s}{4\pi R_c D}\exp\left(\frac{(k_s + 4\pi R_c D)^2}{(4\pi R_c^2\sqrt{D})^2}t\right)\operatorname{erfc}\left(\frac{k_s + 4\pi R_c D}{4\pi R_c^2\sqrt{D}}\sqrt{t}\right)\right] \tag{9.33}$$

It is seen that the effective rate constant depends on the diffusion constant D and the intrinsic rate constant k_s in a rather complicated way, and that it is a function of time. The time dependence is a consequence of the transient approach to stationarity of the concentration profile of B (see Fig. 9.2.2). At stationarity, the rate constant is independent of time, which is also seen from the asymptotic expansion

$$\lim_{z\to\infty} \sqrt{\pi}z\exp(z^2)\operatorname{erfc}(z) = 1 + \cdots \tag{9.34}$$

Thus, for $t \to \infty$ we obtain the steady-state limit for the effective rate constant:

$$\boxed{k(t \to \infty) \equiv k = \frac{4\pi R_c D k_s}{k_s + 4\pi R_c D}} \tag{9.35}$$

The relation for k is often written as

$$\frac{1}{k} = \frac{1}{k_s} + \frac{1}{4\pi R_c D} \tag{9.36}$$

For the case where $k_s \to \infty$, which means that every physical encounter at R_c leads to a reaction, we find

$$\lim_{k_s \to \infty} k = 4\pi R_c D \equiv k_D \tag{9.37}$$

which is the rate constant in the *diffusion-controlled limit*, where the diffusion constant determines the overall rate of the reaction. This relation identifies k_D of the simplified picture in Eq. (9.13) with $4\pi R_c D$. Diffusion-controlled reactions typically include recombination reactions of atoms and radicals that have small, zero, or even negative activation energies. For example, for the association/recombination reaction of iodine atoms in hexane at 298 K, $I + I \to I_2$, the rate constant is found to be 1.3×10^7 m^3 mol^{-1} s^{-1}.

In the *activation-controlled limit* where $k_s \ll k_D$, the rate constant is seen to be k_s, that is, the reaction rate is determined by the intrinsic rate constant alone. The majority of chemical reactions in liquid solution are activation controlled.

Example 9.1: Diffusion and chemical reaction in an ideal gas

Here we estimate the relative magnitude of k_s and k_D in an ideal gas at various pressures. An upper limit to k_s, where it is assumed that every collision leads to reaction, was given by Eq. (4.14), $k_s = \sigma\langle v\rangle$, where $\sigma = \pi d^2$ is the reaction cross-section and $\langle v\rangle = \sqrt{8k_BT/(\mu\pi)}$ is the average velocity at the temperature T. The diffusion constant is (from kinetic gas theory) given by $D = (1/3)\lambda\langle v\rangle$, where $\lambda = k_BT/(\sqrt{2}\sigma p)$ is the mean-free path of the molecule at the pressure p. That is,

$$\begin{aligned}\frac{k_s}{k_D} &= \frac{\sigma\langle v\rangle}{4\pi R_c D}\\ &= \frac{3\sigma}{4\pi R_c\lambda}\\ &\sim \frac{R_c}{\lambda}\end{aligned}$$

where in the last line it was assumed that $d \sim R_c$. Thus, as long as the mean-free path is large compared to the internuclear distance where reaction takes place, R_c, we are in the activation-controlled limit.

Consider, as an example, molecules at $p = 1$ atm and $T = 298$ K. Typically d ($\sim R_c$) is 0.3 nm, which implies that $\lambda \sim 102$ nm. Thus, at moderate—not too high—pressures the mean-free path is several hundred times larger than the molecular dimensions. Reactions in an ideal gas at standard conditions accordingly take place such that 'free flight' prevails between the collisions. Diffusional motion only plays a role at very high pressures.

Before closing, let us add a few comments about the diffusion constant D in the equations above. For diffusion of solutes like A and B in a solvent, it is customary to introduce diffusion constants D_A and D_B such that the associated fluxes J'_A and J'_B are given relative to the flux of the solvent molecules. Since they are present in an overwhelming quantity as compared to the solutes, this will be equivalent to a center-of-mass reference frame, and if the system is stationary, to a laboratory reference frame. In this calculation we have used another reference frame in which the flux of B is measured *relative* to the flux of A, so we need to find a relation between D and the ordinary diffusion constants $D_{\rm A}$ and $D_{\rm B}$ as they may be found in tables. We write

$$\begin{aligned} J'_{\rm A} &= -D_{\rm A}\nabla C_{\rm A} \\ J'_{\rm B} &= -D_{\rm B}\nabla C_{\rm B} \end{aligned} \tag{9.38}$$

so the flux of B relative to the flux of A is given by

$$J_{\rm B} \equiv J'_{\rm B} - J'_{\rm A} = -D_{\rm B}\nabla C_{\rm B} + D_{\rm A}\nabla C_{\rm A} \tag{9.39}$$

The gradients are not independent but coupled through the Gibbs–Duhem equation

$$C_{\rm A}\nabla\mu_{\rm A} + C_{\rm B}\nabla\mu_{\rm B} + C_w\nabla\mu_w = 0 \tag{9.40}$$

Here C_w is the concentration of the solvent and μ_i is the chemical potential of component i. For the solvent we have that $\nabla\mu_w \sim 0$, so

$$C_{\rm A}\nabla\mu_{\rm A} = -C_{\rm B}\nabla\mu_{\rm B} \tag{9.41}$$

and with

$$\mu_i = \mu_i^0 + RT\ln(C_i/C_i^0) \tag{9.42}$$

we find that

$$\nabla C_{\rm A} = -\nabla C_{\rm B} \tag{9.43}$$

Substitution into Eq. (9.39) gives

$$J_{\rm B} = -(D_{\rm A} + D_{\rm B})\nabla C_{\rm B} \tag{9.44}$$

and a comparison with Eq. (9.18) finally shows that

$$D = D_{\rm A} + D_{\rm B} \tag{9.45}$$

Example 9.2: Diffusion and chemical reaction in liquid water

In the diffusion-controlled limit we found that $k = k_D = 4\pi R_c D$, so from the diffusion constant D and a reasonable estimate of R_c it is possible to determine the order of magnitude of k_D. In Table 9.1, we have listed the diffusion constants for a series of ions and molecules in water at room temperature. It is seen that

Table 9.1 Diffusion constants in water at 298 K.

Substance	D m^2 s^{-1}	Substance	D m^2 s^{-1}
H^+	9.1×10^{-9}	Glycin	1.06×10^{-9}
Li^+	1.0×10^{-9}	Dextrose	0.67×10^{-9}
Na^+	1.3×10^{-9}	Sucrose	0.52×10^{-9}
OH^-	5.2×10^{-9}	Methanol	1.58×10^{-9}
Cl^-	2.0×10^{-9}	Ethanol	1.24×10^{-9}

typical values of D are 10^{-8}–10^{-10} m^2 s^{-1}. When the distance between reactants is small enough and equal to R_c, then a reaction may occur as it was assumed in the derivation given above. A reasonable estimate of that distance would be that $R_c \approx 5.0 \times 10^{-10}$ m, so with a diffusion constant $D \approx 10^{-9}$ m^2 s^{-1} we find $k_D = 4\pi \times 5.0 \times 10^{-10} \times 10^{-9}$ m^3 molecule^{-1} s^{-1}, and after multiplication by Avogadros number $N_A = 6.023 \times 10^{23}$ we find $k_D \approx 10^6$–10^7 m^3 mol^{-1} s^{-1}. Note that $k = k_D = 4\pi R_c D$ is the upper limit for the magnitude of a rate constant associated with a bimolecular reaction in solution.

Thus, from the discussion of the rate constant in Eq. (9.35), we have in water the following conditions:

- $k_s \ll 10^6$ m^3 mol^{-1} s^{-1}, the reaction will be *activation controlled*;
- $k_s \gg 10^7$ m^3 mol^{-1} s^{-1}, the reaction will be *diffusion controlled.*

An example of a diffusion-controlled reaction is the acid–base reaction $H^+ + OH^- \rightarrow H_2O$. The experimentally determined rate constant at 298 K is 1.4×10^8 m^3 mol^{-1} s^{-1}, which is very large even for a diffusion-controlled reaction. Although, as observed from the table, the diffusion constants for the ions are large, $k_D = 4\pi R_c D = 5.4 \times 10^7$ m^3 mol^{-1} s^{-1} assuming $R_c = 5.0 \times 10^{-10}$ m. Thus, the predicted value is about a factor of two smaller than the experimentally observed value. This suggests, not surprisingly, that a modified treatment of diffusion is required for reactions involving ions (see the following subsection and Problem 9.2).

The diffusion constant D_i of a particle in a solvent is related to the *viscosity* of the solvent by the *Stokes–Einstein relation* known from hydrodynamics:

$$D_i = \frac{k_B T}{n_h \pi \eta R_i} \tag{9.46}$$

where η is the viscosity, R_i the hydrodynamic radius of the diffusing particle, and n_h is a numerical constant. The equation was derived for a continuum solvent (that is, $R_i \gg$ radius of the solvent molecules), in which case $n_h = 6$. The validity of the equation for molecular diffusion is somewhat questionable; in spite of this it is also used for this case but with a change of the numerical constant n_h from 6 to 4. When we apply the equation to our case, we find the following relation between D and the viscosity of the solvent:

$$D = \frac{k_B T}{n_h \pi \eta}\left[\frac{1}{R_\mathrm{A}} + \frac{1}{R_\mathrm{B}}\right] \tag{9.47}$$

There is rarely much information available on either the hydrodynamic radius of a species and the encounter radius R_c. It is often assumed that $R_c = R_\mathrm{A} + R_\mathrm{B}$, and if R_A and R_B are also approximately equal, Eqs (9.37) and (9.47) may be combined to give a simple expression for the rate constant:

$$k_D = \frac{16 k_B T}{n_h \eta} \approx \frac{4 k_B T}{\eta} \tag{9.48}$$

where we have used $n_h = 4$. This shows that the rate constant in the diffusion-controlled limit is given as a function of the viscosity and the temperature of the solvent, and is independent of the properties of the reactants. The viscosity of water at 298 K is $\eta = 0.8904 \times 10^{-3}$ kg m^{-1} s^{-1}, so for water at 298 K the rate constant will be $k_D = 1.1 \times 10^7$ m^3 mol^{-1} s^{-1}, in rough agreement with the result in Example 9.2.

9.2.3 Stochastic description of diffusion and chemical reaction

In this section, we will describe the results of the previous section within a probabilistic framework based on stochastic dynamics [8,9]. Furthermore, the discussion will be extended to the diffusion of particles with an interaction potential $U(r)$ depending on the distance r between the particles. An example will be the electrostatic potential associated with the interaction of ions.

Diffusion can be considered as a stochastic or random process and described by the so-called *Fokker–Planck equation* adapted to Brownian motion. This equation is also known as the *Smoluchowski equation*. We consider the description of stochastic processes and Brownian motion in more detail in Section 11.1 and Appendix H.

The Fokker–Planck equation, Eq. (H.24), describing one-dimensional diffusive motion in a potential $U(x)$ in terms of the probability $P(x,t)dx$ of finding a particle at the position $[x, x+dx]$, is

$$\begin{aligned}\frac{\partial P(x,t)}{\partial t} &= D\frac{\partial^2 P(x,t)}{\partial x^2} + \frac{D}{k_B T}\frac{\partial}{\partial x}\left(\frac{\partial U}{\partial x}P(x,t)\right)\\ &= -\frac{\partial}{\partial x}\left(-D\frac{\partial P(x,t)}{\partial x} - D\frac{\partial U}{\partial x}\frac{P(x,t)}{k_B T}\right)\\ &\equiv -\frac{\partial}{\partial x}J(x,t)\end{aligned} \tag{9.49}$$

In the second line, the equation is written in the form of a continuity equation that, formally, is identical to the continuity equations in quantum mechanics, Eq. (4.106), and in classical statistical mechanics, Eq. (5.17). The probability flux density is identified as

$$J(x,t) = -D\left(\frac{\partial P(x,t)}{\partial x} + \frac{\partial U}{\partial x}\frac{P(x,t)}{k_B T}\right) \tag{9.50}$$

The first term is equivalent to the flux due to concentration gradients (i.e., Fick's first law) and the second term represents the flux due to gradients in the potential.

Equation (9.49) can be generalized to spherically-symmetric diffusion in three dimensions, and the probability of finding the AB pair at the distance $[r, r+dr]$ is $P(r,t)dr$, and is given by

$$\frac{\partial P(r,t)}{\partial t} = D\left[\frac{\partial^2}{\partial r^2} + \frac{2}{r}\frac{\partial}{\partial r}\right]P(r,t) + \frac{D}{k_B T}\left[\frac{\partial}{\partial r}\left(\frac{\partial U}{\partial r}P(r,t)\right) + \frac{2}{r}\frac{\partial U}{\partial r}P(r,t)\right] \tag{9.51}$$

Note that, if the potential $U(r)$ is independent of r, then the second term disappears, and the equation becomes equivalent to Fick's second law in Eq. (9.17) (furthermore, the two terms containing $2/r$ are absent in the one-dimensional description).

Equation (9.51) is solved with the initial condition (replacing boundary condition (ii) in Section 9.2.2)

$$P(r, t = 0) = C \exp(-U(r)/k_B T) \tag{9.52}$$

which is the probability density, irrespective of the momentum, of finding the AB pair at distance r according to the equilibrium Boltzmann distribution (C is a normalization constant). The boundary condition at $r = R_c$, replacing (iv) above, is

$$4\pi R_c^2 D \left(\frac{\partial P(r,t)}{\partial r} + \frac{\partial U}{\partial r} \frac{P(r,t)}{k_B T} \right)_{r=R_c} = k_s P(R_c, t) \tag{9.53}$$

following Eq. (9.50).

For $U(r) = 0$, the equations are mathematically equivalent to the equations in Section 9.2.2, and Eq. (9.33) is the exact solution for $k(t)$. For $U(r) \neq 0$, it can be shown [8] that an approximate expression for the time-dependent rate constant is given by Eq. (9.33), provided R_c is replaced by R_c^{eff}, where the effective radius is given by

$$1/R_c^{\text{eff}} = \int_{R_c}^{\infty} e^{U(r)/k_B T} r^{-2} dr \tag{9.54}$$

Note that for $U(r) = 0$, $R_c^{\text{eff}} = R_c$.

When $U(r) \neq 0$, the approximate expression for $k(t)$ behaves correctly at $t \to \infty$, and the diffusion-controlled rate constant becomes

$$k_D = 4\pi R_c^{\text{eff}} D \tag{9.55}$$

i.e., the same form as in Eq. (9.37).

For reactions involving ions, it is relevant to consider a Coulomb potential, Eq. (9.2). The integral in Eq. (9.54) is easily evaluated (using the substitution $x = 1/r$) and

$$R_c^{\text{eff}} = R_c (V(R_c)/k_B T)[\exp(V(R_c)/k_B T) - 1]^{-1} \tag{9.56}$$

where $V(R_c) = q_{\text{A}} q_{\text{B}}/(4\pi\epsilon_r \epsilon_0 R_c)$ is the Coulomb potential evaluated at $R = R_c$, and $q_{\text{A}} = z_{\text{A}} e$ and $q_{\text{B}} = z_{\text{B}} e$. For ions of opposite charge, $R_c^{\text{eff}} > R_c$, and the charges give rise to an increased value of the diffusion-controlled rate constant.

The Coulomb potential describes the interaction between two isolated charged particles, i.e., ions. At finite concentrations, a better description is obtained using the so-called (Debye–Hückel) screened or shielded Coulomb potential.

Further reading/references

[1] L. Onsager, *J. Am. Chem. Soc.* **58**, 1486 (1936)

[2] C.J.F. Böttcher, *Theory of electric polarisation* (Elsevier, Amsterdam, 1952) Chapter 3

[3] M.W. Wong and K.B. Wiberg, *J. Chem. Phys.* **95**, 8991 (1991)

[4] J. Tomasi and M. Persico, *Chem. Rev.* **94**, 2027 (1994)

[5] M. Cossi, G. Scalmani, N. Rega, and V. Barone, *J. Chem. Phys.* **117**, 43 (2002)

[6] F.C. Collins and G.E. Kimball, *J. Colloid. Sci.* **4**, 425 (1949)

[7] R.M. Noyes, *Prog. Reaction Kinetics* **1**, 129 (1961)

[8] A. Szabo, *J. Phys. Chem.* **93**, 6929 (1989)
[9] B. Cohen, D. Huppert, and N. Agmon, *J. Phys. Chem.* **105**, 7165 (2001)

Problems

9.1 Show that the function in Eq. (9.52) is a stationary solution to Eq. (9.51), i.e., it satisfies the equation for $\partial P(r,t)/\partial t = 0$.

9.2 Calculate the diffusion-controlled rate constant for $H^+ + OH^- \rightarrow H_2O$ at $T = 298$ K, using Eqs (9.55) and (9.56), and relevant data from Example 9.2.

10

Static solvent effects, transition-state theory

Key ideas and results

In this chapter we consider static solvent effects on the rate constant for chemical reactions in solution. The static equilibrium structure of the solvent will modify the potential energy surface for the chemical reaction. This effect can be analyzed within the framework of transition-state theory. The results are as follows.

- The rate constant can be expressed in terms of the potential of mean force at the activated complex. This potential may, e.g., be defined such that the gradient of the potential gives the average force on an atom in the activated complex due to the solvent molecules, Boltzmann averaged over all configurations.
- A relation between the rate constants in the gas phase and in solution can be derived. The effect can, approximately, be expressed in a form where the relation between the rate constants in the gas phase and in solution is given in terms of the potential of mean force at the transition state.

We shall begin our studies of the transition-state theory for reactions in solution by considering the influence of the static structure of the solvent on the rate constant for a reaction. This corresponds to an evaluation of the average effect of the solvent on the rate constants; that is, we consider all possible configurations of the solvent molecules around the reactants and the activated complex, determine the effect on them for each particular solvent molecule configuration, and evaluate the overall effect as a statistical weighted sum of the particular effects. This approach is therefore often referred to as a *mean field* determination of the solvent effect, since the influence of spontaneous fluctuations in solvent molecule positions and velocities is neglected. The effect of these will be considered in the next chapter using stochastic dynamics for the effect of solvents on the rate constants.

The basic expressions for the rate constant within a fully classical version of conventional transition-state theory were derived in Chapter 5. According to Eq. (5.53), we may write

$$k_{\mathrm{TST}}(T) = \frac{k_{\mathrm{B}}T}{h} V^{\nu-1} \frac{\mathcal{Q}^{\ddagger}}{\mathcal{Q}_r} \tag{10.1}$$

where $\mathcal{Q}^\ddagger$ and $\mathcal{Q}_r$ are the partition functions of the activated complex and the reactants, respectively, *including* the surrounding solvent molecules, and where $\nu = 1$ for a unimolecular reaction and $\nu = 2$ for a bimolecular reaction. Before we go into the more detailed atomic-level description of solvent effects, let us briefly consider the *thermodynamic formulation* of Eq. (10.1). We showed in Section 6.6 that in the gas phase the transition-state theory rate constant given in terms of molecular partition functions can be transcribed into a thermodynamic formulation. In a condensed phase, we can again transcribe the basic expression for the rate constant into a thermodynamic formulation. To that end, we note that the Helmholtz (free) energy is related to the canonical partition function according to [1] $A = -k_B T \ln \mathcal{Q}$, where $\mathcal{Q}$ here is the partition function of N interacting molecules. We define the *Helmholtz energy of activation*, say for a unimolecular reaction A $\rightarrow$ product, as $\Delta A_\ddagger^\ominus \equiv A^\ddagger - A_{\mathrm{A}} = -k_B T \ln(\mathcal{Q}^\ddagger/\mathcal{Q}_{\mathrm{A}})$, and Eq. (10.1) takes the form

$$\begin{aligned} k_{\mathrm{TST}}(T) &= \frac{k_B T}{h} \exp(-\Delta A_\ddagger^\ominus / k_B T) \\ &= \frac{k_B T}{h} \exp(\Delta S_\ddagger^\ominus / k_B) \exp(-\Delta E_\ddagger^\ominus / k_B T) \end{aligned} \tag{10.2}$$

From this equation originates the statement that the rate constant is determined by the free energy of activation. Furthermore, as also discussed in Section 6.6, the pre-exponential factor in front of the activation energy $\Delta E_\ddagger^\ominus$ is related to the entropy of activation. For reactions in solution, it is important to notice that the values of $\Delta A_\ddagger^\ominus$, $\Delta S_\ddagger^\ominus$, and $\Delta E_\ddagger^\ominus$ are determined by the activated complex and the reactant *as well as* by the surrounding solvent molecules.

The interaction potential between solvent molecules and between solvent and solute molecules may, *in principle*, be found by solving the electronic Schrödinger equation, Eq. (3.1), for the solute/solvent system. Thus, the electronic energy (the potential for the nuclear motion) is, for a system with N nuclei, a function of the position vectors of each nucleus, i.e., $U_N(\boldsymbol{r}_1, \boldsymbol{r}_2, \ldots, \boldsymbol{r}_N)$. Note that if the whole system is invariant to translation and rotation, the electronic energy can be expressed in terms of internuclear distances, $r_{ij} = |\boldsymbol{r}_i - \boldsymbol{r}_j|$. With the exception of the case of a few solvent molecules, this approach is, however, not feasible because of the complexity of such calculations.

In practice, empirical or semi-empirical interaction potentials are used. These potential energy functions are often parameterized as *pairwise additive atom–atom interactions*, i.e., $U_N(\boldsymbol{r}_1, \boldsymbol{r}_2, \ldots, \boldsymbol{r}_N) = \sum_{i<j} u(r_{ij})$, where the sum runs over all atom–atom distances. An all-atom model usually requires a substantial amount of computation. This may be reduced by estimating the electronic energy via a continuum solvation model like the Onsager reaction-field model, discussed in Section 9.1.

10.1 An introduction to the potential of mean force

In this section, we will give a short introduction to some important results of this chapter, in particular, to the concept of a *potential of mean force*.

For any given reaction, the rate constant according to transition-state theory (TST) is given by the average flux across the transition state (TS) from reactants to products. Denoting the reaction coordinate by q_1 and the TS by $q^\ddagger$, and asserting (arbitrarily) that the reactants are 'to the left' of the TS ($q_1 < q^\ddagger$), the TST rate constant can, according to Eq. (5.49), be written in the form

$$k_{\text{TST}}(T) = V^{\nu-1} \frac{\int \cdots \int e^{-H_N/k_BT} \delta(q_1 - q^\ddagger)\, \dot{q}_1\, \theta(\dot{q}_1)\, d\boldsymbol{r}_1 \cdots d\boldsymbol{r}_N d\boldsymbol{p}_1 \cdots d\boldsymbol{p}_N}{\int \cdots \int \theta(q^\ddagger - q_1)\, e^{-H_N/k_BT} d\boldsymbol{r}_1 \cdots d\boldsymbol{r}_N d\boldsymbol{p}_1 \cdots d\boldsymbol{p}_N} \tag{10.3}$$

where $\nu = 1$ for a unimolecular reaction and $\nu = 2$ for a bimolecular reaction, $\delta(x)$ is the Dirac delta function, and $\theta(x)$ is the Heaviside step function.

To obtain a simplified expression for k_{TST}, it is convenient to make a coordinate transformation to a set of coordinates that includes the reaction coordinate:

$$(\boldsymbol{r}_1, \ldots, \boldsymbol{r}_N, \boldsymbol{p}_1, \ldots, \boldsymbol{p}_N) \rightarrow (q_1, \ldots, q_{3N}, p_1, \ldots, p_{3N}) \tag{10.4}$$

Such a transformation will typically be a transformation to normal-mode coordinates around the TS for the activated complex leaving the solvent coordinates unchanged (see Sections 10.2.1 and 10.2.2 for details). This implies that

$$d\boldsymbol{r}_1 \cdots d\boldsymbol{r}_N d\boldsymbol{p}_1 \cdots d\boldsymbol{p}_N = dq_1 \cdots dq_{3N} dp_1 \cdots dp_{3N} \tag{10.5}$$

and that the kinetic energy function remains of the same (diagonal) form, although now with reduced masses. Hence, k_{TST} can now be written

$$\begin{aligned} k_{\text{TST}}(T) = V^{\nu-1} & \frac{\int_0^\infty e^{-p_1^2/2\mu_1 k_BT} \frac{p_1}{\mu_1} dp_1}{\int e^{-p_1^2/2\mu_1 k_BT} dp_1} \\ & \times \frac{\int \delta(q_1 - q^\ddagger) \left[\int \cdots \int e^{-U_N(q_1, q_2, \ldots)/k_BT} dq_2 \cdots dq_{3N}\right] dq_1}{\int \theta(q^\ddagger - q_1) \left[\int \cdots \int e^{-U_N(q_1, q_2, \ldots)/k_BT} dq_2 \cdots dq_{3N}\right] dq_1} \end{aligned} \tag{10.6}$$

The first term in this expression, as we have seen several times in Chapters 5 and 6, is easily evaluated:

$$\frac{\int_0^\infty e^{-p_1^2/2\mu_1 k_BT} \frac{p_1}{\mu_1} dp_1}{\int e^{-p_1^2/2\mu_1 k_BT} dp_1} = \sqrt{\frac{k_BT}{2\pi\mu_1}} \tag{10.7}$$

and is the average speed along the reaction coordinate.

With regard to the second term, the quantity $P_1(q_1)dq_1$, where

$$P_1(q_1) = \frac{1}{Z_N} \int \cdots \int e^{-U_N(q_1, q_2, \ldots)/k_BT} dq_2 \cdots dq_{3N} \tag{10.8}$$

and

$$Z_N = \int \cdots \int e^{-U_N(q_1, q_2, \ldots)/k_BT} dq_1 \cdots dq_{3N} \tag{10.9}$$

gives the probability of finding values of the reaction coordinate between q_1 and $q_1 + dq_1$. The dimension of $P_1(q_1)$ is inverse length, and a dimensionless distribution function and a potential $w(q_1)$ may be introduced according to

$$g_1(q_1) \equiv V^{1/3} P_1(q_1) \equiv e^{-w(q_1)/k_BT} \tag{10.10}$$

The physical significance of the potential $w(q_1)$ is easy to establish. We take the logarithm of Eq. (10.10):

$$-w(q_1)/k_BT = \ln\left\{\int \cdots \int e^{-U_N(q_1,q_2,\ldots)/k_BT} dq_2 \cdots dq_{3N}\right\} + \text{const.} \tag{10.11}$$

and take the derivative $\partial/\partial q_1$:

$$\begin{aligned} -\frac{\partial w(q_1)}{\partial q_1} &= \frac{\int \cdots \int -\frac{\partial U_N}{\partial q_1} e^{-U_N(q_1,q_2,\ldots)/k_BT} dq_2 \cdots dq_{3N}}{\int \cdots \int e^{-U_N(q_1,q_2,\ldots)/k_BT} dq_2 \cdots dq_{3N}} \\ &= \langle -\partial U_N/\partial q_1 \rangle \end{aligned} \tag{10.12}$$

that is, $-\partial w(q_1)/\partial q_1$ is the mean value of the force acting on the reaction coordinate, averaged over all other coordinates. These coordinates include the position of all the solvent molecules as well as all the configurations of the reacting molecules, with the exception of the reaction coordinate. Thus, $w(q_1)$ is the potential that gives the mean force acting on the reaction coordinate and $w(q_1)$ is referred to as a *potential of mean force*.

With the definition in Eq. (10.10), the integrals over $q_2, \ldots, q_{3N}$ in Eq. (10.6) equal $Z_N V^{-1/3} \exp[-w(q_1)/k_BT]$, and the TST rate constant can be expressed as [2,3]

$$\boxed{k_{\mathrm{TST}}(T) = V^{\nu-1}\sqrt{\frac{k_BT}{2\pi\mu_1}}\frac{e^{-w(q^\ddagger)/k_BT}}{\int_{-\infty}^{q^\ddagger} e^{-w(q_1)/k_BT} dq_1}} \tag{10.13}$$

Note that, in this expression, the square root has the dimension of speed and the dimension of the integral in the denominator is length. Hence, the expression indeed gives the correct unit for a unimolecular rate constant (where $\nu = 1$) as well as for a bimolecular rate constant (where $\nu = 2$). Equation (10.13) gives an expression for the intrinsic rate constant, k_s of Section 9.2, in the activation-controlled limit.

The multidimensional integrals in the definition of the potential of mean force can be evaluated directly using the Monte Carlo method (see Appendix I).

Example 10.1: Potential of mean force

Consider the S_N2 reaction

$$\mathrm{Cl^- + CH_3Cl' \longrightarrow CH_3Cl + Cl'^-}$$

The dashed line in Fig. 10.1.1 shows, in the *gas phase*, a calculation at the Hartree–Fock (HF) level with a 6-31G* basis set. The electronic energies correspond to a minimum-energy path with a reaction coordinate defined as $r_c = r_{\mathrm{CCl'}} - r_{\mathrm{CCl}}$, where $r_c = 0$ corresponds to the activated complex $(\mathrm{Cl \cdots CH_3 \cdots Cl'})^\ddagger$. The minimum-energy path is optimized in C_{3v} symmetry for fixed values of r_c.

In *aqueous solution*, the intermolecular interactions were assumed to be pairwise additive and described by Lennard–Jones (12–6) potentials with added interactions corresponding to point charges. The solid line in Fig. 10.1.1 shows the potential of mean force $w(r_c)$ evaluated in a solution of 250 water molecules at $T = 25°\mathrm{C}$ [4].

In solution, the pronounced ion–dipole complex is not observed and the overall potential energy barrier is significantly higher. Note that this calculation agrees

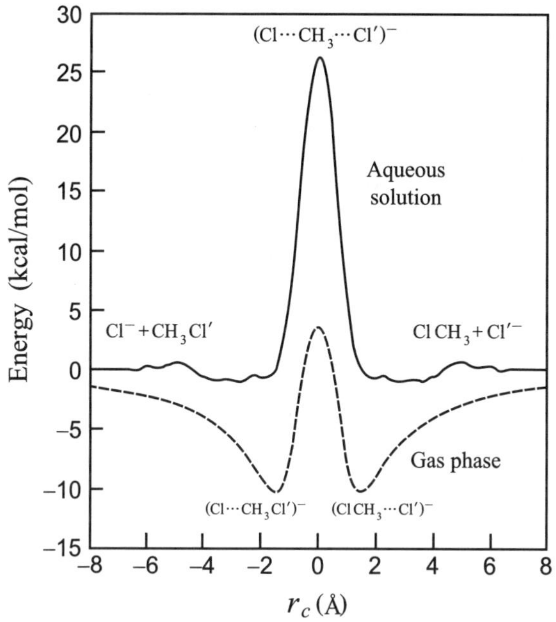

Fig. 10.1.1 Electronic energies in the gas phase (dashed curve) and the potential of mean force in water (solid curve) as a function of the reaction coordinate r_C. [Adapted from J. Chandrasekhar, S.F. Smith, and W.L. Jorgensen, *J. Am. Chem. Soc.* **107**, 154 (1985).]

with the schematic energy diagram in Fig. 9.1. However, in order to facilitate the comparison in Fig. 10.1.1, the curve corresponding to the aqueous solution has been shifted in energy to the same asymptotic value as in the gas phase.

10.2 Transition-state theory and the potential of mean force

The purpose of this section is to give a detailed discussion of the material in Section 10.1, as well as to elaborate on the results. Equation (10.13) is a convenient expression from a computational point of view, but the simplicity of the expression is at the cost of 'hiding' the complexity of the terms involved, including physical insights concerning solvent effects.

We will have a closer look at the coordinate transformations leading to Eq. (10.13) from Eq. (10.3). Furthermore, we will consider some alternative forms of Eq. (10.13). In Eq. (10.13), the potential of mean force is related to the average force on the reaction coordinate. An alternative definition of the potential of mean force, which we will consider, is related to the average force on the atoms of the activated complex exerted by the solvent molecules.

The structure of the activated complex may be perturbed compared to the gas phase when it is placed in a solvent. We derive an exact expression for the static solvent effect on the rate constant and an exact expression for the relation between the solution and gas-phase rate constants. Then, assuming that the activated complex is not perturbed by the solvent, we show that the exact expression for the rate constant in solution simplifies to the well-known gas-phase result in the absence of solvent molecules. Furthermore, when the internal degrees of freedom of the reactants are neglected, the solvent effect on the rate constant is given by the simple expression in Eq. (10.57), where the relation between the rate constant k_s in solution and the rate constant k_g in the gas phase is given in terms of the potential of mean force. This is, however, an approximation and only valid for atomic reactants A and B with no internal degrees of freedom, but can still be used for molecules with internal degrees of freedom as a crude estimate.

We consider in the following a bimolecular reaction, and the starting point for the evaluation of the solvent effects on the reaction is the situation where the reactants A and B have been brought together by diffusion in the same solvent 'cage'. We then consider the reaction

$$\mathrm{A} + \mathrm{B} \rightleftharpoons \mathrm{AB}^{\ddagger} \rightarrow \text{products} \tag{10.14}$$

just like in the gas phase. The Hamiltonian H for a system of reactants A and B and solvent molecules may in general be written as

$$\begin{aligned} H_N &= H_{\mathrm{A}}(p_{\mathrm{A}}, q_{\mathrm{A}}) + H_{\mathrm{B}}(p_{\mathrm{B}}, q_{\mathrm{B}}) + H_{\mathrm{sol}}(P, R) + V_{\mathrm{int}}(q_{\mathrm{A}}, q_{\mathrm{B}}, R) \\ &= T_{\mathrm{A}}(p_{\mathrm{A}}) + V_{\mathrm{A}}(q_{\mathrm{A}}) + T_{\mathrm{B}}(p_{\mathrm{B}}) + V_{\mathrm{B}}(q_{\mathrm{B}}) + T_{\mathrm{sol}}(P) + V_{\mathrm{sol}}(R) + V_{\mathrm{int}}(q_{\mathrm{A}}, q_{\mathrm{B}}, R) \end{aligned} \tag{10.15}$$

Here H_{A} and H_{B} are the Hamiltonians of the isolated reactant molecules, H_{sol} is the Hamiltonian of the pure solvent, and V_{int} is the interaction energy between reactants and between reactant and solvent molecules, i.e., it contains the *solute–solute* as well as the *solute–solvent* interactions. q_{A} and q_{B} are the (for example, Cartesian) coordinates of the atoms in reactant molecules A and B, respectively, and p_{A} and p_{B} are the conjugated momenta. If there are n_{A} atoms in molecule A and n_{B} atoms in molecule B, then there will be, respectively, $3n_{\mathrm{A}}$ coordinates q_{A} and $3n_{\mathrm{B}}$ coordinates q_{B}. Similarly, R are the coordinates for the solvent molecules and P are the conjugated momenta. In the second line of the equation, we have partitioned the Hamiltonians H_i into a kinetic energy part T_i and a potential energy part V_i.

The Hamiltonian of Eq. (10.15) is, of course, valid for any configuration of the system, and also when an activated complex $\mathrm{AB}^{\ddagger}$ is formed and the identity of the reactants is lost. It will then be natural to restructure the terms in Eq. (10.15), so the Hamiltonian will be a sum of a Hamiltonian for the activated complex, a Hamiltonian for the solvent, and an interaction energy term between the activated complex and the solvent:

$$\begin{aligned} H_N^{\ddagger} &= H_{\mathrm{AB}}^{\ddagger}(p^{\ddagger}, q^{\ddagger}) + H_{\mathrm{sol}}(P, R) + V_{\mathrm{int}}^{\ddagger}(q^{\ddagger}, R) \\ &= T_{\mathrm{AB}}^{\ddagger}(p^{\ddagger}) + V_{\mathrm{AB}}^{\ddagger}(q^{\ddagger}) + T_{\mathrm{sol}}(P) + V_{\mathrm{sol}}(R) + V_{\mathrm{int}}^{\ddagger}(q^{\ddagger}, R) \end{aligned} \tag{10.16}$$

Here $q^{\ddagger}$ are the coordinates of the atoms in the activated complex and $p^{\ddagger}$ are the conjugated momenta. There will be a total of $3n = 3n_{\mathrm{A}}+3n_{\mathrm{B}}$ coordinates $q^{\ddagger}$. $V_{\mathrm{AB}}^{\ddagger}$ is the intramolecular *gas-phase potential* and $V_{\mathrm{int}}^{\ddagger}$ describes the *solute–solvent* interaction. The potential energy terms are illustrated in Fig. 10.2.1.

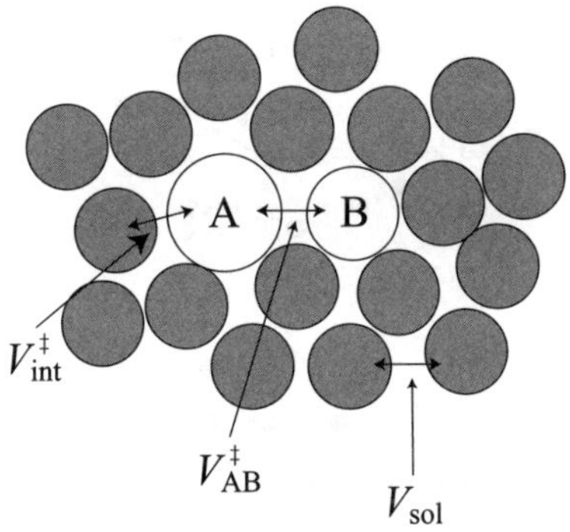

Fig. 10.2.1 Reactants A and B in a solvent 'cage'. $V_{\rm AB}^{\ddagger}$ is the intramolecular gas-phase potential of the activated complex. $V_{\rm sol}$ is the intermolecular potential of the pure solvent. $V_{\rm int}^{\ddagger}$ is the intermolecular potential that describes the interaction between the activated complex and all the solvent molecules.

In a system with many degrees of freedom there may be several activated complexes depending on the constraints imposed on the degrees of freedom. As an example, look at the energy plot for the $H_2 + H$ system in Fig. 3.1.3. The energy of the activated complex at the saddle point of the potential energy surface depends on the angle of approach of the H atom with respect to the H_2 bond. So here and in the following we always refer to the activated complex with the lowest energy, that is, to the complex with all degrees of freedom relaxed except for the reaction coordinate $q_1^{\ddagger}$ to give the activated complex with the lowest energy.

According to Eq. (10.3), we can now write the rate constant in the form

$$k_{\rm TST} = V\frac{\int\int dq^{\ddagger}dp^{\ddagger}dR\exp(-(H_{\rm AB}^{\ddagger}+V_{\rm sol}+V_{\rm int}^{\ddagger})/k_BT)\delta(q_1^{\ddagger}-q^{\ddagger})\,\dot{q}_1^{\ddagger}\,\theta(\dot{q}_1^{\ddagger})}{\int_{\rm React}dqdp\int dR\exp(-(H_{\rm A}+H_{\rm B}+V_{\rm sol}+V_{\rm int})/k_BT)} \tag{10.17}$$

where we have integrated over all solvent momenta P. The integration in the denominator is over all configurations associated with the reactant space. As emphasized previously, the *reaction coordinate* $q_1^{\ddagger}$ is, in general, a superposition of the absolute Cartesian position coordinates for the atoms of the activated complex. We consider in the following the identification of the reaction coordinate.

10.2.1 Solvent-perturbed activated complex

The structure of the activated complex may or may not be perturbed when it is placed in a solvent, depending on the relative strength of the complex–solvent interactions and the intramolecular interactions. Typically, when stable molecules are placed in an external field used to probe some molecular property, it is assumed that the molecular structure is not perturbed. This is often justified, but whether or not that is the case here is more difficult to decide since at least one intramolecular mode becomes soft compared to what it is in a stable molecule, namely the mode associated with the reaction coordinate. This is illustrated in Fig. 10.2.2. In Fig. 10.2.2(a) we see an activated complex consisting of two groups G1 and G2 separated by a bond, which

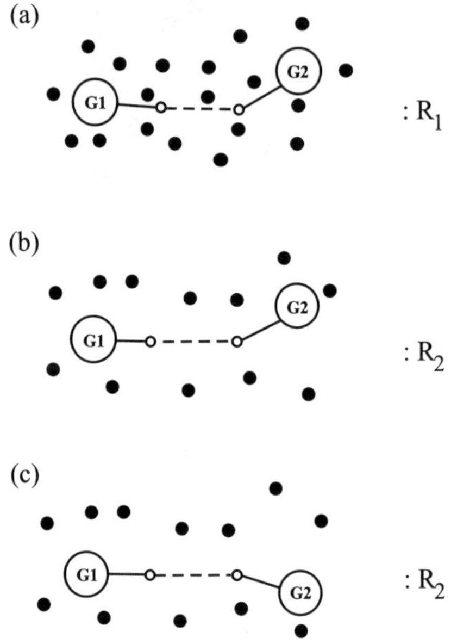

Fig. 10.2.2 Activated complex consisting of two groups G1 and G2 surrounded by solvent molecules (filled circles). The two groups are connected by a bond (shown here as a dashed line), which is the reaction coordinate. (a) The activated complex in a solvent with configuration R_1. (b) Another solvent configuration R_2 with an unperturbed activated complex identical to the one in (a). (c) Same solvent configuration as in (b) but with a perturbed activated complex.

is the reaction coordinate. The complex has been immersed in a solvent with a given configuration of the solvent molecules, referred to as configuration R_1. A different solvent configuration R_2 is shown in Fig. 10.2.2(b), where the structure of the activated complex is identical to the one in Fig. 10.2.2(a), and in Fig. 10.2.2(c) with a different structure of the activated complex. These are the two cases that may occur: one where the structure remains unperturbed by the solvent and one where it is perturbed.

It will therefore be natural to first derive a general expression for the rate constant that includes the possibility that the structure of the activated complex is different from the structure in the gas phase, and then specialize this result to the case where it is assumed that the structure will not be perturbed. We will see that the general result does not allow us to extract a simple physical picture of the solvent effect on the rate constant, whereas more insight may be obtained with the simplifying assumption that the structure of the complex is not perturbed.

The gas-phase potential energy surface for the complex, $V_{AB}^{\ddagger}(q^{\ddagger})$, is perturbed by the interaction potential, $V_{\text{int}}^{\ddagger}(q^{\ddagger}, R)$, when the complex is placed in the solvent. This means that both the *position* of the saddle point on the potential energy surface for the complex and the shape of the potential energy surface around the saddle point may differ from that in the gas phase and depend on the particular configuration R of the solvent molecules around the complex. For a given configuration R of the solvent, we determine the coordinates, $S^\ddagger(R)$, of the saddle point associated with the lowest energy of the *total potential energy surface* $V^\ddagger_{\mathrm{AB,sol}}(q^\ddagger, R)$, where

$$V^\ddagger_{\mathrm{AB,sol}}(q^\ddagger, R) = V^\ddagger_{\mathrm{AB}}(q^\ddagger) + V^\ddagger_{\mathrm{int}}(q^\ddagger, R) \tag{10.18}$$

They will in general be a function of the solvent configuration, as indicated and illustrated in Fig. 10.2.3.

$V^\ddagger_{\mathrm{AB,sol}}(q^\ddagger, R)$ is expanded around the saddle point $S^\ddagger(R)$ to second order in the displacement coordinates

$$\Delta q_i^\ddagger = q_i^\ddagger - S_i^\ddagger(R) \tag{10.19}$$

and the harmonic approximation to the potential surface around the saddle point will be

$$\begin{aligned} V^\ddagger_{\mathrm{harm}}(\Delta q^\ddagger, S^\ddagger(R), R) &= V^\ddagger_{\mathrm{AB,sol}}(S^\ddagger(R), R) \\ &\quad + \frac{1}{2}\sum_i\sum_j \left(\frac{\partial^2 V^\ddagger_{\mathrm{AB,sol}}(q^\ddagger, R)}{\partial q_i^\ddagger \partial q_j^\ddagger}\right)_{q^\ddagger = S^\ddagger(R)} \Delta q_i^\ddagger \Delta q_j^\ddagger \\ &= V^\ddagger_{\mathrm{AB,sol}}(S^\ddagger(R), R) + \frac{1}{2}\sum_i\sum_j \phi^\ddagger_{ij}\Delta q_i^\ddagger \Delta q_j^\ddagger \end{aligned} \tag{10.20}$$

The standard procedure for introducing normal-mode coordinates (see Appendix E) is followed and we define a set of mass-weighted displacement coordinates $\eta_i^\ddagger$:

$$\eta_i^\ddagger = \sqrt{m_i}\Delta q_i^\ddagger \tag{10.21}$$

In these coordinates, the potential has the form

$$\begin{aligned} V^\ddagger_{\mathrm{harm}}(\eta^\ddagger, S^\ddagger(R), R) &= V^\ddagger_{\mathrm{AB,sol}}(S^\ddagger(R), R) + \frac{1}{2}\sum_i\sum_j F^\ddagger_{ij}\eta_i^\ddagger\eta_j^\ddagger \\ &= V^\ddagger_{\mathrm{AB,sol}}(S^\ddagger(R), R) + \frac{1}{2}\boldsymbol{\eta}^{\ddagger T}\boldsymbol{F}^\ddagger\boldsymbol{\eta}^\ddagger \end{aligned} \tag{10.22}$$

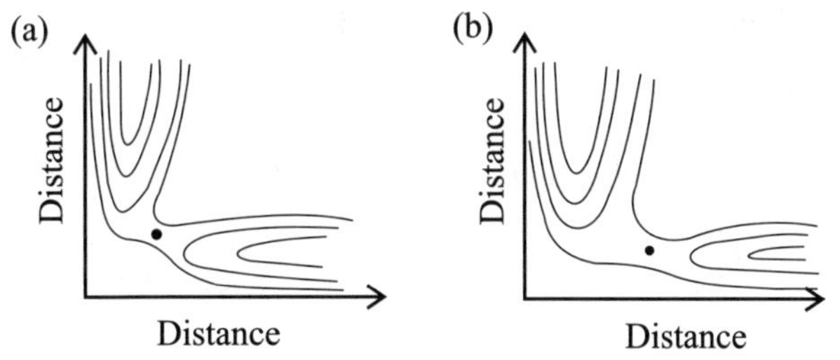

Fig. 10.2.3 (a) Potential energy surface for a chemical reaction in solution with solvent configuration R_1. (b) Potential energy surface for a chemical reaction in solution with another solvent configuration R_2.

where $F^{\ddagger}_{ij} = \phi^{\ddagger}_{ij}/(\sqrt{m_i m_j})$ is the mass-weighted force constant matrix. The normal-mode coordinates $Q^{\ddagger}$ are introduced by the linear transformation of the mass-weighted displacement coordinates:

$$\boldsymbol{\eta}^{\ddagger} = \boldsymbol{L}^{\ddagger}(R)\boldsymbol{Q}^{\ddagger} \tag{10.23}$$

where $\boldsymbol{L}^{\ddagger}(R)$ is a $3n \times 3n$ matrix that depends on the solvent configuration R. The potential energy can then be written as

$$\begin{aligned} V^{\ddagger}_{\text{harm}}(Q^{\ddagger}, S^{\ddagger}(R), R) &= V^{\ddagger}_{\text{AB,sol}}(S^{\ddagger}(R), R) + \frac{1}{2}\boldsymbol{Q}^{\ddagger T}[\boldsymbol{L}^{\ddagger}(R)^T \boldsymbol{F}^{\ddagger} \boldsymbol{L}^{\ddagger}(R)]\boldsymbol{Q}^{\ddagger} \\ &= V^{\ddagger}_{\text{AB,sol}}(S^{\ddagger}(R), R) + \frac{1}{2}\boldsymbol{Q}^{\ddagger T}\boldsymbol{\Omega}^{\ddagger}(R)\boldsymbol{Q}^{\ddagger} \\ &= V^{\ddagger}_{\text{AB,sol}}(S^{\ddagger}(R), R) + \frac{1}{2}\sum_{i=1}^{3n} \omega^2_{ii} Q^{\ddagger 2}_i \end{aligned} \tag{10.24}$$

The $\boldsymbol{L}^{\ddagger}(R)$ matrix is chosen such that the matrix $\boldsymbol{L}^{\ddagger}(R)^T \boldsymbol{F}^{\ddagger} \boldsymbol{L}^{\ddagger}(R) = \boldsymbol{\Omega}^{\ddagger}(R)$ is diagonal with elements ω^2_{ii}. The eigenvectors of $\boldsymbol{F}^{\ddagger}$ are arranged as columns in the $\boldsymbol{L}^{\ddagger}(R)$ matrix. It should also be noticed that $V^{\ddagger}_{\text{AB,sol}}(S^{\ddagger}(R), R)$ can no longer be written as a sum of an intramolecular (gas-phase potential) and an intermolecular part as in Eq. (10.18), because the harmonic expansion of the potential around the saddle point is based on the *total potential energy* surface and not just on the intramolecular part. By combining Eqs (10.19), (10.21), and (10.23) we see that the absolute position coordinates of the atoms in the activated complex around the saddle point of the *total potential energy surface* can be written as

$$q^{\ddagger}_i = S^{\ddagger}_i(R) + \Delta q^{\ddagger}_i = S^{\ddagger}_i(R) + \frac{1}{\sqrt{m_i}}\sum_{j=1}^{3n} L^{\ddagger}_{ij}(R) Q^{\ddagger}_j \tag{10.25}$$

in terms of the normal coordinates $Q^{\ddagger}$ and the position of the saddle point $S^{\ddagger}(R)$.

The salient feature of the normal coordinate description is that there is no coupling between the various normal modes, so the Hamiltonian can be written as a sum of Hamiltonians for each normal mode. The reaction coordinate is defined, like in gas-phase transition-state theory, as the normal mode for which the associated frequency is imaginary. The Hamiltonian for the activated complex may be written as

$$\begin{aligned} H^{\ddagger}_{\text{AB,sol}}(P^{\ddagger}, Q^{\ddagger}, R) &= T^{\ddagger}_{\text{AB}}(P^{\ddagger}) + V^{\ddagger}_{\text{harm}}(Q^{\ddagger}, S^{\ddagger}(R), R) \\ &= H^{\ddagger}_{\text{trans}}(P^{\ddagger}_1, Q^{\ddagger}_1) + H^{\ddagger}_{\text{intra}}(\tilde{P}^{\ddagger}, \tilde{Q}^{\ddagger}, R) \\ &= \frac{P^{\ddagger 2}_1}{2\mu_1} + \frac{1}{2}\omega^2_{11} Q^{\ddagger 2}_1 + \sum_{i=2}^{3n} \frac{P^{\ddagger 2}_i}{2\mu_i} + \frac{1}{2}\sum_{i=2}^{3n} \omega^2_{ii} Q^{\ddagger 2}_i + V^{\ddagger}_{\text{AB,sol}}(S^{\ddagger}(R), R) \end{aligned} \tag{10.26}$$

where we have pulled out the Hamiltonian for the reaction coordinate in the second line of the equation and written the explicit expressions for the Hamiltonians in the following lines. The terms on the right-hand side of the equation will be discussed in detail in the following.

The motion in the reaction coordinate $Q_1^\ddagger$ is, like in gas-phase transition-state theory, described as a free translational motion in a very narrow range of the reaction coordinate at the transition state, that is, for $Q_1^\ddagger = 0$; hence the subscript 'trans' on the Hamiltonian. The potential may be considered to be constant and with zero slope in the direction of the reaction coordinate (that is, zero force in that direction) at the transition state. The *central assumption* in the theory is now that the flow about the transition state is given solely by the free motion at the transition state with no recrossings. So when we associate a free translational motion with that coordinate, it does not mean that the interaction potential energy is independent of the reaction coordinate, but rather that it has been set to its value at the transition state, $Q_1^\ddagger = 0$, because we only consider the motion at that point. The Hamiltonian $H_{\text{trans}}^\ddagger$ accordingly only depends on $P_1^\ddagger$, as for a free translational motion, so

$$H_{\text{trans}}^\ddagger = \frac{P_1^{\ddagger 2}}{2\mu_1} \tag{10.27}$$

with μ_1 being the reduced mass for that particular mode. It is equal to one when mass-weighted coordinates are used, as shown in Appendix E. In spite of this, we have chosen to include the reduced mass in this and the following expression for clarity.

The second term in the last line of Eq. (10.26) includes the contributions from the other normal modes, $\tilde{Q}^\ddagger, \tilde{P}^\ddagger$, and may be written as

$$\begin{aligned} H_{\text{intra}}^\ddagger(\tilde{P}^\ddagger, \tilde{Q}^\ddagger, R) &= T_{\text{intra}}^\ddagger(\tilde{P}^\ddagger) + V_{\text{AB,sol}}^\ddagger(S^\ddagger(R), R) + \frac{1}{2}\sum_{i=2}^{3n} \omega_{ii}^2 Q_i^{\ddagger 2} \\ &\equiv T_{\text{intra}}^\ddagger(\tilde{P}^\ddagger) + V_{\text{intra,sol}}^\ddagger(\tilde{Q}^\ddagger, S^\ddagger(R), R) \end{aligned} \tag{10.28}$$

where we have used Eq. (10.24) and defined the term $V_{\text{intra,sol}}^\ddagger(\tilde{Q}^\ddagger, S^\ddagger(R), R)$ in the second line of the equation. The only difference between that term and the $V_{\text{harm}}^\ddagger(Q^\ddagger, S^\ddagger, R)$ term in Eq. (10.24) is that we have pulled out the contribution from the reaction coordinate in the former. The ~ on $\tilde{P}^\ddagger$ and $\tilde{Q}^\ddagger$ implies all momenta and coordinates except for $P_1^\ddagger$ and $Q_1^\ddagger$ which are associated with the reaction coordinate. Note also that the $V_{\text{intra,sol}}^\ddagger$ term includes both the intramolecular potential of the complex and the complex–solvent interaction energy, and cannot be written as a sum of an intramolecular and intermolecular part for the same reason as mentioned for the term $V_{\text{AB,sol}}^\ddagger(S^\ddagger(R), R)$ in Eq. (10.24).

We now return to the expression for the rate constant in Eq. (10.17). This was, however, expressed in terms of the Cartesian coordinates $p^\ddagger, q^\ddagger$ and not in terms of the normal coordinates $P^\ddagger, Q^\ddagger$ and the *reaction coordinate* $Q_1^\ddagger$ introduced above.

From the normal coordinate analysis in Appendix E, it is clear that the relation between the volume elements in the two sets of coordinate systems is given by

$$dp^\ddagger dq^\ddagger = dP^\ddagger dQ^\ddagger \tag{10.29}$$

The term $H_{\text{AB}}^\ddagger + V_{\text{int}}^\ddagger$ in the argument of the exponential function in Eq. (10.17) may, according to Eq. (10.16), be written as

$$
\begin{aligned}
H^{\ddagger}_{\mathrm{AB}}(p^{\ddagger},q^{\ddagger}) + V^{\ddagger}_{\mathrm{int}}(q^{\ddagger},R) &= T^{\ddagger}_{\mathrm{AB}}(p^{\ddagger}) + V^{\ddagger}_{\mathrm{AB}}(q^{\ddagger}) + V^{\ddagger}_{\mathrm{int}}(q^{\ddagger},R)\\
&= T^{\ddagger}_{\mathrm{AB}}(P^{\ddagger}) + V^{\ddagger}_{\mathrm{harm}}(Q^{\ddagger},S^{\ddagger}(R),R)\\
&= H^{\ddagger}_{\mathrm{AB,sol}}(P^{\ddagger},Q^{\ddagger},R) \qquad (10.30)
\end{aligned}
$$

where we have introduced the normal coordinate expansion of the last two potential energy terms following the derivations in Eqs (10.18)–(10.24) and the conjugated momenta $P^{\ddagger}$. The last relation in the equation follows from Eq. (10.26). In Eq. (10.17), we now introduce this relation and the normal-mode coordinates including $q_1^{\ddagger} \equiv Q_1^{\ddagger}$:

$$
k_{\mathrm{TST}} = V\frac{\int\!\!\int \dot{Q}_1^{\ddagger}\theta(\dot{Q}_1^{\ddagger})dP_1^{\ddagger}d\tilde{P}^{\ddagger}d\tilde{Q}^{\ddagger}dR\exp(-(H^{\ddagger}_{\mathrm{AB,sol}}+V_{\mathrm{sol}})/k_BT)}{\int_{\mathrm{React}} dqdp\int dR\exp(-(H_{\mathrm{A}}+H_{\mathrm{B}}+V_{\mathrm{sol}}+V_{\mathrm{int}})/k_BT)} \qquad (10.31)
$$

The integration over $Q_1^{\ddagger}$ has been carried out, which implies that $H^{\ddagger}_{\mathrm{AB,sol}}$ is evaluated at $Q_1^{\ddagger}=0$. From the Hamiltonian in Eq. (10.27), we find, using Hamilton's equations of motion,

$$
\begin{aligned}
\frac{dQ_1^{\ddagger}}{dt} &= \frac{\partial H^{\ddagger}_{\mathrm{trans}}}{\partial P_1^{\ddagger}}\\
&= \frac{P_1^{\ddagger}}{\mu_1} \qquad (10.32)
\end{aligned}
$$

This is introduced into Eq. (10.31), and with Eqs (10.26) and (10.27) we get

$$
\begin{aligned}
k_{\mathrm{TST}} = V &\int_0^{\infty}\frac{P_1^{\ddagger}}{\mu_1}\exp(-P_1^{\ddagger 2}/(2\mu_1k_BT))dP_1^{\ddagger}\\
&\times\frac{\int\!\!\int d\tilde{P}^{\ddagger}d\tilde{Q}^{\ddagger}dR\exp(-(H^{\ddagger}_{\mathrm{intra}}+V_{\mathrm{sol}})/k_BT)}{\int_{\mathrm{React}} dqdp\int dR\exp(-(H_{\mathrm{A}}+H_{\mathrm{B}}+V_{\mathrm{sol}}+V_{\mathrm{int}})/k_BT)} \qquad (10.33)
\end{aligned}
$$

Note that the integration over all positive $P_1^{\ddagger}$ leading from the reactant part of phase space to the product part comes from the assumption of no recrossing of the transition state. Integration over all momenta $P_1^{\ddagger}$ in the first line of Eq. (10.33) gives

$$
\int_0^{\infty}\frac{P_1^{\ddagger}}{\mu_1}\exp(-P_1^{\ddagger 2}/(2\mu_1k_BT))dP_1^{\ddagger} = k_BT \qquad (10.34)
$$

When Eqs (10.15), (10.28), and (10.34) are introduced into Eq. (10.33), we may write

$$
\begin{aligned}
k_{\mathrm{TST}} = k_BT\,V &\frac{\int d\tilde{P}^{\ddagger}\exp(-T^{\ddagger}_{\mathrm{intra}}/k_BT)}{\int_{\mathrm{React}} dp\exp(-(T_{\mathrm{A}}+T_{\mathrm{B}})/k_BT)}\\
&\times\frac{\int\!\!\int d\tilde{Q}^{\ddagger}dR\exp(-(V^{\ddagger}_{\mathrm{intra,sol}}+V_{\mathrm{sol}})/k_BT)}{\int_{\mathrm{React}} dq\int dR\exp(-(V_{\mathrm{A}}+V_{\mathrm{B}}+V_{\mathrm{sol}}+V_{\mathrm{int}})/k_BT)} \qquad (10.35)
\end{aligned}
$$

where we have pulled out the kinetic energy term $T^{\ddagger}_{\mathrm{intra}}$.

The integrals over momenta may easily be evaluated, since the kinetic energy terms T_{A} and T_{B} are a sum of terms $p_i^2/(2m_i)$. Likewise, $T^{\ddagger}_{\mathrm{intra}}$ is a sum of terms $P_j^{\ddagger 2}/2$, since the reduced masses for the normal modes are all equal to one. The total number of degrees of freedom for the activated complex is $3n$, and the ~ indicates that one degree of freedom $P_1^\ddagger$ has already been considered. We then find

$$\begin{aligned}\Omega &\equiv \frac{\int d\tilde{P}^\ddagger \exp(-T^\ddagger_{\text{intra}}/k_BT)}{\int_{\text{React}} dp \exp(-(T_{\text{A}}+T_{\text{B}})/k_BT)} \\ &= \frac{(\sqrt{2\pi k_BT})^{(3n-1)}}{\prod_{i=1}^{3n_a}\sqrt{2\pi m_{\text{A},i}k_BT}\prod_{j=1}^{3n_b}\sqrt{2\pi m_{\text{B},j}k_BT}} \\ &= \frac{1}{\sqrt{2\pi k_BT}}\frac{1}{\prod_{i=1}^{3n_a}\sqrt{m_{\text{A},i}}\prod_{j=1}^{3n_b}\sqrt{m_{\text{B},j}}}\end{aligned} \tag{10.36}$$

$m_{\text{A},i}$ is the mass of atom i in molecule A and $m_{\text{B},j}$ is the mass of atom j in molecule B. We have then

$$\boxed{k_{\text{TST}} = k_BT\, V\, \Omega \frac{\int\int d\tilde{Q}^\ddagger dR \exp(-(V^\ddagger_{\text{intra,sol}}+V_{\text{sol}})/k_BT)}{\int_{\text{React}} dq \int dR \exp(-(V_{\text{A}}+V_{\text{B}}+V_{\text{sol}}+V_{\text{int}})/k_BT)}} \tag{10.37}$$

This is an exact *mean field* expression for the rate constant when the reaction takes place in a solution. The result incorporates the possibility that the structure of the activated complex may be *perturbed* by the interaction with the solvent molecules. The cost we must pay for the general validity of the expression is that we cannot reduce the complicated integrals further, so the result is as such not very informative and does not provide us with a simple physical picture of how the rate constant is affected by the solvent.

The integrals in Eq. (10.37) are referred to as configuration integrals. Thus, the denominator and the numerator are configurational integrals of, respectively, the reactants including the solvent and the activated complex including the solvent (with the reaction coordinate fixed at $Q_1^\ddagger = 0$). The unit of k_{TST} is m^3 s^{-1} and this is consistent with the expression in Eq. (10.37). The unit of the k_BTV factor is J m^3, the unit of the Ω factor is J$^{1/2}$ kg$^{-3n/2}$, and the unit of the last factor is kg$^{(3n-1)/2}$ m^{-1}. Put together, the unit is J$^{1/2}$ kg$^{-1/2}$ m^2 = m^3 s^{-1}.

10.2.2 Unperturbed activated complex

When we introduce the simplifying assumption that the structure of the activated complex will *not* be perturbed when exposed to the solvent molecules, then it is possible to reduce the integrals and obtain more physical insight into the effect of the solvent. The transition-state coordinates $S^\ddagger$ are obtained from the *gas-phase potential* $V^\ddagger_{\text{AB}}(q^\ddagger)$ alone and are therefore independent of the solvent configuration. The eigenvectors in the $\boldsymbol{L}^\ddagger$ matrix are likewise determined from the gas-phase potential alone and therefore are also independent of the solvent configuration.

This implies that $V^\ddagger_{\text{AB,sol}}(S^\ddagger(R), R)$ in Eq. (10.24) may now be written as a sum of an intramolecular and an intermolecular part:

$$V^\ddagger_{\text{AB,sol}}(S^\ddagger(R), R) = V^\ddagger_{\text{AB}}(S^\ddagger) + V^\ddagger_{\text{int}}(S^\ddagger, Q^\ddagger, R) \tag{10.38}$$

as in Eq. (10.18), and $S^\ddagger$ is no longer a function of the coordinates of the solvent molecules. We have expressed the absolute coordinates $q^\ddagger$ in the $V^\ddagger_{\text{int}}$ term using the transition-state coordinates and the normal coordinates. The relation between the two sets of coordinates is given in Eq. (10.25). When Eq. (10.38) is introduced into the expression for $V^{\ddagger}_{\text{intra,sol}}(S^{\ddagger}, R, \tilde{Q}^{\ddagger})$ in Eq. (10.28), we may write

$$\begin{aligned} V^{\ddagger}_{\text{intra,sol}}(S^{\ddagger}, R, \tilde{Q}^{\ddagger}) &= V^{\ddagger}_{\text{AB}}(S^{\ddagger}) + \frac{1}{2}\sum_{i=2}^{3n} \omega_{ii}^2 Q_i^{\ddagger 2} + V^{\ddagger}_{\text{int}}(S^{\ddagger}, Q_1^{\ddagger} = 0, \tilde{Q}^{\ddagger}, R) \\ &\equiv V^{\ddagger}_{\text{intra}}(S^{\ddagger}, Q_1^{\ddagger} = 0, \tilde{Q}^{\ddagger}) + V^{\ddagger}_{\text{int}}(S^{\ddagger}, Q_1^{\ddagger} = 0, \tilde{Q}^{\ddagger}, R) \end{aligned} \tag{10.39}$$

Note the difference in the definitions of $V^{\ddagger}_{\text{intra,sol}}$ in Eq. (10.28) and $V^{\ddagger}_{\text{intra}}$ in Eq. (10.39). The interactions between the activated complex and the solvent are included in the former but not in the latter. This is used in the expression for the rate constant k_{TST} in Eq. (10.37), and we find

$$\boxed{\begin{aligned} k_{\text{TST}} &= k_B T\, V\, \Omega \frac{\int d\tilde{Q}^{\ddagger} \exp(-V^{\ddagger}_{\text{intra}}/k_B T) \int dR \exp(-(V_{\text{sol}} + V^{\ddagger}_{\text{int}})/k_B T)}{\int_{\text{React}} dq \int dR \exp(-(V_{\text{A}} + V_{\text{B}} + V_{\text{sol}} + V_{\text{int}})/k_B T)} \\ &= k_B T\, V\, \Omega \int d\tilde{Q}^{\ddagger} P^{(n)}(S^{\ddagger}, Q_1^{\ddagger} = 0, Q_2^{\ddagger}, \ldots, Q_{3n-r}^{\ddagger}) \end{aligned}} \tag{10.40}$$

where we have introduced the function

$$P^{(n)}(S^{\ddagger}, Q_1^{\ddagger} = 0, Q_2^{\ddagger}, \ldots, Q_{3n-r}^{\ddagger}) = \frac{\exp(-V^{\ddagger}_{\text{intra}}/k_B T) \int dR \exp(-(V_{\text{sol}} + V^{\ddagger}_{\text{int}})/k_B T)}{\int_{\text{React}} dq \int dR \exp(-(V_{\text{A}} + V_{\text{B}} + V_{\text{sol}} + V_{\text{int}})/k_B T)} \tag{10.41}$$

This function is the n-particle distribution function. For $n = 2$, it is related to the well-known pair-distribution function. It gives the probability of finding the atoms in the activated complex in a given configuration $S^{\ddagger}, Q_1^{\ddagger} = 0, Q_2^{\ddagger}, \ldots, Q_{3n-r}^{\ddagger}$ (represented here by the normal-mode coordinates and the position of the saddle point) averaged over all solvent molecule configurations. $P^{(n)}$ can only depend on the internal coordinates of the activated complex when we average over all possible configurations of the solvent molecules, because that averaging will produce configurations that may be obtained by a rigid translation (center-of-mass translation) and a rigid rotation of the complex. For a non-linear molecule there will be $r = 6$ coordinates that describe the center-of-mass position and the orientation of the molecule, and $r = 5$ coordinates for a linear molecule. Equation (10.40) is an exact *mean field* expression for the rate constant of a reaction taking place in a solvent, when the structure of the activated complex is *not perturbed* by the solvent.

The n-particle distribution function $P^{(n)}$ in Eq. (10.41) is often replaced by the so-called *potential of mean force* function W_{mean} for the activated complex, defined as

$$\begin{aligned} P^{(n)} &\equiv \exp(-V^{\ddagger}_{\text{intra}}/k_B T) \exp(-W_{\text{mean}}/k_B T)/V^n \\ &\equiv g^{(n)}(S^{\ddagger}, Q_1^{\ddagger} = 0, Q_2^{\ddagger}, \ldots, Q_{3n-r}^{\ddagger})/V^n \\ &= \frac{\exp(-V^{\ddagger}_{\text{intra}}/k_B T) \int dR \exp(-(V_{\text{sol}} + V^{\ddagger}_{\text{int}})/k_B T)}{\int_{\text{React}} dq \int dR \exp(-(V_{\text{A}} + V_{\text{B}} + V_{\text{sol}} + V_{\text{int}})/k_B T)} \end{aligned} \tag{10.42}$$

where the factor V^{-n} is introduced, since the unit of the probability density function $P^{(n)}$ is $(m)^{-3n}$ according to Eq. (10.41). $g^{(n)}$ is the so-called n-particle correlation function, and for $n = 2$ it is identical to the pair-correlation function $g(r)$, where r is the internuclear distance. For an infinite separation there will be no correlation between the atoms and $g(r \to \infty) \to 1$, whereas it will be different from one at smaller separations.

It is easy to verify that W_{mean} is the potential of the mean force exerted by the solvent molecules on the activated complex. From the definition in Eq. (10.42), we find, by taking the natural logarithm on both sides of the expression,

$$-W_{\text{mean}}/k_BT = \ln\left[\int dR \exp(-(V_{\text{sol}} + V_{\text{int}}^{\ddagger})/k_BT)\right] - \text{const.} \tag{10.43}$$

Then we take the gradient with respect to the position vector of the jth atom in the activated complex:

$$\begin{aligned} -\nabla_j W_{\text{mean}}/k_BT &= \frac{-\int \nabla_j V_{\text{int}}^{\ddagger}/k_BT \exp(-(V_{\text{sol}} + V_{\text{int}}^{\ddagger})/k_BT)dR}{\int dR \exp(-(V_{\text{sol}} + V_{\text{int}}^{\ddagger})/k_BT)} \\ &\equiv \langle \boldsymbol{F}_j \rangle / k_BT \end{aligned} \tag{10.44}$$

which is just the definition of the average force on the jth atom in the activated complex due to the solvent molecules. The situation is illustrated in Fig. 10.2.4. Here we have shown the atoms (open circles) in an activated complex and the average force $\langle \boldsymbol{F}_i \rangle$ on each of the atoms from the solvent molecules.

Equation (10.40) is sometimes found in a simpler form at the cost of 'hiding' the complexity of the terms involved. This form is based on the introduction of a probability density function for the reaction coordinate and the associated potential of mean force, in contrast to previously, where we considered the probability density of a particular arrangement of n atoms. Let $\Pi(Q_1^{\ddagger})dQ_1^{\ddagger}$ be the probability of finding the reaction coordinate in the range $Q_1^{\ddagger}, Q_1^{\ddagger} + dQ_1^{\ddagger}$. Then, from equilibrium statistical mechanics (see Appendix A.2), the probability density function $\Pi(Q_1^{\ddagger})$ is given by

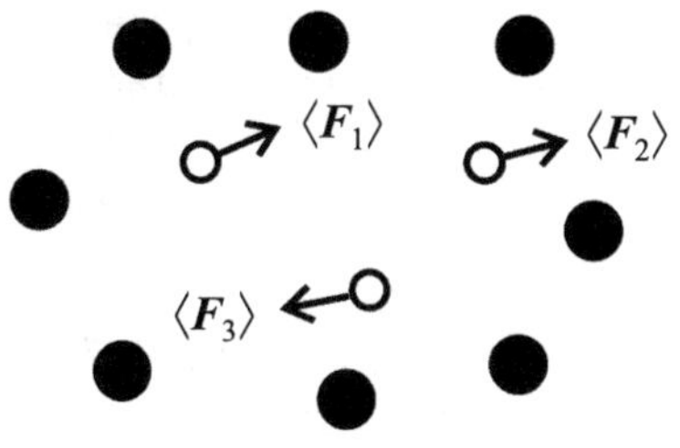

Fig. 10.2.4 Activated complex atoms (open circles) in a cage of solvent molecules (filled circles). The mean force $\langle \boldsymbol{F}_i \rangle$ on the atoms in the activated complex from the solvent molecules is shown.

$$\Pi(Q_1^\ddagger) = \frac{\int d\tilde{Q}^\ddagger \exp(-V_{\text{intra}}^\ddagger/k_BT) \int dR \exp(-(V_{\text{sol}} + V_{\text{int}}^\ddagger)/k_BT)}{\int dQ^\ddagger \exp(-V_{\text{intra}}^\ddagger/k_BT) \int dR \exp(-(V_{\text{sol}} + V_{\text{int}}^\ddagger)/k_BT)}$$
$$\equiv V^{-1/3} \exp(-w(Q_1^\ddagger)/k_BT) \tag{10.45}$$

where in the second line we have introduced the potential of mean force, $w(Q_1^\ddagger)$, for the reaction coordinate in analogy to Eq. (10.42). Here the volume is raised to the power of $-1/3$, since we consider just one coordinate and not n atoms. The derivative of $w(Q_1^\ddagger)$ with respect to $Q_1^\ddagger$ is easily seen to give the average force on the reaction coordinate from the solvent molecules.

The numerator in Eq. (10.40), with $Q_1^\ddagger = 0$, may then be written, using Eq. (10.45):

$$\int d\tilde{Q}^\ddagger \exp(-V_{\text{intra}}^\ddagger/k_BT) \int dR \exp(-(V_{\text{sol}} + V_{\text{int}}^\ddagger)/k_BT) =$$
$$\frac{1}{V^{1/3}} \exp(-w(Q_1^\ddagger)/k_BT) \int dQ^\ddagger \exp(-V_{\text{intra}}^\ddagger/k_BT) \int dR \exp(-(V_{\text{sol}} + V_{\text{int}}^\ddagger)/k_BT) \tag{10.46}$$

In the denominator of the expression in Eq. (10.40), we change integration variables from dq to $dQ^\ddagger$ using the transformation Eq. (10.25), to have the same integration variables in the numerator and denominator. From Appendix E.3, we have the following relation between the volume elements:

$$dq = \frac{1}{\prod_{i=1}^{3n_a} \sqrt{m_{\text{A},i}} \prod_{j=1}^{3n_b} \sqrt{m_{\text{B},j}}} dQ^\ddagger$$
$$= \Omega \sqrt{2\pi k_BT}\, dQ^\ddagger \tag{10.47}$$

using the same notation as in Eq. (10.36), and we find

$$\int_{\text{React}} dq \int dR \exp(-(V_{\text{A}} + V_{\text{B}} + V_{\text{sol}} + V_{\text{int}})/k_BT) =$$
$$\Omega \sqrt{2\pi k_BT} \int_{\text{React}} dQ^\ddagger \int dR \exp(-(V_{\text{A}} + V_{\text{B}} + V_{\text{sol}} + V_{\text{int}})/k_BT) \tag{10.48}$$

The integration over the whole reactant space is equivalent to an integration of Eq. (10.46) over all $Q_1^\ddagger$ restricted to the reactant space.

The combination of the expressions for the numerator, Eq. (10.46), and the denominator, Eq. (10.48), then gives the following simple expression for the rate constant in Eq. (10.40):

$$\boxed{k_{\text{TST}} = V\sqrt{\frac{k_BT}{2\pi}} \frac{e^{-w(Q_1^\ddagger=0)/k_BT}}{\int e^{-w(Q_1^\ddagger)/k_BT} dQ_1^\ddagger}} \tag{10.49}$$

This is identical to the expression in Eq. (10.13). The factor $\sqrt{k_BT/(2\pi)}$ should really be $\sqrt{k_BT/(2\pi\mu_1)}$, where μ_1 is the effective mass associated with the motion in the reaction coordinate. Since, however, we have used mass-weighted coordinates in the construction of the normal-mode coordinates, all effective masses are equal to one (see Appendix E), so it does not appear explicitly in the expression. The expression in

Eq. (10.49) has a simple form and by its derivation we have shown what is involved in the calculation of the potential of mean force $w(Q_1^\ddagger)$.

Before we derive a relation between the rate constants in solution and gas phase, let us first verify that the expression for the rate constant in Eq. (10.40) simplifies to the well-known result for the gas-phase rate constant, when there is no solvent present. With no solvent, we have $V_{\mathrm{sol}} = 0$. Also $V_{\mathrm{int}}^\ddagger = 0$, since this is the interaction energy between solvent and activated complex. The n-particle distribution function in Eq. (10.41) simplifies to

$$P^{(n)}(S^\ddagger, Q_1^\ddagger = 0, Q_2^\ddagger, \ldots, Q_{3n-r}^\ddagger) = \frac{\exp(-V_{\mathrm{intra}}^\ddagger/k_B T)}{\int_{\mathrm{React}} dq \exp(-(V_{\mathrm{A}} + V_{\mathrm{B}} + V_{\mathrm{int}})/k_B T)} \tag{10.50}$$

When introduced into the expression for the rate constant, we find

$$\begin{aligned} k_g = k_B T V \frac{\int d\tilde{P}^\ddagger \exp(-T_{\mathrm{intra}}^\ddagger/k_B T)}{\int_{\mathrm{React}} dp \exp(-(T_{\mathrm{A}} + T_{\mathrm{B}})/k_B T)} \\ \times \frac{\int d\tilde{Q}^\ddagger \exp(-V_{\mathrm{intra}}^\ddagger/k_B T)}{\int_{\mathrm{React}} dq \exp(-(V_{\mathrm{A}} + V_{\mathrm{B}} + V_{\mathrm{int}})/k_B T)} \end{aligned} \tag{10.51}$$

where we have used the definition of the Ω factor from Eq. (10.36). If we also make the assumption that $V_{\mathrm{int}} = 0$, that is, we neglect interactions between the reactant molecules except when they form an activated complex, then the denominator in Eq. (10.51) is simply proportional to the molecular partition functions Z_{A} and Z_{B}. Please note that here we use Z to denote a partition function to avoid confusion with the normal coordinates Q. We have

$$\begin{aligned} \int_{\mathrm{React}} & dp \exp(-(T_{\mathrm{A}} + T_{\mathrm{B}})/k_B T) \int_{\mathrm{React}} dq \exp(-(V_{\mathrm{A}} + V_{\mathrm{B}})/k_B T) \\ &= \int_{\mathrm{React}} dp_{\mathrm{A}} dq_{\mathrm{A}} \exp(-H_{\mathrm{A}}/k_B T) \int_{\mathrm{React}} dp_{\mathrm{B}} dq_{\mathrm{B}} \exp(-H_{\mathrm{B}}/k_B T) \\ &= h^{3n_{\mathrm{A}}} Z_{\mathrm{A}} h^{3n_{\mathrm{B}}} Z_{\mathrm{B}} \\ &= h^{3n} Z_{\mathrm{A}} Z_{\mathrm{B}} \\ &= h^{3n} V^2 (Z_{\mathrm{A}}/V)(Z_{\mathrm{B}}/V) \end{aligned} \tag{10.52}$$

In the same way, the numerator may be written

$$\begin{aligned} \int d\tilde{P}^\ddagger \exp(-T_{\mathrm{intra}}^\ddagger/k_B T) \int d\tilde{Q}^\ddagger \exp(-V_{\mathrm{intra}}^\ddagger/k_B T) &= \int d\tilde{P}^\ddagger d\tilde{Q}^\ddagger \exp(-H_{\mathrm{intra}}^\ddagger/k_B T) \\ &= h^{3n-1} Z^\ddagger \exp(-E_{\mathrm{cl}}/k_B T) \\ &= h^{3n-1} V (Z^\ddagger/V) \exp(-E_{\mathrm{cl}}/k_B T) \end{aligned} \tag{10.53}$$

where it is noticed that the partition function for the activated complex does not include the reaction coordinate, so the phase-space dimension is two less than that of the reactants, and hence the power of h is one less than for the reactants. The exponential function ensures that we refer to the same zero point for the energy for both reactants and the activated complex. E_{cl} is the saddle-point energy, and is given by

$$E_{\mathrm{cl}} = V^{\ddagger}_{\mathrm{intra}}(S^{\ddagger}, Q^{\ddagger}_1 = 0, \tilde{Q}^{\ddagger} = 0) \tag{10.54}$$

When this is introduced into the expression for the gas-phase rate constant, we get the familiar expression

$$k_g = \frac{k_B T}{h} \frac{(Z^{\ddagger}/V)}{(Z_{\mathrm{A}}/V)(Z_{\mathrm{B}}/V)} \exp(-E_{\mathrm{cl}}/k_B T) \tag{10.55}$$

Having established that the general expression for the rate constant in Eq. (10.40) simplifies to the well-known expression from the gas phase in the absence of a solvent, let us then write down a general expression for the relation between the two rate constants, when the activated complex is *unperturbed* by the solvent. We find, using Eqs (10.40) and (10.51),

$$\begin{aligned}\frac{k_s}{k_g} = \int d\tilde{Q}^{\ddagger} P^{(n)}(S^{\ddagger}, Q^{\ddagger}_1 = 0, Q^{\ddagger}_2, \ldots, Q^{\ddagger}_{3n-r}) \\ \times \frac{\int_{\mathrm{React}} dq \exp(-(V_{\mathrm{A}} + V_{\mathrm{B}} + V_{\mathrm{int}})/k_B T)}{\int d\tilde{Q}^{\ddagger} \exp(-V^{\ddagger}_{\mathrm{intra}}/k_B T)}\end{aligned} \tag{10.56}$$

This is not a simple expression that allows us to immediately evaluate the effect of the solvent molecules on the rate constant. The integrals have to be evaluated by, for example, the Monte Carlo technique. Both $P^{(n)}$ and $V^{\ddagger}_{\mathrm{intra}}$ only depend on the internal $3n - r$ coordinates and not on the center-of-mass coordinates and rotational coordinates, as explained above. Integration over these coordinates therefore always cancels in this expression.

A special case: reactants without internal degrees of freedom

The expression in Eq. (10.56) only becomes simple in the special case when the reactants A and B are atoms with no internal degrees of freedom. Then, in the second factor which is related to the gas-phase reaction, $V_{\mathrm{A}} = V_{\mathrm{B}} = 0$, and with no interactions between the reactant molecules, $V_{\mathrm{int}} = 0$. The integral over reactant states is therefore simply equal to the square of the volume, V^2. The reaction coordinate $Q^{\ddagger}_1$ is the distance, r, between the atoms, so the five coordinates $\tilde{Q}^{\ddagger}$ involve the center-of-mass coordinates and the rotational coordinates for the linear activated complex. The integrals over them cancel, since both $P^{(n)}$ and $V^{\ddagger}_{\mathrm{intra}}$ are independent of them. There are no other integration variables, so the integrands just take the value they have for $Q^{\ddagger}_1 = 0$. We find

$$\boxed{\begin{aligned}\frac{k_s}{k_g} &= V^2 P^{(2)}(S^{\ddagger}, Q^{\ddagger}_1 = 0) \exp(V^{\ddagger}_{\mathrm{intra}}(S^{\ddagger}, Q^{\ddagger}_1 = 0)/k_B T) \\ &= g(r^{\ddagger}) \exp(V^{\ddagger}_{\mathrm{intra}}(r^{\ddagger})/k_B T) \\ &= \exp(-W_{\mathrm{mean}}(r^{\ddagger})/k_B T)\end{aligned}} \tag{10.57}$$

where $r^\ddagger$ is the distance between the atoms at the saddle point (i.e., $Q_1^\ddagger = 0$), and we have used Eq. (10.42) to introduce the pair-distribution function $g^{(2)}(r^\ddagger) \equiv g(r^\ddagger)$ and the potential of mean force $W_{\text{mean}}(r^\ddagger)$, i.e.,

$$\begin{aligned} g(r^\ddagger) &= \exp(-V_{\text{intra}}^\ddagger(r^\ddagger)/k_BT)\exp(-W_{\text{mean}}(r^\ddagger)/k_BT) \\ &= \exp(-E_{\text{cl}}/k_BT)\exp(-W_{\text{mean}}(r^\ddagger)/k_BT) \\ &= \frac{\exp(-E_{\text{cl}}/k_BT)\int dR\exp(-(V_{\text{sol}}+V_{\text{int}}^\ddagger)/k_BT)}{\int_{\text{React}} dq_{\text{A}}dq_{\text{B}}\int dR\exp(-(V_{\text{A}}+V_{\text{B}}+V_{\text{sol}}+V_{\text{int}})/k_BT)} \end{aligned} \tag{10.58}$$

Let us emphasize the physical meaning of the result in Eq. (10.57). If we introduce the expression for the gas-phase rate constant k_g in Eq. (10.55), we find

$$k_s = \frac{k_BT}{h}\frac{(Z^\ddagger/V)}{(Z_{\text{A}}/V)(Z_{\text{B}}/V)}\exp(-(E_{\text{cl}}+W_{\text{mean}})/k_BT) \tag{10.59}$$

From this it is clear that the solvent effect on the rate constant amounts to a renormalization of the energy barrier from being E_{cl} in the gas phase to $E_{\text{cl}} + W_{\text{mean}}$ in solution. This represents the net effect of stabilization of the activated complex and the reactants in the solution.

To see this, let us consider the definition of W_{mean}. From Eq. (10.58) we have

$$\exp(-W_{\text{mean}}/k_BT) = V^2\frac{\int dR\exp(-(V_{\text{sol}}+V_{\text{int}}^\ddagger)/k_BT)}{\int_{\text{React}} dq_{\text{A}}dq_{\text{B}}\int dR\exp(-(V_{\text{sol}}+V_{\text{int}})/k_BT)} \tag{10.60}$$

where we have set $V_{\text{A}} = V_{\text{B}} = 0$ in the integral over reactants, since we only consider atoms. The influence of the solvent on the activated complex is embedded in the $V_{\text{int}}^\ddagger$ term and on the reactants in the V_{int} term, since that term not only includes the interaction between reactant molecules but also between reactant and solvent molecules. So, even when we neglect interactions between reactant molecules, there is an extra contribution to V_{int} from the reactant–solvent interaction, i.e., *solute–solvent* interactions, that is absent in the gas phase. If we assume that the stabilization of the activated complex and reactant molecules is the same, that is, $V_{\text{int}}^\ddagger = V_{\text{int}}$, then

$$\exp(-W_{\text{mean}}/k_BT) = V^2\frac{1}{\int_{\text{React}} dq_{\text{A}}dq_{\text{B}}} = 1 \tag{10.61}$$

that is, $W_{\text{mean}} = 0$. This is illustrated in Fig. 10.2.5.

In the central part of the figure, we have shown the energy levels for the activated complex and reactants in the gas phase. If the stabilization of the two components is the same, as illustrated by the energy diagram to the left, then both levels are displaced by the same amount and the rate constant will be the same in solution as in the gas phase. If the stabilization is different, as illustrated in the energy diagram to the right, the two levels are displaced differently, and we will have a different rate constant in solution. The interaction energies V_{int} and $V_{\text{int}}^\ddagger$ can, e.g., be estimated within a continuum solvation model like the Onsager model, Eq. (9.6).

The idea, therefore, is to use this expression on 'real' chemical reactions to give, if not an exact account of the effect, then at least a qualitative account. We then

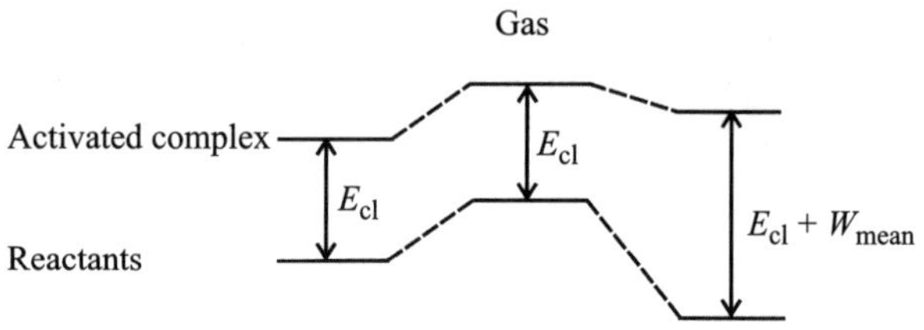

Fig. 10.2.5 Energy diagram for the activated complex and reactants in the gas phase and solution. To the left is shown the case where the stabilization of the two compounds in solution is the same, and to the right where it differs.

need to understand the two functions $g(r)$ and $W_{mean}(r)$ for such a system. The pair-distribution function is proportional to the probability of finding a separation of r between two atoms in a liquid, and here between two groups of atoms along the reaction coordinate. That function looks like the pair-distribution function for a pure liquid as sketched in Fig. 10.2.6. Note that a high value of $g(r)$ implies a low value of the potential of mean force. In the liquid, as well as in the reacting system, it may, for example, be determined by Monte Carlo simulations. Also shown is the potential of mean force based on the relation for a pure liquid:

$$g(r) = \exp(-W_{mean}(r)/k_BT) \Longrightarrow W_{mean}(r) = -k_BT \ln g(r) \tag{10.62}$$

This expression is a little simpler than the expression in Eq. (10.58), since we only consider a one-component liquid and therefore do not have to distinguish between a V_{intra} and a V_{int} contribution to the potential energy.

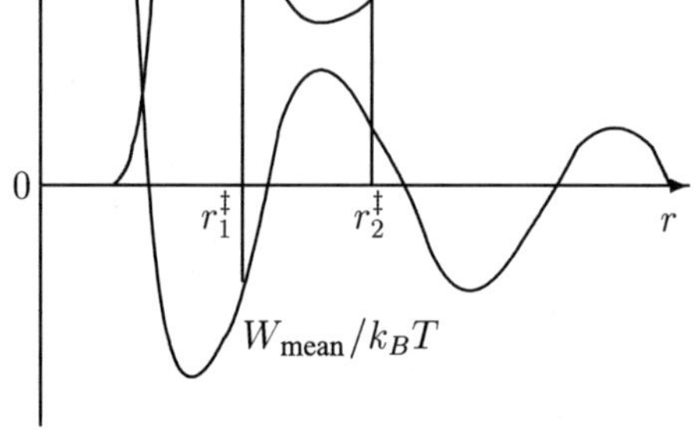

Fig. 10.2.6 Sketch of the pair-distribution function and potential of mean force.

In practice, the value of the reaction coordinate $r^\ddagger$ is determined from the gas-phase potential energy surface of the complex. Then we use the pair-distribution function for the system (for example, determined by a Monte Carlo simulation) and the intramolecular potential energy $V_{\text{intra}}^\ddagger$ to calculate the relation between the two rate constants. Alternatively, one may determine the potential of mean force directly in a Monte Carlo simulation. With the example in Fig. 10.2.6 and a reaction coordinate at $r_1^\ddagger$, we see that the potential of mean force is negative, which implies that the rate constant in solution is larger than in the gas phase. Physically, this means that the transition state is more stabilized (has a lower energy) than in the gas phase. If the reaction coordinate is at $r_2^\ddagger$, then the potential of mean force is positive and the rate constant in solution is smaller than in the gas phase.

The determination of potentials of mean force has, e.g., been applied to nucleophilic substitution (S_N2) reactions in solution, as illustrated in Example 10.1.

Further reading/references

[1] D.A. McQuarrie, *Statistical mechanics* (University Science Books, 2000)

[2] B.J. Berne, M. Borkovec, and J.E. Straub, *J. Phys. Chem.* **92**, 3711 (1988)

[3] G. Ciccotti, M. Ferrario, J.T. Hynes, and R. Kapral, *J. Chem. Phys.* **93**, 7137 (1990)

[4] J. Chandrasekhar, S.F. Smith, and W.L. Jorgensen, *J. Am. Chem. Soc.* **107**, 154 (1985)

[5] D.G. Truhlar, B.C. Garrett, and S.J. Klippenstein, *J. Phys. Chem.* **100**, 12771 (1996)

11

Dynamic solvent effects, Kramers theory

Key ideas and results

In this chapter we consider dynamical solvent effects on the rate constant for chemical reactions in solution. Solvent dynamics may enhance or impede molecular motion. The effect is described by stochastic dynamics, where the influence of the solvent on the reaction dynamics is included by considering the motion along the reaction coordinate as (one-dimensional) Brownian motion. The results are as follows.

- In Kramers theory that is based on the Langevin equation with a constant time-independent friction constant, it is found that the rate constant may be written as a product of the result from conventional transition-state theory and a transmission factor. This factor depends on the ratio of the solvent friction (proportional to the solvent viscosity) and the curvature of the potential surface at the transition state. In the high friction limit the transmission factor goes toward zero, and in the low friction limit the transmission factor goes toward one.
- Grote–Hynes theory is a generalization of Kramers theory, based on the generalized Langevin equation with a time-dependent solvent friction coefficient on the dynamics of the reaction coordinate. The Kramers result is found in the limit where the motion in the reaction coordinate is slow enough for the solvent molecules to adjust to the changes and re-establish equilibrium conditions. In the other limit, the non-adiabatic limit, where the motion in the reaction coordinate is so fast that the solvent molecules are 'frozen', the result may differ from the Kramers result by several orders of magnitude.

In our discussion of the transition-state theory with static solvent effects, it was noticed that it is a *mean field* description where the effects of dynamical fluctuations in the solvent molecule positions and velocities were excluded.

Most modern investigations of the effects of a solvent on the rate constant, where dynamical fluctuations are included, are based on a classical paper by Kramers from 1940 [1]. His theory is based on the transition-state theory approach where we have identified the reaction coordinate as the normal mode of the activated complex that has an imaginary frequency. In ordinary transition-state theory, we assume that the motion in that coordinate is like a free translational motion with no recrossings. This may be partially justified in the gas phase, but hardly in solution. Typically, the activated complex is positioned in a 'cage' in the solvent, which means that the motion in the reaction coordinate cannot be considered as free due to interactions with the surrounding solvent molecules, and it indeed appears that recrossings may be important.

The differences between gas-phase reaction dynamics and reaction dynamics in a solvent are illustrated in the sketches in Fig. 11.0.1. In Fig. 11.0.1(a), a potential energy surface (PES) in the gas phase is shown with a trajectory that passes the saddle point and leads to a chemical reaction. In Fig. 11.0.1(b), the PES is shown for the reaction in solution. It may or may not be perturbed by the interactions with the solvent; in the figure we have shown a slight distortion. The trajectory would have led to a reaction as indicated by the dot–dashed part of it, had it not been for a dynamical interaction with the solvent, where energy is exchanged between reactants and solvent briefly before the saddle-point region is passed. This causes the trajectory to turn back into the phase space of the reactants. Another possibility is shown in Fig. 11.0.1(c), where the trajectory has passed the saddle point before the energy exchange with the solvent takes place and sends it back into the phase space of the reactants. This is an example of recrossing.

It should be emphasized that these dynamical effects can lead to significant corrections to conventional transition-state theory where recrossings are neglected. However,

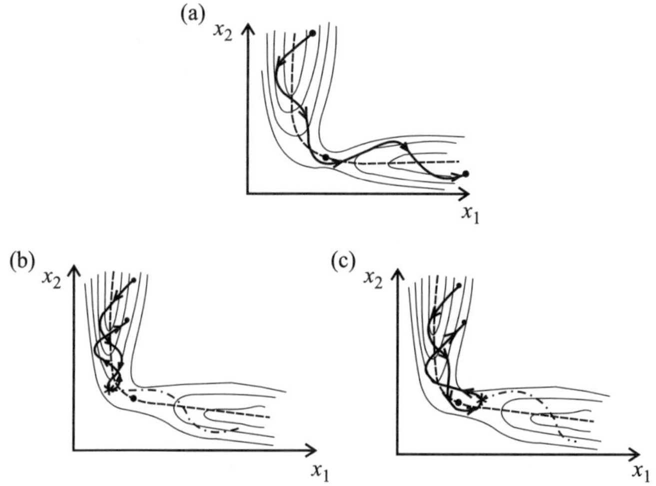

Fig. 11.0.1 (a) PES for a reaction in the gas phase with a trajectory leading to a reaction since it passes the saddle-point region. (b) A slightly perturbed PES for the same reaction as in (a) carried out in solution. The dot–dashed part of the trajectory is the one followed when the solvent is absent. At the 'star', just before the saddle point, there is a dynamical energy exchange between reactants and solvent that causes the trajectory to move back into the phase space of the reactants. (c) As in (b) except for the position along the trajectory where the energy exchange with the solvent takes place; here it is after passing the saddle point, resulting in a recrossing of the barrier.

in general, the corrections are very small compared to the static solvent effects discussed in the previous chapter.

Kramers idea was to give a more realistic description of the dynamics in the reaction coordinate by including dynamical effects of the solvent. Instead of giving a deterministic description, which is only possible in a large-scale molecular dynamics simulation, he proposed to give a stochastic description of the motion similar to that of the Brownian motion of a heavy particle in a solvent. From the normal coordinate analysis of the activated complex, a reduced mass μ_1 has been associated with the motion in the reaction coordinate, so the proposal is to describe the motion in that coordinate as that of a Brownian particle of mass μ_1 in the solvent.

The one-dimensional Brownian motion takes place in a potential, as sketched in Fig. 11.0.2, where the well at r_a refers to the reactants, r_b to the transition state, and r_c to the products.

The influence of the solvent may be described by the *Langevin equation.* Since it is a probabilistic description, we want to determine the probability density $P(r, v; t)$, where $P(r, v; t)dvdr$ is the probability of finding the particle in the position interval $(r, r + dr)$ with velocity in the interval $(v, v + dv)$ at time t.

If n_a is the probability of finding reactants at the a-well, and k_s is the rate constant for going from the a-well to the c-well, then the probability flux j across the barrier is

$$j = k_s n_a \tag{11.1}$$

which defines the rate constant k_s. At stationary conditions $P(r, v; t) = P(r, v)$, so the probability flux j across the transition state at $r = r_b$ may be determined as

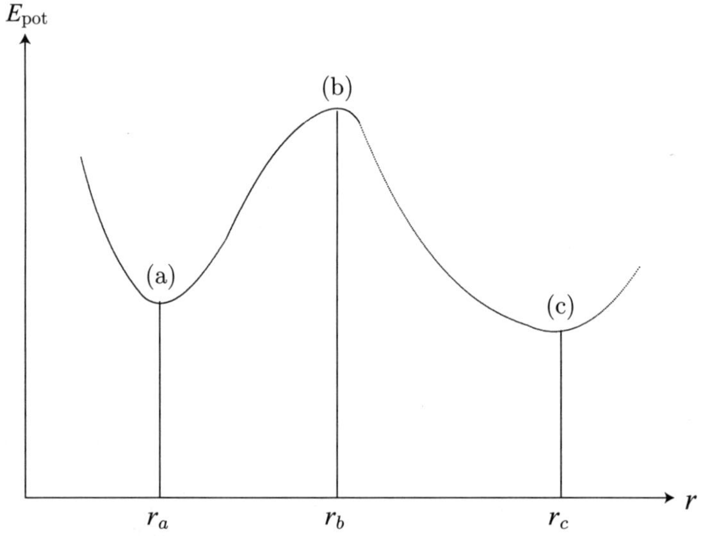

Fig. 11.0.2 Sketch of the potential energy as a function of the reaction coordinate r. The state around (a) represents the reactant state, the transition state is at (b), and the product state is at (c).

$$j = \int_{-\infty}^{\infty} dv\, vP(r_b, v) \tag{11.2}$$

and n_a according to

$$n_a = \int_{a_{\text{well}}} drdvP(r, v) \tag{11.3}$$

So the rate constant k_s may be found as the ratio between j and n_a. This expression for the rate constant is equivalent to a one-dimensional version of Eq. (5.49) when the Boltzmann distribution is replaced by $P(r, v)$. It is, however, important to notice that Eq. (11.2) allows for recrossings across the barrier, since the lower integration limit is $-\infty$ and not 0 as in the previous presentations of transition-state theory where the possibility of recrossings was neglected.

We note that $P(r, v; t)$ has the dimension s m^{-2}, so j has the dimension s^{-1}, n_a is dimensionless, and k_s has the dimension s^{-1}. If both n_a and j are multiplied by the number of reactant molecules, we will get, respectively, the number of molecules in the a-well and the flux of reactants across the transition-state barrier.

Since probabilistic dynamics is central to an understanding of Kramers theory for the influence of solvents on the rate constant, we shall first summarize some of the essential features in such a description.

11.1 Brownian motion, the Langevin equation

The problem of Brownian motion relates to the motion of a heavy colloidal particle immersed in a fluid made up of light particles. In Fig. 11.1.1 the trajectory of a Brownian particle is shown. The coordinates of a particle with diameter 2 μm moving in water are observed every 30 s for 135 min. At the very first step in the argument one renounces an exact deterministic description of the motion and replaces it with a probabilistic description.

Let us consider a Brownian particle of mass M immersed in a fluid. Macroscopically, the laws of hydrodynamics would tell us that during its motion the particle

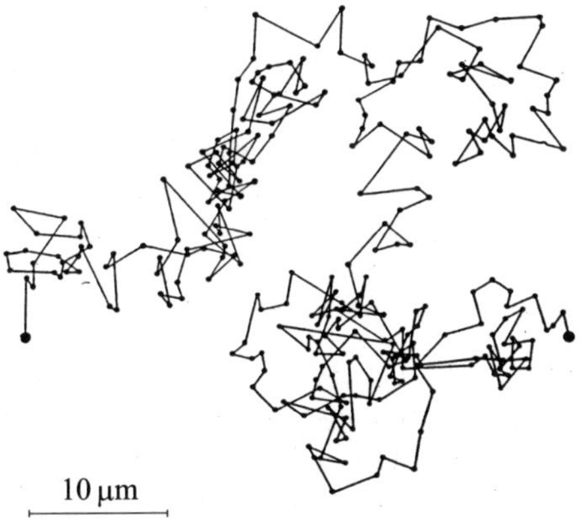

Fig. 11.1.1 The trajectory of a Brownian particle with diameter 2 μm in water. The coordinates of the particle are observed every 30 s for 135 min.

undergoes a friction force due to the viscosity of the fluid that is proportional to its velocity v; hence Newton's equation of motion has the form

$$\frac{d\langle v\rangle}{dt} = -\frac{g}{M}\langle v\rangle \equiv -\gamma\langle v\rangle \tag{11.4}$$

where g is the friction, assumed to be a constant, and $\langle v\rangle$ is the average macroscopic velocity as opposed to the instantaneous microscopic velocity v. γ is defined as the friction constant divided by the mass M. We restrict ourselves to one-dimensional notation for simplicity and because we are going to use the theory on a one-dimensional problem.

We need, however, a more refined and realistic description, since this equation predicts an exponential decay of the initial velocity to zero, in contrast to the observed incessant motion of a Brownian particle. Therefore, we must add to the systematic friction force the action of all individual solvent molecules on the Brownian particle, which results in an additional term $F(t)$:

$$\frac{dv}{dt} = -\gamma v + F(t) \tag{11.5}$$

Note that in this equation we consider the actual velocity v. The effect of each individual collision is to produce a tiny deflection of the particle away from its macroscopic trajectory. Since we do not want to go into the details of this many-body problem, we can only say that the collisions with solvent molecules are very numerous and very irregular as to their strength and direction. This leads to a probabilistic description. We cannot specify the force $F(t)$ as a given function of time, but we can make reasonable assumptions about the average effect of the collisions over a large number of identical macroscopic situations (like an ensemble). Similarly, we cannot predict the velocity of the Brownian particle at every time t, but we may be able to predict the average outcome of a large number of experiments performed under identical conditions. Therefore, the whole idea of solving Eq. (11.5) is a typical example of the class of so-called *stochastic* (or *random*) equations of motion. It is called, after its discoverer, the *Langevin equation*.

Let us see how such an equation is solved. First we must define the random function $F(t)$ quantitatively. The average of $F(t)$ over an ensemble of Brownian particles vanishes. This condition ensures that the average velocity of the Brownian particle obeys the macroscopic law (Eq. (11.4)), that is, that the fluctuations cancel each other on average. This is written as follows:

$$\langle F(t)\rangle = 0 \tag{11.6}$$

We may express the idea of irregularity by assuming that collisions that are well separated in time are statistically independent. In other words, the correlation between values of $F(t)$ at two times t_1 and t_2 is different from zero only for time intervals of the order of the duration of a collision τ_c. Explicitly,

$$\langle F(t_1)F(t_2)\rangle = \phi(t_1 - t_2) \simeq f\delta(t_1 - t_2) \tag{11.7}$$

where $\phi(t)$ is a function that is sharply peaked at $t = 0$ and that vanishes for $t > \tau_c$. Often ϕ is approximated by a delta function, as indicated in the equation. We are now in a position to solve the Langevin equation in Eq. (11.5). Suppose the Brownian particle starts at $t = 0$ with velocity v_0. If we consider Eq. (11.5) as an ordinary differential equation, it is seen to be an inhomogeneous linear first-order differential equation with the solution

$$v = v_0 \exp(-\gamma t) + \exp(-\gamma t) \int_0^t d\tau \exp(\gamma \tau) F(\tau) \tag{11.8}$$

This result is useless, since we do not know $F(t)$. If, however, we average all its terms over the ensemble defined above and use assumption (11.6), we obtain

$$\langle v \rangle = v_0 \exp(-\gamma t) \tag{11.9}$$

This is just the solution of the macroscopic equation (11.4), a result that is not surprising. More interesting is the square of the velocity, as for the Maxwell–Boltzmann velocity distribution law. If we square the right-hand side of Eq. (11.8) we find

$$\begin{aligned} v^2 = v_0^2 \exp(-2\gamma t) &+ v_0 \exp(-2\gamma t) \int_0^t d\tau_2 \exp(\gamma \tau_2) F(\tau_2) \\ &+ v_0 \exp(-2\gamma t) \int_0^t d\tau_1 \exp(\gamma \tau_1) F(\tau_1) \\ &+ \exp(-2\gamma t) \int_0^t d\tau_1 \int_0^t d\tau_2 \exp(\gamma(\tau_1 + \tau_2)) F(\tau_1) F(\tau_2) \end{aligned} \tag{11.10}$$

Then, taking the ensemble average using Eq. (11.6), we get

$$\langle v^2 \rangle = v_0^2 \exp(-2\gamma t) + \exp(-2\gamma t) \int_0^t d\tau_1 \int_0^t d\tau_2 \exp(\gamma(\tau_1 + \tau_2)) \langle F(\tau_1) F(\tau_2) \rangle \tag{11.11}$$

The two-dimensional integral may be reduced to two one-dimensional integrals by a change of variables. Details may be found in Appendix F. The result is

$$\langle v^2 \rangle = v_0^2 \exp(-2\gamma t) + \frac{f}{2\gamma}(1 - \exp(-2\gamma t)) \tag{11.12}$$

This tells us how the square of the velocity of the particle, on the average, develops in time given that it had the value v_0^2 at time $t = 0$. In other words, the average is a conditional average, a point of view that will be emphasized in the following section, where a statistical approach to the Brownian motion will be given. The expression shows that, for short times $t \ll (2\gamma)^{-1}$, the velocity fluctuations are mainly determined by the initial value v_0^2; but for larger times the initial value is progressively forgotten, and the average square of the velocity approaches the value of $f/2\gamma$, which is solely determined by the mechanism of collision and is independent of the initial velocity.

To complete the theory we need to specify f, given the friction constant γ. There is, however, nothing in the theory that allows us to determine this function from microscopic molecular data. However, we may get around the problem if we believe, with good reason, that the end point of the evolution of particle motions will lead to a state of thermal equilibrium at temperature T, no matter the original perturbation of the particle motion. In this state the mean square velocity of the particle is determined by the theorem of equipartition of energy. Hence, we may require that

$$\lim_{t\to\infty} \langle v^2 \rangle = \frac{f}{2\gamma} = \frac{k_B T}{M} \tag{11.13}$$

so

$$f = 2\gamma \frac{k_B T}{M} \tag{11.14}$$

This is a specific example of the fundamental *fluctuation–dissipation* theorem that relates the random force f (*fluctuation*) to the friction constant γ (*dissipation*) to ensure that any initial state eventually evolves into a state in thermal equilibrium with the fluid at temperature T. With this requirement, we obtain from Eq. (11.12) the final result

$$\langle v^2 \rangle = \frac{k_B T}{M} + \left[v_0^2 - \frac{k_B T}{M} \right] \exp(-2\gamma t) \tag{11.15}$$

11.2 Kramers theory for the rate constant

The theory of Brownian motion is a particular example of an application of the general theory of *random* or *stochastic* processes [2]. Since Kramers approach is based on a more general stochastic equation than the Langevin equation, we have reviewed some of the fundamental ideas and methods of the theory of stochastic processes in Appendix H.

Kramers theory is based on the *Fokker–Planck equation* for the position and velocity of a particle. The Fokker–Planck equation is based on the concept of a Markov process and in its generic form it contains no specific information about any particular process. In the case of Brownian motion, where it is sometimes simply called the Kramers equation, it takes the form

$$\frac{\partial P(r,v;t)}{\partial t} = -v\frac{\partial P(r,v;t)}{\partial r} + \frac{1}{M}\frac{\partial U}{\partial r}\frac{\partial P(r,v;t)}{\partial v} + \gamma\frac{\partial}{\partial v}(vP(r,v;t)) + \frac{\gamma k_B T}{M}\frac{\partial^2 P(r,v;t)}{\partial v^2} \tag{11.16}$$

where $P(r, v; t)drdv$ is the probability of finding a particle at $r, r + dr$ with a velocity $v, v + dv$ at time t, M is the mass of the particle, and γ is the friction coefficient known from the Langevin equation. The motion takes place in an external potential $U(r)$, as sketched in Fig. 11.0.2. This is the potential from the gas phase modified by the interactions from the solvent molecules, as determined in Chapter 10. The equation describes the diffusion of a point in phase space for a one-dimensional Brownian particle.

We shall now determine a solution to Eq. (11.16) with proper boundary conditions and use the result to determine the rate constant in a fluid. It is assumed that the barrier in going from reactant states at well (a) in Fig. 11.0.2 to the transition state at (b) is large compared to $k_B T$; the probability of being at (b) is therefore small and it will be reasonable to seek a *steady-state* solution to Eq. (11.16). Similarly, around the a-well, we also assume *stationary conditions* with reactants in thermal equilibrium with the solvent. Thus, we seek a solution to the equation with $(\partial P/\partial t) = 0$.

At equilibrium, stationary conditions exist where $P(r,v;t) = P(r,v)$ and $P(r,v)$ is given by equilibrium statistical mechanics:

$$P(r,v) = Q^{-1}\exp(-(Mv^2/2 + U(r))/k_BT) \tag{11.17}$$

with Q as the partition function:

$$Q = \int_{\text{React}} drdv\,\exp(-(Mv^2/2 + U(r))/k_BT) \tag{11.18}$$

It is easy to show that the function in Eq. (11.17) is indeed a solution to the stationary equation. We have

$$\begin{aligned}
P(r,v) &= Q^{-1}\exp(-(Mv^2/2 + U(r))/k_BT)\\
\left(\frac{\partial P}{\partial r}\right) &= -\frac{1}{k_BT}\left(\frac{\partial U}{\partial r}\right)P\\
\left(\frac{\partial P}{\partial v}\right) &= -\frac{Mv}{k_BT}P\\
\left(\frac{\partial^2 P}{\partial v^2}\right) &= -\frac{M}{k_BT}P + \left(\frac{Mv}{k_BT}\right)^2 P
\end{aligned} \tag{11.19}$$

This is introduced into Eq. (11.16) and all terms do indeed cancel, corresponding to stationarity, that is, $\partial P/\partial t = 0$:

$$\frac{Pv}{k_BT}\left(\frac{\partial U}{\partial r}\right) - \frac{Pv}{k_BT}\left(\frac{\partial U}{\partial r}\right) + \gamma P - \frac{M\gamma v^2}{k_BT}P + \frac{M\gamma v^2}{k_BT}P - \gamma P = 0 \tag{11.20}$$

We will now search for a non-trivial solution to the stationary Fokker–Planck equation.

To find a *steady-state* solution at the transition state, $r \sim r_b$, we make the substitution

$$P(r,v) = Y(r,v)Q^{-1}\exp(-(Mv^2/2 + U(r))/k_BT) \tag{11.21}$$

where $Y(r,v)$ has to be determined with the following boundary conditions:

(i) $r \sim r_a$: $Y(r,v) = 1$;

(ii) $r \sim r_c$: $Y(r,v) = 0$.

Boundary condition (i) ensures that the reactants will be in thermal equilibrium with the solvent, and boundary condition (ii) expresses a sink condition at the product side, that is, the products are removed as soon as they are formed or, equivalently, the concentration of them remains negligible. The shape of the external potential near the transition state is approximated by a parabolic shape:

$$U(r) = U(r_b) - \frac{1}{2}M\omega_b^2(r - r_b)^2 \tag{11.22}$$

and when introduced into Eq. (11.16) we get

$$0 = -v\frac{\partial P(r,v;t)}{\partial r} - \omega_b^2(r - r_b)\frac{\partial P(r,v;t)}{\partial v} + \gamma\frac{\partial}{\partial v}(vP(r,v;t)) + \frac{\gamma k_BT}{M}\frac{\partial^2 P(r,v;t)}{\partial v^2} \tag{11.23}$$

The introduction of Eq. (11.21) gives, after some lengthy but straightforward manipulations, the following equation for $Y(r, v)$:

$$-v\frac{\partial Y}{\partial r} - (\omega_b^2(r - r_b) + \gamma v)\frac{\partial Y}{\partial v} + \frac{\gamma k_B T}{M}\frac{\partial^2 Y}{\partial v^2} = 0 \tag{11.24}$$

where we have used the result in Eq. (11.20). In order to convert this partial differential equation in the two variables r and v into an ordinary second-order differential equation in one variable u, Kramers introduced the substitution

$$u = v - a(r - r_b) \tag{11.25}$$

and assumed that Y depends on r and v only in this combination, that is, $Y(r, v) = Y(u)$. The idea with this transformation is that the partial derivatives with respect to the two variables r and v may be expressed as the partial derivatives with respect to a single variable u, since

$$\begin{aligned}\frac{\partial Y}{\partial r} &= \left(\frac{dY}{du}\right)\left(\frac{\partial u}{\partial r}\right) = -a\left(\frac{dY}{du}\right)\\ \frac{\partial Y}{\partial v} &= \left(\frac{dY}{du}\right)\left(\frac{\partial u}{\partial v}\right) = \left(\frac{dY}{du}\right)\end{aligned} \tag{11.26}$$

So the differential equation becomes

$$(av - \omega_b^2(r - r_b) - \gamma v)\left(\frac{dY}{du}\right) + \frac{\gamma k_B T}{M}\frac{d^2Y}{du^2} = 0 \tag{11.27}$$

This is now an ordinary second-order differential equation. The solution $Y =$ const. corresponds to thermal equilibrium according to Eq. (11.21). There is, however, another steady-state solution to the equation, if it can be brought into the standard form

$$A\frac{\mathrm{d}^2Y}{\mathrm{d}u^2} + Bu\frac{\mathrm{d}Y}{\mathrm{d}u} = 0 \tag{11.28}$$

This can be done if

$$A = \frac{\gamma k_B T}{M} \tag{11.29}$$

and the coefficient of the first derivative can be brought into the form Bu, that is,

$$\begin{aligned}Bu &= Bv - aB(r - r_b)\\ &= (a - \gamma)v - \omega_b^2(r - r_b)\end{aligned} \tag{11.30}$$

From this it follows that

$$B = a - \gamma \tag{11.31}$$

and

$$a(a - \gamma) = \omega_b^2 \tag{11.32}$$

or

$$a = \frac{\gamma \pm \sqrt{\gamma^2 + 4\omega_b^2}}{2}$$
$$= \frac{\gamma}{2} \pm \sqrt{\frac{\gamma^2}{4} + \omega_b^2} \tag{11.33}$$

The solution to the standard equation is

$$Y(u) = K \int_{u_c}^{u} dz \exp(-Bz^2/(2A))$$
$$= K \int_{u_c}^{u} dz \exp(-(a-\gamma)z^2 M/(2\gamma k_B T)) \tag{11.34}$$

where $u_c \equiv u(r = r_c) = v - a(r_c - r_b)$. This can easily be checked by substitution of the expression into the standard differential equation.

In Eq. (11.33) we choose the upper sign, so that the argument to the exponential will be negative and Y will therefore be well behaved. The lower limit in the integration is chosen such that boundary condition (ii) is satisfied. K may be determined from boundary condition (i). We have

$$Y(u_a) = 1 = K \int_{u_c}^{u_a} dz \exp(-(a-\gamma)z^2 M/(2\gamma k_B T))$$
$$= K \int_{-\infty}^{\infty} dz \exp(-(a-\gamma)z^2 M/(2\gamma k_B T))$$
$$= K \sqrt{\frac{2\pi\gamma k_B T}{(a-\gamma)M}} \tag{11.35}$$

Thus

$$K = \sqrt{\frac{(a-\gamma)M}{2\pi\gamma k_B T}} \tag{11.36}$$

We have replaced the integration limits $u_a = v - a(r_a - r_b)$ and $u_c = v - a(r_c - r_b)$ by ∞ and $-\infty$, respectively. This can be done because $u_a > u_c$ and because the integrand only differs from zero in a relatively narrow zone about $z = 0$.

We may now determine the probability flux j at $r = r_b$, where $u = v$:

$$j = \int_{-\infty}^{\infty} dv\, v P(r_b, v)$$
$$= K Q^{-1} \exp(-U(r_b)/k_B T) \int_{-\infty}^{\infty} dv\, v \exp(-v^2 M/(2k_B T))$$
$$\times \int_{-\infty}^{v} dz \exp(-(a-\gamma)z^2 M/(2\gamma k_B T)) \tag{11.37}$$

The order of integration is interchanged, as sketched in Fig. 11.2.1. The region covered in the integration is marked by the hatched region between the 45° line and the abscissa axis. This region may be spanned in two ways. We may choose an interval

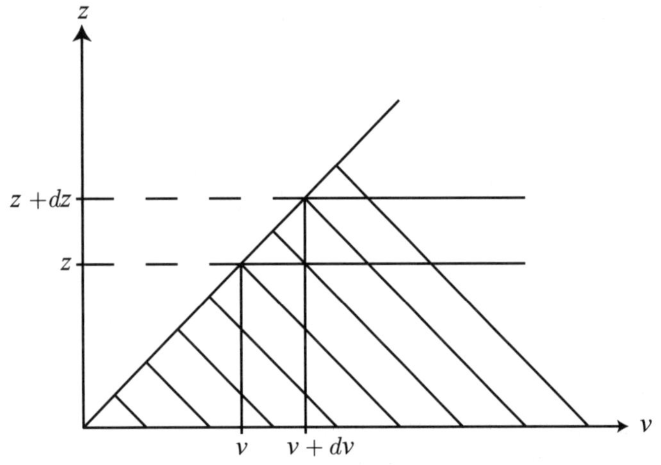

Fig. 11.2.1 A sketch of the order of integration in a double integration.

$(v, v + dv)$ along the v-axis and let z vary from the lower bound in the integral to the value of v on the 45° line. This is how the integral is performed in the equation as written. The alternative is to choose an interval $(z, z + dz)$ along z and then let v span the region z to ∞. With the integrations done in that order, Eq. (11.37) will be

$$j = KQ^{-1}\exp(-U(r_b)/k_BT)\int_{-\infty}^{\infty} dz\,\exp(-(a-\gamma)z^2M/(2\gamma k_BT)) \times \int_{z}^{\infty} dv\,v\exp(-v^2M/(2k_BT)) \tag{11.38}$$

The integral over the variable v is now easily done, and we get

$$\begin{aligned} j &= KQ^{-1}\exp(-U(r_b)/k_BT)\frac{k_BT}{M} \\ &\quad\times \int_{-\infty}^{\infty} dz\,\exp(-(a-\gamma)z^2M/(2\gamma k_BT))\exp(-z^2M/(2k_BT)) \\ &= KQ^{-1}\exp(-U(r_b)/k_BT)\frac{k_BT}{M}\int_{-\infty}^{\infty} dz\,\exp(-az^2M/(2\gamma k_BT)) \\ &= KQ^{-1}\exp(-U(r_b)/k_BT)\frac{k_BT}{M}\sqrt{\frac{2\pi\gamma k_BT}{aM}} \end{aligned} \tag{11.39}$$

In order to obtain an expression for the population of reactant states around well (a), we expand the potential around (a) to second order:

$$U(r) = U(r_a) + \frac{1}{2}M\omega_a^2(r - r_a)^2 \tag{11.40}$$

and find the population of reaction states in well (a) according to Eq. (11.3) using that $Y = 1$ in that region:

$$\begin{aligned}
n_a &= \int_{-\infty}^{\infty} dv \int_{a_{\text{well}}} dr P(r,v) \\
&= Q^{-1}\exp(-U(r_a)/k_BT)\int_{-\infty}^{\infty} dv \exp(-v^2M/(2k_BT)) \\
&\qquad \times \int_{-\infty}^{\infty} dr \exp(-M\omega_a^2(r-r_a)^2/(2k_BT)) \\
&= Q^{-1}\exp(-U(r_a)/k_BT)\sqrt{\frac{2\pi k_BT}{M}}\sqrt{\frac{2\pi k_BT}{M\omega_a^2}} \\
&= Q^{-1}\exp(-U(r_a)/k_BT)\frac{2\pi k_BT}{M\omega_a}
\end{aligned} \tag{11.41}$$

We may now use the expressions for j and n_a to determine the rate of transition across the transition state according to Eq. (11.2):

$$\begin{aligned}
k_s = \frac{j}{n_a} &= \frac{K\exp(-U(r_b)/k_BT)k_BT/M\sqrt{2\pi\gamma k_BT/(aM)}}{\exp(-U(r_a)/k_BT)2\pi k_BT/(M\omega_a)} \\
&= \sqrt{\frac{a-\gamma}{a}}\frac{\omega_a}{2\pi}\exp(-(U(r_b)-U(r_a))/k_BT)
\end{aligned} \tag{11.42}$$

where we have used the expression for K in Eq. (11.36). From Eq. (11.32) we have

$$a = \frac{\omega_b^2}{a-\gamma} \tag{11.43}$$

so

$$\boxed{\begin{aligned}
k_s &= \sqrt{\frac{(a-\gamma)^2}{\omega_b^2}}\frac{\omega_a}{2\pi}\exp(-(U(r_b)-U(r_a))/k_BT) \\
&= \frac{\omega_a}{2\pi\omega_b}\left(\sqrt{\frac{\gamma^2}{4}+\omega_b^2}-\frac{\gamma}{2}\right)\exp(-(U(r_b)-U(r_a))/k_BT)
\end{aligned}} \tag{11.44}$$

This is Kramers equation for the rate constant of a chemical reaction in a solution. The influence of the solvent is represented by the friction coefficient, which may be directly related to the viscosity η of the solvent via Stokes law:

$$g = 6\pi\eta R \tag{11.45}$$

or

$$\gamma = \frac{6\pi\eta R}{M} \tag{11.46}$$

where R is the hydrodynamic radius of the particle.

For $\gamma/2 \gg \omega_b$ we find that

$$k_s = \frac{\omega_a}{2\pi}\frac{\omega_b}{\gamma}\exp(-(U(r_b)-U(r_a))/k_BT) \tag{11.47}$$

which is the high friction or high viscosity limit. It shows that the rate constant goes toward zero for an infinite friction constant. In the other limit, where $\gamma/2 \ll \omega_b$, we find

$$k_s = \frac{\omega_a}{2\pi} \exp(-(U(r_b) - U(r_a))/k_B T) \tag{11.48}$$

which is just the ordinary (gas-phase) transition-state theory result. This is seen by realizing that our reactant state is represented by just a one-dimensional oscillator at the a-well; the expression for the ordinary (gas-phase) transition-state rate constant, where recrossings of the transition state are neglected, is then (Eq. (6.8))

$$\begin{aligned} k_{\text{TST}} &= \frac{k_B T}{h} \frac{Q^\ddagger}{Q_{\text{vib}}} \exp(-E_{\text{cl}}/k_B T) \\ &= \frac{k_B T}{h} \frac{1}{k_B T/(h\nu_a)} \exp(-E_{\text{cl}}/k_B T) \\ &= \nu_a \exp(-E_{\text{cl}}/k_B T) \\ &= \frac{\omega_a}{2\pi} \exp(-E_{\text{cl}}/k_B T) \end{aligned} \tag{11.49}$$

which is identical to the expression in Eq. (11.48) with $E_{\text{cl}} = U(r_b) - U(r_a)$ being the energy barrier for the reaction along the reaction coordinate. Note that we have used the classical expression for E_0 and the vibrational partition function, since we neglect any quantum effects in this theory, and that the partition function of the transition state is equal to one, since the number of degrees of freedom in this state is one less than in the reactant state, that is, equal to zero.

We may therefore write the expression for the rate constant as a product of the conventional transition-state rate constant k_{TST} and a *transmission factor* κ_{KR}:

$$\boxed{k_s = \kappa_{\text{KR}} k_{\text{TST}}} \tag{11.50}$$

with

$$\begin{aligned} \kappa_{\text{KR}} &= \left(\sqrt{\gamma^2/4 + \omega_b^2} - \gamma/2\right) \Big/ \omega_b \\ &= \sqrt{1 + (\gamma/(2\omega_b))^2} - \gamma/(2\omega_b) \end{aligned} \tag{11.51}$$

The departure from ordinary transition-state theory is seen to be determined by the relation between the friction γ and the frequency ω_b representing the curvature of the potential surface along the reaction coordinate at the transition state. For $\gamma/(2\omega_b) \to \infty$, $\kappa_{\text{KR}} \to 0$, whereas for $\gamma/(2\omega_b) \to 0$, $\kappa_{\text{KR}} \to 1$, the ordinary result. κ_{KR} is shown in Fig. 11.2.2 as a function of $\gamma/(2\omega_b)$.

The result in Eq. (11.50) is only valid for a one-dimensional system. However, if we identify this degree of freedom with the reaction coordinate, it can also be used in multidimensional cases as an estimate of the dynamical influence of a solvent on the conventional transition-state rate constant. The theory can be tested experimentally by studying the kinetics of a reaction in a series of solvents with varying viscosity.

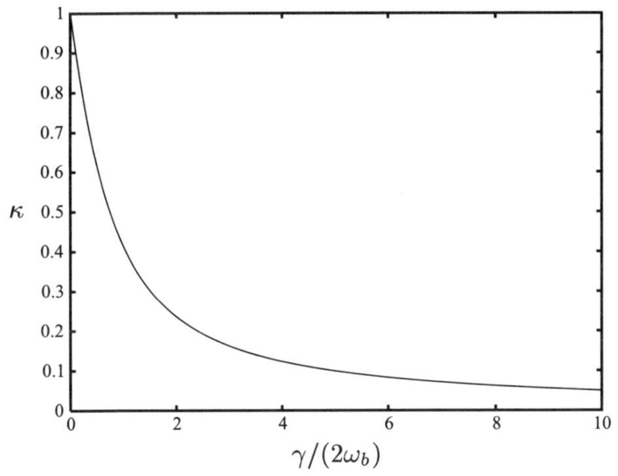

Fig. 11.2.2 The correction factor $\kappa_{\rm KR}$ to the ordinary (gas-phase) transition-state rate constant according to Kramers theory, as a function of $\gamma/(2\omega_b)$.

Example 11.1: Validity of Kramers theory

In order to evaluate the validity of Kramers theory, it is natural to focus on unimolecular reactions since in this case diffusive motion does not come into play.

Kramers theory in the high-viscosity regime, Eq. (11.47), predicts that $k_s \propto \eta^{-1}$. Experimental results for the rate of excited-state isomerization dynamics (i.e., thermally-activated barrier crossing in the lowest excited singlet state) in the high-viscosity regime agree qualitatively with this prediction. However, the experimental results show a much weaker viscosity dependence. Thus, the rate constant can be fitted to $k_s \propto \eta^{-\alpha}$, where $\alpha = 0.32$ for trans-stilbene [G. Rothenberger, D.K. Negus, and R.M. Hochstrasser, *J. Chem. Phys.* **79**, 5360 (1983)], and $\alpha = 0.2$ for cis-2-vinylanthracene [K. Hara, H. Kiyotani, and D.S. Bulgarevich, *Chem. Phys. Lett.* **242**, 455 (1995)].

This discrepancy can be due to a breakdown in the hydrodynamics friction law, Eq. (11.45), i.e., Stokes law, and/or a breakdown of the basic assumptions of Kramers theory. As we will see in the following section, a problem with Kramers theory is that the Langevin equation does not provide a sufficiently accurate description of the dynamics associated with the reaction coordinate.

11.3 Beyond Kramers, Grote–Hynes theory and MD

The Langevin equation, Eq. (11.5), that was used in Kramers calculation of the dynamical effects on the rate constant, is only valid in the limit of long times, where an equilibrium situation may be established. The reaction coordinate undergoes many collisions with the atoms in the solvent due to thermal agitation. From the Langevin equation of motion and Eq. (11.9), we obtained an expression for the autocorrelation function of the velocity:

$$\langle v(0)v(t)\rangle = v_0^2 \exp(-\gamma|t|) \tag{11.52}$$

This result cannot hold for small times t. The function in Eq. (11.52) has a discontinuity in the first derivative at $t = 0$ that the exact correlation function will not have.

To overcome this difficulty Kubo, Mori, Zwanzig, and others introduced a time-dependent friction coefficient, writing, instead of the 'simple' Langevin equation, the *generalized Langevin equation* (GLE)

$$M\,\dot{v}(t) = -\int_{\infty}^{\infty} g(t-t')\,v(t')\,dt' - f(t) \tag{11.53}$$

In this equation $g(t)$ represents the retarded effect of the frictional force, and $f(t)$ is an external force including the random force from the solvent molecules. We see, in contrast to the simple Langevin equation with a constant friction coefficient, that the friction force at a given time t depends on all previous velocities along the trajectory. The friction force is no longer local in time and does not depend on the current velocity alone. The time-dependent friction coefficient is therefore also referred to as a 'memory kernel'. A short-time expansion of the velocity correlation function based on the GLE gives $(k_B T/M)(1 - (g/M)t^2/(2\tau) + \cdots)$, where τ is the decay time of $g(t)$, and it therefore does not have a discontinuous first derivative at $t = 0$. The discussion of the properties of the GLE is most easily accomplished by using so-called *linear response theory*, which forms the theoretical basis for the equation and is a powerful method that allows us to determine non-equilibrium transport coefficients from equilibrium properties of the systems. A discussion of this is, however, beyond the scope of this book.

In this section, we will describe the *Grote–Hynes theory* for the calculation of the dynamical effect on the rate constant. The theory generalizes the Kramers result by using the GLE to describe the dynamics of the reaction coordinate instead of the 'simple' Langevin equation used in the derivation of Kramers result. The Grote–Hynes theory 'lifts' the requirement of equilibrium at all times, a requirement that is not fulfilled at small times, in particular in cases where the curvature at the top of the barrier in the direction of the reaction coordinate is large, corresponding to very fast motion in the reaction coordinate, much faster than the relaxation time of the surrounding solvent molecules. The limit of this *non-equilibrium solvation* will be the non-adiabatic limit, where all solvent molecules do not have time to move (are 'frozen') in the very short time when the destiny of a trajectory is determined, that is, whether it is reactive or not. In this limit there may be an order of magnitude difference between the Grote–Hynes result and the Kramers result.

11.3.1 Generalized Langevin equation

The GLE for the reaction coordinate $y(t)$ has the form

$$\mu\,\ddot{y}(t) = -\frac{\partial U[y(t)]}{\partial y(t)} - \int_0^t d\tau\,\eta(\tau)\,\dot{y}(t-\tau) + F(t) \tag{11.54}$$

$F(t)$ is assumed to be a random force from the solvent molecules on the reaction coordinate with zero average, $\langle F(t)\rangle = 0$, and describes together with the *time-dependent friction coefficient* $\eta(t)$ the effect of the solvent molecules on the reaction coordinate.

If we take an equilibrium ensemble average in Eq. (11.54), the second as well as the last term on the right-hand side of the equation disappears because $\langle \dot{y} \rangle = 0$ and because $\langle F(t) \rangle = 0$. Thus, the potential $U(y)$ is the equilibrium potential of mean force along the reaction coordinate, and it is assumed to be parabolic ($U(0) - (1/2)k_x y^2$, as in Eq. (11.22)) in the region of interest near the transition state, where $y = 0$.

It will be convenient to introduce the mass-weighted coordinate x according to

$$x(t) = \sqrt{\mu}\, y(t) \tag{11.55}$$

where μ is the effective mass associated with the motion in the reaction coordinate. After division by $\sqrt{\mu}$, Eq. (11.54) takes the form

$$\ddot{x}(t) = \omega_{b,\text{eq}}^2\, x(t) - \int_0^t d\tau\, \xi(\tau)\, \dot{x}(t-\tau) + R(t) \tag{11.56}$$

with

$$\begin{aligned} \omega_{b,\text{eq}}^2 &= k_x/\mu \\ \xi(t) &= \eta(t)/\mu \\ R(t) &= F(t)/\sqrt{\mu} \end{aligned} \tag{11.57}$$

The subscript 'eq' on the frequency ω_b stresses that it is the curvature of the equilibrium potential of mean force at the top of the energy barrier in the direction of the reaction coordinate.

The important *fluctuation–dissipation theorem* that expresses a balance between the dissipation and fluctuating forces to ensure the approach to equilibrium at a specified temperature is generalized to the expression that in the mass-weighted coordinates has the form

$$k_B T\, \xi(t) = \langle R(0) R(t) \rangle \tag{11.58}$$

Before we continue with the derivation of the Grote–Hynes expression for the transmission coefficient, it may be instructive to study the GLE, if not from the basic *linear response theory* point of view, then for a simple system where the GLE can be derived from the Hamiltonian of the system. For the special case where all forces are linear, that is, a parabolic reaction barrier and a 'harmonic' solvent, it is possible to derive the GLE directly from the Hamiltonian. This allows us to identify and express the various terms in the GLE by system parameters, which helps to clarify the origin of the various terms in the equation.

Derivation of the GLE for a harmonic system

We expand the potential energy surface at the saddle point to second order in the coordinates at the top of the barrier and determine the normal modes of the activated complex; one of them is the reaction coordinate y identified as the mode with an imaginary frequency. Since the other normal modes of the activated complex are not coupled to the reaction coordinate in the harmonic approximation, we do not consider them here because they are irrelevant. For the harmonic solvent, we may likewise find the normal modes S. We use these normal modes to write down the Hamiltonian, and then add a *linear coupling term* representing the coupling between the reaction coordinate and the solvent atoms. Let us start with the Lagrange function for the system:

$$L = \frac{1}{2}\mu\dot{y}^2 + \sum_i \frac{1}{2}\mu_i\dot{S}^2 + \frac{1}{2}k_x y^2 - \sum_i \frac{1}{2}k_i S_i^2 + \sum_i k_{ix}\, y\, S \tag{11.59}$$

Here μ and μ_i are the reduced masses for the normal mode associated with the reaction coordinate and for the ith normal mode of the solvent, respectively. Note the different signs of the potential energy terms for the 'unstable' reaction coordinate and the 'stable' solvent coordinates.

Introducing mass-weighted coordinates, the Lagrangian takes the form

$$\begin{aligned} x &= \sqrt{\mu}\, y \\ Q_i &= \sqrt{\mu_i}\, S_i \end{aligned} \tag{11.60}$$

$$L = \frac{1}{2}\dot{x}^2 + \sum_i \frac{1}{2}\dot{Q}_i^2 + \frac{1}{2}\omega_b^2 x^2 - \sum_i \frac{1}{2}\omega_i^2 Q_i^2 + \sum_i \omega_{ix}^2\, x\, Q_i \tag{11.61}$$

with

$$\begin{aligned} \omega_b^2 &= k_x/\mu \\ \omega_i^2 &= k_i/\mu_i \\ \omega_{ix}^2 &= k_{ix}/(\sqrt{\mu}\sqrt{\mu_i}) \end{aligned} \tag{11.62}$$

It is important to notice that the ω_b frequency is not, as we shall see, identical to the curvature of the potential of mean force $\omega_{b,\mathrm{eq}}$, but is identical to the *non-adiabatic* curvature, and the coupling term, ω_{ix}^2, may be positive or negative.

The Hamiltonian for the system when we use mass-weighted coordinates is then, using $p_i = (\partial L/\partial \dot{q}_i)$,

$$H = \frac{1}{2}\, p_x^2 - \frac{1}{2}\omega_b^2 x^2 + \frac{1}{2}\sum_i P_i^2 + \frac{1}{2}\sum_i \omega_i^2 Q_i^2 - \sum_i \omega_{ix}^2\, Q_i\, x \tag{11.63}$$

Hamilton's equations of motion are then given by

$$\begin{aligned} \dot{x} &= \left(\frac{\partial H}{\partial p_x}\right) = p_x, & \dot{Q}_i &= \left(\frac{\partial H}{\partial P_i}\right) = P_i \\ \dot{p}_x &= -\left(\frac{\partial H}{\partial x}\right) = \omega_b^2\, x + \sum_i \omega_{ix}^2\, Q_i, & \dot{P}_i &= -\left(\frac{\partial H}{\partial Q_i}\right) = -\omega_i^2\, Q_i + \omega_{ix}^2\, x \end{aligned} \tag{11.64}$$

and combining the equations of motion for the coordinates and momenta gives equations of motion for the coordinates, that is,

$$\begin{aligned} \ddot{x} &= \dot{p}_x = \omega_b^2\, x + \sum_i \omega_{ix}^2\, Q_i \\ \ddot{Q}_i &= \dot{P}_i = -\omega_i^2\, Q_i + \omega_{ix}^2\, x \end{aligned} \tag{11.65}$$

These coupled linear second-order differential equations may be converted into simple algebraic equations by a Laplace transformation. The Laplace transformation is defined as (see Section 9.2.2)

$$\hat{f}(\lambda) = \int_0^\infty dt\, \exp(-\lambda t)\, f(t) \tag{11.66}$$

By repeated partial integration, we find the expression for the Laplace transform of the accelerations:

$$\begin{aligned}\hat{\ddot{x}}(\lambda) &= \int_0^\infty dt\, \exp(-\lambda t)\, \ddot{x} \\ &= [\exp(-\lambda t)\, \dot{x}]_0^\infty + \lambda \int_0^\infty dt\, \exp(-\lambda t)\, \dot{x} \\ &= -\dot{x}(0) + \lambda \left\{ [\exp(-\lambda t)\, x]_0^\infty + \lambda \int_0^\infty dt\, \exp(-\lambda t)\, x \right\} \\ &= -\dot{x}(0) - \lambda\, x(0) + \lambda^2\, \hat{x}(\lambda)\end{aligned} \tag{11.67}$$

So, the Laplace transforms of Eqs (11.65) with $x(0) = 0$ (start at the top of the barrier) are

$$\begin{aligned}\lambda^2\, \hat{x} - p_x(0) &= \omega_b^2\, \hat{x} + \sum_i \omega_{ix}^2\, \hat{Q}_i \\ \lambda^2 \hat{Q}_i - P_i(0) - \lambda\, Q_i(0) &= -\omega_i^2\, \hat{Q}_i + \omega_{ix}^2\, \hat{x}\end{aligned} \tag{11.68}$$

Here we have used Eqs (11.64) to replace $\dot{x}(0)$ with the momentum $p_x(0)$, and $\dot{Q}_i(0)$ with $P_i(0)$. We are only interested in the motion in the reaction coordinate, so we solve the second equation for $\hat{Q}_i$ and substitute the solution into the first equation. We find

$$\lambda^2\, \hat{x} - p_x(0) = \omega_b^2\, \hat{x} + \sum_i \frac{\omega_{ix}^4}{\lambda^2 + \omega_i^2}\, \hat{x} + \sum_i \frac{\omega_{ix}^2 [P_i(0) + \lambda\, Q_i(0)]}{\lambda^2 + \omega_i^2} \tag{11.69}$$

This is the equation for the Laplace transform of the reaction coordinate, and we could then take the inverse transform of this equation to show that it is identical to the GLE in Eq. (11.56).

Alternatively, we may take the Laplace transform of the GLE and compare with Eq. (11.69), and since this is easier, we follow this strategy. The transformation is straightforward except perhaps for the second term on the right-hand side, which is a convolution term. The Laplace transform of this is found as

$$\begin{aligned}\int_0^\infty dt\, \exp(-\lambda t) \int_0^\infty d\tau\, \xi(\tau)\, \dot{x}(t-\tau) &= \int_0^\infty d\tau\, \xi(\tau) \int_0^\infty dt\, \exp(-\lambda t)\, \dot{x}(t-\tau) \\ &= \int_0^\infty d\tau\, \xi(\tau) \exp(-\lambda \tau) \int_0^\infty dt'\, \exp(-\lambda t')\, \dot{x}(t') \\ &= \hat{\xi}(\lambda)\, \lambda \hat{x}(\lambda)\end{aligned} \tag{11.70}$$

where we have replaced the upper integration limit t with ∞, which is permissible since the friction coefficient is usually only non-zero over relatively small times.

Thus the Laplace transform of the GLE is

$$\lambda^2\, \hat{x} - p_x(0) = \omega_{b,\text{eq}}^2 \hat{x} - \hat{\xi}(\lambda)\, \lambda\, \hat{x} + \hat{R}(\lambda) \tag{11.71}$$

From a comparison with Eq. (11.69), it is clear that the last term on the right-hand side is related to the random force $\hat{R}$. At the same time it is also clear that the second term on the right-hand side of the equation still does not have a form that makes it possible to identify $\omega_{b,\text{eq}}^2$ and $\hat{\xi}$. A factor of λ is missing in Eq. (11.69).

However, it is possible to recast the second term in such a way that the missing factor of λ appears. Consider a fraction, $a/(b+c)$, like the second term in Eq. (11.69). It may be broken up in the following way:

$$\frac{a}{b+c} = \frac{a}{b}\left[1 + \frac{z}{b+c}\right] \tag{11.72}$$

where the unknown z is determined by the equation and gives $z = -c$. With $a = \omega_{ix}^4$, $b = \omega_i^2$, and $c = \lambda^2$, we have that

$$\frac{\omega_{ix}^4}{\lambda^2+\omega_i^2}\hat{x} = \frac{\omega_{ix}^4}{\omega_i^2}\left[1 - \frac{\lambda^2}{\lambda^2+\omega_i^2}\right]\hat{x} = \frac{\omega_{ix}^4}{\omega_i^2}\hat{x} - \frac{\omega_{ix}^4}{\omega_i^2}\frac{\lambda}{\lambda^2+\omega_i^2}\lambda\,\hat{x} \tag{11.73}$$

When introduced into Eq. (11.69), we finally arrive at a form that makes it possible to identify the terms in the GLE with the parameters of our simple harmonic model. We find

$$\lambda^2\hat{x} - p_x(0) = \left[\omega_b^2 + \sum_i \frac{\omega_{ix}^4}{\omega_i^2}\right]\hat{x} - \sum_i\left[\frac{\omega_{ix}^4}{\omega_i^2}\frac{\lambda}{\lambda^2+\omega_i^2}\right]\lambda\hat{x} + \sum_i \frac{\omega_{ix}^2[P_i(0)+\lambda\,Q_i(0)]}{\lambda^2+\omega_i^2} \tag{11.74}$$

This has exactly the same form as the Laplace transform of the GLE in Eq. (11.71), and the different terms may be identified according to

$$\begin{aligned}\omega_{b,\text{eq}}^2 &= \omega_b^2 + \sum_i \frac{\omega_{ix}^4}{\omega_i^2} = \omega_b^2 + \xi(0)\\ \hat{\xi}(\lambda) &= \sum_i \frac{\omega_{ix}^4}{\omega_i^2}\frac{\lambda}{\lambda^2+\omega_i^2}\\ \hat{R}(\lambda) &= \sum_i \frac{\omega_{ix}^2[P_i(0)+\lambda\,Q_i(0)]}{\lambda^2+\omega_i^2}\end{aligned} \tag{11.75}$$

The following important features are noticed.

- The curvature of the barrier potential along the reaction coordinate, ω_b^2, in the Hamiltonian is different from the equilibrium curvature $\omega_{b,\text{eq}}^2$. The physical reason for this is clear. If we go back to the Hamiltonian in Eq. (11.63), then we see that

$$\omega_b^2 = -\left(\frac{\partial^2 H}{\partial^2 x}\right) = -\frac{\partial}{\partial x}\left(\frac{\partial H}{\partial x}\right) = \frac{\partial}{\partial x}\dot{p}_x = \left(\frac{\partial \ddot{x}}{\partial x}\right)_{x=0} \tag{11.76}$$

that is, it describes the change in the force on the reaction coordinate when it moves along the reaction path with *all* solvent molecules 'frozen' at their position, that is, it is the *non-adiabatic* curvature of the potential. The second relation between the non-adiabatic and equilibrium curvatures of the barrier potential is derived in the following way. Assume that the friction kernel is a constant and equal to its value at $t = 0$, that is, $\xi(t) = \xi(t = 0)$; then the Laplace transform is simply

$$\hat{\xi}(\lambda) = \int_0^\infty dt \, \exp(-\lambda \, t) \, \xi(0) = \xi(0)/\lambda \tag{11.77}$$

Therefore, if we multiply the expression for $\hat{\xi}(\lambda)$ in Eq. (11.75) by λ we should get a constant, independent of λ. This is obtained when the fraction $1/(1 + \omega_i^2/\lambda^2) \sim 1$, that is, when $\omega_i^2/\lambda^2 \ll 1$, corresponding to 'frozen' solvent molecules thanks to the very small ω_i^2 compared to λ, and therefore 'slow' solvent molecule motions. This is the background for the second relation in the first equation in Eq. (11.75).

- We see how the fluctuating forces depend on the initial positions and momenta of the solvent molecules—a result that is meaningful.

We may also use the results to check that the expressions satisfy the fluctuation–dissipation theorem. The Laplace transform of the autocorrelation function for the fluctuating forces is

$$\hat{C}(\lambda) = \int_0^\infty dt \, \exp(-\lambda \, t) \langle R(0) R(t) \rangle = \langle R(0) \hat{R}(\lambda) \rangle \tag{11.78}$$

Inserting the relation $R(0) = \lambda \, \hat{R}(\lambda)$, equivalent to the result in Eq. (11.77), we find using Eq. (11.75) that

$$\hat{C}(\lambda) = \left\langle \sum_i \frac{\lambda \, \omega_{ix}^4}{(\lambda^2 + \omega_i^2)^2} \left[P_i(0)^2 + \lambda^2 \, Q_i^2(0) + 2\lambda P_i(0) \, Q_i(0) \right] \right\rangle \tag{11.79}$$

The bracket indicates a Boltzmann average over the solvent degrees of freedom with the reaction coordinate at the top of the barrier ($x = 0$). We have only included diagonal terms in the sum, since the ensemble average will be zero for all off-diagonal terms. Then

$$\begin{aligned} \langle P_i^2(0) \rangle &= \frac{\int_{-\infty}^\infty dP_i(0) \, P_i^2(0) \exp(-P_i^2(0)/(2k_B T))}{\int_{-\infty}^\infty dP_i \, \exp(-P_i^2(0)/(2k_B T))} = k_B T \\ \langle Q_i^2(0) \rangle &= \frac{\int_{-\infty}^\infty dQ_i(0) \, Q_i^2(0) \exp(-\omega_i^2 \, Q_i^2(0)/(2k_B T))}{\int_{-\infty}^\infty dQ_i \, \exp(-\omega_i^2 \, Q_i^2(0)/(2k_B T))} = \frac{k_B T}{\omega_i^2} \end{aligned} \tag{11.80}$$

This is then introduced into Eq. (11.79) and we get

$$\begin{aligned} \hat{C}(\lambda) = \langle R(0) \, \hat{R}(\lambda) \rangle &= \sum_i \frac{\lambda \, \omega_{ix}^4}{(\lambda^2 + \omega_i^2)^2} \left[k_B T + k_B T \, \frac{\lambda^2}{\omega_i^2} \right] \\ &= k_B T \sum_i \frac{\lambda \, \omega_{ix}^4}{(\lambda^2 + \omega_i^2)^2} \left[1 + \frac{\lambda^2}{\omega_i^2} \right] \\ &= k_B T \sum_i \frac{\lambda \, \omega_{ix}^4}{(\lambda^2 + \omega_i^2) \omega_i^2} = k_B T \, \hat{\xi}(\lambda) \end{aligned} \tag{11.81}$$

in perfect agreement with the *fluctuation–dissipation theorem* in Eq. (11.58).

11.3.2 The Grote–Hynes equation

Here we will give a simple derivation of the Grote–Hynes equation based on the GLE [7]. Alternative derivations can be found in the literature [4,5].

The trajectory $x(t)$ is reactive if $x(t) > 0$ for $t \to \infty$. Let us therefore write the solution that diverges at large t as

$$x(t) = C\exp(\lambda_r t), \quad t \to \infty \tag{11.82}$$

where $\lambda_r > 0$ and the subscript r refers to a reactive trajectory. In order to determine λ_r, we introduce this equation into Eq. (11.56) and take the average for an ensemble of trajectories initiated at the barrier top. Then, since $\langle R(t)\rangle = 0$, we find

$$\lambda_r^2 C\exp(\lambda_r t) = \omega_{b,\text{eq}}^2 C\exp(\lambda_r t) - \int_0^\infty d\tau\, \xi(\tau)\, \lambda_r\, C\exp(\lambda_r(t-\tau)) \tag{11.83}$$

which may be rearranged to

$$C\exp(\lambda_r t)\,\lambda_r^2 = C\exp(\lambda_r t)\left[\omega_{b,\text{eq}}^2 - \lambda_r \int_0^\infty d\tau\, \xi(\tau)\, \exp(-\lambda_r \tau)\right] \tag{11.84}$$

or

$$\boxed{\lambda_r^2 = \omega_{b,\text{eq}}^2 - \lambda_r \int_0^\infty d\tau\, \xi(\tau)\, \exp(-\lambda_r \tau) = \omega_{b,\text{eq}}^2 - \lambda_r\, \hat{\xi}(\lambda_r)} \tag{11.85}$$

which is the *Grote–Hynes equation* for the frequency λ_r by which the trajectories cross the top of the barrier to form products.

We see that a calculation of λ_r involves a Laplace transform of the time-dependent friction kernel. This may typically be determined in a molecular dynamics (MD) simulation, where the autocorrelation function of the random force $\langle R(0)R(t)\rangle$ may be determined, which then allows us to determine $\xi(t)$ using the fluctuation–dissipation theorem in Eq. (11.58). Note that Eq. (11.85) is an implicit equation for λ_r that in general must be solved by iteration. In the absence of friction we see from Eq. (11.85) that $\lambda_r = \omega_{b,\text{eq}}$.

The *transmission coefficient* κ_{GH} is, as previously, a measure of the departure of the rate constant from transition-state theory. κ_{GH} is given as the ratio between the frequencies with and without friction, that is,

$$\boxed{\kappa_{\text{GH}} = \frac{\lambda_r}{\omega_{b,\text{eq}}}} \tag{11.86}$$

as can indeed be shown more formally [4,7].

In practical calculations it is faster to calculate Fourier transforms (with fast Fourier transform (FFT) routines), so usually the Fourier transform of the friction kernel is introduced:

$$\begin{aligned}\tilde{\xi}(\omega) &= \int_{-\infty}^{\infty} dt\, \exp(i\omega t)\xi(t)\\ \xi(t) &= \int_{-\infty}^{\infty} \frac{d\omega}{2\pi}\, \exp(-i\omega t)\,\tilde{\xi}(\omega)\end{aligned} \tag{11.87}$$

If we introduce the second of the equations in Eq. (11.87) into Eq. (11.85), we find

$$\begin{aligned}\int_0^\infty dt\,\exp(-\lambda_r t)\,\xi(t) &= \int_0^\infty dt\,\exp(-\lambda_r t)\int_{-\infty}^\infty \frac{d\omega}{2\pi}\exp(-i\omega t)\,\tilde{\xi}(\omega)\\ &= \int_{-\infty}^\infty \frac{d\omega}{2\pi}\tilde{\xi}(\omega)\int_0^\infty dt\,\exp(-(i\omega+\lambda_r)\,t)\\ &= \int_{-\infty}^\infty \frac{d\omega}{2\pi}\,\tilde{\xi}(\omega)\frac{1}{i\omega+\lambda_r}\\ &= \int_{-\infty}^\infty \frac{d\omega}{2\pi}\,\tilde{\xi}(\omega)\frac{\lambda_r - i\omega}{\lambda_r^2+\omega^2}\\ &= \int_{-\infty}^\infty \frac{d\omega}{2\pi}\,\tilde{\xi}(\omega)\frac{\lambda_r}{\lambda_r^2+\omega^2}\end{aligned} \tag{11.88}$$

The imaginary term on the right-hand side of the equation vanishes since the left-hand side of the equation is real and because $\tilde{\xi}(\omega)$ is real according to Eq. (11.87), where $\xi(t)$ is an even function in t. When Eq. (11.88) is introduced into Eq. (11.85) we find

$$\lambda_r^2 - \omega_{b,\text{eq}}^2 + \lambda_r^2 \int_{-\infty}^\infty \frac{d\omega}{2\pi}\frac{\tilde{\xi}(\omega)}{\lambda_r^2+\omega^2} = 0 \tag{11.89}$$

Discussion of the Grote–Hynes equation

When friction is present, there will be solvent-induced recrossings of the barrier and non-equilibrium solvation, and the frequency λ_r by which products are formed depends both on the equilibrium curvature $\omega_{b,\text{eq}}^2$ of the barrier and on the dynamics of the solvent as expressed by the time-dependent friction kernel. The following two limiting cases should be noticed.

- In the *adiabatic regime*, the solvent relaxation time $\tau_c \ll \lambda_r^{-1}$, that is, the solvent responds instantaneously to any change in the reaction coordinate. This limit corresponds to $\xi(t) = \xi\,\delta(t)$, so the power spectrum (Eq. (11.87)) is equal to ξ, that is, to 'white noise'. The GLE is reduced to the simple Langevin equation with a time-local friction force $-\xi\,\dot{x}$. λ_r is found from Eq. (11.85):

$$\lambda_r^2 - \omega_{b,\text{eq}}^2 + \xi\,\lambda_r = 0 \tag{11.90}$$

  with the (positive) solution

$$\lambda_r = \frac{-\xi + \sqrt{\xi^2 + 4\,\omega_{b,\text{eq}}^2}}{2} = \sqrt{\omega_{b,\text{eq}}^2 + \frac{\xi^2}{4}} - \frac{\xi}{2} \tag{11.91}$$

  which leads to a transmission coefficient, $\kappa_{\text{GH}} = \lambda_r/\omega_{b,\text{eq}} = \kappa_{\text{KR}}$, i.e., identical to Kramers result in Eq. (11.51).
- In the *non-adiabatic regime*, the time scale $\lambda_r^{-1} \ll \tau_c$ of the reaction is so short compared to the relaxation times of the solvent molecules that they have no chance to respond to the motion in the reaction coordinate; said differently, the solvent molecules are effectively 'frozen' in their positions during the rapid passage of the

barrier. This limit corresponds to the limit with $\xi(t) = \xi(0)$, that is, solvent dynamics is ignored, so the power spectrum $\tilde{\xi}(\omega) \propto \delta(\omega)$ (see Eq. (11.87)) is peaked around zero frequency. λ_r is found from Eq. (11.85):

$$\lambda_r^2 = \omega_{b,\mathrm{eq}}^2 - \xi(0) \tag{11.92}$$

The motion is not determined by the equilibrium barrier $-\omega_{b,\mathrm{eq}}^2 x^2/2$, since the solvent molecules cannot provide an equilibrium solvation. Instead, the motion is along a different non-adiabatic barrier $-\omega_{b,\mathrm{na}}^2 x^2/2$ (the $\omega_b^2 x^2/2$ barrier in the harmonic model system considered above) that depends on the initial configuration of the solvent molecules. We have seen in Eq. (11.75) that

$$\omega_{b,\mathrm{na}}^2 = \omega_{b,\mathrm{eq}}^2 - \xi(0) \tag{11.93}$$

in agreement with Eq. (11.92). When the solvent molecules cannot yield to the motion in the reaction coordinate, they tend to offset the equilibrium potential barrier, and hence to reduce the reaction rate, giving the transmission coefficient

$$\kappa_{\mathrm{na}} = \frac{\omega_{b,\mathrm{na}}}{\omega_{b,\mathrm{eq}}} = \sqrt{\omega_{b,\mathrm{eq}}^2 - \xi(0)} \Big/ \omega_{b,\mathrm{eq}} \tag{11.94}$$

Let us in the following derive a relation between the actual transmission coefficient, κ_{GH}, and the non-adiabatic coefficient, κ_{na}, since it will show why $\kappa_{\mathrm{GH}} > \kappa_{\mathrm{na}}$ and which part of the power spectrum for the solvent motion that is responsible for this.

We see from Eq. (11.87) that

$$\xi(0) = \int_{-\infty}^{\infty} \frac{d\omega}{2\pi} \tilde{\xi}(\omega) \tag{11.95}$$

Substitution of Eqs (11.93) and (11.95) into Eq. (11.89) gives

$$\begin{aligned} \lambda_r^2 - \omega_{b,\mathrm{na}}^2 &= \int_{-\infty}^{\infty} \frac{d\omega}{2\pi} \tilde{\xi}(\omega) - \lambda_r^2 \int_{-\infty}^{\infty} \frac{d\omega}{2\pi} \frac{\tilde{\xi}(\omega)}{\lambda_r^2 + \omega^2} \\ &= \int_{-\infty}^{\infty} \frac{d\omega}{2\pi} \left[1 - \frac{\lambda_r^2}{\lambda_r^2 + \omega^2} \right] \tilde{\xi}(\omega) \end{aligned} \tag{11.96}$$

Division by $\omega_{b,\mathrm{eq}}^2$ gives

$$\kappa_{\mathrm{GH}}^2 - \kappa_{\mathrm{na}}^2 = (\kappa_{\mathrm{GH}} - \kappa_{\mathrm{na}})(\kappa_{\mathrm{GH}} + \kappa_{\mathrm{na}}) = \int_{-\infty}^{\infty} \frac{d\omega}{2\pi} \frac{\omega^2}{\lambda_r^2 + \omega^2} \frac{\tilde{\xi}(\omega)}{\omega_{b,\mathrm{eq}}^2} \tag{11.97}$$

and finally

$$\kappa_{\mathrm{GH}} - \kappa_{\mathrm{na}} = \frac{1}{\omega_{b,\mathrm{eq}}(\lambda_r + \omega_{b,\mathrm{na}})} \int_{-\infty}^{\infty} \frac{d\omega}{2\pi} \frac{\omega^2 \tilde{\xi}(\omega)}{\lambda_r^2 + \omega^2} \tag{11.98}$$

Since κ_{na} is equal to the transmission coefficient when the solvent molecules do not respond to the motion in the reaction coordinate ('frozen' solvent molecules), the right-hand side of the equation represents the dynamical response of the solvent. We see that, with the factor $\omega^2/(\lambda_r^2+\omega^2)$ in the integrand, the high frequency part of the friction kernel power spectrum is emphasized since the factor is small for small frequencies and approaches the value of one at high frequencies. This is exactly what is expected, since it will be the high frequency modes that may respond to the fast motion in the reaction coordinate. We also note that, provided the integral is convergent, $\kappa_{\text{GH}} > \kappa_{\text{na}}$ since the right-hand side of the equation will always be positive and, therefore, any dynamical response of the solvent will always enhance the reaction rate.

The integral on the right-hand side of Eq. (11.98) diverges unless $\tilde{\xi}(\omega)$ falls off faster than $1/\omega$ at infinity. Thus, the inequality established above might not be valid close to the adiabatic limit. In the adiabatic limit where $\tilde{\xi}(\omega) = \xi$, we use instead Eqs (11.90) and (11.92) to determine the difference between the transmission coefficients in the adiabatic (Kramers) limit and in the non-adiabatic limit. We find

$$\lambda_{r,\text{KR}}^2 - \lambda_{r,\text{na}}^2 = \xi(0) - \xi\,\lambda_{r,\text{KR}} \tag{11.99}$$

After division by $\omega_{b,\text{eq}}^2$ and factorization of $\kappa_{\text{KR}}^2 - \kappa_{\text{na}}^2$, we find

$$\kappa_{\text{KR}} - \kappa_{\text{na}} = \frac{\xi(0) - \xi\,\lambda_{r,\text{KR}}}{\omega_{b,\text{eq}}(\lambda_{r,\text{KR}} + \lambda_{r,\text{na}})} \tag{11.100}$$

So, $\kappa_{\text{KR}} > \kappa_{\text{na}}$ if $\xi(0) > \xi\,\lambda_{r,\text{KR}}$, and $\kappa_{\text{KR}} < \kappa_{\text{na}}$ when $\xi(0) < \xi\,\lambda_{r,\text{KR}}$. The latter case is found in Example 11.2.

Finally, it will also be interesting to develop a general relation between the actual transmission coefficient κ_{GH} and the limiting value of one corresponding to no friction. This will show that $\kappa_{\text{GH}} < 1$ in a solution and which part of the power spectrum for the solvent that is responsible for this. We have from Eq. (11.89) that

$$\lambda_r^2 - \omega_{b,\text{eq}}^2 = -\lambda_r^2 \int_{-\infty}^{\infty} \frac{d\omega}{2\pi} \frac{\tilde{\xi}(\omega)}{\lambda_r^2 + \omega^2} \tag{11.101}$$

and after division by $\omega_{b,\text{eq}}^2$ and factorization of $\kappa_{\text{GH}}^2 - 1^2$,

$$\kappa_{\text{GH}} - 1 = -\frac{\lambda_r^2}{\omega_{b,\text{eq}}(\lambda_r + \omega_{b,\text{eq}})} \int_{-\infty}^{\infty} \frac{d\omega}{2\pi} \frac{\tilde{\xi}(\omega)}{\lambda_r^2 + \omega^2} \tag{11.102}$$

This relation shows that $\kappa_{\text{GH}} < 1$ and, in contrast to above, the factor $1/(\lambda_r^2 + \omega^2)$ emphasizes the low frequency part of the power spectrum of the friction kernel, since it approaches zero at high frequencies. The low frequency modes cannot follow the fast reaction mode and therefore a non-equilibrium solvent distribution is produced that causes the reaction rate to decrease.

Example 11.2: Comparison of Kramers, Grote–Hynes, and MD

As in Example 10.1, we consider the S_N2 reaction

$$\text{Cl}^- + \text{CH}_3\text{Cl}' \longrightarrow \text{CH}_3\text{Cl} + \text{Cl}'^-$$

in water at $T = 298$ K. The transmission coefficient for this reaction has been evaluated at different levels of approximation [6,7].

The multidimensional potential energy surface was written as the sum of a gas-phase (LEPS) energy surface incorporating the main features of the one-dimensional double-well potential in Example 10.1, solvent–solute interactions described by Lennard–Jones potentials with added (Coulomb) interactions corresponding to point charges, and solvent–solvent interactions including intermolecular degrees of freedom. The solvent consisted of 64 water molecules.

The exact transmission factor (within classical molecular dynamics) κ_{MD} was calculated using the approach described in Section 5.1.2; that is, trajectories were sampled from the thermal equilibrium distribution at a dividing surface. Good agreement between κ_{MD} and κ_{GH} was found (with a transmission coefficient ~ 0.5), whereas κ_{KR} severely underestimates the transmission (with a transmission coefficient < 0.05). The transmission coefficient in the non-adiabatic (frozen solvent) regime gives, on the other hand, a description that is in much better agreement with the numerical value of κ_{MD}.

In full *molecular dynamics computer simulations*, the time evolution of all atomic nuclei (including solvent atoms) is followed. This approach has been applied to several examples of chemical reactions in solution, including bimolecular reactions (as in Example 11.2), and unimolecular reactions like photodissociation and isomerization.

Further reading/references

[1] H.A. Kramers, *Physica* **7**, 284 (1940)

[2] N.G. van Kampen, *Stochastic processes in physics and chemistry* (North-Holland, 1981)

[3] D.J. Tannor and D. Kohen, *J. Chem. Phys.* **100**, 4932 (1994)

[4] R.F. Grote and J.T. Hynes, *J. Chem. Phys.* **73**, 2715 (1980)

[5] D. Kohen and D.J. Tannor, *J. Chem. Phys.* **103**, 6013 (1995)

[6] J.P. Bergsma, B.J. Gertner, K.R. Wilson, and J.T. Hynes, *J. Chem. Phys.* **86**, 1356 (1987)

[7] B.J. Gertner, K.R. Wilson, and J.T. Hynes, *J. Chem. Phys.* **90**, 3537 (1989)

[8] P. Hänggi, P. Talkner, and M. Borkovec, *Rev. Mod. Phys.* **62**, 251 (1990)

[9] D.G. Truhlar, B.C. Garrett, and S.J. Klippenstein, *J. Phys. Chem.* **100**, 12771 (1996)

Problems

11.1 We consider a bimolecular reaction with two product channels:

$$
\begin{array}{ccc}
 & & \mathrm{C_1 + D_1} \\
 & \overset{k_1}{\nearrow} & \\
\mathrm{A + B} & & \\
 & \overset{k_2}{\searrow} & \\
 & & \mathrm{C_2 + D_2}
\end{array}
$$

(a) Write down the rate laws for the formation of the products C_1 and C_2, respectively, for the parallel bimolecular reactions. Show that $[C_1]/[C_2] = k_1/k_2$, when the initial concentrations are $[C_1]_0 = [C_2]_0 = 0$.

First, we assume that the reaction takes place in the gas phase.

(b) Using transition-state theory, write down an expression for $[C_1]/[C_2]$.

The reactants are now surrounded by a solvent.

(c) How is the equation for the effective rate constant, Eq. (9.35), modified when it is taken into consideration that the reactants can disappear via two channels? Under what condition is the reaction 'activation controlled'?

(d) The 'potentials of mean force' for the two activated complexes are $W_{\mathrm{mean}}^{(1)}$ and $W_{\mathrm{mean}}^{(2)}$, respectively. When, in addition, dynamic solvent effects are included according to Kramers theory, write down an expression for $[C_1]/[C_2]$.

Part III

Appendices

A

Statistical mechanics

The main objective of *statistical mechanics* is to calculate macroscopic (thermodynamic) properties from a knowledge of microscopic information like quantum mechanical energy levels. The purpose of the present appendix is merely to present a selection[1] of the results that are most relevant in the context of reaction dynamics.

Consider a macroscopic system with N particles, a volume given by V, and an energy given by E. This is the so-called *microcanonical* system. There will be an enormous number of quantum states consistent with the fixed macroscopic properties N, V, and E. The energy E must be one of the eigenvalues of the N-body Hamiltonian H_N, and the number of states of energy E is equal to the degeneracy, which we denote by $\Omega(N, V, E)$. This number is also called the microcanonical *partition function*. A basic postulate of statistical mechanics is that all states at the same energy are equally probable (the assumption of *equal a priori probabilities*). That is, the probability of finding the system in any of the states is $P = 1/\Omega(N, V, E)$.

For most practical applications, we consider not isolated systems but systems in which the temperature is fixed. This is the *canonical* system consistent with the fixed macroscopic properties (N, V, T), where T is the absolute temperature.

We consider a canonical system of N *interacting* particles described by the N-body Hamiltonian H_N. In such a system the energy is not fixed. We shall not give the detailed derivation here but merely state that the probability that the system is found in the ith energy state $\mathcal{E}_i$, given as an eigenvalue of the N-body Hamiltonian, is

$$\mathcal{P}_i = \frac{\exp(-\mathcal{E}_i/k_B T)}{\mathcal{Q}} \tag{A.1}$$

where the normalization factor

$$\begin{aligned} \mathcal{Q}(N, V, T) &= \sum_{\text{states}, i} \exp(-\mathcal{E}_i/k_B T) \\ &= \sum_{\text{levels}, \mathcal{E}_i} \Omega(N, V, \mathcal{E}_i) \exp(-\mathcal{E}_i/k_B T) \end{aligned} \tag{A.2}$$

is the canonical partition function. Note that the summation (i.e., the enumeration) can be written as a sum over all the energy states or, alternatively, as a sum over the energy levels, when the degeneracy is included in the sum. Likewise, the probability that the system is found in the energy level $\mathcal{E}_i$ is given by $\mathcal{P}(\mathcal{E}_i) = \Omega(N, V, \mathcal{E}_i)\mathcal{P}_i$.

[1]For a detailed discussion of this subject see, e.g., D.A. McQuarrie, *Statistical mechanics* (University Science Books, 2000).

A.1 A system of non-interacting molecules

Now, at sufficiently high temperatures, for N *non-interacting* identical and *indistinguishable* particles (e.g., an ideal gas), the partition function can be written in the form

$$\mathcal{Q}(N, V, T) = \frac{Q(V, T)^N}{N!} \tag{A.3}$$

where Q is the partition function of the individual particles. In the following it will be referred to as the molecular partition function. When Eq. (A.3) is valid, the molecules are said to obey Boltzmann statistics. All results that are given below fall within this 'ideal gas' limit (except when it is explicitly stated that the equation is valid for interacting molecules).

The probability of finding a molecule in the energy level E_i is

$$P(E_i) = \frac{\omega_i}{Q} \exp(-E_i/k_B T) \tag{A.4}$$

where Q is the molecular partition function,

$$Q = \sum_i \omega_i \exp(-E_i/k_B T) \tag{A.5}$$

and E_i and ω_i are the energy and the degeneracy (the microcanonical partition function) of the ith quantum level. Note that the summation here runs over all the energy levels. The probability distribution in Eq. (A.4) is referred to as the *Boltzmann distribution*. If the energy is continuous (e.g., the free particle), then the probability that a molecule has energy in the energy interval $E \to E + dE$ is

$$P(E)dE = \frac{N(E)}{Q} \exp(-E/k_B T)dE \tag{A.6}$$

where

$$Q = \int_0^\infty N(E) \exp(-E/k_B T)dE \tag{A.7}$$

and $N(E)dE$ is the number of states in the energy interval $E \to E + dE$ (the microcanonical partition function). Thus, $N(E)$ is the density of states.

The Boltzmann distribution is illustrated in Figs 1.2.1 to 1.2.3 of Chapter 1.

A.1.1 The molecular partition function

The evaluation of the molecular partition function can be simplified by noting that the total energy of the molecule may be written as a sum of the center-of-mass translational energy and the internal energy, $E = E_{\text{trans}} + E_{\text{int}}$, which implies

$$Q = Q_{\text{trans}} Q_{\text{int}} \tag{A.8}$$

since a product of exponentials is equal to an exponential with an argument that equals the sum of the arguments of the exponentials, and since $E_{\text{int}} \sim E_{\text{vib}} + E_{\text{rot}} + E_{\text{elec}}$, the partition function for the internal degrees of freedom can be written in the form

$$Q_{\text{int}} \sim Q_{\text{vib}} Q_{\text{rot}} Q_{\text{elec}} \tag{A.9}$$

These partition functions can be evaluated quite readily. In the following it turns out to be convenient to choose the zero of energy such that it coincides with the zero-point level of the quantized energy levels. To that end, we note that, if we subtract the zero-point energy E_0 from all energy levels, the partition function Eq. (A.5) takes the form

$$Q = \exp(-E_0/k_BT) \sum_i \omega_i \exp(-[E_i - E_0]/k_BT) \tag{A.10}$$

Here the sum is the partition function with the energy measured relative to the zero-point level and Q is obtained after multiplication by $\exp(-E_0/k_BT)$. The Boltzmann distribution Eq. (A.4) can be written in the form

$$P(E_i - E_0) = \frac{\omega_i \exp(-[E_i - E_0]/k_BT)}{\sum_i \omega_i \exp(-[E_i - E_0]/k_BT)} = P(E_i) \tag{A.11}$$

Thus, the partition functions differ by the factor $\exp(-E_0/k_BT)$, whereas the Boltzmann distribution is invariant to such a shift of the energy scale in the standard expressions for the energy levels.

For the *translational partition function*, we first consider a particle in a one-dimensional box of length l. The energy levels, with the zero of energy as the zero-point level, are $E_n = (n^2 - 1)h^2/(8ml^2)$, with $n = 1, 2, \ldots$, and degeneracy $\omega_n = 1$. The partition function takes the form

$$Q_{\text{trans}} = \sum_{n=1}^{\infty} \exp\left(-\frac{(n^2 - 1)h^2}{8ml^2 k_BT}\right) \tag{A.12}$$

This sum cannot be evaluated analytically. However, when the energy difference between subsequent levels can be considered as small, then the sum can be replaced by an integral. Thus, when $E_{n+1} - E_n = (2n + 1)h^2/(8ml^2) \ll k_BT$, that is, at high temperatures,[2] we have

$$\begin{aligned} Q_{\text{trans}} &\sim \int_0^{\infty} \exp\left(-\frac{n^2h^2}{8ml^2 k_BT}\right) dn \\ &= (2\pi m k_BT)^{1/2} l/h \end{aligned} \tag{A.13}$$

The energy levels for a particle in a three-dimensional box are given as the sum of the energies for each dimension, and the partition function for the three-dimensional box is simply a product of the partition functions for each dimension; that is,

$$Q_{\text{trans}} = (2\pi m k_BT)^{3/2} V/h^3 \tag{A.14}$$

where $V = l^3$ is the volume of the box.

In order to evaluate the *vibrational partition function* we consider a single harmonic oscillator. The energy levels, with the zero of energy as the zero-point level, are $E_n = h\nu_s n$, with $n = 0, 1, \ldots$, and degeneracy $\omega_n = 1$. The partition function takes the form

[2] The inequality cannot hold as n increases; but, by the time n is large enough to contradict this, the terms are so small that they make no contribution to the sum.

$$\begin{aligned} Q_{\text{vib}} &= \sum_{n=0}^{\infty} e^{-h\nu_s n/k_B T} \\ &= \sum_{n=0}^{\infty} \left(e^{-h\nu_s/k_B T}\right)^n \\ &= 1 + e^{-h\nu_s/k_B T} + \left(e^{-h\nu_s/k_B T}\right)^2 + \cdots \\ &= (1 - \exp(-h\nu_s/k_B T))^{-1} \end{aligned} \tag{A.15}$$

since $1 + x + x^2 + \cdots = (1 - x)^{-1}$. For a set of s harmonic oscillators (e.g., normal modes, see Appendix E), the total energy is the sum of the energies for each oscillator and the partition function becomes, accordingly, a product of partition functions of the form given in Eq. (A.15); that is,

$$Q_{\text{vib}} = \prod_{i=1}^{s} (1 - \exp(-h\nu_i/k_B T))^{-1} \tag{A.16}$$

For the *rotational partition function*, we first consider a linear rigid rotor. The energy levels are $E_J = J(J+1)\hbar^2/(2I)$, with $J = 0, 1, \ldots$, and I is the moment of inertia. Each energy level has a degeneracy of $m_J = 2J + 1$. The partition function takes the form

$$Q_{\text{rot}} = \sum_{J=0}^{\infty} (2J+1) \exp\{-J(J+1)\hbar^2/(2Ik_B T)\} \tag{A.17}$$

This sum cannot be evaluated analytically. However, when the energy difference between subsequent levels can be considered as small ($E_{J+1} - E_J = \hbar^2(J+1)/I \ll k_B T$, that is, at high temperatures) then the sum can be replaced by an integral:

$$\begin{aligned} Q_{\text{rot}} &\sim \int_0^{\infty} (2J+1) \exp\{-J(J+1)\hbar^2/(2Ik_B T)\} dJ \\ &= \int_0^{\infty} \exp\{-J(J+1)\hbar^2/(2Ik_B T)\} d\{J(J+1)\} \end{aligned} \tag{A.18}$$

The last integral is easily evaluated, and

$$Q_{\text{rot}} = 8\pi^2 I k_B T/h^2 \tag{A.19}$$

This is the correct expression for the rotational partition function of a heteronuclear diatomic molecule. For a homonuclear diatomic molecule, however, it must be taken into account that the total wave function must be either symmetric or antisymmetric under the interchange of the two identical nuclei: symmetric if the nuclei have integral spins or antisymmetric if they have half-integral spins. The effect on Q_{rot} is that it should be replaced by Q_{rot}/σ, where σ is a *symmetry number* that represents the number of indistinguishable orientations that the molecule can have (i.e., the number of ways the molecule can be rotated 'into itself'). Thus, Q_{rot} in Eq. (A.19) should be replaced by Q_{rot}/σ, where $\sigma = 1$ for a heteronuclear diatomic molecule and $\sigma = 2$

for a homonuclear diatomic molecule. When Eq. (A.19) is extended to a non-linear molecule with moments of inertia I_a, I_b, and I_c about its principal axes, the rotational partition function takes the form

$$Q_{\text{rot}} = \pi^{1/2}\sqrt{8\pi^2 I_a k_B T/h^2}\sqrt{8\pi^2 I_b k_B T/h^2}\sqrt{8\pi^2 I_c k_B T/h^2}/\sigma \tag{A.20}$$

Finally, the *electronic partition function* is considered. The zero of energy is chosen as the electronic ground-state energy. The spacings between the electronic energy levels are, normally, large and only the first term in the partition function will make a significant contribution; that is,

$$\begin{aligned} Q_{\text{elec}} &= \sum_i \omega_i \exp(-E_i/k_B T) \\ &\sim \omega_0 \end{aligned} \tag{A.21}$$

Thus, the partition function is simply the degeneracy of the electronic ground state.

A.1.2 Macroscopic properties

When we know the partition function, we can calculate thermodynamic quantities from a knowledge of the quantum mechanical energy levels. Consider, as an example, the energy E. A basic postulate of statistical mechanics is that such an energy is the average value of all the quantum mechanical energy levels with the weights given by the Boltzmann distribution; that is,

$$\begin{aligned} \langle E \rangle &= \sum_i E_i P(E_i) \\ &= \frac{k_B T^2}{Q} dQ/dT \\ &= k_B T^2 d\ln Q/dT \end{aligned} \tag{A.22}$$

using Eqs (A.4) and (A.5).

In a similar manner, the equilibrium constant of a chemical reaction can be related to the quantum mechanical energy levels of the reactants and products. Consider, as an example, a mixture of A and B molecules in equilibrium:

$$\mathrm{A} \rightleftarrows \mathrm{B} \tag{A.23}$$

The equilibrium constant is

$$\begin{aligned} K_c(T) &= [\mathrm{B}]/[\mathrm{A}] \\ &= N_{\mathrm{B}}/N_{\mathrm{A}} \end{aligned} \tag{A.24}$$

that is, equal to the ratio of the number of products and reactants.

According to the Boltzmann distribution, Eq. (A.4), the probability of finding an A molecule in the energy level E_a is

$$\begin{aligned} P(E_a) &= n_a/N \\ &= \frac{\omega_a}{Q}\exp(-E_a/k_B T) \end{aligned} \tag{A.25}$$

where n_a is the number of A molecules in the energy level E_a, $N = N_\mathrm{A} + N_\mathrm{B}$ is the total number of molecules, and Q is the total partition function. Similarly,

$$\begin{aligned} P(E_b) &= n_b/N \\ &= \frac{\omega_b}{Q}\exp(-E_b'/k_BT) \end{aligned} \tag{A.26}$$

is the probability of finding a B molecule in the energy level E_b', where $E_b' = E_b + E_0$, and E_0 is the difference between the zero-point levels of the products and the reactants. Thus, E_b denotes the energy levels of molecule B measured relative to the zero-point level of the molecule. Note that $Q = \sum_a \omega_a \exp(-E_a/k_BT) + \sum_b \omega_b \exp(-E_b'/k_BT)$. Now,

$$\begin{aligned} N_\mathrm{A} &= \sum_a n_a \\ &= (N/Q)\sum_a \omega_a \exp(-E_a/k_BT) \\ &= NQ_\mathrm{A}/Q \end{aligned} \tag{A.27}$$

and

$$\begin{aligned} N_\mathrm{B} &= \sum_b n_b \\ &= (N/Q)\sum_b \omega_b \exp(-E_b'/k_BT) \\ &= (N/Q)\exp(-E_0/k_BT)\sum_b \omega_b \exp(-E_b/k_BT) \\ &= NQ_\mathrm{B}\exp(-E_0/k_BT)/Q \end{aligned} \tag{A.28}$$

and the equilibrium constant is given by

$$K_c(T) = \frac{Q_\mathrm{B}}{Q_\mathrm{A}}\exp(-E_0/k_BT) \tag{A.29}$$

that is, it can be calculated from the energy levels of the molecules.

For a general equilibrium of the form

$$\nu_\mathrm{A}\mathrm{A} + \nu_\mathrm{B}\mathrm{B} \rightleftharpoons \nu_\mathrm{C}\mathrm{C} + \nu_\mathrm{D}\mathrm{D} \tag{A.30}$$

where the νs are stoichiometric coefficients, the result is

$$K_c(T) = \frac{(Q_\mathrm{C}/V)^{\nu_\mathrm{C}}(Q_\mathrm{D}/V)^{\nu_\mathrm{D}}}{(Q_\mathrm{A}/V)^{\nu_\mathrm{A}}(Q_\mathrm{B}/V)^{\nu_\mathrm{B}}}e^{-E_0/k_BT} \tag{A.31}$$

where Q_A is the partition function for molecule A, E_0 is the difference between the zero-point levels of the products and the reactants, and the partition functions are evaluated such that the zero of energy is the zero-point level.

A.2 Classical statistical mechanics

It is often impossible to obtain the quantized energies of a complicated system and therefore the partition function. Fortunately, a classical mechanical description will often suffice. Classical statistical mechanics is valid at sufficiently high temperatures. The classical treatment can be derived as a limiting case of the quantum version for cases where energy differences become small compared with $k_B T$.

The state of a classical system can be completely described by coordinates and momenta, i.e., a point in phase space; see Fig. A.2.1. For a system with s degrees of freedom (that is, s coordinates are required to completely describe its position), the phase space has the dimension $2s$. When the system evolves in time, its dynamics is described by the trajectory of the phase-space point through phase space. The trajectory is given by Hamilton's equations of motion, Eq. (4.63), and the total energy for the molecule is given by the classical Hamiltonian H. The summations over states (or energy levels) in quantum statistical mechanics are replaced by integrations over regions of phase space.

The problem now is how to count the number of states in this continuous description. For proper counting, the use of quantum mechanics (Planck's constant) is unavoidable. For a system with one degree of freedom, the number of states in the area element $dqdp$ is

$$dqdp/h \tag{A.32}$$

or, equivalently, a 'phase-space cell' of area h contains one state. The *sum of states* $G(E)$, that is, the total number of states having energy in the range from 0 to E, is then the total phase-space area occupied at the energy E divided by h. For a multidimensional system with s degrees of freedom, a phase-space cell of volume h^s contains one state (see Fig. A.2.1), and the sum of states is then given by

$$G(E) = \frac{1}{h^s} \int_{H=0}^{H=E} \cdots \int dq_1 \cdots dq_s dp_1 \cdots dp_s \tag{A.33}$$

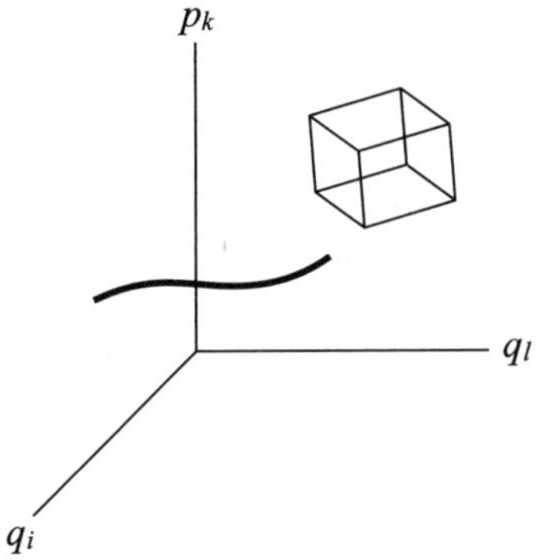

Fig. A.2.1 An illustration of the $2s$-dimensional phase space of a system with s degrees of freedom. The solid path describes the motion of the system according to Hamilton's equations of motion. The cube illustrates a phase-space cell of volume h^s that contains one state.

where all possible combinations of coordinates and momenta are included with the restriction that $H(q_1, \ldots, q_s, p_1, \ldots, p_s)$ lies between 0 and E. The *density of states* $N(E)$ is given by

$$
\begin{aligned}
N(E) &= (G(E + dE) - G(E))/dE \\
&= \frac{1}{h^s} \int_{H=E} \cdots \int dq_1 \cdots dq_s dp_1 \cdots dp_s \\
&= \frac{1}{h^s} \int \cdots \int \delta(E - H) dq_1 \cdots dq_s dp_1 \cdots dp_s
\end{aligned} \qquad \text{(A.34)}
$$

using that $G(E + dE) - G(E)$ is a multidimensional volume of phase space enclosed by the two hypersurfaces defined by $H = E$ and $H = E + dE$, and this integral can be expressed as the area of the hypersurface for $H = E$ multiplied by dE. The condition that the range of integration is restricted to this hypersurface is in the last integral expressed in terms of a delta function. Note that the definition in Eq. (A.34) implies

$$N(E) = dG(E)/dE \qquad \text{(A.35)}$$

The Boltzmann statistics of particles described by classical mechanics is obtained from Eqs (A.6) and (A.34):

$$P(E)dE = \frac{N(E)}{Q} \exp(-E/k_B T) dE \qquad \text{(A.36)}$$

where

$$
\begin{aligned}
Q &= \int_0^\infty N(E) \exp(-E/k_B T) dE \\
&= \frac{1}{h^s} \int \cdots \int dq_1 \cdots dq_s dp_1 \cdots dp_s \exp(-H(q, p)/k_B T)
\end{aligned} \qquad \text{(A.37)}
$$

The expectation value of the energy in Eq. (A.22) is given by

$$
\begin{aligned}
\langle E \rangle &= \frac{1}{Q} \int E N(E) \exp(-E/k_B T) dE \\
&= \frac{\int \cdots \int dq_1 \cdots dq_s dp_1 \cdots dp_s H(q, p) \exp(-H(q, p)/k_B T)}{\int \cdots \int dq_1 \cdots dq_s dp_1 \cdots dp_s \exp(-H(q, p)/k_B T)}
\end{aligned} \qquad \text{(A.38)}
$$

Finally, the classical partition function for N *interacting* molecules, that is, the classical limit of Eq. (A.2), takes the form

$$\mathcal{Q}(N, V, T) = \frac{1}{N! h^{sN}} \int \cdots \int dq_1 \cdots dq_{sN} dp_1 \cdots dp_{sN} \exp(-H_N(q, p)/k_B T) \qquad \text{(A.39)}$$

where $H_N(q, p)$ is the classical N-body Hamiltonian with each molecule described by s coordinates in configuration space. The probability of finding the system at the point $q_1, \ldots, q_{sN}, p_1, \ldots, p_{sN}$ in a volume element $dq_1 \cdots dq_{sN} dp_1 \cdots dp_{sN}$ is given by

$$
\begin{aligned}
&\mathcal{P}(q_1, \ldots, q_{sN}, p_1, \ldots, p_{sN}) dq_1 \cdots dq_{sN} dp_1 \cdots dp_{sN} = \\
&\qquad \frac{dq_1 \cdots dq_{sN} dp_1 \cdots dp_{sN} \exp(-H_N(q, p)/k_B T)}{\int \cdots \int dq_1 \cdots dq_{sN} dp_1 \cdots dp_{sN} \exp(-H_N(q, p)/k_B T)}
\end{aligned} \qquad \text{(A.40)}
$$

A.2.1 Applications of classical statistical mechanics

In the following we derive some results that are used in various parts of this book.

Example A.1: The free particle

We consider the sum of states, density of states, and energies of an ideal gas in a box of volume V. The Hamiltonian for a free particle of mass m is

$$H(p,q) = (p_x^2 + p_y^2 + p_z^2)/2m \tag{A.41}$$

We first consider the sum of states. Now, in Eq. (A.33) the integration over coordinates gives the volume of the container, and the integral over the momenta is the momentum-space volume for H having values between 0 and E. Equation (A.41) is the equation for a sphere in momentum space with radius $\sqrt{2mH}$. Thus, the volume of the sphere is $4\pi(\sqrt{2mH})^3/3$ and

$$G(E) = \frac{1}{h^3}\frac{4\pi}{3}(\sqrt{2mE})^3 V \tag{A.42}$$

and after differentiation with respect to E,

$$N(E) = (2^{1/2}\pi^2\hbar^3)^{-1} m^{3/2} E^{1/2} V \tag{A.43}$$

(note that the same result is obtained from the quantum solution to a 'particle in a box', for high quantum numbers or, equivalently, high energies [1]). The partition function can then be evaluated:

$$\begin{aligned} Q_{\text{trans}} &= \int_0^\infty N(E)\exp(-E/k_BT)dE \\ &= (2\pi m k_B T)^{3/2} V/h^3 \end{aligned} \tag{A.44}$$

Note that this expression is identical to the approximate form of the quantum mechanical partition function in Eq. (A.14). The energy distribution in Eq. (A.36) becomes

$$P(E)dE = 2\pi\left(\frac{1}{\pi k_B T}\right)^{3/2}\sqrt{E}\exp\left\{-\frac{E}{k_BT}\right\}dE \tag{A.45}$$

This result was given in Eq. (2.28). The well-known Maxwell–Boltzmann distribution of molecular speeds, Eq. (2.27), is obtained after substitution of $E = mv^2/2$, $dE = mvdv$.

Example A.2: s uncoupled harmonic oscillators

We consider the sum of states, density of states, and energies for a set of harmonic oscillators. The Hamiltonian for s harmonic oscillators of unit mass is

$$H(p,q) = \sum_{i=1}^{s} \left(\frac{p_i^2}{2} + \frac{\omega_i^2 q_i^2}{2} \right) \tag{A.46}$$

where $\omega_i = 2\pi\nu_i$ and ν_i is the frequency of the ith oscillator. First consider a single harmonic oscillator ($s = 1$). Equation (A.46) can be written in the form

$$\frac{p_1^2}{(\sqrt{2H})^2} + \frac{q_1^2}{(\sqrt{2H/\omega_1^2})^2} = 1 \tag{A.47}$$

This is the equation of an ellipse with semiaxes $\sqrt{2H}$ and $\sqrt{2H/\omega_1^2}$. The integral in Eq. (A.33) is simply the area of this ellipse, which is equal to π multiplied by the product of the two semiaxes. The area is, accordingly, H/ν_1, and $G(E) = E/(h\nu_1)$. For s harmonic oscillators, Eq. (A.46) is the equation for a $2s$-dimensional ellipsoid and the integral in Eq. (A.33) is simply the volume of this ellipsoid. The volume of a $2s$-dimensional ellipsoid is given by $(\pi^s/s!) \prod_{i=1}^{2s} a_i$, where a_i are the semiaxes. Thus, the sum of states takes the form

$$G(E) = \frac{E^s}{s! \prod_{i=1}^{s} h\nu_i} \tag{A.48}$$

and the density of states is

$$N(E) = \frac{E^{s-1}}{(s-1)! \prod_{i=1}^{s} h\nu_i} \tag{A.49}$$

which implies that the partition function (Eq. (A.37)) is

$$Q_{\text{vib}} = \frac{(k_B T)^s}{\prod_{i=1}^{s} h\nu_i} \tag{A.50}$$

Note that this expression differs from the quantum mechanical partition function in Eq. (A.16); it is the high-temperature limit, where $h\nu_i/k_B T$ is small. The Boltzmann energy distribution function, as given by Eq. (A.36), takes the form

$$P(E)dE = \frac{1}{(s-1)!} \left(\frac{E}{k_B T} \right)^{s-1} \exp\left\{ -\frac{E}{k_B T} \right\} \frac{dE}{k_B T} \tag{A.51}$$

Finally, since Eq. (A.48) is based on a classical evaluation of the sum of states, the fact that according to quantum mechanics no vibrational states exist at energies below the zero-point energy E_z is clearly violated. Thus, we can anticipate that a better estimate of the sum of states at the vibrational energy E, defined as the energy in excess of the lowest possible vibrational energy, is $G(E) = G(E + E_z) - G(E_z)$.

Example A.3: The rigid rotor

We consider the sum of states and density of states for a rigid rotor. The Hamiltonian for a *linear* rigid rotor with the moment of inertia I is

$$H(\theta, \phi, p_\theta, p_\phi) = \frac{p_\theta^2}{2I} + \frac{p_\phi^2}{2I\sin^2\theta} \tag{A.52}$$

where (θ, ϕ) are polar angles that specify the orientation of the rotor ($\theta \in [0, \pi]$ and $\phi \in [0, 2\pi]$) and (p_θ, p_ϕ) are the conjugate momenta. Equation (A.52) can be written in the form

$$\frac{p_\theta^2}{2IH} + \frac{p_\phi^2}{2IH\sin^2\theta} = 1 \tag{A.53}$$

For a fixed value of θ, this is the equation of an ellipse with semiaxes $\sqrt{2IH}$ and $\sqrt{2IH\sin^2\theta}$. The integral in Eq. (A.33) takes the form

$$\begin{aligned} G(E) &= \frac{1}{h^2}\int_0^\pi\int_0^{2\pi} d\theta d\phi \int_0^{H=E}\int dp_\theta dp_\phi \\ &= \frac{1}{h^2}\int_0^\pi\int_0^{2\pi} \pi 2IE\sin\theta\, d\theta d\phi \end{aligned} \tag{A.54}$$

since the integral over p_θ and p_ϕ with the restriction that the energy is between 0 and E is simply the area of the ellipse defined by Eq. (A.53), which is equal to π multiplied by the product of the two semiaxes.

Thus, the sum of states takes the form

$$G(E) = \frac{8\pi^2 IE}{h^2} \tag{A.55}$$

and the density of states is

$$N(E) = \frac{8\pi^2 I}{h^2} \tag{A.56}$$

The partition function (Eq. (A.37)) is then

$$\begin{aligned} Q_{\text{rot}} &= \int_0^\infty N(E)e^{-E/k_BT}dE \\ &= \frac{8\pi^2 Ik_BT}{h^2} \end{aligned} \tag{A.57}$$

Note that this expression is identical to the (approximate) quantum mechanical partition function, Eq. (A.19), for a linear rigid rotor.

The Hamiltonian for a *non-linear* rigid rotor is quite complicated [1] and the derivation of the expressions for the sum and density of states is cumbersome. We know, however, that the partition function is given by Eq. (A.20), and it is quite easy to find the expressions for the sum and density of states that are consistent with Eq. (A.20).

Thus, for the sum of states we use a form that is analogous to Eq. (A.55):

$$G(E) = \frac{4}{3}\sqrt{8\pi^2 I_a/h^2}\sqrt{8\pi^2 I_b/h^2}\sqrt{8\pi^2 I_c/h^2}\,E^{3/2} \tag{A.58}$$

and the density of states is then

$$N(E) = 2\sqrt{8\pi^2 I_a/h^2}\sqrt{8\pi^2 I_b/h^2}\sqrt{8\pi^2 I_c/h^2}\,E^{1/2} \tag{A.59}$$

The partition function (Eq. (A.37)) is then

$$\begin{aligned}
Q_{\text{rot}} &= \int_0^\infty N(E)e^{-E/k_BT}dE \\
&= 2\sqrt{8\pi^2 I_a/h^2}\sqrt{8\pi^2 I_b/h^2}\sqrt{8\pi^2 I_c/h^2}\int_0^\infty E^{1/2}e^{-E/k_BT}dE \\
&= 2\sqrt{8\pi^2 I_a/h^2}\sqrt{8\pi^2 I_b/h^2}\sqrt{8\pi^2 I_c/h^2}\,(k_BT)^{3/2}\int_0^\infty E'^{1/2}e^{-E'}dE' \\
&= \pi^{1/2}\sqrt{8\pi^2 I_a k_BT/h^2}\sqrt{8\pi^2 I_b k_BT/h^2}\sqrt{8\pi^2 I_c k_BT/h^2}
\end{aligned} \tag{A.60}$$

since the integral is the Gamma function with the argument 3/2, and $\Gamma(3/2) = \sqrt{\pi}/2$. This expression is identical to the (approximate) quantum mechanical partition function in Eq. (A.20).

Example A.4: The Maxwell–Boltzmann distribution for N *interacting* molecules

The well-known Maxwell–Boltzmann distribution for the velocity or momentum distribution associated with the translational motion of a molecule is valid not only for free molecules but also for interacting molecules, say in a liquid phase. We start with the general expression, Eq. (A.40), i.e., the Boltzmann distribution for N identical molecules each with s degrees of freedom:

$$\mathcal{P}(q_1,\ldots,q_{sN},p_1,\ldots,p_{sN})dq_1\cdots dq_{sN}dp_1\cdots dp_{sN} = \frac{dq_1\cdots dq_{sN}dp_1\cdots dp_{sN}\exp(-H_N(q,p)/k_BT)}{\int\cdots\int dq_1\cdots dq_{sN}dp_1\cdots dp_{sN}\exp(-H_N(q,p)/k_BT)} \tag{A.61}$$

where q,p is a short-hand notation for $q_1,\ldots,q_{sN},p_1,\ldots,p_{sN}$ and the Hamiltonian can be written in the form

$$H_N(q,p) = T_{\text{trans}}(p_{\text{trans}}) + T_{\text{rot,vib}}(p_{\text{rot,vib}}) + U(q) \tag{A.62}$$

We choose the first $3N$ momenta to be associated with the Cartesian coordinates of the centers of mass of the N molecules, that is, the kinetic energy associated with the translational motion of the N molecules is

$$T_{\text{trans}}(p_{\text{trans}}) = \sum_{n=1}^{N} (p_{xn}^2 + p_{yn}^2 + p_{zn}^2)/2m \tag{A.63}$$

$T_{\text{rot,vib}}(p_{\text{rot,vib}})$ is the kinetic energy associated with rotational and vibrational motion, and $U(q)$ is the potential energy describing the interaction between all molecules.

We want the probability distribution irrespective of all the position coordinates, q, as well as all the momenta associated with the rotational and vibrational motion, $p_{\text{rot,vib}}$. Thus, in Eq. (A.61), we integrate over all these coordinates. The form of the Hamiltonian, Eq. (A.62), implies that the exponentials can be factorized into three terms, and integration over the three sets of coordinates can be carried out separately, i.e.,

$$P_{\text{trans}}(p_1, \ldots, p_{3N})dp_1 \cdots dp_{3N} = \frac{dp_1 \cdots dp_{3N} \exp(-T_{\text{trans}}(p_{\text{trans}})/k_B T)}{\int \cdots \int dp_1 \cdots dp_{3N} \exp(-T_{\text{trans}}(p_{\text{trans}})/k_B T)} \tag{A.64}$$

Finally, we integrate over the three momentum components for $N-1$ of the molecules, i.e., all the molecules except one, and obtain the well-known result

$$P_{\text{trans}}(p_{x1}, p_{y1}, p_{z1})dp_{x1}dp_{y1}dp_{z1} = (2\pi m k_B T)^{-3/2} \exp[-(p_{x1}^2 + p_{y1}^2 + p_{z1}^2)/(2mk_B T)]dp_{x1}dp_{y1}dp_{z1} \tag{A.65}$$

The corresponding velocity distribution is obtained after substitution of $p_{x1} = mv_{x1}$, $p_{y1} = mv_{y1}$, and $p_{z1} = mv_{z1}$.

Further reading/references

[1] D.A. McQuarrie, *Statistical mechanics* (University Science Books, 2000)

B

Microscopic reversibility and detailed balance

Both Newton's equation of motion for a classical system and Schrödinger's equation for a quantum system are unchanged by *time reversal*, i.e., when the sign of the time is changed. Due to this symmetry under time reversal, the transition probability for a forward and the reverse reaction is the same, and consequently a definite relationship exists between the cross-sections for forward and reverse reactions. This relationship, based on the reversibility of the equations of motion, is known as the principle of *microscopic reversibility*, sometimes also referred to as the *reciprocity theorem*. The statistical relationship between rate constants for forward and reverse reactions at equilibrium is known as the principle of *detailed balance*, and we will show that this principle is a consequence of microscopic reversibility. These relations are very useful for obtaining information about reverse reactions once the forward rate constants or cross-sections are known. Let us begin with a discussion of microscopic reversibility.

B.1 Microscopic reversibility

B.1.1 Transition probability

The trajectory of a classical particle may be found by integrating Newton's equation of motion

$$\boldsymbol{F} = m\frac{d}{dt}\left(\frac{d\boldsymbol{r}}{dt}\right) \tag{B.1}$$

from say time t_0 to time t_1. Let us then introduce the substitution

$$\tau = -t + (t_0 + t_1) \tag{B.2}$$

into Eq. (B.1). We find

$$\begin{aligned} \boldsymbol{F} &= m\frac{d}{d(-\tau)}\left(\frac{d\boldsymbol{r}}{d(-\tau)}\right) \\ &= m\frac{d}{d\tau}\left(\frac{d\boldsymbol{r}}{d\tau}\right) \end{aligned} \tag{B.3}$$

As t varies from t_0 to t_1, τ is seen from Eq. (B.2) to vary from t_1 to t_0, so the substitution is equivalent to a time reversal. Since the equations of motion are identical, the system will follow exactly the same trajectory; the only difference will be that in one case it is followed forward in time from t_0 to t_1, and in the other case backward in time from t_1 to t_0. In other words, it is not possible from the equation of motion to decide whether it describes a forward or a reverse propagation in time. This is summarized by saying that the classical equation of motion has time-reversal symmetry. As a consequence, the probability for a forward scattering process must equal the probability for the reverse scattering process, where all velocities $d\boldsymbol{r}/dt$ and time have changed sign (see Eq. (B.3)).

The Schrödinger equation for a quantum system also has time-reversal symmetry. The solution to the time-dependent Schrödinger equation

$$i\hbar\frac{\partial\psi}{\partial t} = \hat{H}\psi \tag{B.4}$$

may be formally written as

$$\begin{aligned}\psi(t_1) &= \exp(-i\hat{H}(t_1 - t_0)/\hbar)\psi(t_0) \\ &\equiv \hat{U}(t_1 - t_0)\psi(t_0)\end{aligned} \tag{B.5}$$

where the propagator

$$\begin{aligned}\hat{U}(t_1 - t_0) &= \exp(-i\hat{H}(t_1 - t_0)/\hbar) \\ &\equiv 1 - i\hat{H}(t_1 - t_0)/\hbar - \hat{H}^2(t_1 - t_0)^2/2\hbar^2 + \cdots\end{aligned} \tag{B.6}$$

(formally defined by its Taylor expansion) propagates the wave function from time t_0 to time t_1. From this definition it is clear that

$$\hat{U}(t_0 - t_1) = \hat{U}^\dagger(t_1 - t_0) \tag{B.7}$$

where $\dagger$ indicates Hermitian conjugation. Suppose now that we propagate the wave function $\psi(t_1)$ at time t_1 backward in time to time t_0; then we have

$$\psi(t_0) = \hat{U}(t_0 - t_1)\psi(t_1) \tag{B.8}$$

or

$$\psi^*(t_0) = \hat{U}(t_1 - t_0)\psi^*(t_1) \tag{B.9}$$

where we have used Eq. (B.6). The propagators in Eqs (B.5) and (B.9) for, respectively, a forward and a backward propagation in time are identical, and if the wave functions are real, it is immediately obvious that the same equation may be used for forward and backward propagation in time as in a classical system; hence the Schrödinger equation has time-reversal symmetry. In general, however, the wave functions are complex, so the two equations of motion differ in the sense that it is the wave function itself that is propagated in the forward direction of time, whereas it is the complex conjugate wave function that is propagated in the reverse direction of time. The complex conjugation of the wave packet is equivalent to a change in sign of the momentum of the wave packet (the Gaussian wave packet, Eq. (4.113), has this property), just like time reversal in a classical system, where the velocities change sign.

Then it is remembered that the wave function itself does not have a physical meaning. So let us, for example, determine the transition probability from say state $|k(t_0)\rangle$ at time t_0 to state $|m(t_1)\rangle$ at time t_1. Then we have

$$|k(t_1)\rangle = \hat{U}(t_1 - t_0)|k(t_0)\rangle \tag{B.10}$$

and the transition probability is

$$P_{|k(t_0)\rangle \to |m(t_1)\rangle} = |\langle m(t_1)|\hat{U}(t_1 - t_0)|k(t_0)\rangle|^2 \tag{B.11}$$

This transition probability is now compared to the transition probability for the reverse process, where we consider the transition from $|m(t_1)\rangle$ to $|k(t_0)\rangle$. We find

$$\begin{aligned} P_{|m(t_1)\rangle \to |k(t_0)\rangle} &= |\langle k(t_0)|\hat{U}(t_0 - t_1)|m(t_1)\rangle|^2 \\ &= |\langle m(t_1)|\hat{U}(t_1 - t_0)|k(t_0)\rangle|^2 \end{aligned} \tag{B.12}$$

where we have used the relation $\langle k(t_0)|\hat{U}(t_0 - t_1)|m(t_1)\rangle^* = \langle m(t_1)|\hat{U}^\dagger(t_0 - t_1)|k(t_0)\rangle$ and Eq. (B.7). In other words, the transition probability from state $|k(t_0)\rangle$ at time t_0 to state $|m(t_1)\rangle$ at time t_1 is equal to the transition probability for the reverse process, that is, from $|m(t_1)\rangle$ to $|k(t_0)\rangle$. This is a manifestation of the time-reversal symmetry of the equation of motion.

The transition probabilities per unit time and per scattering center in Eqs (B.11) and (B.12) are for a collision process, often written in a more explicit form to emphasize the particular transition considered:

$$\boxed{P(ml, ij; v_i, \Omega) = P(ij, ml; v_f, \Omega)} \tag{B.13}$$

where v_i and v_f are the initial and final relative speeds, respectively, of the colliding particles, ij are the initial internal states of the two colliding molecules, ml are the final internal states of the product molecules, and Ω is the solid angle into which the particle of interest is scattered. This angle, as measured from the direction of the initial relative velocities, respectively, $\boldsymbol{v}_i$ and $\boldsymbol{v}_f$, is the same for both the forward and the reverse scattering process. Finally, let us emphasize that the transition probability P refers to a well-specified initial and final state and that it gives us the probability per unit time and per scattering center for the scattering process indicated.

B.1.2 Cross-sections

From the fundamental relation in Eqs (B.12) or (B.13), we are now going to derive a relation between the differential cross-section for the forward and the reverse scattering process.

As defined previously in Eq. (2.7), differential cross-sections are defined as the flux of particles scattered into a range of solid angle $d\Omega$ around the solid angle Ω per unit initial flux and per scattering center. The state-to-state transition probabilities are already per unit scattering center and, since we do not scatter into one single final state but a range of final states as given by $d\Omega$, we need to determine how many final states are consistent with that uncertainty in the final solid angle. It is then assumed that the transition probability to that small range of final states is the same, irrespective of the state; then the product of the transition probability and the number of final states will be the total transition probability. The scattering into the solid angle $d\Omega$ around Ω with a relative final velocity $\boldsymbol{v}_f$ is equivalent to scattering into a final state with momentum $\boldsymbol{p}_f, \boldsymbol{p}_f + d\boldsymbol{p}_f$. The number of final states in this range may be determined in the following way. We consider a free particle in one dimension where the momentum eigenfunction is given by $\psi_p(x) = c(p)\exp(ipx/\hbar)$. In a (macroscopic) box of length L, periodic boundary conditions on the wave function $\psi_p(x+L) = \psi_p(x)$ imply that $\exp(ipL/\hbar) = 1$ or $p/\hbar = 2n\pi/L$, where n is an integer. That is, the spacing between momentum values is h/L, which implies that the density is L/h. This argument can be extended to three dimensions, and it can be shown that the density, $\rho(\boldsymbol{p})$, is $(L/h)^3$ and

$$\begin{aligned}\rho(\boldsymbol{p}_f)d\boldsymbol{p}_f &= \frac{V}{h^3}d\boldsymbol{p}_f \\ &= \frac{V}{h^3}p_f^2 \sin\theta d\theta d\phi dp_f \\ &= \frac{V}{h^3}p_f^2 dp_f d\Omega\end{aligned} \tag{B.14}$$

where we have changed from Cartesian coordinates to spherical coordinates and used the definition of the solid angle $d\Omega$.

If the density of particles in the incident beam of A(i) molecules with relative velocity $\boldsymbol{v}_i$ with respect to the B molecules is $n_{\mathrm{A}(i)}$, then the magnitude of the incoming flux is

$$J_{\mathrm{A}} = v_i n_{\mathrm{A}(i)} = v_i N_{\mathrm{A}(i)}/V \tag{B.15}$$

and the number of scattered molecules in the solid angle $d\Omega$ per unit time is

$$\begin{aligned}J_{\mathrm{C}} &= N_{\mathrm{A}(i)} P(ml, ij; v_i, \Omega)\rho(\boldsymbol{p}_f)d\boldsymbol{p}_f \\ &= N_{\mathrm{A}(i)} P(ml, ij; v_i, \Omega)\frac{V}{h^3}p_f^2 dp_f d\Omega\end{aligned} \tag{B.16}$$

From the definition of the differential cross-section, we obtain[1]

$$\sigma_R(ml|ij; v_i, \Omega)d\Omega\delta(E_f - E_i)dE_f = P(ml, ij; v_i, \Omega)\frac{V}{v_i}\frac{Vp_f^2 dp_f d\Omega}{h^3} \tag{B.17}$$

The factor $\delta(E_f - E_i)dE_f$ on the left-hand side of the equation has been introduced to emphasize energy conservation for the process indicated in the argument to σ_R. The transition probability P on the right-hand side is determined directly from the equation of motion, so it is therefore not necessary to include the delta function on the right-hand side.

[1]Note that, in this appendix, we use a notation for the differential cross-sections that differs somewhat from the one in Chapter 2. Thus, $\sigma_R(ml|ij; v_i, \Omega)$ is $\frac{d\sigma_R}{d\Omega}(ij, v|ml, \Omega)$ in the notation of Chapter 2.

The cross-section for the reverse reaction may be written analogously:

$$\sigma_R(ij|ml; v_f, \Omega)d\Omega\delta(E_i - E_f)dE_i = P(ij, ml; v_f, \Omega)\frac{V}{v_f}\frac{Vp_i^2 dp_i d\Omega}{h^3} \quad \text{(B.18)}$$

When we use Eq. (B.13) and cancel common factors, we find from Eqs (B.17) and (B.18) that

$$\sigma_R(ml|ij; v_i, \Omega)\frac{v_i}{p_f^2 dp_f} = \sigma_R(ij|ml; v_f, \Omega)\frac{v_f}{p_i^2 dp_i} \quad \text{(B.19)}$$

From the energy balance

$$E_i = E_{i,\text{internal}} + \frac{p_i^2}{2\mu_i} = E_f = E_{f,\text{internal}} + \frac{p_f^2}{2\mu_f} \quad \text{(B.20)}$$

where μ_i is the reduced mass for the relative motion of the reactants and μ_f for the products, we find the following relation between p_i and p_f for fixed internal states:

$$\frac{p_i}{\mu_i}dp_i = \frac{p_f}{\mu_f}dp_f$$
$$\Rightarrow \quad v_i dp_i = v_f dp_f \quad \text{(B.21)}$$

The desired relation between cross-sections for the forward and the reverse reactions is thus found to be

$$\boxed{p_i^2\sigma_R(ml|ij; v_i, \Omega) = p_f^2\sigma_R(ij|ml; v_f, \Omega)} \quad \text{(B.22)}$$

This relation expresses the principle of *microscopic reversibility* for the differential cross-sections.

Alternatively, using Eq. (4.173) and the symmetry of the S-matrix element, using Eq. (B.12), we immediately obtain

$$k_n^2\sigma_R(n, E|m) = k_m^2\sigma_R(m, E|n) \quad \text{(B.23)}$$

where, on the left-hand side, n and m refer to the quantum state of the reactant and product and, on the right-hand side, the reverse reaction is considered, i.e., the initial and final quantum states are m and n, respectively. E is the total energy.

Equation (B.23) is the desired relation (at the level of integrated cross-sections), and it is equivalent to Eq. (B.22), since $p = \hbar k$.

If the system has an internal angular momentum (associated with rotational states of molecules) there will, in the absence of an external field, be degeneracies in the system that will be practical to display explicitly in the expression for microscopic reversibility in Eq. (B.22). For systems with angular momenta, time reversal of the quantum equations of motion reverses the signs of both the momenta and their projections on a given direction, just like in a classical system. To express this explicitly, Eq. (B.13) is written as

$$P(ml, ij; v_i, \Omega) = P(i^*j^*, m^*l^*; v_f, \Omega) \quad \text{(B.24)}$$

The 'starred' quantum states differ from the 'unstarred' ones by the sign of the projection of their angular momenta, so if the projection of the angular momentum of state i on the z-axis is m_i then the projection of i^* on the z-axis is $-m_i$, etc. By an analysis similar to that used to derive Eq. (B.22), we obtain

$$p_i^2 \sigma_R(ml|ij; v_i, \Omega) = p_f^2 \sigma_R(i^* j^* | m^* l^*; v_f, \Omega) \tag{B.25}$$

Experimental and calculated cross-sections are usually averaged over the degenerate quantum states associated with the angular momenta when there is no external field. A more useful statement of microscopic reversibility can therefore be obtained by working with average cross-sections $\bar{\sigma}_R$, which refer to transitions between sets of degenerate levels in the initial and final states. Let J_i denote the total angular momentum of state i and m_i its projection on the z-axis. Then by summing Eq. (B.24) over all degenerate states of the reactants and products one obtains

$$\sum_{m_i, m_j, m_m, m_l} P(ml, ij; v_i, \Omega) = \sum_{m_i, m_j, m_m, m_l} P(ij, ml; v_f, \Omega) \tag{B.26}$$

where the summations over the m_k are from $-J_k$ to J_k. Note that we have dropped the star on the quantum states because the range of the m_k extends over both positive and negative values.

Since the p_i and p_f are constant when summed over degenerate states, we may now use Eqs (B.17) and (B.26) to introduce a cross-section, $\sigma_{R(s)}$, where we have summed over all degenerate states:

$$\sigma_{R(s)}(\bar{m}\bar{l}|\bar{i}\bar{j}; v_i, \Omega) d\Omega \delta(E_f - E_i) dE_f = \sum_{m_i, m_j, m_m, m_l} P(ml, ij; v_i, \Omega) \frac{V}{v_i} \frac{V p_f^2 dp_f d\Omega}{h^3} \tag{B.27}$$

So the principle of microscopic reversibility may be rewritten as

$$p_i^2 \sigma_{R(s)}(\bar{m}\bar{l}|\bar{i}\bar{j}; v_i, \Omega) = p_f^2 \sigma_{R(s)}(\bar{i}\bar{j}|\bar{m}\bar{l}; v_f, \Omega) \tag{B.28}$$

The $\sigma_{R(s)}$ refers to the sum of cross-sections for the transition from the degenerate set of states $\bar{i}, \bar{j}$ to $\bar{l}, \bar{m}$ defined for unit flux from each initial state (see Eqs (B.15)–(B.17)). We may instead operate with an average cross-section $\bar{\sigma}_R$ based on an initial flux for the total set of degenerate states:

$$\bar{\sigma}_R(\bar{m}\bar{l}|\bar{i}\bar{j}; v_i, \Omega) = \frac{1}{g_i g_j} \sigma_{R(s)}(\bar{m}\bar{l}|\bar{i}\bar{j}; v_i, \Omega) \tag{B.29}$$

where g_i is the number of degenerate states, $g_i = 2J_i + 1$. Substitution of Eq. (B.29) into Eq. (B.28) leads to yet another expression of microscopic reversibility, based here on the average cross-sections for the degenerate states:

$$p_i^2 g_i g_j \bar{\sigma}_R(\bar{m}\bar{l}|\bar{i}\bar{j}; v_i, \Omega) = p_f^2 g_m g_l \bar{\sigma}_R(\bar{i}\bar{j}|\bar{m}\bar{l}; v_f, \Omega) \tag{B.30}$$

It is noted that this relation only holds in the absence of an external field, when the sets of states are degenerate. Other degeneracies may also occur in the internal molecular states. In this case, the g factors should include those degeneracies.

B.2 Detailed balance

Detailed balance provides a relation between the macroscopic rate constants k_f and k_r for the forward and reverse reactions, respectively. On a macroscopic level, the relation is derived by equating the rates of the forward and reverse reactions at equilibrium. Here it will be shown that the principle of detailed balance can be readily obtained as a direct consequence of the microscopic reversibility of the fundamental equations of motion.

On a macroscopic level in the absence of an external field, we cannot distinguish between sets of degenerate states in the reactants and products; so the most detailed relation between macroscopic rate constants and microscopic cross-sections will be one where we have summed over all degenerate states as in Eq. (B.27). The macroscopic rate constant for a particular transition between degenerate states is then given by

$$k_{\sigma(s)}(\bar{m}\bar{l}, \bar{i}\bar{j}) = \iint v_i \sigma_{R(s)}(\bar{m}\bar{l}|\bar{i}\bar{j}; v_i, \Omega) f_{\mathrm{A}(i)}(\boldsymbol{v}_\mathrm{A}) f_{\mathrm{B}(j)}(\boldsymbol{v}_\mathrm{B}) d\Omega d\boldsymbol{p}_\mathrm{A} d\boldsymbol{p}_\mathrm{B} \tag{B.31}$$

analogous to the relation in Eq. (2.18). This relation is general in the sense that it applies to both equilibrium and non-equilibrium systems. We now assume that the velocity distributions for $\boldsymbol{v}_\mathrm{A}$ and $\boldsymbol{v}_\mathrm{B}$ (or momenta distributions $\boldsymbol{p}_\mathrm{A}$ and $\boldsymbol{p}_\mathrm{B}$) are given by the equilibrium Maxwell–Boltzmann distribution at temperature T. We obtain, as in Section 2.2 (see Eq. (2.29)), the following result:

$$k_{\sigma(s)}(\bar{m}\bar{l}, \bar{i}\bar{j}) = \frac{1}{(2\pi\mu_i k_B T)^{3/2}} \int v_i \sigma_{R(s)}(\bar{m}\bar{l}|\bar{i}\bar{j}; v_i) \exp\left(-\frac{p_i^2}{2\mu_i k_B T}\right) d\boldsymbol{p}_i \tag{B.32}$$

where μ_i is the reduced mass and $\boldsymbol{p}_i = \mu_i \boldsymbol{v}_i$ is the momentum associated with the relative translation of the reactants. Similarly, the expression for the rate constant for the reverse reaction is

$$k_{\sigma(s)}(\bar{i}\bar{j}, \bar{m}\bar{l}) = \frac{1}{(2\pi\mu_f k_B T)^{3/2}} \int v_f \sigma_{R(s)}(\bar{i}\bar{j}|\bar{m}\bar{l}; v_f) \exp\left(-\frac{p_f^2}{2\mu_f k_B T}\right) d\boldsymbol{p}_f \tag{B.33}$$

where μ_f refers to the reduced mass and $\boldsymbol{p}_f = \mu_f \boldsymbol{v}_f$ is the relative momentum of the products. Note that $\sigma_{R(s)}(\bar{m}\bar{l}|\bar{i}\bar{j}; v_i)$ is the integrated cross-section (see Eq. (2.12)) where $\sigma_{R(s)}(\bar{m}\bar{l}|\bar{i}\bar{j}; v_i, \Omega)$ is integrated over all space angles.

To relate the rate constants in Eqs (B.32) and (B.33), we substitute Eq. (B.28) and use the conservation of energy to relate the differentials and limits of integration. Conservation of energy requires

$$\frac{p_i^2}{2\mu_i} = \frac{p_f^2}{2\mu_f} + \Delta E_{\mathrm{int}} \tag{B.34}$$

where

$$\Delta E_{\mathrm{int}} = E_m + E_l - (E_i + E_j) + E_{0,p} - E_{0,r} \tag{B.35}$$

is the change in the internal energies associated with the reaction. $E_{0,p}$ is the zero-point energy of the products and $E_{0,r}$ is the zero-point energy of the reactants. We then have for the reverse rate constant, using Eqs (B.21), (B.28), and $d\boldsymbol{p} = 4\pi p^2 dp$,

$$k_{\sigma(s)}(\bar{i}\bar{j},\bar{m}\bar{l}) =$$
$$\frac{4\pi}{(2\pi\mu_f k_B T)^{3/2}}\int_0^\infty v_i p_i^2 \sigma_{R(s)}(\bar{m}\bar{l}|\bar{i}\bar{j};v_i)\exp\left(-\left[\frac{p_i^2}{2\mu_i k_B T}-\frac{\Delta E_{\text{int}}}{k_B T}\right]\right)dp_i \tag{B.36}$$

From Eq. (B.32), we then obtain the following relation between the forward and reverse rate constants:

$$\boxed{\frac{k_{\sigma(s)}(\bar{m}\bar{l},\bar{i}\bar{j})}{k_{\sigma(s)}(\bar{i}\bar{j},\bar{m}\bar{l})} = \left(\frac{\mu_f}{\mu_i}\right)^{3/2}\exp(-\Delta E_{\text{int}}/k_B T) \equiv K(\bar{m}\bar{l},\bar{i}\bar{j})} \tag{B.37}$$

It is noted that the right-hand side is the ratio of the translational partition functions of products and reactants times the Boltzmann factor for the internal energy change. In the derivation of this expression we have only used that the translational degrees of freedom have been equilibrated at T through the use of the Maxwell–Boltzmann velocity distribution. No assumption about the internal degrees of freedom has been used, so they may or may not be equilibrated at the temperature T. The quantity $K(\bar{m}\bar{l},\bar{i}\bar{j})$ may therefore be considered as a partial equilibrium constant for reactions in which the reactants and products are in translational but not necessarily internal equilibrium.

To obtain the statement of detailed balance for complete equilibrium, with both translational and internal degrees of freedom in thermal equilibrium, we must sum over the rate constants in Eqs (B.32) and (B.33), weighting each by its equilibrium Boltzmann distribution; that is (as in Eq. (2.18)),

$$k_f = \sum_{\bar{i}\bar{j}}\sum_{\bar{m}\bar{l}} k_{\sigma(s)}(\bar{m}\bar{l},\bar{i}\bar{j})p_{\text{A}(\bar{i})}p_{\text{B}(\bar{j})} \tag{B.38}$$

and

$$k_r = \sum_{\bar{i}\bar{j}}\sum_{\bar{m}\bar{l}} k_{\sigma(s)}(\bar{i}\bar{j},\bar{m}\bar{l})p_{\text{C}(\bar{m})}p_{\text{D}(\bar{l})} \tag{B.39}$$

Here k_f is the rate constant for the forward reaction and k_r for the reverse reaction. $p_{\text{A}(\bar{i})}$ is the probability (mole fraction) of A in any of the set of states $\bar{i}$, which is given by statistical mechanics according to Eq. (A.4), and similarly for the other constituents. If we substitute these relations into Eqs (B.38) and (B.39), we obtain

$$k_f = \frac{1}{Q_\text{A}Q_\text{B}}\sum_{\bar{i}\bar{j}}\sum_{\bar{m}\bar{l}} k_{\sigma(s)}(\bar{m}\bar{l},\bar{i}\bar{j})\exp(-[E_i+E_j+E_{0,r}]/k_B T) \tag{B.40}$$

and for the reverse reaction

$$k_r = \frac{1}{Q_\text{C}Q_\text{D}}\sum_{\bar{i}\bar{j}}\sum_{\bar{m}\bar{l}} k_{\sigma(s)}(\bar{i}\bar{j},\bar{m}\bar{l})\exp(-[E_l+E_m+E_{0,p}]/k_B T) \tag{B.41}$$

Note that the degeneracies g_i do not appear explicitly in the equations, because we use 'barred' quantities as indices in the sums. They imply a sum over degenerate states; had we used 'unbarred' indices, then the degeneracy factor g_i should be included explicitly. From Eqs (B.40) and (B.41) and using Eq. (B.37), we find the following relation between the rate constants:

$$
\begin{aligned}
k_r &= \left(\frac{\mu_i}{\mu_f}\right)^{3/2} \left(\frac{Q_\mathrm{A} Q_\mathrm{B}}{Q_\mathrm{C} Q_\mathrm{D}}\right) k_f \\
&= \left(\frac{\mu_i}{\mu_f}\right)^{3/2} \left(\frac{Q_\mathrm{A} Q_\mathrm{B}}{Q_\mathrm{C} Q_\mathrm{D}}\right)_{\mathrm{int}} \exp([E_{0,p} - E_{0,r}]/k_B T) k_f
\end{aligned}
\tag{B.42}
$$

where the subscript 'int' has been added in order to emphasize that the partition functions refer to internal (non-translational) degrees of freedom. Furthermore, in the second line, following Eq. (A.10), the partition functions are evaluated with the energies measured relative to the zero-point levels of the reactants and products, respectively. Rearranged, the general statement of detailed balance at equilibrium may be written

$$
\boxed{\frac{k_f}{k_r} = \left(\frac{\mu_f}{\mu_i}\right)^{3/2} \left(\frac{Q_\mathrm{C} Q_\mathrm{D}}{Q_\mathrm{A} Q_\mathrm{B}}\right)_{\mathrm{int}} \exp(-[E_{0,p} - E_{0,r}]/k_B T) \equiv K(T)}
\tag{B.43}
$$

where $K(T)$ is the equilibrium constant for the reaction. This is the usual statistical mechanical expression for the equilibrium constant in terms of the molecular partition functions.

In summary, we have seen that the application of microscopic reversibility for the forward and reverse cross-sections and the use of complete equilibrium distributions for the evaluation of the statistical rate constant lead to the usual results known from equilibrium statistical mechanics. If one knows the cross-section for a forward reaction, one can always determine the inverse cross-section through the principle of microscopic reversibility. Also, if one knows the cross-section for the forward reaction, and in addition one knows that the translational and internal distribution functions of reactants and products have reached equilibrium, one can calculate the rate constant. Detailed balance then permits the calculation of the reverse rate constant.

Further reading/references

[1] J.C. Light, J. Ross, and K.E. Shuler, in *Kinetic processes in gases and plasmas* (Academic Press, 1969) Chapter 8

C

Cross-sections in various frames

We consider the scattering process between two molecules in the gas phase. In the laboratory, the experiments are usually done by letting two beams, one consisting of one kind of molecules and another consisting of the other kind of molecules, cross in a small volume element, where the collision takes place. The result of the scattering experiment is usually reported as a scattering angle formed by the velocity of some molecule after the scattering and one of the incident molecules, and as a differential cross-section with respect to some space angle, both given in the laboratory coordinate system. The scattering angle alone does not uniquely specify the direction in which the scattered molecules move, only that the velocity is somewhere on a cone with an opening angle that is twice the scattering angle. We also need to specify an azimuthal angle for rotation around the incident molecular velocity to uniquely determine the position of the velocity vector on the cone and thereby its direction.

It is intuitively clear that the outcome of the scattering event only depends on the relative motion of the colliding molecules rather than on the overall motion of the molecules. In a theoretical calculation of the scattering event, it is therefore natural to describe it in terms of the center-of-mass motion of the colliding molecules and their relative motion. The center-of-mass motion, representing the overall motion of the system, is irrelevant for the scattering event and therefore not followed further, whereas the relative motion determines the outcome of the scattering event and is followed in detail.

Since the center-of-mass coordinate system used in a theoretical calculation is different from the laboratory coordinate system used in an experiment, we need to determine a relation between the scattering angle χ and the azimuthal angle ϕ in the center-of-mass coordinate system, and the scattering angle Θ and the azimuthal angle ξ in the laboratory coordinate system when theoretical and experimental results are compared. We also need a relation between the differential scattering cross-sections $(d\sigma/d\Omega)_{\text{c.m.}}$ in the center-of-mass coordinate system and $(d\sigma/d\Omega)_{\text{lab}}$ in the laboratory coordinate system, where $d\Omega$ is a space angle.

In the following we first derive a general relation between χ, ϕ and Θ, ξ for an elastic or inelastic scattering event between two molecules. If the molecules are two atoms, they will interact with central forces acting along the relative position vector of the two atoms, and then the scattering event takes place in the plane spanned by the initial velocities of the molecules. Otherwise, the forces will be non-central and the angular momentum is therefore not conserved, which infers that there will be out-of-plane scattering, and the velocities of the scattered molecules will no longer be in the plane spanned by the initial velocities of the molecules. From the relations between the scattering angles and azimuthal angles we derive an expression for the relation between the differential scattering cross-sections in the two coordinate systems. Finally, the results are generalized to a reactive scattering event.

C.1 Elastic and inelastic scattering of two molecules

Let us consider the collision between two atoms with masses m_A and m_B. The center-of-mass coordinate $\boldsymbol{R}$ and center-of-mass velocity $\boldsymbol{V}$ of the two-atom system are given by

$$\begin{aligned} \boldsymbol{R} &= \left(\frac{m_A}{M}\right)\boldsymbol{r}_A + \left(\frac{m_B}{M}\right)\boldsymbol{r}_B \\ \boldsymbol{V} &= \left(\frac{m_A}{M}\right)\boldsymbol{v}_A + \left(\frac{m_B}{M}\right)\boldsymbol{v}_B \\ &= \left(\frac{m_A}{M}\right)\boldsymbol{v}_A^0 + \left(\frac{m_B}{M}\right)\boldsymbol{v}_B^0 \end{aligned} \tag{C.1}$$

$M = m_A + m_B$ is the total mass of the system and the superscript 0 on the velocities $\boldsymbol{v}_i$ implies the start velocities before the scattering event. The last identity in Eq. (C.1) follows from the fact that the center-of-mass velocity $\boldsymbol{V}$ is constant, since the action forces only depend on the distance between the atoms, making the total force on the system equal to zero.

The relative coordinate $\boldsymbol{r}$ and relative velocity $\boldsymbol{v}$ are given by

$$\begin{aligned} \boldsymbol{r} &= \boldsymbol{r}_A - \boldsymbol{r}_B \\ \boldsymbol{v} &= \boldsymbol{v}_A - \boldsymbol{v}_B \\ \boldsymbol{v}^0 &= \boldsymbol{v}_A^0 - \boldsymbol{v}_B^0 \\ |\boldsymbol{v}| &= \beta|\boldsymbol{v}^0| \end{aligned} \tag{C.2}$$

We have also given the initial relative velocity $\boldsymbol{v}^0$ and the β parameter determines the magnitude of the relative velocity after the scattering event. For an elastic scattering event $\beta = 1$, while it will be different from 1 in an inelastic scattering event, where relative translational energy may be lost to internal degrees of freedom ($0 < \beta < 1$) or gained from the internal degrees of freedom ($\beta > 1$).

The definitions in Eqs (C.1) and (C.2) for two atoms may easily be generalized to two molecules. m_A and m_B are then the masses, $\boldsymbol{r}_A$ and $\boldsymbol{r}_B$ the center-of-mass positions, and $\boldsymbol{v}_A$ and $\boldsymbol{v}_B$ the center-of-mass velocities of the two molecules. The interactions between the two molecules are usually more complex than between atoms and will most likely not be given in terms of central forces. This means that the angular momentum for the relative motion will change both in magnitude and direction during the scattering event, so velocities like $\boldsymbol{v}_A$ and $\boldsymbol{v}$ after the scattering will no longer be in the plane spanned by the initial velocities $\boldsymbol{v}_A^0$ and $\boldsymbol{v}_B^0$. However, the center-of-mass velocity, $\boldsymbol{V}$, of the two molecules will not change during the scattering event because there are no external forces acting on the system.

The scattering geometry is illustrated in Fig. C.1.1, often referred to as a *Newton diagram*. The χ angle is given as the angle between $\boldsymbol{v}$ and $\boldsymbol{v}^0$, and is the angle usually determined in a theoretical calculation of the scattering event, whereas the experimentally determined scattering angle Θ in the laboratory coordinate system is

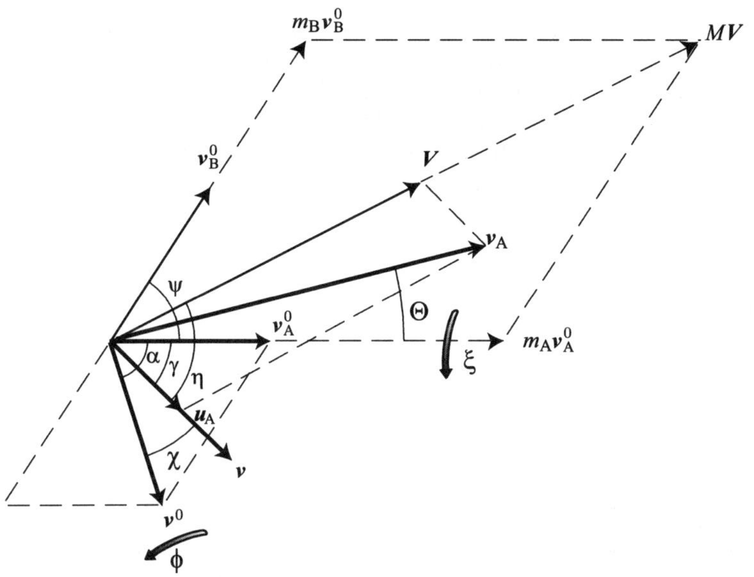

Fig. C.1.1 The geometry for the scattering between two molecules. ψ is the angle between the two molecular beams. α, γ, and η are auxiliary angles used in the final expressions. χ is the scattering angle and ϕ is the azimuthal angle in the center-of-mass coordinate system. Θ is the scattering angle and ξ is the azimuthal angle in the laboratory coordinate system. All velocities are in the plane spanned by the initial velocities $\boldsymbol{v}_A^0$ and $\boldsymbol{v}_B^0$ of the molecules, when only central forces are acting. Otherwise, $\boldsymbol{v}$ and $\boldsymbol{v}_A$ will in general be out of the plane as given by the azimuthal angles ϕ and ξ.

given by the angle between the $\boldsymbol{v}_A$ and $\boldsymbol{v}_A^0$ vectors, the velocities of particle A in the laboratory coordinate system after and before the scattering event, respectively.

The χ angle is, however, not sufficient to specify the orientation of the relative velocity $\boldsymbol{v}$. It only limits $\boldsymbol{v}$ to be somewhere on a cone with opening angle 2χ and generated by rotation of $\boldsymbol{v}$ through 2π around $\boldsymbol{v}^0$. We therefore introduce an azimuthal angle ϕ that gives the rotation angle of $\boldsymbol{v}$ around $\boldsymbol{v}^0$, with $\phi = 0$ when $\boldsymbol{v}$ is in the plane spanned by the initial velocities of the molecules. The angle ϕ is also determined in a simulation of the scattering event, and when $\phi = 0$ the scattering takes place in the plane spanned by the initial velocities of the molecules, so all velocities after the scattering will still be in that plane. This occurs when only central forces are acting, since the angular momentum of the relative motion is then conserved. Otherwise, $\boldsymbol{v}$ and $\boldsymbol{v}_A$ will be out of that plane when non-central forces are acting and in the plane spanned by $\boldsymbol{v}$ and $\boldsymbol{V}$.

Likewise, $\boldsymbol{v}_A$ is bound to be somewhere on a cone with opening angle 2Θ and generated by rotation of $\boldsymbol{v}_A$ through 2π around $\boldsymbol{v}_A^0$, and we introduce the azimuthal angle ξ to give the rotation angle of $\boldsymbol{v}_A$ around $\boldsymbol{v}_A^0$ to completely specify the orientation of $\boldsymbol{v}_A$. Then $\xi = 0$ when $\boldsymbol{v}_A$ is in the plane spanned by the initial velocities of the molecules. The ξ and Θ angles determine the position of the detector in the laboratory coordinate system for monitoring the intensity of a scattering experiment that in the center-of-mass coordinate system leads to scattering through angles χ and ϕ.

$\boldsymbol{v}_{\mathrm{A}}^{0}$ and $\boldsymbol{v}_{\mathrm{B}}^{0}$ are the velocities of the molecules in the two incident beams. From them we may determine the center-of-mass velocity $\boldsymbol{V}$ as shown. Also, the relative velocity $\boldsymbol{v}^{0} = \boldsymbol{v}_{\mathrm{A}}^{0} - \boldsymbol{v}_{\mathrm{B}}^{0}$ is found as illustrated in Fig. C.1.1. The relative velocity $\boldsymbol{v}$ is at an angle χ with respect to $\boldsymbol{v}^{0}$, rotated through an angle ϕ around $\boldsymbol{v}^{0}$, and of magnitude $\beta|\boldsymbol{v}^{0}|$ after the scattering event.

The velocity $\boldsymbol{u}_{\mathrm{A}}$ of molecule A with respect to the center-of-mass velocity is given by

$$\begin{aligned} \boldsymbol{u}_{\mathrm{A}} &= \boldsymbol{v}_{\mathrm{A}} - \boldsymbol{V} \\ &= \boldsymbol{v}_{\mathrm{A}} - \left(\frac{m_{\mathrm{A}}}{M}\right)\boldsymbol{v}_{\mathrm{A}} - \left(\frac{m_{\mathrm{B}}}{M}\right)\boldsymbol{v}_{\mathrm{B}} \\ &= \left(\frac{m_{\mathrm{B}}}{M}\right)\boldsymbol{v} \end{aligned} \tag{C.3}$$

that is, in the same direction as the relative velocity $\boldsymbol{v}$, as shown in Fig. C.1.1. By combining $\boldsymbol{u}_{\mathrm{A}}$ with $\boldsymbol{V}$, we may determine the velocity $\boldsymbol{v}_{\mathrm{A}}$ of particle A after the scattering event in the laboratory coordinate system. $\boldsymbol{v}_{\mathrm{A}}$ is also shown in the figure. All these velocities will be in the same plane during the scattering event when only central forces are acting along $\boldsymbol{r}$, since the angular momentum of the relative motion is then conserved. In general, with non-central forces acting, velocities $\boldsymbol{v}$ and $\boldsymbol{v}_{\mathrm{A}}$ after the scattering will be in a plane that is rotated around the center-of-mass velocity $\boldsymbol{V}$, which is invariant during the scattering since no external forces are acting.

C.1.1 Scattering angle Θ

The scattering angle Θ in the laboratory coordinate system is the angle between $\boldsymbol{v}_{\mathrm{A}}^{0}$ and $\boldsymbol{v}_{\mathrm{A}}$, as shown in Fig. C.1.1. We have

$$\begin{aligned} \boldsymbol{v}_{\mathrm{A}} \cdot \boldsymbol{v}_{\mathrm{A}}^{0} &= (\boldsymbol{V} + \boldsymbol{u}_{\mathrm{A}}) \cdot \boldsymbol{v}_{\mathrm{A}}^{0} \\ &= \left(\boldsymbol{V} + \frac{m_{\mathrm{B}}}{M}\boldsymbol{v}\right) \cdot \boldsymbol{v}_{\mathrm{A}}^{0} \\ &= |\boldsymbol{v}_{\mathrm{A}}||\boldsymbol{v}_{\mathrm{A}}^{0}| \cos\Theta \end{aligned} \tag{C.4}$$

where we have used the relation for $\boldsymbol{u}_{\mathrm{A}}$ in Eq. (C.3). When we introduce the definition of the center-of-mass velocity $\boldsymbol{V}$ from Eq. (C.1), we get

$$|\boldsymbol{v}_{\mathrm{A}}||\boldsymbol{v}_{\mathrm{A}}^{0}| \cos\Theta = \left(\frac{m_{\mathrm{A}}}{M}\right)\boldsymbol{v}_{\mathrm{A}}^{0} \cdot \boldsymbol{v}_{\mathrm{A}}^{0} + \left(\frac{m_{\mathrm{B}}}{M}\right)\boldsymbol{v}_{\mathrm{B}}^{0} \cdot \boldsymbol{v}_{\mathrm{A}}^{0} + \left(\frac{m_{\mathrm{B}}}{M}\right)\boldsymbol{v} \cdot \boldsymbol{v}_{\mathrm{A}}^{0} \tag{C.5}$$

It will be convenient to introduce the angles ψ, α, and γ as defined by their cosines in the following. ψ is the angle between the molecular beams and defines the geometry of the scattering experiment. α is an auxiliary angle that is determined by the set-up of the scattering experiment (see Eq. (C.20)), and γ is another auxiliary angle that is determined by the set-up (see Eq. (C.11)). All three angles are shown in Fig. C.1.1. We have

$$\begin{aligned}
\cos\psi &= \frac{\boldsymbol{v}_{\mathrm{B}}^{0}\cdot\boldsymbol{v}_{\mathrm{A}}^{0}}{|\boldsymbol{v}_{\mathrm{B}}^{0}||\boldsymbol{v}_{\mathrm{A}}^{0}|}\\
\cos\alpha &= \frac{\boldsymbol{v}_{\mathrm{A}}^{0}\cdot\boldsymbol{v}^{0}}{|\boldsymbol{v}_{\mathrm{A}}^{0}||\boldsymbol{v}^{0}|}\\
\cos\gamma &= \frac{\boldsymbol{v}\cdot\boldsymbol{v}_{\mathrm{A}}^{0}}{|\boldsymbol{v}||\boldsymbol{v}_{\mathrm{A}}^{0}|}
\end{aligned}\tag{C.6}$$

When the expressions in Eq. (C.6) are introduced into Eq. (C.5), it may be written

$$|\boldsymbol{v}_{\mathrm{A}}||\boldsymbol{v}_{\mathrm{A}}^{0}|\cos\Theta = \left(\frac{m_{\mathrm{A}}}{M}\right)|\boldsymbol{v}_{\mathrm{A}}^{0}|^{2} + \left(\frac{m_{\mathrm{B}}}{M}\right)|\boldsymbol{v}_{\mathrm{B}}^{0}||\boldsymbol{v}_{\mathrm{A}}^{0}|\cos\psi + \left(\frac{m_{\mathrm{B}}}{M}\right)|\boldsymbol{v}||\boldsymbol{v}_{\mathrm{A}}^{0}|\cos\gamma \tag{C.7}$$

or

$$\boxed{\cos\Theta = \frac{1}{|\boldsymbol{v}_{\mathrm{A}}|}\left[\left(\frac{m_{\mathrm{A}}}{M}\right)|\boldsymbol{v}_{\mathrm{A}}^{0}| + \left(\frac{m_{\mathrm{B}}}{M}\right)|\boldsymbol{v}_{\mathrm{B}}^{0}|\cos\psi + \left(\frac{m_{\mathrm{B}}}{M}\right)|\boldsymbol{v}|\cos\gamma\right]} \tag{C.8}$$

The angle γ will be given once the χ and ϕ angles have been specified, so let us determine the relation between γ, χ, and ϕ.

We write $\boldsymbol{v}$ as a sum of three vectors, one along $\boldsymbol{v}^0$, one along the perpendicular direction to $\boldsymbol{v}^0$ in the plane spanned by the initial velocities of the molecules and specified by the unit vector $\hat{\mathbf{v}}_{\perp}^{0}$, and one along the perpendicular direction to the plane and specified by the unit vector $\hat{\mathbf{k}}$. The projections of $\boldsymbol{v}$ onto the three directions are easily expressed in terms of the angles χ and ϕ, and we get

$$\boldsymbol{v} = \frac{|\boldsymbol{v}|}{|\boldsymbol{v}^{0}|}\cos\chi\,\boldsymbol{v}^{0} + |\boldsymbol{v}|\sin\chi\,\cos\phi\,\hat{\mathbf{v}}_{\perp}^{0} + |\boldsymbol{v}|\sin\chi\,\sin\phi\,\hat{\mathbf{k}} \tag{C.9}$$

We then take the dot product of $\boldsymbol{v}$ and $\boldsymbol{v}_{\mathrm{A}}^{0}$ to determine the angle γ, and use Eq. (C.6) to find

$$\begin{aligned}
\boldsymbol{v}\cdot\boldsymbol{v}_{\mathrm{A}}^{0} &= \left[\frac{|\boldsymbol{v}|}{|\boldsymbol{v}^{0}|}\cos\chi\,\boldsymbol{v}^{0} + |\boldsymbol{v}|\sin\chi\,\cos\phi\,\hat{\mathbf{v}}_{\perp}^{0} + |\boldsymbol{v}|\,\sin\chi\,\sin\phi\,\hat{\mathbf{k}}\right]\cdot\boldsymbol{v}_{\mathrm{A}}^{0}\\
&= \frac{|\boldsymbol{v}|}{|\boldsymbol{v}^{0}|}|\boldsymbol{v}_{\mathrm{A}}^{0}||\boldsymbol{v}^{0}|\cos\chi\,\cos\alpha + |\boldsymbol{v}||\boldsymbol{v}_{\mathrm{A}}^{0}|\sin\chi\,\cos\phi\,\cos(90-\alpha)\\
&= |\boldsymbol{v}||\boldsymbol{v}_{\mathrm{A}}^{0}|\cos\chi\,\cos\alpha + |\boldsymbol{v}||\boldsymbol{v}_{\mathrm{A}}^{0}|\sin\chi\,\cos\phi\,\sin\alpha\\
&= |\boldsymbol{v}||\boldsymbol{v}_{\mathrm{A}}^{0}|\cos\gamma
\end{aligned}\tag{C.10}$$

We have used that, if the angle between two vectors is α, then the angle between one of the vectors and a vector perpendicular to the other in the plane spanned by the two vectors is $90 - \alpha$. Thus, we have

$$\cos\gamma = \cos\chi\,\cos\alpha + \sin\chi\,\sin\alpha\,\cos\phi \tag{C.11}$$

All quantities on the right-hand side of Eq. (C.8) are now known except for $|\boldsymbol{v}_{\mathrm{A}}|$. It may be determined from the triangle with sides $\boldsymbol{V}$, $\boldsymbol{u}_{\mathrm{A}}$, and $\boldsymbol{v}_{\mathrm{A}}$, and we find

$$\begin{aligned}
|\boldsymbol{v}_{\mathrm{A}}| &= \sqrt{|\boldsymbol{u}_{\mathrm{A}}|^{2} + |\boldsymbol{V}|^{2} - 2|\boldsymbol{u}_{\mathrm{A}}||\boldsymbol{V}|\cos(180-\eta)}\\
&= \sqrt{\left(\frac{m_{\mathrm{B}}}{M}\right)^{2}|\boldsymbol{v}|^{2} + |\boldsymbol{V}|^{2} + 2\left(\frac{m_{\mathrm{B}}}{M}\right)|\boldsymbol{v}||\boldsymbol{V}|\cos\eta}
\end{aligned}\tag{C.12}$$

where η is the angle between the center-of-mass velocity $\boldsymbol{V}$ and $\boldsymbol{v}$. This angle is, like γ, given when χ and ϕ are specified. To obtain the relation to those angles, take the dot product of $\boldsymbol{V}$ and $\boldsymbol{v}$. We find

$$\begin{aligned}
\boldsymbol{v}\cdot\boldsymbol{V} &= \left[\frac{|\boldsymbol{v}|}{|\boldsymbol{v}^0|}\cos\chi\,\boldsymbol{v}^0 + |\boldsymbol{v}|\sin\chi\,\cos\phi\,\hat{\mathbf{v}}_\perp^0 + |\boldsymbol{v}|\,\sin\chi\,\sin\phi\,\hat{\mathbf{k}}\right]\cdot\left[\frac{m_\mathrm{A}}{M}\boldsymbol{v}_\mathrm{A}^0 + \frac{m_\mathrm{B}}{M}\boldsymbol{v}_\mathrm{B}^0\right] \\
&= \frac{m_\mathrm{A}}{M}|\boldsymbol{v}_\mathrm{A}^0||\boldsymbol{v}|\cos\chi\,\cos\alpha + \frac{m_\mathrm{B}}{M}|\boldsymbol{v}_\mathrm{B}^0||\boldsymbol{v}|\cos\chi\,\cos(\alpha+\psi) \\
&\quad + \frac{m_\mathrm{A}}{M}|\boldsymbol{v}_\mathrm{A}^0||\boldsymbol{v}|\sin\chi\,\cos\phi\,\cos(90-\alpha) + \frac{m_\mathrm{B}}{M}|\boldsymbol{v}_\mathrm{B}^0||\boldsymbol{v}|\sin\chi\,\cos\phi\,\cos(90-(\alpha+\psi)) \\
&= |\boldsymbol{v}||\boldsymbol{V}|\,\cos\eta
\end{aligned} \tag{C.13}$$

which leads to the following relation:

$$\begin{aligned}
\cos\eta = \frac{m_\mathrm{B}|\boldsymbol{v}_\mathrm{A}^0|}{M|\boldsymbol{V}|}\Bigg[&\left(\frac{m_\mathrm{A}}{m_\mathrm{B}}\cos\alpha + \frac{|\boldsymbol{v}_\mathrm{B}^0|}{|\boldsymbol{v}_\mathrm{A}^0|}\cos(\alpha+\psi)\right)\cos\chi \\
&+\left(\frac{m_\mathrm{A}}{m_\mathrm{B}}\sin\alpha + \frac{|\boldsymbol{v}_\mathrm{B}^0|}{|\boldsymbol{v}_\mathrm{A}^0|}\sin(\alpha+\psi)\right)\cos\phi\,\sin\chi\Bigg]
\end{aligned} \tag{C.14}$$

C.1.2 Azimuthal angle ξ

The azimuthal angle ξ for rotation of $\boldsymbol{v}_\mathrm{A}$ around $\boldsymbol{v}_\mathrm{A}^0$ is related to χ and ϕ in the following way. It is clear from the construction of $\boldsymbol{v}_\mathrm{A}$ that it has the same component perpendicular to the plane spanned by the initial velocities of the molecules as the $\boldsymbol{u}_\mathrm{A}$ vector has because $\boldsymbol{V}$ does not change during the scattering event. This component may be found by analogy with the construction in Eq. (C.9), and is given by the component of the unit vector $\hat{\mathbf{k}}$ with $|\boldsymbol{v}|$ replaced by $|\boldsymbol{u}_\mathrm{A}|$ since the two vectors are collinear. We may, however, also express the component in terms of the azimuthal angle ξ for the rotation of $\boldsymbol{v}_\mathrm{A}$ around $\boldsymbol{v}_\mathrm{A}^0$ by a similar construction as in Eq. (C.9), and we get

$$|\boldsymbol{u}_\mathrm{A}|\sin\chi\,\sin\phi = |\boldsymbol{v}_\mathrm{A}|\sin\Theta\,\sin\xi \tag{C.15}$$

which leads to the relation

$$\boxed{\sin\xi = \frac{m_\mathrm{B}}{M}\frac{|\boldsymbol{v}|}{|\boldsymbol{v}_\mathrm{A}|}\frac{\sin\chi\,\sin\phi}{\sin\Theta}} \tag{C.16}$$

where we have used Eq. (C.3). A combination of Eqs (C.8), (C.12), and (C.14) allows us to determine the scattering angle Θ, and Eq. (C.16) gives the azimuthal angle ξ in the laboratory coordinate system as a function of the theoretically calculated scattering angles χ and ϕ in the center-of-mass coordinate system, since all terms in the equations are known, remembering that $|\boldsymbol{v}| = \beta|\boldsymbol{v}^0|$ with $\beta = 1$ for an elastic scattering event.

C.1.3 Further development of expressions for Θ and ξ

By algebraic manipulations it is possible to show that the scattering angles Θ and ξ in the laboratory coordinate system may be expressed as functions of the scattering angles χ, ϕ in the center-of-mass coordinate system in terms of the following parameters that characterize the initial geometry and set-up of the scattering event:

(i) ψ: defined by the geometry of the beam experiment as the angle between the velocities $\boldsymbol{v}_{\mathrm{A}}^0$ and $\boldsymbol{v}_{\mathrm{B}}^0$ in the beams, $0 \leqslant \psi \leqslant \pi$;

(ii) $m_{\mathrm{A}}/m_{\mathrm{B}}$: the relative mass of the two molecules in the beams;

(iii) $|\boldsymbol{v}_{\mathrm{B}}^0|/|\boldsymbol{v}_{\mathrm{A}}^0|$: the relative initial speed of the two molecules in the beams.

Let us in the following give explicit expressions for all the relevant variables in Eqs (C.8), (C.12), and (C.16) and introduce a shorthand notation f_i $(i = 1, 2, 3)$ for certain combinations of terms that occur repeatedly:

$$
\begin{aligned}
|\boldsymbol{V}| &= \sqrt{\left(\frac{m_{\mathrm{A}}}{M}\right)^2 |\boldsymbol{v}_{\mathrm{A}}^0|^2 + \left(\frac{m_{\mathrm{B}}}{M}\right)^2 |\boldsymbol{v}_{\mathrm{B}}^0|^2 + 2\,\frac{m_{\mathrm{A}} m_{\mathrm{B}}}{M^2}\,\boldsymbol{v}_{\mathrm{A}}^0 \cdot \boldsymbol{v}_{\mathrm{B}}^0} \\
&= \frac{m_{\mathrm{B}}|\boldsymbol{v}_{\mathrm{A}}^0|}{M} \sqrt{\left(\frac{m_{\mathrm{A}}}{m_{\mathrm{B}}}\right)^2 + \frac{|\boldsymbol{v}_{\mathrm{B}}^0|^2}{|\boldsymbol{v}_{\mathrm{A}}^0|^2} + 2\left(\frac{m_{\mathrm{A}}}{m_{\mathrm{B}}}\right)\frac{|\boldsymbol{v}_{\mathrm{B}}^0|\cos\psi}{|\boldsymbol{v}_{\mathrm{A}}^0|}} \\
&= \frac{m_{\mathrm{B}}|\boldsymbol{v}_{\mathrm{A}}^0|}{M}\sqrt{f_1}
\end{aligned}
\tag{C.17}
$$

$$
\begin{aligned}
|\boldsymbol{v}^0| &= \sqrt{|\boldsymbol{v}_{\mathrm{A}}^0|^2 + |\boldsymbol{v}_{\mathrm{B}}^0|^2 - 2|\boldsymbol{v}_{\mathrm{A}}^0||\boldsymbol{v}_{\mathrm{B}}^0|\cos\psi} \\
&= |\boldsymbol{v}_{\mathrm{A}}^0|\sqrt{1 + \frac{|\boldsymbol{v}_{\mathrm{B}}^0|^2}{|\boldsymbol{v}_{\mathrm{A}}^0|^2} - 2\frac{|\boldsymbol{v}_{\mathrm{B}}^0|\cos\psi}{|\boldsymbol{v}_{\mathrm{A}}^0|}} \\
&= |\boldsymbol{v}_{\mathrm{A}}^0|\sqrt{f_2}
\end{aligned}
\tag{C.18}
$$

$$
|\boldsymbol{v}| = \beta|\boldsymbol{v}^0| = \beta|\boldsymbol{v}_{\mathrm{A}}^0|\sqrt{f_2}
\tag{C.19}
$$

$$
\begin{aligned}
\cos\alpha &= \frac{\boldsymbol{v}_{\mathrm{A}}^0 \cdot \boldsymbol{v}^0}{|\boldsymbol{v}_{\mathrm{A}}^0||\boldsymbol{v}^0|} = \frac{\boldsymbol{v}_{\mathrm{A}}^0 \cdot [\boldsymbol{v}_{\mathrm{A}}^0 - \boldsymbol{v}_{\mathrm{B}}^0]}{|\boldsymbol{v}_{\mathrm{A}}^0||\boldsymbol{v}^0|} \\
&= \frac{|\boldsymbol{v}_{\mathrm{A}}^0|}{|\boldsymbol{v}^0|}\left[1 - \frac{|\boldsymbol{v}_{\mathrm{B}}^0|\cos\psi}{|\boldsymbol{v}_{\mathrm{A}}^0|}\right] \\
&= \frac{1 - \frac{|\boldsymbol{v}_{\mathrm{B}}^0|\cos\psi}{|\boldsymbol{v}_{\mathrm{A}}^0|}}{\sqrt{1 + \frac{|\boldsymbol{v}_{\mathrm{B}}^0|^2}{|\boldsymbol{v}_{\mathrm{A}}^0|^2} - 2\frac{|\boldsymbol{v}_{\mathrm{B}}^0|\cos\psi}{|\boldsymbol{v}_{\mathrm{A}}^0|}}} \\
&= \frac{f_3}{\sqrt{f_2}}
\end{aligned}
\tag{C.20}
$$

$$
\cos\gamma = \cos\alpha\cos\chi + \sin\alpha\,\sin\chi\,\cos\phi
\tag{C.21}
$$

$$
\begin{aligned}
\cos\eta = \frac{m_{\mathrm{B}}|\boldsymbol{v}_{\mathrm{A}}^0|}{M|\boldsymbol{V}|}\Bigg[&\left(\frac{m_{\mathrm{A}}}{m_{\mathrm{B}}}\cos\alpha + \frac{|\boldsymbol{v}_{\mathrm{B}}^0|}{|\boldsymbol{v}_{\mathrm{A}}^0|}\cos(\alpha+\psi)\right)\cos\chi \\
&+ \left(\frac{m_{\mathrm{A}}}{m_{\mathrm{B}}}\sin\alpha + \frac{|\boldsymbol{v}_{\mathrm{B}}^0|}{|\boldsymbol{v}_{\mathrm{A}}^0|}\sin(\alpha+\psi)\right)\cos\phi\,\sin\chi\Bigg]
\end{aligned}
$$

$$= \frac{1}{\sqrt{f_1}} \left[\left(\frac{m_A}{m_B} \cos\alpha + \frac{|\boldsymbol{v}_B^0|}{|\boldsymbol{v}_A^0|} [\cos\alpha \cos\psi - \sin\alpha \sin\psi] \right) \cos\chi \right.$$
$$\left. + \left(\frac{m_A}{m_B} \sin\alpha + \frac{|\boldsymbol{v}_B^0|}{|\boldsymbol{v}_A^0|} [\sin\alpha \cos\psi + \cos\alpha \sin\psi] \right) \cos\phi \sin\chi \right] \tag{C.22}$$

where the shorthand notations f_i are defined as

$$\begin{aligned} f_1 &= \left(\frac{m_A}{m_B}\right)^2 + \frac{|\boldsymbol{v}_B^0|^2}{|\boldsymbol{v}_A^0|^2} + 2\left(\frac{m_A}{m_B}\right)\frac{|\boldsymbol{v}_B^0| \cos\psi}{|\boldsymbol{v}_A^0|} \\ f_2 &= 1 + \frac{|\boldsymbol{v}_B^0|^2}{|\boldsymbol{v}_A^0|^2} - 2\frac{|\boldsymbol{v}_B^0| \cos\psi}{|\boldsymbol{v}_A^0|} \\ f_3 &= 1 - \frac{|\boldsymbol{v}_B^0| \cos\psi}{|\boldsymbol{v}_A^0|} \end{aligned} \tag{C.23}$$

There is an ambiguity in the definition of the sine of the angle α, but from the usage in Eq. (C.13), where the angle between $\boldsymbol{v}^0$ and $\boldsymbol{v}_B^0$ is given as $\psi + \alpha$, it follows that the signs of both angles are positive, that is, we do not distinguish between 'clockwise' and 'counterclockwise' rotations, so

$$\sin\alpha = \sqrt{1 - \cos^2\alpha} \tag{C.24}$$

The introduction of Eqs (C.17) and (C.19) into Eq. (C.12) gives the following expression for $|\boldsymbol{v}_A|$:

$$\begin{aligned} |\boldsymbol{v}_A| &= \sqrt{\left(\frac{m_B}{M}\right)^2 |\boldsymbol{v}|^2 + |\boldsymbol{V}|^2 + 2\left(\frac{m_B}{M}\right)|\boldsymbol{v}||\boldsymbol{V}| \cos\eta} \\ &= \frac{m_B}{M}|\boldsymbol{v}_A^0|\sqrt{\beta^2 f_2 + f_1 + 2\beta\sqrt{f_1 f_2}\cos\eta} \end{aligned} \tag{C.25}$$

If we introduce the relations defined above into the expression for Θ, Eq. (C.8), we get the final expression for the scattering angle Θ:

$$\boxed{\cos\Theta = \frac{\frac{m_A}{m_B} + \frac{|\boldsymbol{v}_B^0| \cos\psi}{|\boldsymbol{v}_A^0|} + \beta\sqrt{f_2}\cos\gamma}{\sqrt{\beta^2 f_2 + f_1 + 2\beta\sqrt{f_1 f_2}\cos\eta}}} \tag{C.26}$$

and for the angle ξ in Eq. (C.16):

$$\boxed{\sin\xi = \frac{\beta\sqrt{f_2}\sin\chi \sin\phi}{\sin\Theta\sqrt{\beta^2 f_2 + f_1 + 2\beta\sqrt{f_1 f_2}\cos\eta}}} \tag{C.27}$$

with $\cos\gamma$ and $\cos\eta$ given by the expressions in Eq. (C.21) and Eq. (C.22). These expressions relate the laboratory scattering angles Θ, ξ to the center-of-mass scattering angles χ, ϕ for the scattering of two molecules, including elastic ($\beta = 1$), inelastic ($\beta \neq 1$), in-plane ($\phi = 0$), and out-of-plane ($\phi \neq 0$) scattering.

C.1.4 Relation between differential cross-sections

The differential cross-section is not invariant when we change our description from one coordinate system to another, since the space angle $d\Omega$ is not. To find a relation between the differential cross-section in the laboratory coordinate system, $(d\sigma/d\Omega)_{\text{lab}}$, and in the center-of-mass coordinate system, $(d\sigma/d\Omega)_{\text{c.m.}}$, we write

$$\boxed{\begin{aligned}\left(\frac{d\sigma}{d\Omega}\right)_{\text{lab}} &= \left(\frac{d\sigma}{d\Omega}\right)_{\text{c.m.}} \left|\frac{d\Omega_{\text{c.m.}}}{d\Omega_{\text{lab}}}\right| \\ &= \left(\frac{d\sigma}{d\Omega}\right)_{\text{c.m.}} \left|\frac{\sin\chi\, d\chi\, d\phi}{\sin\Theta\, d\Theta\, d\xi}\right| \\ &= \left(\frac{d\sigma}{d\Omega}\right)_{\text{c.m.}} \left|\frac{-\sin\chi}{(d\cos\Theta/d\chi)(d\xi/d\phi)}\right| \\ &= \left(\frac{d\sigma}{d\Omega}\right)_{\text{c.m.}} \left|\frac{-\sin\chi\,\cos\xi}{(d\cos\Theta/d\chi)(d\sin\xi/d\phi)}\right|\end{aligned}} \tag{C.28}$$

Note that we have taken the absolute value of the second factor on the right-hand side of the equations since Θ and ξ are, in general, not monotonic functions of χ and ϕ, and a negative differential scattering cross-section is meaningless.

An expression for the denominator in the last relation in Eq. (C.28) is easily derived from Eqs (C.26) and (C.27), using Eqs (C.21) and (C.22). It will be convenient to introduce a shorthand notation for the numerator and denominator in Eq. (C.26) since they occur repeatedly in the final expressions. This will simplify the equations. We introduce

$$\begin{aligned} f_4 &= \frac{m_{\text{A}}}{m_{\text{B}}} + \frac{|\boldsymbol{v}_{\text{B}}^0|\cos\psi}{|\boldsymbol{v}_{\text{A}}^0|} + \beta\sqrt{f_2}\cos\gamma \\ f_5 &= \beta^2 f_2 + f_1 + 2\beta\sqrt{f_1 f_2}\cos\eta \end{aligned} \tag{C.29}$$

and then find

$$\begin{aligned}\frac{d\cos\Theta}{d\chi} &= \frac{1}{\sqrt{f_5}}\left(\frac{df_4}{d\chi}\right) - \frac{f_4}{2f_5^{3/2}}\left(\frac{df_5}{d\chi}\right) \\ &= \frac{\beta\sqrt{f_2}}{\sqrt{f_5}}\left(\frac{d\cos\gamma}{d\chi}\right) - \frac{\beta\, f_4\sqrt{f_1 f_2}}{f_5^{3/2}}\left(\frac{d\cos\eta}{d\chi}\right) \\ &= \frac{\beta\sqrt{f_2}}{\sqrt{f_5}}\left[\sin\alpha\,\cos\chi\,\cos\phi - \cos\alpha\,\sin\chi - \frac{f_4\sqrt{f_1}}{f_5}\left(\frac{d\cos\eta}{d\chi}\right)\right]\end{aligned} \tag{C.30}$$

where we have used Eq. (C.21), and from Eq. (C.22) we find that

$$\begin{aligned}\frac{d\cos\eta}{d\chi} = \frac{1}{\sqrt{f_1}}\Bigg[&-\left(\frac{m_{\text{A}}}{m_{\text{B}}}\cos\alpha + \frac{|\boldsymbol{v}_{\text{B}}^0|}{|\boldsymbol{v}_{\text{A}}^0|}[\cos\alpha\,\cos\psi - \sin\alpha\,\sin\psi]\right)\sin\chi \\ &+ \left(\frac{m_{\text{A}}}{m_{\text{B}}}\sin\alpha + \frac{|\boldsymbol{v}_{\text{B}}^0|}{|\boldsymbol{v}_{\text{A}}^0|}[\sin\alpha\,\cos\psi + \cos\alpha\,\sin\psi]\right)\cos\phi\,\cos\chi\Bigg]\end{aligned} \tag{C.31}$$

For the azimuthal angle ξ we find

$$\left(\frac{d\sin\xi}{d\phi}\right) = \frac{\beta\sqrt{f_2}\sin\chi\,\cos\phi}{\sqrt{f_5}\sin\Theta} - \frac{\beta\sqrt{f_2}\sin\chi\,\sin\phi}{f_5\sin^2\Theta}\left[\sqrt{f_5}\left(\frac{d\sin\Theta}{d\phi}\right) + \frac{\sin\Theta}{2\sqrt{f_5}}\left(\frac{df_5}{d\phi}\right)\right]$$
$$= \frac{\beta\sqrt{f_2}\sin\chi}{\sqrt{f_5}\sin\Theta}\left[\cos\phi + \left(\frac{\cos\Theta}{\sin^2\Theta}\left(\frac{d\cos\Theta}{d\phi}\right) - \frac{\beta\sqrt{f_1 f_2}}{f_5}\left(\frac{d\cos\eta}{d\phi}\right)\right)\sin\phi\right] \tag{C.32}$$

with

$$\left(\frac{d\cos\Theta}{d\phi}\right) = -\frac{\beta\sqrt{f_2}}{\sqrt{f_5}}\left(\frac{d\cos\gamma}{d\chi}\right) - \frac{\beta f_4\sqrt{f_1 f_2}}{f_5^{3/2}}\left(\frac{d\cos\eta}{d\phi}\right)$$
$$= -\frac{\beta\sqrt{f_2}}{\sqrt{f_5}}\left[\sin\alpha\,\sin\chi\,\sin\phi + \frac{f_4\sqrt{f_1}}{f_5}\left(\frac{d\cos\eta}{d\phi}\right)\right] \tag{C.33}$$

where we have used Eq. (C.21), and

$$\frac{d\cos\eta}{d\phi} = -\frac{1}{\sqrt{f_1}}\left[\frac{m_A}{m_B}\sin\alpha + \frac{|\boldsymbol{v}_B^0|}{|\boldsymbol{v}_A^0|}[\sin\alpha\,\cos\psi + \cos\alpha\,\sin\psi]\,\sin\chi\,\sin\phi\right] \tag{C.34}$$

where we have used Eq. (C.22). Equation (C.30) with Eq. (C.31), and Eq. (C.32) with Eqs (C.33) and (C.34) are introduced into Eq. (C.28) to give the relation between the differential cross-sections.

In summary, we have derived general relations between the center-of-mass scattering angles, χ and ϕ, and the laboratory scattering angles, Θ and ξ, that are valid for elastic and inelastic as well as in-plane and out-of-plane scattering events. They are also used to derive a general relation between the differential scattering cross-sections in the two coordinate systems. The central results are given in Eqs (C.26)–(C.28) with Eqs (C.20)–(C.24) and Eqs (C.30)–(C.34). The expressions may be used to convert theoretical results for the scattering event of two molecules to results that may be compared directly with experimental results.

C.1.5 The special case with $|\boldsymbol{v}_B^0|/|\boldsymbol{v}_A^0| = 0$

For the special case where $|\boldsymbol{v}_B^0|/|\boldsymbol{v}_A^0| = 0$ or is very small, that is, the target molecules are at rest before the scattering event, the relations are greatly simplified since $\boldsymbol{V}$, $\boldsymbol{v}_A^0$, and $\boldsymbol{v}^0$ are collinear. We have

$$\begin{gathered} f_1 = \left(\frac{m_A}{m_B}\right)^2, \quad f_2 = 1, \quad f_3 = 1 \\ \cos\alpha = 1, \quad \sin\alpha = 0, \quad \cos\gamma = \cos\chi, \quad \cos\eta = \cos\chi \\ f_4 = \frac{m_A}{m_B} + \beta\cos\chi, \quad f_5 = \beta^2 + \left(\frac{m_A}{m_B}\right)^2 + 2\beta\left(\frac{m_A}{m_B}\right)\cos\chi \\ \left(\frac{d\cos\eta}{d\chi}\right) = -\sin\chi, \quad \left(\frac{d\cos\eta}{d\phi}\right) = 0, \quad \left(\frac{d\cos\Theta}{d\phi}\right) = 0 \end{gathered} \tag{C.35}$$

With the target molecule at rest, $\cos\psi$ is not defined, but since it is always multiplied by the ratio of the initial speeds, which is equal to zero, it does not matter. We find

$$\cos\Theta = \frac{\frac{m_A}{m_B} + \beta\cos\chi}{\sqrt{\beta^2 + \left(\frac{m_A}{m_B}\right)^2 + 2\,\beta\left(\frac{m_A}{m_B}\right)\cos\chi}} \tag{C.36}$$

From this we may determine the sine of Θ as

$$\begin{aligned}\sin\Theta &= \sqrt{1-\cos^2\Theta}\\ &= \frac{\beta\sin\chi}{\sqrt{\beta^2 + \left(\frac{m_A}{m_B}\right)^2 + 2\,\beta\left(\frac{m_A}{m_B}\right)\cos\chi}}\end{aligned} \tag{C.37}$$

and therefore

$$\tan\Theta = \frac{\beta\sin\chi}{\frac{m_A}{m_B} + \beta\cos\chi} \tag{C.38}$$

which is identical to the expression derived in Eq. (4.55) for elastic in-plane scattering. However, the derivations show that it is also valid for out-of-plane scattering since ϕ has not been constrained to be zero.

For the azimuthal angle ξ we find

$$\begin{aligned}\sin\xi &= \frac{\beta\sin\chi\,\sin\phi}{\sin\Theta\sqrt{f_5}}\\ &= \frac{\beta\sin\chi\,\sin\phi\sqrt{f_5}}{\beta\sin\chi\sqrt{f_5}} = \sin\phi\\ \Rightarrow\quad \xi &= \phi\end{aligned} \tag{C.39}$$

where we have used Eq. (C.37).

For the relation between the differential scattering cross-sections we find

$$\begin{aligned}\frac{d\,\cos\Theta}{d\chi} &= \frac{\beta\left(\frac{m_A}{m_B}\right)\sin\chi\left[\frac{m_A}{m_B} + \beta\,\cos\chi\right]}{\left(\beta^2 + \left(\frac{m_A}{m_B}\right)^2 + 2\beta\left(\frac{m_A}{m_B}\right)\cos\chi\right)^{3/2}} - \frac{\beta\,\sin\chi}{\sqrt{\beta^2 + \left(\frac{m_A}{m_B}\right)^2 + 2\beta\,\left(\frac{m_A}{m_B}\right)\cos\chi}}\\ &= -\frac{\beta^2\sin\chi\left[\beta + \left(\frac{m_A}{m_B}\right)\cos\chi\right]}{\left(\beta^2 + \left(\frac{m_A}{m_B}\right)^2 + 2\beta\left(\frac{m_A}{m_B}\right)\cos\chi\right)^{3/2}}\end{aligned} \tag{C.40}$$

and

$$\frac{d\sin\xi}{d\phi} = \frac{\beta\sin\chi\cos\phi}{\sqrt{f_5}\sin\Theta} = \frac{\beta\sin\chi\cos\phi}{\sqrt{f_5}}\,\frac{\sqrt{f_5}}{\beta\sin\chi} = \cos\phi = \cos\xi \tag{C.41}$$

Then

$$\begin{aligned}\left(\frac{d\sigma}{d\Omega}\right)_{\text{lab}} &= \left(\frac{d\sigma}{d\Omega}\right)_{\text{c.m.}}\left|\frac{-\sin\chi\cos\xi}{(d\cos\Theta/d\chi)(d\sin\xi/d\phi)}\right|\\ &= \left(\frac{d\sigma}{d\Omega}\right)_{\text{c.m.}}\left|\frac{\left(\beta^2 + \left(\frac{m_A}{m_B}\right)^2 + 2\beta\left(\frac{m_A}{m_B}\right)\cos\chi\right)^{3/2}}{\beta^2\left[\beta + \left(\frac{m_A}{m_B}\right)\cos\chi\right]}\right|\end{aligned} \tag{C.42}$$

which is identical to the expression found in the literature for in-plane scattering. However, the derivations show that it is also valid for out-of-plane scattering since ϕ is not constrained to be zero.

C.2 Reactive scattering between two molecules

We will now show that the expressions for non-reactive scattering may also, with a few changes, be applied to reactive scattering. Let us consider the reactive scattering event

$$\mathrm{A} + \mathrm{B} \rightarrow \mathrm{C} + \mathrm{D} \tag{C.43}$$

The distinction between elastic and inelastic scattering events in non-reactive scattering becomes meaningless in reactive scattering. Still, we may use the parameter β introduced in Eq. (C.2) to relate the relative speed of products C and D after the scattering to the relative speed of the reactants. For non-reactive scattering we may write the total energy of the system before and after the scattering event as

$$\begin{aligned} E_{\text{before}} &= \frac{1}{2}M|\boldsymbol{V}|^2 + \frac{1}{2}\frac{m_\mathrm{A} m_\mathrm{B}}{M}|\boldsymbol{v}^0|^2 + E_\mathrm{A}^0 + E_\mathrm{B}^0 \\ E_{\text{after}} &= \frac{1}{2}M|\boldsymbol{V}|^2 + \frac{1}{2}\frac{m_\mathrm{A} m_\mathrm{B}}{M}|\boldsymbol{v}|^2 + E_\mathrm{A} + E_\mathrm{B} \end{aligned} \tag{C.44}$$

where E_A^0 and E_B^0 are the intramolecular energies (vibrational, rotational, and electronic) before the scattering event and E_A and E_B are the energies after the scattering event. Conservation of energy requires that

$$E_{\text{after}} - E_{\text{before}} = \frac{1}{2}\frac{m_\mathrm{A} m_\mathrm{B}}{M}|\boldsymbol{v}^0|^2\left[\beta^2 - 1\right] + (E_\mathrm{A} - E_\mathrm{A}^0) + (E_\mathrm{B} - E_\mathrm{B}^0) = 0 \tag{C.45}$$

which gives the following equation for β:

$$\beta^2 = 1 - \frac{(E_\mathrm{A} - E_\mathrm{A}^0) + (E_\mathrm{B} - E_\mathrm{B}^0)}{\frac{1}{2}\frac{m_\mathrm{A} m_\mathrm{B}}{M}|\boldsymbol{v}^0|^2} \tag{C.46}$$

We see that $\beta = 1$ for elastic scattering, where $E_\mathrm{A} = E_\mathrm{A}^0$ and $E_\mathrm{B} = E_\mathrm{B}^0$. For the reactive scattering event in Eq. (C.43) we may write

$$\begin{aligned} E_{\text{before}} &= \frac{1}{2}M|\boldsymbol{V}|^2 + \frac{1}{2}\frac{m_\mathrm{A} m_\mathrm{B}}{M}|\boldsymbol{v}^0|^2 + E_\mathrm{A}^0 + E_\mathrm{B}^0 \\ E_{\text{after}} &= \frac{1}{2}M|\boldsymbol{V}|^2 + \frac{1}{2}\frac{m_\mathrm{C} m_\mathrm{D}}{M}|\boldsymbol{v}|^2 + E_\mathrm{C} + E_\mathrm{D} \end{aligned} \tag{C.47}$$

Conservation of energy gives

$$E_{\text{after}} - E_{\text{before}} = \frac{1}{2}\frac{m_\mathrm{A} m_\mathrm{B}}{M}|\boldsymbol{v}^0|^2\left[\frac{m_\mathrm{C} m_\mathrm{D}}{m_\mathrm{A} m_\mathrm{B}}\beta^2 - 1\right] + (E_\mathrm{C} + E_\mathrm{D}) - (E_\mathrm{A}^0 + E_\mathrm{B}^0) = 0 \tag{C.48}$$

which leads to the following equation for β:

$$\beta^2 = \left[1 - \frac{(E_\mathrm{C} + E_\mathrm{D}) - (E_\mathrm{A}^0 + E_\mathrm{B}^0)}{\frac{1}{2}\frac{m_\mathrm{A} m_\mathrm{B}}{M}|\boldsymbol{v}^0|^2}\right]\left[\frac{m_\mathrm{A} m_\mathrm{B}}{m_\mathrm{D} m_\mathrm{C}}\right] \tag{C.49}$$

We note that β for the inelastic non-reactive scattering event differs from β for the reactive scattering event even with $(E_\mathrm{C} + E_\mathrm{D}) - (E_\mathrm{A}^0 + E_\mathrm{B}^0) = (E_\mathrm{A} - E_\mathrm{A}^0) + (E_\mathrm{B} - E_\mathrm{B}^0)$

because of the mass factor $(m_A m_B)/(m_C m_D)$ in Eq. (C.49). For this special situation where the difference in the intramolecular energy of the molecules after and before the scattering is the same, the values of β for the two types of scattering event differ and are related by

$$\beta_{\text{reactive}} = \sqrt{\frac{m_A m_B}{m_C m_D}}\, \beta_{\text{non-reactive}} \tag{C.50}$$

Thus, even in reactive scattering it makes sense to work with the β parameter as defined in Eq. (C.2), although it is no longer related to whether the scattering event is elastic or inelastic. It may always be determined from the simulation of the scattering event as shown in Eq. (C.49).

C.2.1 Scattering angles Θ and ξ

Let us then go through the derivations leading to the results for non-reactive scattering events and detect where changes are necessary to get a result for a reactive scattering event. We begin by extending the definition of the center-of-mass velocity in Eq. (C.1) with the expression

$$\boldsymbol{V} = \left(\frac{m_C}{M}\right) \boldsymbol{v}_C + \left(\frac{m_D}{M}\right) \boldsymbol{v}_D \tag{C.51}$$

and replace the definition of the relative velocity $\boldsymbol{v}$ in Eq. (C.2) by the relative velocity of the product molecules

$$\boldsymbol{v} = \boldsymbol{v}_C - \boldsymbol{v}_D \tag{C.52}$$

We then have to decide how to report the results of the scattering event. Let us choose to follow product C, so the angle Θ will be the angle between $\boldsymbol{v}_A^0$ and $\boldsymbol{v}_C$, with χ still being the angle between $\boldsymbol{v}^0$ and $\boldsymbol{v}$. Then we replace Eq. (C.3) by

$$\begin{aligned} \boldsymbol{u}_C &= \boldsymbol{v}_C - \boldsymbol{V} \\ &= \boldsymbol{v}_C - \left(\frac{m_C}{M}\right) \boldsymbol{v}_C - \left(\frac{m_D}{M}\right) \boldsymbol{v}_D \\ &= \left(\frac{m_D}{M}\right) \boldsymbol{v} \end{aligned} \tag{C.53}$$

where we have used Eq. (C.51). Equation (C.4) is then replaced by

$$\begin{aligned} \boldsymbol{v}_C \cdot \boldsymbol{v}_A^0 &= (\boldsymbol{V} + \boldsymbol{u}_C) \cdot \boldsymbol{v}_A^0 \\ &= \left(\boldsymbol{V} + \frac{m_D}{M} \boldsymbol{v}\right) \cdot \boldsymbol{v}_A^0 \\ &= |\boldsymbol{v}_C||\boldsymbol{v}_A^0| \cos\Theta \end{aligned} \tag{C.54}$$

and Eq. (C.5) is replaced by

$$|\boldsymbol{v}_C||\boldsymbol{v}_A^0| \cos\Theta = \left(\frac{m_A}{M}\right) \boldsymbol{v}_A^0 \cdot \boldsymbol{v}_A^0 + \left(\frac{m_B}{M}\right) \boldsymbol{v}_B^0 \cdot \boldsymbol{v}_A^0 + \left(\frac{m_D}{M}\right) \boldsymbol{v} \cdot \boldsymbol{v}_A^0 \tag{C.55}$$

where we have used the last expression in Eq. (C.1) for the center-of-mass velocity $\boldsymbol{V}$. Equation (C.16) is replaced by

$$\sin\xi = \frac{m_D}{M} \frac{|\boldsymbol{v}|}{|\boldsymbol{v}_C|} \frac{\sin\chi \sin\phi}{\sin\Theta} \tag{C.56}$$

With the definitions in Eq. (C.6), we get the equation for Θ, equivalent to Eq. (C.8):

$$\cos\Theta = \frac{1}{|\boldsymbol{v}_\mathrm{C}|}\left[\left(\frac{m_\mathrm{A}}{M}\right)|\boldsymbol{v}_\mathrm{A}^0| + \left(\frac{m_\mathrm{B}}{M}\right)|\boldsymbol{v}_\mathrm{B}^0|\cos\psi + \left(\frac{m_\mathrm{D}}{M}\right)|\boldsymbol{v}|\cos\gamma\right] \tag{C.57}$$

and the following equation, equivalent to Eq. (C.12):

$$\begin{aligned}|\boldsymbol{v}_\mathrm{C}| &= \sqrt{|\boldsymbol{u}_\mathrm{C}|^2 + |\boldsymbol{V}|^2 - 2|\boldsymbol{u}_\mathrm{C}||\boldsymbol{V}|\cos(180-\eta)} \\ &= \sqrt{\left(\frac{m_\mathrm{D}}{M}\right)^2|\boldsymbol{v}|^2 + |\boldsymbol{V}|^2 + 2\left(\frac{m_\mathrm{D}}{M}\right)|\boldsymbol{v}||\boldsymbol{V}|\cos\eta}\end{aligned} \tag{C.58}$$

Equations (C.17)–(C.24) are the same as before since all are based on the initial set-up of the scattering event and the direction of $\boldsymbol{v}$ as given by the angles χ and ϕ. Equation (C.25) changes to

$$\begin{aligned}|\boldsymbol{v}_\mathrm{C}| &= \sqrt{\left(\frac{m_\mathrm{D}}{M}\right)^2|\boldsymbol{v}|^2 + |\boldsymbol{V}|^2 + 2\left(\frac{m_\mathrm{D}}{M}\right)|\boldsymbol{v}||\boldsymbol{V}|\cos\eta} \\ &= \frac{m_\mathrm{B}}{M}|\boldsymbol{v}_\mathrm{A}^0|\sqrt{\left(\frac{m_\mathrm{D}}{m_\mathrm{B}}\right)^2\beta^2 f_2 + f_1 + 2\left(\frac{m_\mathrm{D}}{m_\mathrm{B}}\right)\beta\sqrt{f_1 f_2}\cos\eta}\end{aligned} \tag{C.59}$$

so Eq. (C.57) becomes

$$\boxed{\cos\Theta = \frac{\frac{m_\mathrm{A}}{m_\mathrm{B}} + \frac{|\boldsymbol{v}_\mathrm{B}^0|\cos\psi}{|\boldsymbol{v}_\mathrm{A}^0|} + \left(\frac{m_\mathrm{D}}{m_\mathrm{B}}\right)\beta\sqrt{f_2}\cos\gamma}{\sqrt{\left(\frac{m_\mathrm{D}}{m_\mathrm{B}}\right)^2\beta^2 f_2 + f_1 + 2\left(\frac{m_\mathrm{D}}{m_\mathrm{B}}\right)\beta\sqrt{f_1 f_2}\cos\eta}}} \tag{C.60}$$

which replaces Eq. (C.26). Similarly, Eq. (C.16) becomes

$$\boxed{\sin\xi = \frac{\frac{m_\mathrm{D}}{m_\mathrm{B}}\beta\sqrt{f_2}\sin\chi\sin\phi}{\sin\Theta\sqrt{\left(\frac{m_\mathrm{D}}{m_\mathrm{B}}\right)^2\beta^2 f_2 + f_1 + 2\frac{m_\mathrm{D}}{m_\mathrm{B}}\beta\sqrt{f_1 f_2}\cos\eta}}} \tag{C.61}$$

and replaces Eq. (C.27).

C.2.2 Relation between differential cross-sections

From Eq. (C.60) it is seen that Eqs (C.29) are replaced by

$$\begin{aligned}f_4 &= \frac{m_\mathrm{A}}{m_\mathrm{B}} + \frac{|\boldsymbol{v}_\mathrm{B}^0|\cos\psi}{|\boldsymbol{v}_\mathrm{A}^0|} + \left(\frac{m_\mathrm{D}}{m_\mathrm{B}}\right)\beta\sqrt{f_2}\cos\gamma \\ f_5 &= \left(\frac{m_\mathrm{D}}{m_\mathrm{B}}\right)^2\beta^2 f_2 + f_1 + 2\left(\frac{m_\mathrm{D}}{m_\mathrm{B}}\right)\beta\sqrt{f_1 f_2}\cos\eta\end{aligned} \tag{C.62}$$

The relations in Eqs (C.30) and (C.31) are then replaced by

$$\frac{d\cos\Theta}{d\chi} = \frac{\frac{m_D}{m_B}\beta\sqrt{f_2}}{\sqrt{f_5}}\left[\sin\alpha\,\cos\chi\,\cos\phi - \cos\alpha\,\sin\chi - \frac{f_4\sqrt{f_1}}{f_5}\left(\frac{d\cos\eta}{d\chi}\right)\right] \tag{C.63}$$

$$\begin{aligned}\frac{d\cos\eta}{d\chi} = \frac{1}{\sqrt{f_1}}\Bigg[&-\left(\frac{m_A}{m_B}\cos\alpha + \frac{|\boldsymbol{v}_B^0|}{|\boldsymbol{v}_A^0|}[\cos\alpha\,\cos\psi - \sin\alpha\,\sin\psi]\right)\sin\chi \\ &+\left(\frac{m_A}{m_B}\sin\alpha + \frac{|\boldsymbol{v}_B^0|}{|\boldsymbol{v}_A^0|}[\sin\alpha\,\cos\psi + \cos\alpha\,\sin\psi]\right)\cos\phi\,\cos\chi\Bigg]\end{aligned} \tag{C.64}$$

$$\begin{aligned}\left(\frac{d\sin\xi}{d\phi}\right) = &\frac{\frac{m_D}{m_B}\beta\sqrt{f_2}\,\sin\chi}{\sqrt{f_5}\,\sin\Theta} \\ &\times\left[\cos\phi + \left(\frac{\cos\Theta}{\sin^2\Theta}\left(\frac{d\cos\Theta}{d\phi}\right) - \frac{\frac{m_D}{m_B}\beta\sqrt{f_1f_2}}{f_5}\left(\frac{d\cos\eta}{d\phi}\right)\right)\sin\phi\right]\end{aligned} \tag{C.65}$$

$$\left(\frac{d\cos\Theta}{d\phi}\right) = -\frac{\frac{m_D}{m_B}\beta\sqrt{f_2}}{\sqrt{f_5}}\left[\sin\alpha\,\sin\chi\,\sin\phi + \frac{f_4\sqrt{f_1}}{f_5}\left(\frac{d\cos\eta}{d\phi}\right)\right] \tag{C.66}$$

$$\frac{d\cos\eta}{d\phi} = -\frac{1}{\sqrt{f_1}}\left[\frac{m_A}{m_B}\sin\alpha + \frac{|\boldsymbol{v}_B^0|}{|\boldsymbol{v}_A^0|}[\sin\alpha\,\cos\psi + \cos\alpha\,\sin\psi]\sin\chi\,\sin\phi\right] \tag{C.67}$$

Equation (C.63) with Eq. (C.64), and Eq. (C.65) with Eqs (C.66) and (C.67) may then be introduced into Eq. (C.28) to determine the relation between the differential cross-sections in the center-of-mass coordinate system and in the laboratory coordinate system.

In summary, we have derived a general relation between the center-of-mass scattering angles, χ, ϕ, and the laboratory scattering angles, Θ, ξ, for reactive scattering events. The central results are given in Eqs (C.60), (C.61), and (C.28) with Eqs (C.20)–(C.24) and Eqs (C.62)–(C.67). We see that, in order to get the results for the reactive scattering process in Eq. (C.43), we may use the general results for an inelastic non-reactive scattering event with β replaced by $\beta\,(m_D/m_B)$, when we follow product molecule C. Had we followed product molecule D, we would just need to replace β by $\beta\,(m_C/m_B)$. The angles χ and ϕ, and the parameter β and differential scattering cross-section in the center-of-mass coordinate system may be obtained in a simulation of the process, and the expressions above may be used to convert the simulation results from the center-of-mass coordinate description to the laboratory coordinate system used in experiments.

C.2.3 Resolution of ambiguity in the determination of ξ

The angle ξ is not given directly in Eqs (C.27) or (C.61) but indirectly as the sine of the angle. This introduces an ambiguity in the determination of ξ, since ξ has the same sine as $\pi - \xi$. The other azimuthal angle ϕ is given in the range from 0 to 2π, and we need to resolve the ambiguity in the determination of ξ to get the correct position for the detector in an experiment. The problem is that the arcsine function returns angles in the range from $-\pi/2$ to $\pi/2$. A simple way to resolve the ambiguity is to determine whether the projection of $\boldsymbol{v}_{\mathrm{A}}$ onto the plane spanned by the initial velocities of the molecules has a component 'above' or 'below' $\boldsymbol{v}_{\mathrm{A}}^0$ in Fig. C.1.1. We introduce a Cartesian coordinate system with the x-axis along $\boldsymbol{v}_{\mathrm{A}}^0$ and the y-axis perpendicular to that vector in the plane of the initial velocities of the molecules and 'pointing' upward in the figure.

We write $\sin\xi = a$ and let the y-component of the projection of $\boldsymbol{v}_{\mathrm{A}}$ onto the plane be $v_{\mathrm{A},y}$. Then the following four cases may arise:

$$\begin{aligned}
&\text{(i) } v_{\mathrm{A},y} > 0 \text{ and } a > 0 \Longrightarrow \ \xi = \arcsin(a); \\
&\text{(ii) } v_{\mathrm{A},y} < 0 \text{ and } a > 0 \Longrightarrow \ \xi = \pi - \arcsin(a); \\
&\text{(iii) } v_{\mathrm{A},y} < 0 \text{ and } a < 0 \Longrightarrow \ \xi = \pi + |\arcsin(a)|; \\
&\text{(iv) } v_{\mathrm{A},y} > 0 \text{ and } a < 0 \Longrightarrow \ \xi = 2\pi - |\arcsin(a)|.
\end{aligned} \tag{C.68}$$

The y-component of $\boldsymbol{v}_{\mathrm{A}}$ is determined in the following way. We have from Eqs (C.1), (C.3), and (C.9) that

$$\begin{aligned}
\boldsymbol{v}_{\mathrm{A}} &= \boldsymbol{V} + \boldsymbol{u}_{\mathrm{A}} \\
&= \frac{m_{\mathrm{A}}}{M}\boldsymbol{v}_{\mathrm{A}}^0 + \frac{m_{\mathrm{B}}}{M}\boldsymbol{v}_{\mathrm{B}}^0 + \frac{m_{\mathrm{B}}}{M}\boldsymbol{v} \\
&= \frac{m_{\mathrm{A}}}{M}\boldsymbol{v}_{\mathrm{A}}^0 + \frac{m_{\mathrm{B}}}{M}\boldsymbol{v}_{\mathrm{B}}^0 + \frac{m_{\mathrm{B}}}{M}\left[\frac{|\boldsymbol{v}|}{|\boldsymbol{v}^0|}\cos\chi\,\boldsymbol{v}^0 + |\boldsymbol{v}|\sin\chi\,\cos\phi\,\hat{\mathbf{v}}_{\perp}^0 + |\boldsymbol{v}|\sin\chi\sin\phi\,\hat{\mathbf{k}}\right]
\end{aligned} \tag{C.69}$$

The y-component of $\boldsymbol{v}_{\mathrm{A}}^0$ is zero and for $\boldsymbol{v}_{\mathrm{B}}^0$, $\boldsymbol{v}^0$, and $\hat{\mathbf{v}}_{\perp}^0$ we find

$$\begin{aligned}
v_{\mathrm{B},y}^0 &= |\boldsymbol{v}_{\mathrm{B}}^0|\sin\psi \\
v_y^0 &= -|\boldsymbol{v}_{\mathrm{B}}^0|\sin\psi \\
\hat{v}_{\perp,y}^0 &= \sin(90 - \alpha) = \cos\alpha = f_3/\sqrt{f_2}
\end{aligned} \tag{C.70}$$

so the y-component of $\boldsymbol{v}_{\mathrm{A}}$ is

$$\begin{aligned}
v_{\mathrm{A},y} &= \frac{m_{\mathrm{B}}}{M}|\boldsymbol{v}_{\mathrm{B}}^0|\,\sin\psi + \frac{m_{\mathrm{B}}}{M}\left[-\frac{|\boldsymbol{v}|}{|\boldsymbol{v}^0|}|\boldsymbol{v}_{\mathrm{B}}^0|\cos\chi\,\sin\psi + |\boldsymbol{v}|\,\sin\chi\,\cos\phi\,\cos\alpha\right] \\
&= \frac{m_{\mathrm{B}}|\boldsymbol{v}_{\mathrm{A}}^0|}{M}\left[\frac{|\boldsymbol{v}_{\mathrm{B}}^0|}{|\boldsymbol{v}_{\mathrm{A}}^0|}\sin\psi - \beta\frac{|\boldsymbol{v}_{\mathrm{B}}^0|}{|\boldsymbol{v}_{\mathrm{A}}^0|}\cos\chi\,\sin\psi + \beta\sqrt{f_2}\sin\chi\,\cos\phi\,\cos\alpha\right]
\end{aligned} \tag{C.71}$$

The sign of $v_{\mathrm{A},y}$ is given by the sign of the expression in the square brackets. A positive sign means that the y-component is positive, whereas a negative sign means that it is negative. The logic presented above is used to decide the proper angle ξ.

In the case of a reactive scattering event, β is replaced by $\beta(m_{\mathrm{D}}/m_{\mathrm{B}})$, as we did above for the scattering angles and differential cross-sections.

D

Classical mechanics, coordinate transformations

When the dynamics of a molecular system is studied, we have to choose a set of coordinates to describe the system. There are many possibilities and in the following we will focus on coordinates that are convenient in the analysis of molecular collisions and chemical reactions.

D.1 Diagonalization of the internal kinetic energy

It is often useful to transform from simple Cartesian coordinates to other sets of coordinates when we study collision processes including chemical reactions. In a collision process, it is obvious that the relative positions of the reactants are relevant and not the absolute positions as given by the simple Cartesian coordinates. It is therefore customary to change from simple Cartesian coordinates to a set describing the relative motions of the atoms and the overall motion of the atoms. For the latter motion the center-of-mass motion is usually chosen. In the following we will describe a general method of transformation from Cartesian coordinates to internal coordinates and determine its effect on the expression for the kinetic energy.

A system of N particles is described by the N position coordinates $\boldsymbol{r}_1, \ldots, \boldsymbol{r}_N$ and N momenta $\boldsymbol{p}_1, \ldots, \boldsymbol{p}_N$. Here $\boldsymbol{p}_i = m_i \dot{\boldsymbol{r}}_i$, since we use Cartesian coordinates.

By a linear transformation, let us introduce a new set of coordinates $\boldsymbol{R}_1, \ldots, \boldsymbol{R}_N$:

$$\boldsymbol{R} = \boldsymbol{A}\boldsymbol{r} \tag{D.1}$$

$\boldsymbol{R}$ is a column vector with N elements $\boldsymbol{R}_i$, $\boldsymbol{r}$ is a column vector with N elements $\boldsymbol{r}_i$, and $\boldsymbol{A}$ is an $N \times N$ square matrix with constant elements A_{ij}. The new momenta $\boldsymbol{P}_i$ may be determined using the definition in Eq. (4.61). From Eq. (D.1) we obtain

$$\dot{\boldsymbol{R}} = \boldsymbol{A}\dot{\boldsymbol{r}} \tag{D.2}$$

and find from Eq. (4.61)

$$\boldsymbol{p}_i = \sum_j \left(\frac{\partial L}{\partial \dot{\boldsymbol{R}}_j}\right)\left(\frac{\partial \dot{\boldsymbol{R}}_j}{\partial \dot{\boldsymbol{r}}_i}\right) = \sum_j \boldsymbol{P}_j A_{ji} \tag{D.3}$$

In matrix form this equation may be written

$$\boldsymbol{p} = \mathbf{A}^T \boldsymbol{P} \tag{D.4}$$

where $\boldsymbol{A}^T$ is the transpose of $\boldsymbol{A}$. In Cartesian laboratory coordinates, the kinetic energy has the form

$$T_{\text{kin}} = \frac{1}{2}\boldsymbol{p}^T \boldsymbol{m}^{-1} \boldsymbol{p} \tag{D.5}$$

where $\boldsymbol{m}^{-1}$ is a diagonal square matrix with elements m_i^{-1}. In the new coordinates, the expression for the kinetic energy may be found by substitution of Eq. (D.4). We find

$$T_{\text{kin}} = \frac{1}{2}\boldsymbol{P}^T \boldsymbol{A} m^{-1} \boldsymbol{A}^T \boldsymbol{P} \tag{D.6}$$

The $N \times N$ matrix $\boldsymbol{A}m^{-1}\boldsymbol{A}^T$ will in general not be diagonal, so there will be cross terms of the kind $P_i P_j$ in the expression for the kinetic energy. This may sometimes be inconvenient, and we shall see in the following how one may choose the matrix $\boldsymbol{A}$ in such a way that the kinetic energy is still diagonal in the new momenta. This leads to the so-called *Jacobi coordinates* that are often used in reaction dynamics calculations.

First we want to single out the overall motion of the system, where all atoms move by the same amount, so all distances will be preserved. This is done by introducing the following condition on the matrix elements in $\boldsymbol{A}$:

$$\sum_i^N A_{ki} = \delta_{k,N} \tag{D.7}$$

If all particles are displaced by the amount $\boldsymbol{u}$ to the position $\boldsymbol{r}_i + \boldsymbol{u}$, then coordinates $\boldsymbol{R}_1, \ldots, \boldsymbol{R}_{N-1}$ are seen from Eq. (D.1) not to change because of the condition in Eq. (D.7), while $\boldsymbol{R}_N$ is displaced by $\boldsymbol{u}$. $\boldsymbol{R}_1, \ldots, \boldsymbol{R}_{N-1}$ are thus internal coordinates, whereas $\boldsymbol{R}_N$ describes the overall position of the system. For the momenta, we get

$$\sum_{i=1}^N \boldsymbol{p}_i = \sum_{i=1}^N \sum_{k=1}^N \boldsymbol{P}_k A_{ki} = \sum_{k=1}^N \boldsymbol{P}_k \delta_{k,N} = \boldsymbol{P}_N \tag{D.8}$$

The moment $\boldsymbol{P}_N$, conjugate to coordinate $\boldsymbol{R}_N$, is therefore the total momentum of the system.

Usually, $\boldsymbol{R}_N$ is chosen as the center-of-mass coordinate:

$$\boldsymbol{R}_N = \sum_{i=1}^N \frac{m_i}{M} \boldsymbol{r}_i, \quad M = \sum_{i=1}^N m_i \tag{D.9}$$

so the elements in the last row of the matrix $\boldsymbol{A}$ are

$$A_{Ni} = \frac{m_i}{M} \tag{D.10}$$

If we develop the matrix product in Eq. (D.6) then we get

$$T_{\text{kin}} = \sum_{k=1}^{N-1} \sum_{l=1}^{N-1} \left(\sum_{i=1}^N \frac{1}{2m_i} A_{ki} A_{li} \right) \boldsymbol{P}_k \cdot \boldsymbol{P}_l + T_{\text{c.m.}} \tag{D.11}$$

The kinetic energy has been divided into two contributions: an internal kinetic energy (the first term) and an external center-of-mass kinetic energy $T_{\text{c.m.}}$, where

$$T_{\text{c.m.}} = \sum_{i=1}^{N} \frac{1}{2m_i} \frac{m_i^2}{M^2} P_N^2 = \frac{P_N^2}{2M} \tag{D.12}$$

The condition for the internal kinetic energy to be 'diagonal', so that there will be no cross term, is seen from Eq. (D.11) to be

$$\sum_{i=1}^{N} \frac{1}{2m_i} A_{ki} A_{li} = 0 \quad \text{for} \quad k \neq l \quad \text{and} \quad k, l \neq N \tag{D.13}$$

If we introduce the notation

$$\sum_{i=1}^{N} \frac{1}{2m_i} A_{ki}^2 = \frac{1}{2M_k} \tag{D.14}$$

then the internal kinetic energy T_{int} will be

$$T_{\text{int}} = \sum_{k=1}^{N-1} \frac{1}{2M_k} P_k^2 \tag{D.15}$$

Such internal coordinates for which the internal kinetic energy is diagonal are called generalized *Jacobi coordinates* when more than two particles are considered.

D.1.1 An example

In this example we shall describe a systematic way to choose the matrix elements such that the kinetic energy will be 'diagonal', that is, with no cross terms. The procedure is to couple the particles with the aid of the center-of-mass coordinates for larger and larger clusters of particles. Let us illustrate the method for a five-particle system. We start with particles one and two. For the first row in $\boldsymbol{A}$ we find, using Eq. (D.7),

$$\text{A}_{1i}: \quad -\alpha_1 \quad \alpha_1 \quad 0 \quad 0 \quad 0$$

where α_1 is a constant. Then, using Eq. (D.13) for rows one and two, and Eq. (D.7) for row two, we get

$$\text{A}_{2i}: \quad -\alpha_2 \tfrac{m_1}{s_2} \quad -\alpha_2 \tfrac{m_2}{s_2} \quad \alpha_2 \quad 0 \quad 0, \qquad s_2 = m_1 + m_2$$

Particles one and two are now coupled in the center-of-mass coordinates for those two particles. This center is now coupled to particle three. Again we use Eq. (D.13) for rows two and three, and Eq. (D.7) for row three to get

$$\text{A}_{3i}: \quad -\alpha_3 \tfrac{m_1}{s_3} \quad -\alpha_3 \tfrac{m_2}{s_3} \quad -\alpha_3 \tfrac{m_3}{s_3} \quad \alpha_3 \quad 0, \qquad s_3 = m_1 + m_2 + m_3$$

The particles one, two, and three are now coupled and we continue with particle four in the same way:

$$\text{A}_{4i}: \quad -\alpha_4 \tfrac{m_1}{s_4} \quad -\alpha_4 \tfrac{m_2}{s_4} \quad -\alpha_4 \tfrac{m_3}{s_4} \quad -\alpha_4 \tfrac{m_4}{s_4} \quad \alpha_4, \qquad s_4 = m_1 + m_2 + m_3 + m_4$$

The fifth row is finally given by

$$\mathrm{A}_{5i}: \quad \frac{m_1}{s_5} \quad \frac{m_2}{s_5} \quad \frac{m_3}{s_5} \quad \frac{m_4}{s_5} \quad \frac{m_5}{s_5}, \qquad s_5 = m_1 + m_2 + m_3 + m_4 + m_5$$

which is the center-of-mass coordinate for the five-particle system. Then, from Eq. (D.14) we find the effective masses for the internal coordinates:

$$\text{Coordinate } \boldsymbol{R}_1: \quad \frac{1}{2M_1} = \frac{\alpha_1^2}{2}\left[\frac{1}{m_1} + \frac{1}{m_2}\right]$$

$$\text{Coordinate } \boldsymbol{R}_2: \quad \frac{1}{2M_2} = \frac{\alpha_2^2}{2}\left[\frac{1}{m_1}\left(\frac{m_1}{s_2}\right)^2 + \frac{1}{m_2}\left(\frac{m_2}{s_2}\right)^2 + \frac{1}{m_3}\right]$$

$$\text{Coordinate } \boldsymbol{R}_3: \quad \frac{1}{2M_3} = \frac{\alpha_3^2}{2}\left[\frac{1}{m_1}\left(\frac{m_1}{s_3}\right)^2 + \frac{1}{m_2}\left(\frac{m_2}{s_3}\right)^2 + \frac{1}{m_3}\left(\frac{m_3}{s_3}\right)^2 + \frac{1}{m_4}\right]$$

Coordinate $\boldsymbol{R}_4$:

$$\frac{1}{2M_4} = \frac{\alpha_4^2}{2}\left[\frac{1}{m_1}\left(\frac{m_1}{s_4}\right)^2 + \frac{1}{m_2}\left(\frac{m_2}{s_4}\right)^2 + \frac{1}{m_3}\left(\frac{m_3}{s_4}\right)^2 + \frac{1}{m_4}\left(\frac{m_4}{s_4}\right)^2 + \frac{1}{m_5}\right]$$

Here we recognize the mass associated with coordinate $\boldsymbol{R}_1$ as the reduced mass M_1 for a two-particle system (as in Eq. (2.25)). With these *generalized Jacobi* coordinates, the internal kinetic energy has the simple form without cross terms according to Eq. (D.15):

$$T_{\text{int}} = \frac{P_1^2}{2M_1} + \frac{P_2^2}{2M_2} + \frac{P_3^2}{2M_3} + \frac{P_4^2}{2M_4} \tag{D.16}$$

Often one chooses to set $\alpha_1 = \alpha_2 = \alpha_3 = \alpha_4 = 1$ in the expressions for the effective masses. Another possibility would be to choose values of α_i so that all the effective masses will be equal to one, that is,

$$\begin{aligned} \alpha_1 &= \left[\frac{1}{m_1} + \frac{1}{m_2}\right]^{-1/2} \\ \alpha_2 &= \left[\frac{1}{m_1}\left(\frac{m_1}{s_2}\right)^2 + \frac{1}{m_2}\left(\frac{m_2}{s_2}\right)^2 + \frac{1}{m_3}\right]^{-1/2} \end{aligned} \tag{D.17}$$

and similarly for the other αs. Then the internal kinetic energy will have the simple form

$$T_{\text{int}} = \frac{P_1^2}{2} + \frac{P_2^2}{2} + \frac{P_3^2}{2} + \frac{P_4^2}{2} \tag{D.18}$$

When the potential energy, $V = V(\boldsymbol{R}_1, \boldsymbol{R}_2, \boldsymbol{R}_3, \boldsymbol{R}_4)$, only depends on the internal coordinates, Hamilton's equations of motion are (use Eq. (4.63))

$$\begin{aligned} \dot{\boldsymbol{P}}_1 &= -\frac{\partial V}{\partial \boldsymbol{R}_1} \\ \dot{\boldsymbol{P}}_2 &= -\frac{\partial V}{\partial \boldsymbol{R}_2} \\ \dot{\boldsymbol{P}}_3 &= -\frac{\partial V}{\partial \boldsymbol{R}_3} \\ \dot{\boldsymbol{P}}_4 &= -\frac{\partial V}{\partial \boldsymbol{R}_4} \end{aligned} \tag{D.19}$$

and

$$\begin{aligned}\dot{\boldsymbol{R}}_1 &= \frac{P_1}{M_1} \quad \text{or} \quad P_1\\ \dot{\boldsymbol{R}}_2 &= \frac{P_2}{M_2} \quad \text{or} \quad P_2\\ \dot{\boldsymbol{R}}_3 &= \frac{P_3}{M_3} \quad \text{or} \quad P_3\\ \dot{\boldsymbol{R}}_4 &= \frac{P_4}{M_4} \quad \text{or} \quad P_4\end{aligned} \tag{D.20}$$

where the first case on the right-hand side of the equality sign refers to the kinetic energy expression in Eq. (D.16) and the second case to the expression in Eq. (D.18).

D.1.2 Mass-weighted skewed angle coordinate systems

Sometimes a mass-weighted skewed angle coordinate system rather than a rectangular system is used to plot the potential energy surface and the trajectories for a simple triatomic reaction like

$$\mathrm{A} + \mathrm{BC} \rightarrow \mathrm{AB} + \mathrm{C} \tag{D.21}$$

For simplicity, we assume that the collision is collinear, say along the x-axis. The atoms A, B, and C are numbered 1, 2, and 3, respectively, to harmonize the notation with the previous section. Then, according to the analysis above, the following internal Jacobi coordinates are consistent with a 'diagonal' kinetic energy:

$$\begin{aligned}X_1 &= \alpha_1(x_2 - x_1)\\ X_2 &= \alpha_2\left[(x_3 - x_2) + \frac{m_1}{s_2}(x_2 - x_1)\right]\end{aligned} \tag{D.22}$$

We note that the coordinate X_1 directly reflects the distance between atoms one and two, whereas the coordinate X_2 reflects a combination of both distances. Therefore, a knowledge of the two coordinates does not directly tell us what the distances are between the involved atoms. Also, for the potential energy function in a collinear collision, the natural variables will be the distances between atoms A and B and atoms B and C. These variables appear as the components along a new set of coordinate axes, if instead of a rectangular coordinate system we use a mass-weighted skewed angle coordinate system.

Let X_1 and X_2 be the coordinates in a rectangular Cartesian coordinate system with X_1 along the ordinate axis and X_2 along the abscissa axis. One of the axes in the new coordinate system is now chosen to be collinear with the abscissa axis in the rectangular coordinate system, and we let the coordinate along this axis be the first term in the expression for X_2 in Eq. (D.22), namely $\alpha_2(x_3 - x_2)$. The situation is sketched in Fig. D.1.1. The other axis with a coordinate proportional to the other distance $x_2 - x_1$ forms an angle ϕ with the first. This angle is determined from the requirements that the projections of this coordinate on the X_1 coordinate axis is $\alpha_1(x_2 - x_1)$ and on the X_2 coordinate axis is $\alpha_2 m_1(x_2 - x_1)/s_2$. If we let the proportionality constant of $x_2 - x_1$ be β, then we have

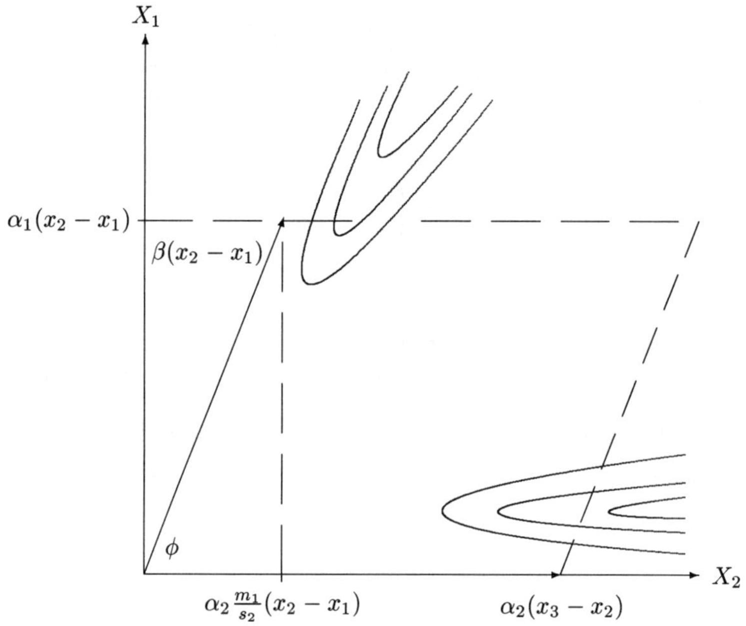

Fig. D.1.1 Sketch of an ordinary Cartesian coordinate system and the associated mass-weighted skewed angle coordinate system.

$$\begin{aligned}\alpha_1(x_2 - x_1) &= \beta(x_2 - x_1)\sin(\phi) \\ \alpha_2\frac{m_1}{s_2}(x_2 - x_1) &= \beta(x_2 - x_1)\cos(\phi)\end{aligned} \tag{D.23}$$

Thus

$$\tan(\phi) = \frac{\alpha_1 s_2}{m_1 \alpha_2} \tag{D.24}$$

and

$$\beta^2 = \alpha_1^2 + \alpha_2^2\left(\frac{m_1}{s_2}\right)^2 \tag{D.25}$$

For the case where $\alpha_1 = \alpha_2 = 1$, we find

$$\begin{aligned}\tan(\phi) &= \frac{s_2}{m_1} \\ \beta &= \left[1 + \left(\frac{m_1}{s_2}\right)^2\right]^{1/2} \\ \alpha_2 &= 1\end{aligned} \tag{D.26}$$

and for the case where all reduced masses are equal to one, the α values are given in Eq. (D.17) and we find

$$\tan(\phi) = \frac{m_2 s_3}{m_1 m_3}$$
$$\beta = \left[\frac{m_1(m_2+m_3)}{s_3}\right]^{1/2} \qquad \text{(D.27)}$$
$$\alpha_2 = \left[\frac{m_3 s_2}{s_3}\right]^{1/2}$$

If we consider the potential energy as a function of the Jacobi coordinates X_1 and X_2 and draw the energy contours in the X_1–X_2 plane, then the entrance and exit valleys will asymptotically be at an angle ϕ to one another and in the mass-weighted skewed angle coordinate system parallel to its axes. So the idea with this coordinate system is that it allows us to directly determine the atomic distances as they develop in time and that it shows us the asymptotic directions of the entrance and exit channels.

It is important to note that the dynamics is done in the Jacobi coordinates, because they make the equations of motion (see Eqs (D.19) and (D.20)) particularly simple. The idea of choosing the αs as in Eq. (D.17) is that all masses M_k equal one, so the equations of motion are analogous to those for a particle of mass one with coordinates X_1 and X_2.

In fact, we can construct a mechanical analogue to the trajectories generated by the equations of motion by rolling a particle of mass one on a hard surface under the influence of gravity. The topography of this analogue surface is closely related to that of the potential energy surface. It is determined by realizing that we have to transform from a potential energy surface to a position surface or 'mountain landscape', so the energy axis is converted into a position axis with a coordinate giving the 'height' of the particle above the X_1–X_2 plane. In Fig. D.1.2(a) a cut through the potential energy surface and the force at some point X_0 are shown. In Fig. D.1.2(b) the same cut through the 'mountain landscape' of the analogue surface is shown. The tangential component F_g of the gravity force mg (here $m = 1$) causes the particle to move on the

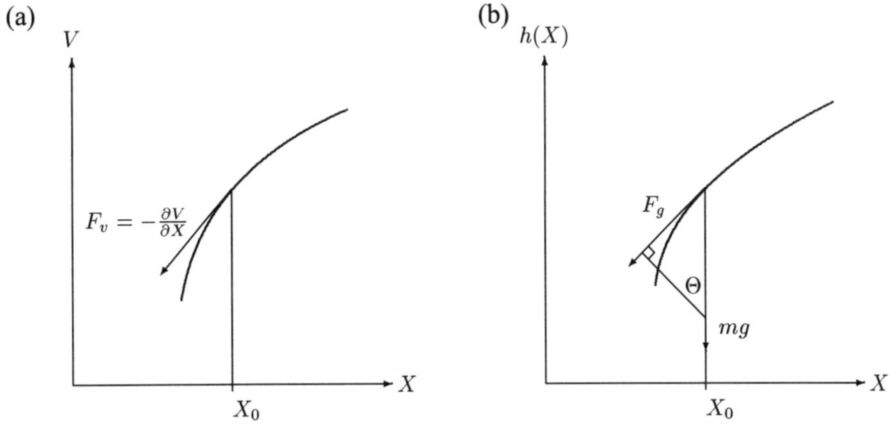

Fig. D.1.2 (a) A cut through the potential energy surface V. (b) The same cut as in (a) through the analogue 'physical surface' $h(X)$.

analogue surface. We can construct the analogue surface, $h(X)$, by requiring that this force shall equal the force derived from the potential energy function at any point; that is,

$$F_g = -mg \sin\theta = -\frac{\partial V}{\partial X} \tag{D.28}$$

or

$$\sin\theta = \frac{1}{mg}\frac{\partial V}{\partial X} = \frac{\partial \frac{V}{mg}}{\partial X} \tag{D.29}$$

Thus, the analogue surface should be constructed in such a way that θ at any point X_1, X_2 satisfies Eq. (D.29).

Further reading/references

[1] J.O. Hirschfelder, *Int. J. Quant. Chem.* **3**, 17 (1969)

E

Small-amplitude vibrations, normal-mode coordinates

In the following we show that a simple description of the (quantum or classical) dynamics can be obtained in a multidimensional system close to a stationary point. Thus, the system can be described by a set of uncoupled *harmonic oscillators*. The formalism is related to the generalization of the harmonic expansion in Eq. (1.7) to multidimensional systems.

E.1 Diagonalization of the potential energy

We consider a potential energy surface expressed in Cartesian laboratory coordinates, q_i $(i = 1, \ldots, n)$, for a system of $n = 3N$ (where N is the number of nuclei) degrees of freedom. A Taylor expansion of the potential V around the point $(q_1^0, \ldots, q_n^0)$ gives

$$V(q_1, \ldots, q_n) = V(q_1^0, \ldots, q_n^0) + \sum_i \left(\frac{\partial V}{\partial q_i}\right)_0 \eta_i' + 1/2 \sum_{i,j} \left(\frac{\partial^2 V}{\partial q_i \partial q_j}\right)_0 \eta_i' \eta_j' + \cdots \tag{E.1}$$

where $\eta_i' = q_i - q_i^0$ is the displacement from the point of expansion.

We now assume that the expansion of the potential is around a *stationary point* (stable or unstable, depending on the sign of the second-order derivatives), that is, all the first-order derivatives vanish. The energy is measured relative to the value at equilibrium, and we obtain

$$\begin{aligned} V(q_1, \ldots, q_n) &= 1/2 \sum_{i,j} \left(\frac{\partial^2 V}{\partial q_i \partial q_j}\right)_0 \eta_i' \eta_j' \\ &= 1/2 \sum_{i,j} \eta_i' V_{ij}' \eta_j' \\ &= 1/2 (\boldsymbol{\eta}')^T \boldsymbol{V}' \boldsymbol{\eta}' \end{aligned} \tag{E.2}$$

where $(\boldsymbol{\eta}')^T$ ('T' for transpose) and $\boldsymbol{\eta}'$ are row and column vectors, respectively, and $\boldsymbol{V}'$ is an $n \times n$ matrix.

We introduce mass-weighted displacement coordinates

$$\eta_i = \eta_i' \sqrt{m_i} \tag{E.3}$$

In these coordinates, the potential takes the form

$$V = 1/2\boldsymbol{\eta}^T \boldsymbol{F}\boldsymbol{\eta} \tag{E.4}$$

where $F_{ij} = V'_{ij}/(\sqrt{m_i}\sqrt{m_j})$ is the symmetric mass-weighted *force constant matrix*.

In the potential of Eq. (E.4), we still find that all the coordinates are coupled, i.e., it contains off-diagonal terms of the form $\eta_i\eta_j$ with $i \neq j$. However, since the potential is a quadratic form, we know from mathematics that it is possible to introduce a linear transformation of the coordinates such that the potential takes a diagonal form in the new coordinates. To that end, *normal-mode coordinates*, $\boldsymbol{Q}$, are introduced by the following linear transformation of the mass-weighted displacement coordinates:

$$\boldsymbol{\eta} = \boldsymbol{L}\boldsymbol{Q} \tag{E.5}$$

where $\boldsymbol{L}$ is an $n \times n$ matrix. The potential can now be written in the form

$$\begin{aligned} V &= 1/2\boldsymbol{\eta}^T \boldsymbol{F}\boldsymbol{\eta} \\ &= 1/2\boldsymbol{Q}^T\boldsymbol{L}^T\boldsymbol{F}\boldsymbol{L}\boldsymbol{Q} \end{aligned} \tag{E.6}$$

and we are going to determine $\boldsymbol{L}$ such that the matrix $\boldsymbol{L}^T\boldsymbol{F}\boldsymbol{L}$ becomes diagonal. The columns of $\boldsymbol{L}$ are now chosen as eigenvectors of the matrix $\boldsymbol{F}$. Thus,

$$\boldsymbol{F}\begin{pmatrix} L_{1j} \\ \vdots \\ L_{nj} \end{pmatrix} = \omega_j^2 \begin{pmatrix} L_{1j} \\ \vdots \\ L_{nj} \end{pmatrix} \tag{E.7}$$

where ω_j^2 $(j = 1, \ldots, n)$ are the corresponding eigenvalues, which are determined as roots to the equation $|\boldsymbol{F} - \omega^2\mathbf{I}| = 0$. Since $\boldsymbol{F}$ is real and symmetric, we know from matrix theory that the column vectors of $\boldsymbol{L}$ are mutually orthogonal. If the column vectors are normalized to unit length, then $\boldsymbol{L}$ becomes an orthogonal matrix. The inverse of an orthogonal matrix is obtained by transponation, $\boldsymbol{L}^{-1} = \boldsymbol{L}^T$, that is, $\boldsymbol{L}^T\boldsymbol{L} = \mathbf{I}$ or equivalently $\sum_i (\boldsymbol{L})_{il}(\boldsymbol{L})_{ik} = \delta_{lk}$.

Equation (E.7) can be written in the form

$$\boldsymbol{F}\boldsymbol{L} = \boldsymbol{L}\boldsymbol{\omega}^2 \tag{E.8}$$

where $\boldsymbol{\omega}^2$ is a diagonal matrix with the n eigenvalues along the diagonal (note that $\boldsymbol{L}\boldsymbol{\omega}^2 \neq \boldsymbol{\omega}^2\boldsymbol{L}$) and

$$\begin{aligned} \boldsymbol{\omega}^2 &= \boldsymbol{L}^{-1}\boldsymbol{F}\boldsymbol{L} \\ &= \boldsymbol{L}^T\boldsymbol{F}\boldsymbol{L} \end{aligned} \tag{E.9}$$

since the inverse of $\boldsymbol{L}$ is obtained by transponation. Using Eqs (E.6) and (E.9), we obtain

$$\begin{aligned} V &= 1/2\boldsymbol{Q}^T\boldsymbol{L}^T\boldsymbol{F}\boldsymbol{L}\boldsymbol{Q} \\ &= 1/2\boldsymbol{Q}^T\boldsymbol{\omega}^2\boldsymbol{Q} \\ &= \sum_{s=1}^{n}(1/2)\omega_s^2 Q_s^2 \end{aligned} \tag{E.10}$$

We have now obtained the desired diagonal form of the potential energy. If all the frequencies, ω_s^2, are positive then the stationary point represents a minimum on

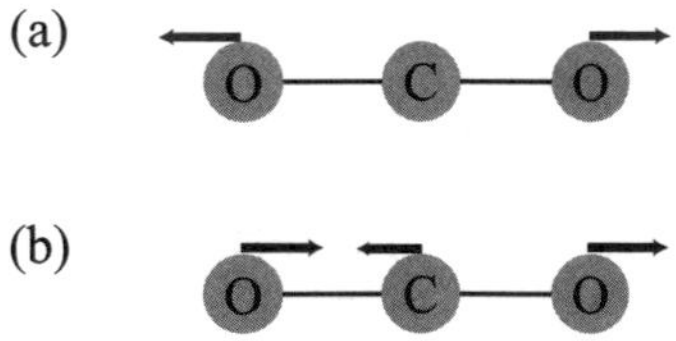

Fig. E.1.1 (a) Symmetric and (b) anti-symmetric stretch in a symmetric molecule like CO_2. Note that the C atom does not participate in the symmetric stretch motion.

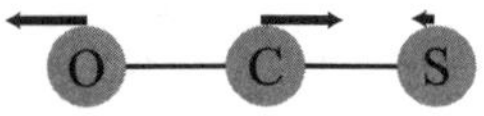

Fig. E.1.2 One of the two stretch modes in OCS. Note that this mode, essentially, corresponds to a pure CO vibration.

the potential energy surface. If, on the other hand, one or more frequencies, ω_k^2, are negative (which implies that ω_k is imaginary) then the potential corresponds to an inverted harmonic potential in that mode and the associated motion is not oscillatory but unbound. A saddle point is an example of such an unstable point.

In practice, from a given potential we first calculate the mass-weighted force constant matrix $\boldsymbol{F}$ (Eq. (E.4)). The eigenvalues of this matrix give the normal-mode frequencies (Eq. (E.7)). The corresponding eigenvectors give, according to Eq. (E.5), the normal-mode coordinates expressed as a linear combination of atomic (mass-weighted) displacement coordinates; thus, $\boldsymbol{Q} = \boldsymbol{L}^T \boldsymbol{\eta}$. The normal modes are often presented in graphical form by 'arrows' that represent (the magnitude and sign of) the coefficients in the linear combinations of the atomic displacement coordinates.

Some examples are given in Figs E.1.1 and E.1.2, for triatomic molecules. A linear triatomic molecule has four $(3 \times 3 - 5)$ vibrational modes: two bond-stretching modes and two (degenerate) bending modes. Fig. E.1.2 shows one of the four normal modes in OCS.

E.2 Transformation of the kinetic energy

As shown above, the potential energy can be expressed as a sum of harmonic potentials. We now consider the expression for the kinetic energy in classical as well as quantum mechanical form. In classical mechanics,

$$\begin{aligned} T &= 1/2 \sum_i m_i \dot{q}_i \dot{q}_i \\ &= 1/2 \sum_i m_i \dot{\eta}_i' \dot{\eta}_i' \\ &= 1/2 \dot{\boldsymbol{\eta}}^T \dot{\boldsymbol{\eta}} \end{aligned} \tag{E.11}$$

where we have introduced mass-weighted coordinates according to Eq. (E.3). Expressed in normal-mode coordinates, the kinetic energy takes the form

$$\begin{aligned} T &= 1/2\dot{\boldsymbol{\eta}}^T\dot{\boldsymbol{\eta}} \\ &= 1/2\dot{\boldsymbol{Q}}^T\boldsymbol{L}^{-1}\boldsymbol{L}\dot{\boldsymbol{Q}} \\ &= 1/2\sum_{s=1}^{n}\dot{Q}_s^2 \end{aligned} \tag{E.12}$$

Thus, from Eqs (E.12) and (E.10) we see that the classical dynamics of the normal modes is just the dynamics of n *uncoupled* harmonic oscillators.

In quantum mechanics the kinetic energy is represented by the operator

$$\begin{aligned} \hat{T} &= -\sum_i \frac{\hbar^2}{2m_i}\frac{\partial^2}{\partial q_i^2} \\ &= -\sum_i \frac{\hbar^2}{2}\frac{\partial^2}{\partial \eta_i^2} \end{aligned} \tag{E.13}$$

where the relation $\partial/\partial q_i = \sqrt{m_i}\partial/\partial\eta_i$ was used in order to derive the second line. Since $Q_l = \sum_k(\boldsymbol{L}^T)_{lk}\eta_k$, the chain rule gives $\partial/\partial\eta_i = \sum_l(\partial Q_l/\partial\eta_i)(\partial/\partial Q_l) = \sum_l(\boldsymbol{L})_{il}\partial/\partial Q_l$ and

$$\begin{aligned} \hat{T} &= -\sum_i \frac{\hbar^2}{2}\frac{\partial^2}{\partial \eta_i^2} \\ &= -\sum_{l,k}\frac{\hbar^2}{2}\left(\sum_i(\boldsymbol{L})_{il}(\boldsymbol{L})_{ik}\right)\frac{\partial}{\partial Q_l}\frac{\partial}{\partial Q_k} \\ &= -\sum_{s=1}^{n}\frac{\hbar^2}{2}\frac{\partial^2}{\partial Q_s^2} \end{aligned} \tag{E.14}$$

Thus, with this result for the kinetic energy and Eq. (E.10) for the potential energy, we conclude that the quantum dynamics of the normal modes is just the dynamics of n *uncoupled* harmonic oscillators; that is,

$$\boxed{\hat{H} = -\sum_{s=1}^{n}\frac{\hbar^2}{2}\frac{\partial^2}{\partial Q_s^2} + \sum_{s=1}^{n}(1/2)\omega_s^2 Q_s^2} \tag{E.15}$$

The total wave function can, accordingly, be written as a product of wave functions corresponding to each mode. The energy eigenfunctions corresponding to each mode are, in particular, just the well-known eigenfunctions for a harmonic oscillator.

E.3 Transformation of phase-space volumes

We consider here the relation between volume elements in phase space; in particular, the relation between $d\boldsymbol{q}d\boldsymbol{p}$ and $d\boldsymbol{Q}d\boldsymbol{P}$, where $d\boldsymbol{q} = dq_1\cdots dq_n$ refers to Cartesian coordinates in a laboratory fixed coordinate system, $d\boldsymbol{Q} = dQ_1\cdots dQ_n$ refers to normal-mode coordinates, and $\boldsymbol{p}$ and $\boldsymbol{P}$ are the associated generalized conjugate momenta.

For coordinate transformations we generally have the following relation between the volume elements: $d\boldsymbol{q} = |J| d\boldsymbol{Q}$, where $|J|$ is the absolute value of the Jacobian J, which is given by the determinant

$$J = \begin{vmatrix} \partial q_1/\partial Q_1 & \partial q_1/\partial Q_2 & \cdots & \partial q_1/\partial Q_n \\ \partial q_2/\partial Q_1 & \partial q_2/\partial Q_2 & \cdots & \cdots \\ \vdots & & & \\ \partial q_n/\partial Q_1 & \cdots & \cdots & \partial q_n/\partial Q_n \end{vmatrix} \tag{E.16}$$

From Eq. (E.5), we get $q_i = q_i^0 + \frac{1}{\sqrt{m_i}} \sum_j L_{ij} Q_j$, and therefore

$$\partial q_i/\partial Q_j = \frac{1}{\sqrt{m_i}} L_{ij} \tag{E.17}$$

so

$$\begin{aligned} J &= \begin{vmatrix} \frac{1}{\sqrt{m_1}} L_{11} & \frac{1}{\sqrt{m_1}} L_{12} & \cdots & \frac{1}{\sqrt{m_1}} L_{1n} \\ \frac{1}{\sqrt{m_2}} L_{21} & \frac{1}{\sqrt{m_2}} L_{22} & \cdots & \cdots \\ \vdots & & & \\ \frac{1}{\sqrt{m_n}} L_{n1} & \cdots & \cdots & \frac{1}{\sqrt{m_n}} L_{nn} \end{vmatrix} \\ &= \left| \begin{pmatrix} \frac{1}{\sqrt{m_1}} & 0 & \cdots & 0 \\ 0 & \frac{1}{\sqrt{m_2}} & 0 & \cdots \\ \vdots & & & \\ 0 & \cdots & \cdots & \frac{1}{\sqrt{m_n}} \end{pmatrix} \begin{pmatrix} L_{11} & L_{12} & \cdots & L_{1n} \\ L_{21} & L_{22} & \cdots & \cdots \\ \vdots & & & \\ L_{n1} & \cdots & \cdots & L_{nn} \end{pmatrix} \right| \\ &= \frac{1}{\sqrt{m_1 m_2 \cdots m_n}} |\boldsymbol{L}| \end{aligned} \tag{E.18}$$

Since $\boldsymbol{L}$ is an orthogonal matrix, $|\boldsymbol{L}\boldsymbol{L}^T| = |\mathbf{I}|$, i.e., $|\boldsymbol{L}| \cdot |\boldsymbol{L}^T| = 1$, which implies that $|\boldsymbol{L}| = \pm 1$. That is,

$$d\boldsymbol{q} = \frac{1}{\sqrt{m_1 m_2 \cdots m_n}} d\boldsymbol{Q} \tag{E.19}$$

is the relation between the volume elements in configuration space.

In momentum space, we have the similar relation $d\boldsymbol{p} = |J| d\boldsymbol{P}$. The Jacobian is given by a determinant similar to the one in Eq. (E.16), now with the elements $\partial p_i/\partial P_j$. The momenta are defined by Eq. (4.61), that is, $p_i = \partial L/\partial \dot{q}_i$ and $P_j = \partial L/\partial \dot{Q}_j$, where L is the Lagrange function. The relation between the two sets of momenta is

$$\begin{aligned} p_i &= \partial L/\partial \dot{q}_i \\ &= \sum_j \frac{\partial L}{\partial \dot{Q}_j} \frac{\partial \dot{Q}_j}{\partial \dot{q}_i} \\ &= \sum_j P_j \frac{\partial \dot{Q}_j}{\partial \dot{q}_i} \end{aligned} \tag{E.20}$$

From Eq. (E.5), we get $\dot{\boldsymbol{Q}} = \boldsymbol{L}^T \dot{\boldsymbol{\eta}}$, that is, $\dot{Q}_j = \sum_{i'} L_{i'j} \sqrt{m_{i'}} \dot{q}_{i'}$. Thus,

$$\partial p_i / \partial P_j = \sqrt{m_i} L_{ij} \tag{E.21}$$

and

$$J = \begin{vmatrix} \sqrt{m_1} L_{11} & \sqrt{m_1} L_{12} & \cdots & \sqrt{m_1} L_{1n} \\ \sqrt{m_2} L_{21} & \sqrt{m_2} L_{22} & \cdots & \cdots \\ \vdots & & & \\ \sqrt{m_n} L_{n1} & \cdots & \cdots & \sqrt{m_n} L_{nn} \end{vmatrix}$$

$$= \left| \begin{pmatrix} \sqrt{m_1} & 0 & \cdots & 0 \\ 0 & \sqrt{m_2} & 0 & \cdots \\ \vdots & & & \\ 0 & \cdots & \cdots & \sqrt{m_n} \end{pmatrix} \begin{pmatrix} L_{11} & L_{12} & \cdots & L_{1n} \\ L_{21} & L_{22} & \cdots & \cdots \\ \vdots & & & \\ L_{n1} & \cdots & \cdots & L_{nn} \end{pmatrix} \right|$$

$$= \sqrt{m_1 m_2 \cdots m_n}\, |\boldsymbol{L}| \tag{E.22}$$

Thus,

$$d\boldsymbol{p} = \sqrt{m_1 m_2 \cdots m_n} d\boldsymbol{P} \tag{E.23}$$

and finally,

$$d\boldsymbol{q} d\boldsymbol{p} = d\boldsymbol{Q} d\boldsymbol{P} \tag{E.24}$$

demonstrating the invariance of the volume element in phase space. It can be shown that this invariance of the volume element in phase space holds in general, i.e., for *any* two sets of coordinates and their conjugate momenta. That is, $\boldsymbol{q}$ and $\boldsymbol{Q}$ can be any two sets of coordinates that describe the same point.

Further reading/references

[1] H. Goldstein, *Classical mechanics*, second edition (Addison–Wesley, 1980)

[2] E.B. Wilson, J.C. Decius, and P.C. Cross, *Molecular vibrations* (Dover, 1980)

F

Quantum mechanics

In the following we present the axioms or basic postulates of quantum mechanics and accompany them by their classical counterparts in the Hamiltonian formalism. We begin the presentation with a brief summary of some of the mathematical background essential for the developments in the following. It is by no means a comprehensive presentation, and the reader is supposed to have some basic knowledge about quantum mechanics that may be obtained from any of the many introductory textbooks in quantum mechanics. The focus here is on results of particular relevance to the subjects of this book. We consider, for example, a derivation of a formal expression for the flux density operator in quantum mechanics and its coordinate representation. A systematic way of generating any representation of any combination of operators is set up, and is of immediate usage for the time autocorrelation function of the flux operator used to determine the rate constants of a chemical process.

F.1 Basic axioms of quantum mechanics

Before describing the axioms of quantum mechanics, one needs some mathematical background in linear vector spaces. Since this may be acquired from any of the introductory textbooks on quantum mechanics, we shall just review some of the main points without going into much detail.

We use the *ket* symbol $|\psi\rangle$ introduced by Dirac to denote the state of a physical system. The scalar product (the 'dot product') between two vectors is a complex number denoted by

$$\langle\phi|\psi\rangle = \langle\psi|\phi\rangle^* \tag{F.1}$$

Physical observables are represented by Hermitian operators $\hat{\Omega}$, where a *Hermitian* operator is defined by $\hat{\Omega} = \hat{\Omega}^\dagger$. Here $\hat{\Omega}^\dagger$ is the so-called adjoint operator defined by the relation

$$\langle\phi|\hat{\Omega}|\psi\rangle = \langle\psi|\hat{\Omega}^\dagger|\phi\rangle^* \tag{F.2}$$

We can expand the state vector $|\psi\rangle$ in any orthonormal basis of basis vectors, e.g., the eigenstates of the position operator:

$$|\psi\rangle = \int dq_i \psi(q_i)|q_i\rangle \tag{F.3}$$

where the eigenstates and eigenvalues associated with the position are given by

$$\hat{q}_i|q_i\rangle = q_i|q_i\rangle \tag{F.4}$$

where $\langle q_i'|q_i\rangle = \delta(q_i' - q_i)$.

To identify the expansion coefficients in Eq. (F.3), $\psi(q_i)$, we multiply from the left with the *bra* $\langle q_i'|$:

$$\langle q_i'|\psi\rangle = \int dq_i \psi(q_i)\langle q_i'|q_i\rangle$$
$$= \psi(q_i') \tag{F.5}$$

and $\psi(q_i)$ is referred to as the representation of $|\psi\rangle$ in coordinate space, i.e., the ordinary wave function in coordinate space. Using this result, we may write Eq. (F.3) as

$$\begin{aligned}|\psi\rangle &= \int dq_i \psi(q_i)|q_i\rangle \\ &= \int dq_i \langle q_i|\psi\rangle|q_i\rangle \\ &= \int dq_i |q_i\rangle\langle q_i|\psi\rangle\end{aligned} \tag{F.6}$$

When written like this, we immediately see that the identity operator $\hat{I}$ can be written in the form

$$\hat{I} = \int dq_i |q_i\rangle\langle q_i| \tag{F.7}$$

This is a very important relation that we will be using over and over again in the following. The operator $\hat{P}_{q_i} = |q_i\rangle\langle q_i|$ is called the *projection* operator for the *ket* $|q_i\rangle$. Equation (F.7), which is called the *completeness relation*, or *closure relation*, expresses the identity operator as a sum over projection operators. The relation is true for any orthonormal basis we may choose.

We consider a system of N particles. To keep the notation simple, we have chosen only to consider the coordinate q_i and conjugate momentum p_i, and we neglect all the other coordinates and momenta in all the derivations to follow. The straightforward generalization to include all coordinates and momenta will be discussed toward the end of the appendix.

(1) In *classical mechanics*, the state of the system is specified by $q(t)$ and $p(t)$, i.e., as a point in the $6N$-dimensional phase space of the N-particle system.

In *quantum mechanics*, the state of the system is represented by a vector $|\psi(t)\rangle$ in Hilbert space.

(2) In *classical mechanics*, every dynamical variable ω is a function of coordinates q and conjugate momenta p, i.e., $\omega = \omega(q, p)$.

In *quantum mechanics*, the independent variables q and p of classical mechanics are represented by the Hermitian operators $\hat{q}$ and $\hat{p}$ with the following matrix elements in the Cartesian coordinate basis $|q\rangle$, here just written for the coordinate q_i and the conjugate momentum p_i:

$$\langle q_i|\hat{q}_i|q_i'\rangle = q_i\delta(q_i - q_i') \tag{F.8}$$

and

$$\langle q_i|\hat{p}_i|q_i'\rangle = -i\hbar\frac{d}{dq_i}\delta(q_i - q_i') = i\hbar\,\frac{d}{dq_i'}\,\delta(q_i - q_i')$$
$$\equiv -i\hbar\delta'_{q_i}(q_i - q_i') = i\hbar\,\delta'_{q_i'}(q_i - q_i') \tag{F.9}$$

The prime on the delta function indicates differentiation with respect to the variable given in the subscript. The prime on the coordinate is just another coordinate value, different from the coordinate without a prime. This prime should not be confused with the prime on the delta function. The operator corresponding to a dependent variable $\omega(q,p)$ is given by a Hermitian operator $\hat{\Omega}(\hat{q},\hat{p}) = \omega(q \to \hat{q}, p \to \hat{p})$. At the end of this section the complete expression for the relations with all coordinates is given. For brevity of notation, we usually only include the coordinate of interest, as in Eqs (F.8) and (F.9).

(3) In *classical mechanics*, if a system is in a given state (q,p), the measurement of the dynamical variable ω will yield a value $\omega(q,p)$. The state of the system will remain unaffected.

In *quantum mechanics*, if a system is in the state $|\psi\rangle$, the measurement of the dynamical variable corresponding to the operator $\hat{\Omega}$ will yield one of the eigenvalues ω with a probability $P(\omega) = |\langle\omega|\psi\rangle|^2$. The state of the system will change from $|\psi\rangle$ to $|\omega\rangle$ as a result of the measurement.
The expectation value $\langle\hat{\Omega}\rangle$ of an operator in a state $|\psi\rangle$ is just the mean value as defined in statistics:

$$\begin{aligned}\langle\hat{\Omega}\rangle &= \sum_{i=1}^{n} P(\omega_i)\,\omega_i = \sum_{i=1}^{n} |\langle\omega_i|\psi\rangle|^2\,\omega_i \\ &= \sum_{i=1}^{n}\langle\psi|\omega_i\rangle\langle\omega_i|\psi\rangle\,\omega_i \\ &= \sum_{i=1}^{n}\langle\psi|\hat{\Omega}|\omega_i\rangle\langle\omega_i|\psi\rangle \\ &= \langle\psi|\hat{\Omega}|\psi\rangle\end{aligned} \tag{F.10}$$

where we have assumed that the eigenstates of $\hat{\Omega}$ are discrete and have used a completeness relation similar to Eq. (F.7) for the eigenstates $|\omega_i\rangle$ of $\hat{\Omega}$. In particular, the expectation value of the projection operator $\hat{P}_{q_i} = |q_i\rangle\langle q_i|$ (see Eq. (F.7)) is

$$\langle\hat{P}_{q_i}\rangle = \langle\psi|q_i\rangle\langle q_i|\psi\rangle = \psi^*(q_i)\psi(q_i) \tag{F.11}$$

which is the probability density for the i*th* coordinate being equal to q_i.

(4) In *classical mechanics*, the state variables change with time according to Hamilton's equations of motion

$$\begin{aligned}\dot{q}_i &= \frac{\partial H}{\partial p_i}, \quad i = 1, 3N \\ \dot{p}_i &= -\frac{\partial H}{\partial q_i}, \quad i = 1, 3N\end{aligned} \tag{F.12}$$

In *quantum mechanics*, the state vector $|\psi\rangle$ obeys the Schrödinger equation

$$i\hbar\frac{d}{dt}|\psi(t)\rangle = \hat{H}|\psi(t)\rangle \tag{F.13}$$

where $\hat{H}(\hat{q},\hat{p}) = H(q \to \hat{q}, p \to \hat{p})$ is the Hamilton operator for the system.

F.2 Application of the axioms—examples

F.2.1 The action of the position operator

Suppose we are in a state $|\psi\rangle$. Then determine the action of the operator $\hat{q}_i$, that is,

$$\hat{q}_i|\psi\rangle = |\psi'\rangle \tag{F.14}$$

From the axioms, we have

$$\langle q_i|\hat{q}_i|q_i'\rangle = q_i\,\delta(q_i - q_i') \tag{F.15}$$

so let us convert Eq. (F.14) to a form that has the matrix element in Eq. (F.15). To do that, we multiply Eq. (F.14) from the left by the bra $\langle q_i|$ and insert the unit operator $\int dq_i'|q_i'\rangle\langle q_i'|$, according to Eq. (F.7), between $\hat{q}_i$ and $|\psi\rangle$ to give

$$\int dq_i'\,\langle q_i|\hat{q}_i|q_i'\rangle\langle q_i'|\psi\rangle = \langle q_i|\psi'\rangle \tag{F.16}$$

and with Eq. (F.15) we have

$$\int dq_i'\,q_i\,\delta(q_i - q_i')\langle q_i'|\psi\rangle = q_i\langle q_i|\psi\rangle = q_i\,\psi(q_i) \tag{F.17}$$

That is, in the coordinate representation the action of $\hat{q}_i$ is simply a multiplication of the coordinate representation of the wave function with q_i.

Then, what is the action of $\hat{q}_i^2$? We write the operator as $\hat{q}_i\hat{q}_i$, multiply from the left by the bra $\langle q_i|$, and insert two unit operators, one between the operators and one between the operator and $|\psi\rangle$ to introduce the matrix elements of the operator in the coordinate representation that are known from the axioms. We get

$$\begin{aligned}\iint dq_i'\,dq_i''\langle q_i|\hat{q}_i|q_i'\rangle\langle q_i'|\hat{q}_i|q_i''\rangle\langle q_i''|\psi\rangle &= \iint dq_i'\,dq_i''\,q_i\,\delta(q_i - q_i')\,q_i'\,\delta(q_i' - q_i'')\,\langle q_i''|\psi\rangle \\ &= \int dq_i'\,q_i\,\delta(q_i - q_i')\,q_i'\,\langle q_i'|\psi\rangle \\ &= q_i^2\langle q_i|\psi\rangle = q_i^2\,\psi(q_i)\end{aligned} \tag{F.18}$$

that is, the wave function is multiplied by q_i^2. This result is easily generalized to any function f of the position operator, i.e., in the coordinate representation it is simply given by $f(q_i)$.

F.2.2 The action of the momentum operator

We now want to determine the action of the momentum operator

$$\hat{p}_i|\psi\rangle = |\psi'\rangle \tag{F.19}$$

In order to use the axioms, we need to recast Eq. (F.19) such that the coordinate representation of the momentum operator appears. This is done by first multiplying the equation from the left by the bra $\langle q_i|$, then introducing the unit operator Eq. (F.7) between the operator and $|\psi\rangle$, and finally using *partial* integration to evaluate the resulting integral:

$$
\begin{aligned}
\langle q_i|\hat{p}_i|\psi\rangle &= \int dq_i' \langle q_i|\hat{p}_i|q_i'\rangle\langle q_i'|\psi\rangle \\
&= \int dq_i'\,(-i\hbar)\,\delta'_{q_i}(q_i - q_i')\,\psi(q_i') = \int dq_i'\,(i\hbar)\,\delta'_{q_i'}(q_i - q_i')\,\psi(q_i') \\
&= i\hbar\left([\delta(q_i - q_i')\,\psi(q_i')]_{-\infty}^{\infty} - \int dq_i'\delta(q_i - q_i')\frac{d\psi(q_i')}{dq_i'}\right) \\
&= -i\hbar\int dq_i'\,\delta(q_i - q_i')\frac{d\psi(q_i')}{dq_i'} = -i\hbar\left(\frac{d\psi(q_i')}{dq_i'}\right)_{q_i=q_i'}
\end{aligned} \tag{F.20}
$$

The term $[\delta(q_i - q_i')\,\psi(q_i')]_{-\infty}^{\infty}$ always equals zero since $\psi(q_i') = 0$ for $q_i' \to \pm\infty$. The momentum operator in the coordinate representation is therefore a differential operator and the result of its action is the derivative of the wave function at q_i multiplied by $-i\hbar$, a standard result given in any textbook on quantum mechanics, here deduced directly from the axioms of quantum mechanics.

This result may be used to determine the eigenstates $|p_i\rangle$ of the momentum operator, defined by

$$\hat{p}_i|p_i\rangle = p_i|p_i\rangle \tag{F.21}$$

in the coordinate representation. We may use the result from above if we replace $|\psi\rangle$ with $|p_i\rangle$ and $|\psi'\rangle$ by $p_i|p_i\rangle$ in Eq. (F.19), and therefore $\psi(q_i) = \langle q_i|\psi\rangle$ by $\langle q_i|p_i\rangle$ and $\psi'(q_i)$ by $p_i\langle q_i|p_i\rangle$ in Eq. (F.20). This gives the following first-order differential equation in $\langle q_i|p_i\rangle$:

$$-i\hbar\frac{d\langle q_i|p_i\rangle}{dq_i} = p_i\langle q_i|p_i\rangle \tag{F.22}$$

where $\langle q_i|p_i\rangle$ is the momentum eigenfunction in the coordinate representation. Equation (F.22) is easily solved, and we get

$$\langle q_i|p_i\rangle = \frac{1}{\sqrt{2\pi\hbar}}\exp(ip_iq_i/\hbar) \tag{F.23}$$

where we have used the normalization $\langle p_i|p_i'\rangle = \delta(p_i - p_i')$, and the representation of the delta function (with $p_i = k_i\hbar$)

$$\delta(k_i) = \frac{1}{2\pi}\int_{-\infty}^{\infty} dx\,\exp(ik_ix) \tag{F.24}$$

F.2.3 The displacement operator

In Section 4.2.2, we used the displacement (translation) operator $\exp(-ib\hat{p}_i/\hbar)$. We consider here this operator and its action on the state $|\phi\rangle$, i.e., we consider the momentum-space and coordinate-space representations of $|\phi_b\rangle = \exp(-ib\hat{p}_i/\hbar)|\phi\rangle$.

In the momentum-space representation, we have

$$\begin{aligned}\langle p_i|\phi_b\rangle &= \langle p_i|e^{-ib\hat{p}_i/\hbar}|\phi\rangle \\ &= \int dp_i' \langle p_i|e^{-ib\hat{p}_i/\hbar}|p_i'\rangle\langle p_i'|\phi\rangle \\ &= \int dp_i' \langle p_i|e^{-ibp_i'/\hbar}|p_i'\rangle\langle p_i'|\phi\rangle \\ &= \int dp_i' e^{-ibp_i'/\hbar}\delta(p_i - p_i')\langle p_i'|\phi\rangle \\ &= e^{-ibp_i/\hbar}\langle p_i|\phi\rangle \end{aligned} \tag{F.25}$$

where Eq. (F.21) was used in the third line. Using this result and Eq. (F.23), we get, in the coordinate-space representation,

$$\begin{aligned}\langle q_i|\phi_b\rangle &= \int dp_i \langle q_i|p_i\rangle\langle p_i|\phi_b\rangle \\ &= (2\pi\hbar)^{-1/2}\int dp_i\, e^{ip_iq_i/\hbar}e^{-ibp_i/\hbar}\langle p_i|\phi\rangle \\ &= (2\pi\hbar)^{-1/2}\int dp_i\, e^{ip_i(q_i-b)/\hbar}\langle p_i|\phi\rangle \\ &= \int dp_i \langle q_i - b|p_i\rangle\langle p_i|\phi\rangle \\ &= \langle q_i - b|\phi\rangle \end{aligned} \tag{F.26}$$

which demonstrates that the operator $\exp(-ib\hat{p}/\hbar)$ generates a displacement of b in coordinate space.

F.2.4 Equivalence of spatial and momentum projection operators

In Section 5.2.1, we used the equivalence of the two projection operators

$$\begin{aligned}\hat{\mathcal{P}}_{r_1} &= \lim_{t\to\infty} \exp(i\hat{H}t/\hbar)h(\hat{r}_1)\exp(-i\hat{H}t/\hbar) \\ \hat{\mathcal{P}}_{p_1} &= \lim_{t\to\infty} \exp(i\hat{H}t/\hbar)h(\hat{p}_1)\exp(-i\hat{H}t/\hbar)\end{aligned} \tag{F.27}$$

where $h(x)$ is the Heaviside step function. In momentum space we defined the reactant region as the one where the translational momentum in the reaction coordinate was negative, while positive in the product region. In coordinate space we define the reactant region as the one with a negative reaction coordinate and the product region as the one with a positive reaction coordinate, with the reaction coordinate $r_1 = 0$ separating the two regions. Intuitively, such a modification should be possible because the definition in momentum space implies an equivalent motion in the reaction coordinate that therefore may also be used to define the reactant and product regions.

We consider in the following the proof that the projection operators in Eq. (F.27) are equivalent. We introduce the unit operator $\exp(-i\hat{H}_0t/\hbar)\exp(i\hat{H}_0t/\hbar)$ on both sides of the Heaviside step functions. We find

$$\hat{\mathcal{P}}_{r_1} = \lim_{t\to\infty} \exp(i\hat{H}t/\hbar)\exp(-i\hat{H}_0t/\hbar)\exp(i\hat{H}_0t/\hbar)h(\hat{r}_1) \times \exp(-i\hat{H}_0t/\hbar)\exp(i\hat{H}_0t/\hbar)\exp(-i\hat{H}t/\hbar)$$
$$= \lim_{t\to\infty} \hat{\Omega}_- \exp(i\hat{H}_0t/\hbar)h(\hat{r}_1)\exp(-i\hat{H}_0t/\hbar)\hat{\Omega}_-^\dagger \tag{F.28}$$

and similarly for $\hat{\mathcal{P}}_{p_1}$ we have

$$\hat{\mathcal{P}}_{p_1} = \lim_{t\to\infty} \hat{\Omega}_- \exp(i\hat{H}_0t/\hbar)h(\hat{p}_1)\exp(-i\hat{H}_0t/\hbar)\hat{\Omega}_-^\dagger \tag{F.29}$$

where $\hat{\Omega}_-$ is the Møller operator in Eq. (4.133). Comparing the operators in Eqs (F.28) and (F.29), we just need to prove that

$$\hat{\mathcal{P}}^0_{r_1} = \lim_{t\to\infty} \exp(i\hat{H}_0t/\hbar)h(\hat{r}_1)\exp(-i\hat{H}_0t/\hbar) \tag{F.30}$$

is equal to

$$\hat{\mathcal{P}}^0_{p_1} = \lim_{t\to\infty} \exp(i\hat{H}_0t/\hbar)h(\hat{p}_1)\exp(-i\hat{H}_0t/\hbar) \tag{F.31}$$

This is done by writing down the coordinate representations of the operators using the methodology presented in this appendix. We introduce unit operators on both sides of the Heaviside step function using the momentum eigenstates and find

$$\langle r_1'|\hat{\mathcal{P}}^0_{p_1}|r_1\rangle = \langle r_1'|\exp(i\hat{H}_0t/\hbar)h(\hat{p}_1)\exp(-i\hat{H}_0t/\hbar)|r_1\rangle$$
$$= \int_{-\infty}^{\infty} dp_1 \int_{-\infty}^{\infty} dp_1' \langle r_1'|\exp(i\hat{p}_1^2t/(2\mu\hbar))|p_1\rangle\langle p_1|h(\hat{p}_1)|p_1'\rangle \times \langle p_1'|\exp(-i\hat{p}_1^2t/(2\mu\hbar))|r_1\rangle$$
$$= \int_{-\infty}^{\infty} dp_1 \int_{-\infty}^{\infty} dp_1' \langle r_1'|\exp(ip_1^2t/(2\mu\hbar))|p_1\rangle h(p_1')\delta(p_1 - p_1') \times \langle p_1'|\exp(-ip_1'^2t/(2\mu\hbar))|r_1\rangle$$
$$= \frac{1}{2\pi\hbar}\int_{-\infty}^{\infty} dp_1 \int_{-\infty}^{\infty} dp_1' \exp(ip_1r_1'/\hbar)\exp(ip_1^2t/(2\mu\hbar))h(p_1')\delta(p_1 - p_1') \times \exp(-ip_1'^2t/(2\mu\hbar))\exp(-ip_1'r_1/\hbar)$$
$$= \frac{1}{2\pi\hbar}\int_{-\infty}^{\infty} dp_1 \exp(ip_1r_1'/\hbar)h(p_1)\exp(-ip_1r_1/\hbar)$$
$$= \frac{1}{2\pi\hbar}\int_{0}^{\infty} dp_1 \exp(ip_1(r_1' - r_1)/\hbar) \tag{F.32}$$

In the second line of the equation we have used that $\hat{H}_0$, in this one degree of freedom case, contains only kinetic energy in the coordinate r_1 since the potential energy term is zero in the reactant/product region. In the sixth line we have introduced the momentum eigenfunctions in the coordinate representation, using Eq. (F.23).

Similarly, for the $\hat{\mathcal{P}}^0_{r_1}$ operator, we introduce unit operators on both sides of the Heaviside step function, but using the coordinate eigenfunctions in this case:

$$\begin{aligned}
\langle r_1'|\hat{\mathcal{P}}^0_{r_1}|r_1\rangle &= \langle r_1'|\exp(i\hat{H}_0t/\hbar)h(\hat{r}_1)\exp(-i\hat{H}_0t/\hbar)|r_1\rangle \\
&= \int_{-\infty}^{\infty} dr_1'' \int_{-\infty}^{\infty} dr_1''' \langle r_1'|\exp(i\hat{H}_0t/\hbar)|r_1''\rangle\langle r_1''|h(\hat{r}_1)|r_1'''\rangle \\
&\qquad \times \langle r_1'''|\exp(-i\hat{H}_0t/\hbar)|r_1\rangle \\
&= \int_{-\infty}^{\infty} dr_1'' \int_{-\infty}^{\infty} dr_1''' \langle r_1'|\exp(i\hat{H}_0t/\hbar)|r_1''\rangle h(r_1''')\delta(r_1''-r_1''') \\
&\qquad \times \langle r_1'''|\exp(-i\hat{H}_0t/\hbar)|r_1\rangle \\
&= \int_{-\infty}^{\infty} dr_1'' \langle r_1'|\exp(i\hat{H}_0t/\hbar)|r_1''\rangle h(r_1'')\langle r_1''|\exp(-i\hat{H}_0t/\hbar)|r_1\rangle
\end{aligned} \tag{F.33}$$

The matrix elements in the integral are evaluated by introducing a unit operator using the momentum eigenfunctions between the operator and the coordinate eigenfunction:

$$\begin{aligned}
\langle r_1'|\exp(i\hat{H}_0t/\hbar)|r_1''\rangle &= \int dp_1 \langle r_1'|\exp(i\hat{p}_1^2t/(2\mu\hbar))|p_1\rangle\langle p_1|r_1''\rangle \\
&= \int dp_1 \langle r_1'|\exp(ip_1^2t/(2\mu\hbar))|p_1\rangle\langle p_1|r_1''\rangle \\
&= \frac{1}{2\pi\hbar}\int dp_1 \exp(ip_1^2t/(2\mu\hbar))\exp(ir_1'p_1/\hbar)\exp(-ir_1''p_1/\hbar) \\
&= \frac{1}{2\pi\hbar}\int dp_1 \exp[ip_1^2t/(2\mu\hbar) + ip_1(r_1'-r_1'')/\hbar] \\
&= \frac{1}{\sqrt{2\pi\hbar}}\sqrt{\frac{\mu}{-it}}\exp(-i\mu(r_1'-r_1'')^2/(2\hbar t))
\end{aligned} \tag{F.34}$$

In the first line we have introduced the unit operator using the momentum eigenstates because we may then evaluate the resulting matrix elements. In the third line we have introduced the momentum eigenfunctions in the coordinate representation, and in the last line we have used the standard integral $\int dx \exp(-p^2x^2 \pm qx) = \exp(q^2/(4p^2))\sqrt{\pi}/p$. For the second matrix element in Eq. (F.33) we find by analogy that

$$\langle r_1''|\exp(-i\hat{H}_0t/\hbar)|r_1\rangle = \frac{1}{\sqrt{2\pi\hbar}}\sqrt{\frac{\mu}{it}}\exp(i\mu(r_1-r_1'')^2/(2\hbar t)) \tag{F.35}$$

The expressions in Eqs (F.34) and (F.35) are finally introduced in Eq. (F.33) and we find

$$\begin{aligned}
\langle r_1'|\hat{\mathcal{P}}^0_{r_1}|r_1\rangle &= \frac{\mu}{t}\frac{1}{2\pi\hbar}\int_0^{\infty} dr_1'' \exp[i\mu((r_1-r_1'')^2-(r_1'-r_1'')^2)/(2\hbar t)] \\
&= \frac{\mu}{2\pi\hbar t}\int_0^{\infty} dr_1'' \exp[i\mu((r_1^2-r_1'^2)+2r_1''(r_1'-r_1))/(2\hbar t)] \\
&= \frac{\mu}{2\pi\hbar t}\exp(i\mu(r_1^2-r_1'^2)/(2\hbar t)) \\
&\qquad \times \int_0^{\infty} dr_1'' \exp[i\mu r_1''(r_1'-r_1)/(\hbar t)]
\end{aligned} \tag{F.36}$$

For comparison with the result in Eq. (F.32), we introduce the substitution $r_1'' = p_1 t/\mu$ in Eq. (F.36) and find

$$\begin{aligned}\langle r_1'|\hat{\mathcal{P}}^0_{r_1}|r_1\rangle &= \frac{1}{2\pi\hbar}\exp(i\mu(r_1^2 - r_1'^2)/(2\hbar t))\int_0^\infty dp_1 \exp[ip_1(r_1' - r_1)/\hbar] \\ &= \frac{1}{2\pi\hbar}\int_0^\infty dp_1 \exp[ip_1(r_1' - r_1)/\hbar] \qquad \text{(F.37)}\end{aligned}$$

when the limit $t \to \infty$ is taken. We then see that the expression is identical to the one in Eq. (F.32) and have thereby proven the equivalence of the two expressions in Eq. (F.27).

F.3 The flux operator

In Section 5.2, we used an expression for the flux operator. In most quantum mechanics textbooks expressions for the probability current density, or probability flux density, are given in terms of the wave function in the coordinate representation. We need an expression for the flux density operator without reference to any particular representation, and since it is rarely found in the textbooks, let us in the following derive this expression.

Consider the Hermitian position projection operator $(\hat{P}_{q_i})_{q_i=r}$:

$$(\hat{P}_{q_i})_{q_i=r} = (|q_i\rangle\langle q_i|)_{q_i=r} \equiv |r\rangle\langle r| \equiv \hat{P}_r \qquad \text{(F.38)}$$

It is related to the probability density for one of the coordinates of a particle to be r. This is seen by forming the expectation value of $\hat{P}_r$ for the system in the state $|\psi\rangle$. We find

$$\langle\psi|\hat{P}_r|\psi\rangle = \langle\psi|r\rangle\langle r|\psi\rangle = \psi^*(r)\psi(r) = \rho(r) \qquad \text{(F.39)}$$

where we have introduced the symbol $\rho(r)$ for the probability density.

We seek an expression for the flux operator by deriving an expression for the time variation of the position projection operator $\hat{P}_r$, and compare the resulting equation with the standard continuity equation known in several branches of physics, for example, in fluid dynamics:

$$\frac{\partial\rho}{\partial t} = -\nabla\cdot\boldsymbol{F}_\rho \qquad \text{(F.40)}$$

which expresses that the rate of change in density is equal to the net flux of mass in and out of the volume element, as given by the divergence of the *flux density* of mass $\boldsymbol{F}_\rho$ as defined in Eq. (F.40).

Let us therefore determine the rate of change of the position projection operator for the coordinate q_i, which in the Heisenberg picture is given by the following commutator equation:

$$\frac{d\hat{P}_r}{dt} = \frac{i}{\hbar}[\hat{H}, \hat{P}_r] \qquad \text{(F.41)}$$

where $\hat{H}$ is the Hamiltonian of the system. Since all momenta operators $\hat{p}_{j\neq i}$ and position operators $\hat{q}_{j\neq i}$ commute with $\hat{p}_i$ and $\hat{q}_i$ we only need to include the momentum operator for the coordinate q_i in the kinetic energy term of the Hamiltonian, and find

$$\frac{d\hat{P}_r}{dt} = \frac{i}{\hbar}[\hat{p}_i^2/(2m_i) + V(\hat{q}), \hat{P}_r]$$
$$= \frac{i}{2m_i\hbar}[\hat{p}_i^2, \hat{P}_r] + \frac{i}{\hbar}[V(\hat{q}), \hat{P}_r] \quad \text{(F.42)}$$

Note that the potential energy term in general cannot be split up into a term only depending on the coordinate q_i like the kinetic energy term. The commutator $[\hat{p}_i^2, \hat{P}_r]$ may be written

$$[\hat{p}_i^2, \hat{P}_r] = \hat{p}_i^2\hat{P}_r - \hat{P}_r\hat{p}_i^2$$
$$= \hat{p}_i^2\hat{P}_r - \hat{p}_i\hat{P}_r\hat{p}_i + \hat{p}_i\hat{P}_r\hat{p}_i - \hat{P}_r\hat{p}_i^2$$
$$= \hat{p}_i(\hat{p}_i\hat{P}_r - \hat{P}_r\hat{p}_i) + (\hat{p}_i\hat{P}_r - \hat{P}_r\hat{p}_i)\hat{p}_i$$
$$= \hat{p}_i[\hat{p}_i, \hat{P}_r] + [\hat{p}_i, \hat{P}_r]\hat{p}_i \quad \text{(F.43)}$$

The commutator $[\hat{p}_i, \hat{P}_r]$ is found in the usual way by forming the matrix element with any two states $|\phi\rangle$ and $|\psi\rangle$:

$$\langle\phi|[\hat{p}_i, \hat{P}_r]|\psi\rangle = \langle\phi|\hat{p}_i\hat{P}_r - \hat{P}_r\hat{p}_i|\psi\rangle$$
$$= \langle\phi|\hat{p}_i|r\rangle\langle r|\psi\rangle - \langle\phi|r\rangle\langle r|\hat{p}_i|\psi\rangle$$
$$= \langle\phi|\hat{p}_i|r\rangle\psi(r) - \phi^*(r)\langle r|\hat{p}_i|\psi\rangle \quad \text{(F.44)}$$

The matrix elements involving the momentum operator were evaluated in Eq. (F.20):

$$\langle r|\hat{p}_i|\psi\rangle = -i\hbar\left(\frac{d\psi(q_i)}{dq_i}\right)_{q_i=r} \quad \text{(F.45)}$$

and similarly for

$$\langle\phi|\hat{p}_i|r\rangle = \langle r|\hat{p}_i|\phi\rangle^* = i\hbar\left(\frac{d\phi^*(q_i)}{dq_i}\right)_{q_i=r} \quad \text{(F.46)}$$

using that the momentum operator is Hermitian. These results are inserted into Eq. (F.44) and we get

$$\langle\phi|[\hat{p}_i, \hat{P}_r]|\psi\rangle = i\hbar\left[\psi(r)\left(\frac{d\phi^*(q_i)}{dq_i}\right)_{q_i=r} + \phi^*(r)\left(\frac{d\psi(q_i)}{dq_i}\right)_{q_i=r}\right]$$
$$= i\hbar\left(\frac{d(\phi^*(q_i)\,\psi(q_i))}{dq_i}\right)_{q_i=r}$$
$$= i\hbar\left(\frac{d}{dq_i}\langle\phi|q_i\rangle\langle q_i|\psi\rangle\right)_{q_i=r}$$
$$= i\hbar\left(\langle\phi|\frac{d}{dq_i}\{|q_i\rangle\langle q_i|\}|\psi\rangle\right)_{q_i=r}$$
$$= i\hbar\left(\langle\phi|\frac{d}{dq_i}\hat{P}_{q_i}|\psi\rangle\right)_{q_i=r} \quad \text{(F.47)}$$

and from this

$$[\hat{p}_i, \hat{P}_r] = i\hbar \left(\frac{d\hat{P}_{q_i}}{dq_i} \right)_{q_i=r} = i\hbar \frac{d\hat{P}_r}{dr} \tag{F.48}$$

The commutator $[V(\hat{q}), \hat{P}_r]$ is found in the same way. We get

$$\begin{aligned}
\langle\phi|[V(\hat{q}), \hat{P}_r]|\psi\rangle &= \langle\phi|V(\hat{q})\hat{P}_r|\psi\rangle - \langle\phi|\hat{P}_r V(\hat{q})|\psi\rangle \\
&= \int dq_i \, \langle\phi|q_i\rangle\langle q_i|V(\hat{q}_i; q_{j\neq i})|r\rangle\langle r|\psi\rangle \\
&\qquad - \int dq_i \, \langle\phi|r\rangle\langle r|V(\hat{q}_i; q_{j\neq j})|q_i\rangle\langle q_i|\psi\rangle \\
&= \int dq_i \, \phi^*(q_i)V(q_i; q_{j\neq i})\delta(q_i - r)\psi(r) \\
&\qquad - \int dq_i \, \phi^*(r)V(q_i; q_{j\neq i})\delta(q_i - r)\psi(q_i) \\
&= \phi^*(r)V(r; q_{j\neq i})\psi(r) - \phi^*(r)V(r; q_{j\neq i})\psi(r) \\
&= 0
\end{aligned} \tag{F.49}$$

that is,

$$[V(\hat{q}), \hat{P}_r] = 0 \tag{F.50}$$

With the results in Eqs (F.48) and (F.50) introduced into Eq. (F.42), we find

$$\begin{aligned}
\frac{d\hat{P}_r}{dt} &= \frac{i}{\hbar}\left[\hat{H}, \hat{P}_r\right] = \frac{i}{2m_i\hbar}\left[\hat{p}_i^2, \hat{P}_r\right] \\
&= \frac{i}{2m_i\hbar}\left[\hat{p}_i\,[\hat{p}_i, \hat{P}_r] + [\hat{p}_i, \hat{P}_r]\,\hat{p}_i\right] \\
&= \frac{i}{2m_i\hbar}\left[\hat{p}_i\, i\hbar\left(\frac{d}{dq_i}\hat{P}_{q_i}\right)_{q_i=r} + i\hbar\left(\frac{d}{dq_i}\hat{P}_{q_i}\right)_{q_i=r}\hat{p}_i\right] \\
&= -\frac{1}{2m_i}\left[\hat{p}_i\left(\frac{d}{dq_i}\hat{P}_{q_i}\right)_{q_i=r} + \left(\frac{d}{dq_i}\hat{P}_{q_i}\right)_{q_i=r}\hat{p}_i\right] \\
&= -\frac{1}{2m_i}\left(\frac{d}{dq_i}\left[\hat{p}_i\hat{P}_{q_i} + \hat{P}_{q_i}\hat{p}_i\right]\right)_{q_i=r}
\end{aligned} \tag{F.51}$$

When we compare with Eq. (F.40), we see that the flux (density) operator related to the coordinate q_i, at $q_i = r$, is given by the expression

$$\hat{F}_r = \frac{1}{2m_i}\left[\hat{p}_i\hat{P}_{q_i} + \hat{P}_{q_i}\hat{p}_i\right]_{q_i=r} = \frac{1}{2m_i}\left[\hat{p}_i\hat{P}_r + \hat{P}_r\hat{p}_i\right] \tag{F.52}$$

We note that the flux is a vector and the expression in Eq. (F.52) is therefore the ith component of the quantum flux operator. The quantum flux of probability through a surface given by $S(q) = 0$ for a system in the quantum state $|\psi\rangle$ may therefore be determined as the dot product of the quantum flux and the normalized gradient vector ∇S, and integrated over the entire surface.

Let us then derive an expression for the matrix element of the flux operator in the coordinate representation, an expression we need in order to develop the time autocorrelation function of the flux operator in the coordinate representation. We use the axiom for the matrix element of the momentum operator in the coordinate representation, and obtain

$$\begin{aligned}
\langle q_i|\hat{F}_r|q_i'\rangle &= \frac{1}{2m_i}\left[\langle q_i|\hat{p}_i\hat{P}_r|q_i'\rangle + \langle q_i|\hat{P}_r\hat{p}_i|q_i'\rangle\right] \\
&= \frac{1}{2m_i}\left[\langle q_i|\hat{p}_i|r\rangle\langle r|q_i'\rangle + \langle q_i|r\rangle\langle r|\hat{p}_i|q_i'\rangle\right] \\
&= \frac{1}{2m_i}\left[-i\hbar\frac{d}{dq_i}\,\delta(q_i-r)\,\delta(q_i'-r) + i\hbar\,\delta(q_i-r)\,\frac{d}{dq_i'}\,\delta(r-q_i')\right] \\
&= \frac{\hbar}{i2m_i}\left[\delta(q_i'-r)\,\frac{d}{dq_i}\delta(q_i-r) - \delta(q_i-r)\,\frac{d}{dq_i'}\,\delta(r-q_i')\right] \\
&= \frac{\hbar}{i2m_i}\left[\delta(q_i'-r)\,\delta'_{q_i}(q_i-r) - \delta(q_i-r)\,\delta'_{q_i'}(r-q_i')\right]
\end{aligned} \tag{F.53}$$

The primes on the delta functions in the last line of the equation indicate differentiation with, respectively, q_i and q_i'.

As an example we use this expression for the matrix element of the flux operator in the coordinate representation to determine the flux at r for a system in the state $|\psi\rangle$. The expectation value of the flux operator is

$$\begin{aligned}
\langle\psi|\hat{F}_r|\psi\rangle &= \int dq_i \int dq_i'\,\langle\psi|q_i\rangle\langle q_i|\hat{F}_r|q_i'\rangle\langle q_i'|\psi\rangle \\
&= \frac{\hbar}{i2m_i}\int dq_i \int dq_i'\psi^*(q_i)[\delta(q_i'-r)\delta'_{q_i}(q_i-r) - \delta(q_i-r)\delta'_{q_i'}(r-q_i')]\psi(q_i')
\end{aligned} \tag{F.54}$$

Partial integration is used to evaluate the terms. Let us begin with the first term:

$$\begin{aligned}
&\int dq_i \int dq_i'\,\psi^*(q_i)\delta(q_i'-r)\,\delta'_{q_i}(q_i-r)\psi(q_i') \\
&\qquad = \int dq_i'\left\{[\delta(q_i-r)\psi^*(q_i)\,\delta(q_i'-r)\,\psi(q_i')\,]_{-\infty}^{\infty}\right. \\
&\qquad\qquad \left. - \int dq_i\,\delta(q_i-r)\,\frac{d}{dq_i}\,(\psi^*(q_i)\,\delta(q_i'-r)\,\psi(q_i'))\right\} \\
&\qquad = -\int dq_i \int dq_i'\,\delta(q_i-r)\,\delta(q_i'-r)\,\psi(q_i')\frac{d\psi^*(q_i)}{dq_i} \\
&\qquad = -\psi(r)\left(\frac{d\psi^*(q_i)}{dq_i}\right)_{q_i=r}
\end{aligned} \tag{F.55}$$

since the first term on the right-hand side of the equation in the second line is zero, because $\psi(q_i) = 0$ for $q_i \to \pm\infty$. Similarly, we find for the second term:

$$
\begin{aligned}
-\int dq_i \int dq_i'\, \psi^*(q_i)\delta(q_i - r)\, \delta'_{q_i'}(r - q_i')\psi(q_i') \\
&= -\int dq_i \Big\{ [\delta(r - q_i')\psi^*(q_i)\,\delta(q_i - r)\,\psi(q_i')\,]_{-\infty}^{\infty} \\
&\qquad - \int dq_i'\,\delta(r - q_i')\,\frac{d}{dq_i'}\,(\psi^*(q_i)\,\delta(q_i - r)\,\psi(q_i'))\Big\} \\
&= \int dq_i \int dq_i'\,\delta(r - q_i')\,\delta(q_i - r)\,\psi^*(q_i)\frac{d\psi(q_i')}{dq_i'} \\
&= \psi^*(r)\left(\frac{d\psi(q_i')}{dq_i'}\right)_{q_i'=r}
\end{aligned} \tag{F.56}
$$

that is, we have

$$
\langle\psi|\hat{F}_r|\psi\rangle = \frac{\hbar}{i2m_i}\left[\psi^*(r)\left(\frac{d\psi(q_i')}{dq_i'}\right)_{q_i'=r} - \psi(r)\left(\frac{d\psi^*(q_i)}{dq_i}\right)_{q_i=r}\right] \tag{F.57}
$$

which is the expression for the ith component of the flux density found in many textbooks. It is immediately evident from this expression that the flux density in a system where the wave function is real will be zero.

F.4 Time-correlation function of the flux operator

In Section 5.2, we have seen how the rate constant for a chemical reaction may be determined as a time integral of the auto-time-correlation function of the flux operator given by

$$
C_F(t) = \mathrm{Tr}[\hat{F}\hat{U}^\dagger\hat{F}\hat{U}] \tag{F.58}
$$

with

$$
\begin{aligned}
\hat{U} &= \exp(-i\hat{H}t/\hbar)\exp(-\hat{H}/2k_BT) \\
\hat{U}^\dagger &= \exp(i\hat{H}t/\hbar)\exp(-\hat{H}/2k_BT)
\end{aligned} \tag{F.59}
$$

In order to use this formal expression in a calculation of the rate constant we need to choose a representation. In the following we will determine the coordinate representation of the correlation function. We use the coordinate representation of the flux operator as derived above. It is introduced in the expression for the time-correlation function by introducing three unit operators like

$$
\begin{aligned}
C_F(t) &= \int dq_i \int dq_i' \int dq_i'' \int dq_i'''\langle q_i|\hat{F}|q_i'\rangle\langle q_i'|\hat{U}^\dagger|q_i''\rangle\langle q_i''|\hat{F}|q_i'''\rangle\langle q_i'''|\hat{U}|q_i\rangle \\
&= -\frac{\hbar^2}{4m^2}\int dq_i \int dq_i' \int dq_i'' \int dq_i'''[\cdots]_{q_iq_i'}\, U^\dagger_{q_i'q_i''}\,[\cdots]_{q_i''q_i'''}\, U_{q_i'''q_i}
\end{aligned} \tag{F.60}
$$

where the expressions for the matrix element of the flux operator in Eq. (F.53) are used. We have introduced the following shorthand notations:

$$\begin{aligned} U^{\dagger}_{q'_i q''_i} &= \langle q'_i|\hat{U}^{\dagger}|q''_i\rangle = \langle q''_i|\hat{U}|q'_i\rangle^* = U^*_{q''_i q'_i} \\ U_{q'''_i q_i} &= \langle q'''_i|\hat{U}|q_i\rangle \\ [\cdots]_{q_i q'_i} &= [\delta(q'_i - r)\,\delta'_{q_i}(q_i - r) - \delta(q_i - r)\,\delta'_{q'_i}(r - q'_i)] \\ [\cdots]_{q''_i q'''_i} &= [\delta(q'''_i - r)\,\delta'_{q''_i}(q''_i - r) - \delta(q''_i - r)\,\delta'_{q'''_i}(r - q'''_i)] \end{aligned} \tag{F.61}$$

We see that there are four terms to be evaluated, and each term is evaluated by partial integration with respect to, respectively, q_i, q'_i, q''_i, and q'''_i. They may look a little complicated, so let us take them term by term. Let us start with an evaluation of the two terms associated with $[\cdots]_{q_i q'_i}$. The first term in $[\cdots]_{q_i q'_i}$ in Eq. (F.61) gives

$$\begin{aligned} &\int dq_i \int dq'_i \int dq''_i \int dq'''_i\, \delta'_{q_i}(q_i - r)\,\delta(q'_i - r)\,U^*_{q''_i q'_i}\,[\cdots]_{q''_i q'''_i}\,U_{q'''_i q_i} \\ &\quad = \int dq'_i \int dq''_i \int dq'''_i \Big\{ [\delta(q_i - r)\,\delta(q'_i - r)\,U^*_{q''_i q'_i}\,[\cdots]_{q''_i q'''_i}\,U_{q'''_i q_i}]^{\infty}_{-\infty} \\ &\qquad\qquad - \int dq_i\,\delta(q_i - r)\frac{d}{dq_i}\left(\delta(q'_i - r)\,U^*_{q''_i q'_i}\,[\cdots]_{q''_i q'''_i}\,U_{q'''_i q_i}\right)\Big\} \\ &\quad = -\int dq_i \int dq'_i \int dq''_i \int dq'''_i\,\delta(q_i - r)\,\delta(q'_i - r)\,U^*_{q''_i q'_i}\,[\cdots]_{q''_i q'''_i}\,\frac{dU_{q'''_i q_i}}{dq_i} \end{aligned} \tag{F.62}$$

using that $[\delta(q_i - r)\cdots]^{q_i=\infty}_{q_i=-\infty}$ equals zero since r takes a finite value. Similarly, the next term in $[\cdots]_{q_i q'_i}$ in Eq. (F.61) gives

$$\begin{aligned} &-\int dq_i \int dq'_i \int dq''_i \int dq'''_i\, \delta'_{q'_i}(r - q'_i)\,\delta(q_i - r)\,U^*_{q''_i q'_i}\,[\cdots]_{q''_i q'''_i}\,U_{q'''_i q_i} \\ &\quad = -\int dq_i \int dq''_i \int dq'''_i \Big\{ [\delta(r - q'_i)\,\delta(q_i - r)\,U^*_{q''_i q'_i}\,[\cdots]_{q''_i q'''_i}\,U_{q'''_i q_i}]^{\infty}_{-\infty} \\ &\qquad\qquad - \int dq'_i\,\delta(r - q'_i)\frac{d}{dq'_i}\left(\delta(q_i - r)\,U^*_{q''_i q'_i}\,[\cdots]_{q''_i q'''_i}\,U_{q'''_i q_i}\right)\Big\} \\ &\quad = \int dq_i \int dq'_i \int dq''_i \int dq'''_i\,\delta(q_i - r)\,\delta(r - q'_i)\,[\cdots]_{q''_i q'''_i}\,U_{q'''_i q_i}\,\frac{dU^*_{q''_i q'_i}}{dq'_i} \end{aligned} \tag{F.63}$$

Then we need to evaluate the two terms associated with the $[\cdots]_{q''_i q'''_i}$ operator in each of the equations above. We begin with the result in Eq. (F.62), combine it with the first term in $[\cdots]_{q''_i q'''_i}$ in Eq. (F.61), and use partial integration over q''_i:

$$\begin{aligned} &-\int dq_i \int dq'_i \int dq''_i \int dq'''_i\,\delta(q_i - r)\,\delta(q'_i - r)\,U^*_{q''_i q'_i}\,\delta'_{q''_i}(q''_i - r)\,\delta(q'''_i - r)\,\frac{dU_{q'''_i q_i}}{dq_i} \\ &\quad = -\int dq_i \int dq'_i \int dq'''_i \Big\{ \Big[\delta(q''_i - r)\,\delta(q_i - r)\,\delta(q'_i - r)\,\delta(q'''_i - r)U^*_{q''_i q'_i}\,\frac{dU_{q'''_i q_i}}{dq_i}\Big]^{\infty}_{-\infty} \\ &\qquad - \int dq''_i\,\delta(q''_i - r)\frac{d}{dq''_i}\Big(\delta(q_i - r)\,\delta(q'_i - r)\,\delta(q'''_i - r)\,U^*_{q''_i q'_i}\,\frac{dU_{q'''_i q_i}}{dq_i}\Big)\Big\} \end{aligned}$$

$$= \int dq_i \int dq_i' \int dq_i'' \int dq_i''' \, \delta(q_i - r)\, \delta(q_i' - r)\, \delta(q_i'' - r)\, \delta(q_i''' - r) \, \frac{dU^*_{q_i'' q_i'}}{dq_i''} \frac{dU_{q_i''' q_i}}{dq_i} \tag{F.64}$$

and finally combine the result in Eq. (F.62) with the last term in $[\cdots]_{q_i'' q_i'''}$ in Eq. (F.61), and use partial integration over q_i''':

$$\begin{aligned}
&\int dq_i \int dq_i' \int dq_i'' \int dq_i''' \, \delta(q_i - r)\, \delta(q_i' - r)\, U^*_{q_i'' q_i'} \, \delta'_{q_i'''}(r - q_i''')\, \delta(q_i'' - r) \, \frac{dU_{q_i''' q_i}}{dq_i} \\
&= \int dq_i \int dq_i' \int dq_i'' \Bigg\{ \left[\delta(r - q_i''')\, \delta(q_i - r)\, \delta(q_i' - r)\, \delta(q_i'' - r) U^*_{q_i'' q_i'} \frac{dU_{q_i''' q_i}}{dq_i} \right]_{-\infty}^{\infty} \\
&\qquad - \int dq_i''' \, \delta(r - q_i''') \frac{d}{dq_i'''} \left(\delta(q_i - r)\, \delta(q_i' - r)\, \delta(q_i'' - r)\, U^*_{q_i'' q_i'} \frac{dU_{q_i''' q_i}}{dq_i} \right) \Bigg\} \\
&= - \int dq_i \int dq_i' \int dq_i'' \int dq_i''' \, \delta(q_i - r)\, \delta(q_i' - r)\, \delta(q_i'' - r)\, \delta(r - q_i''')\, U^*_{q_i'' q_i'} \frac{d^2 U_{q_i''' q_i}}{dq_i''' \, dq_i}
\end{aligned} \tag{F.65}$$

Similarly, the result in Eq. (F.63) is combined with the first term in $[\cdots]_{q_i'' q_i'''}$ in Eq. (F.61), and we use partial integration over q_i'':

$$\begin{aligned}
&\int dq_i \int dq_i' \int dq_i'' \int dq_i''' \, \delta(q_i - r)\, \delta(r - q_i')\, \delta'_{q_i''}(q_i'' - r)\, \delta(q_i''' - r)\, U_{q_i''' q_i} \frac{dU^*_{q_i'' q_i'}}{dq_i'} \\
&= \int dq_i \int dq_i' \int dq_i''' \Bigg\{ \left[\delta(q_i'' - r)\, \delta(q_i - r)\, \delta(r - q_i')\, \delta(q_i''' - r) U_{q_i''' q_i} \frac{dU^*_{q_i'' q_i'}}{dq_i'} \right]_{-\infty}^{\infty} \\
&\qquad - \int dq_i'' \, \delta(q_i'' - r) \frac{d}{dq_i''} \left(\delta(q_i - r)\, \delta(r - q_i')\, \delta(q_i''' - r)\, U_{q_i''' q_i} \frac{dU^*_{q_i'' q_i'}}{dq_i'} \right) \Bigg\} \\
&= - \int dq_i \int dq_i' \int dq_i'' \int dq_i''' \, \delta(q_i - r)\, \delta(q_i' - r)\, \delta(q_i'' - r)\, \delta(q_i''' - r)\, U_{q_i''' q_i} \frac{d^2 U^*_{q_i'' q_i'}}{dq_i' \, dq_i''}
\end{aligned} \tag{F.66}$$

and finally combine the result in Eq. (F.63) with the last term in $[\cdots]_{q_i'' q_i'''}$ in Eq. (F.61), and use partial integration over q_i''':

$$\begin{aligned}
&- \int dq_i \int dq_i' \int dq_i'' \int dq_i''' \, \delta(q_i - r)\, \delta(r - q_i')\, \delta'_{q_i'''}(r - q_i''')\, \delta(q_i'' - r)\, U_{q_i''' q_i} \frac{dU^*_{q_i'' q_i'}}{dq_i'} \\
&= - \int dq_i \int dq_i' \int dq_i'' \Bigg\{ \left[\delta(r - q_i''')\, \delta(q_i - r)\, \delta(r - q_i')\, \delta(r - q_i'') U_{q_i''' q_i} \frac{dU^*_{q_i'' q_i'}}{dq_i'} \right]_{-\infty}^{\infty} \\
&\qquad + \int dq_i''' \, \delta(r - q_i''') \frac{d}{dq_i'''} \left(\delta(q_i - r)\, \delta(r - q_i')\, \delta(q_i'' - r)\, U_{q_i''' q_i} \frac{dU^*_{q_i'' q_i'}}{dq_i'} \right) \Bigg\} \\
&= \int dq_i \int dq_i' \int dq_i'' \int dq_i''' \, \delta(q_i - r)\, \delta(r - q_i')\, \delta(q_i'' - r)\, \delta(r - q_i''') \frac{dU^*_{q_i'' q_i'}}{dq_i'} \frac{dU_{q_i''' q_i}}{dq_i'''}
\end{aligned} \tag{F.67}$$

These results are now introduced into Eq. (F.60), and we have

$$\frac{\hbar^2}{4m^2}\int dq_i \int dq_i' \int dq_i'' \int dq_i'''\, \delta(q_i - r)\, \delta(q_i' - r)\, \delta(q_i'' - r)\, \delta(r - q_i''')$$
$$\times \left[U^*_{q_i''q_i'} \frac{d^2 U_{q_i'''q_i}}{dq_i'''\, dq_i} + U_{q_i'''q_i} \frac{d^2 U^*_{q_i''q_i'}}{dq_i'\, dq_i''} - \frac{dU^*_{q_i''q_i'}}{dq_i''} \frac{dU_{q_i'''q_i}}{dq_i} - \frac{dU^*_{q_i''q_i'}}{dq_i'} \frac{dU_{q_i'''q_i}}{dq_i'''} \right]$$
$$= \frac{\hbar^2}{4m^2} \left[U^*_{rr} \left(\frac{d^2 U_{q_i'''q_i}}{dq_i'''\, dq_i} \right)_{q_i = q_i''' = r} + U_{rr} \left(\frac{d^2 U^*_{q_i''q_i'}}{dq_i'\, dq_i''} \right)_{q_i' = q_i'' = r} \right.$$
$$\left. - \left(\frac{dU^*_{q_i''r}}{dq_i''} \right)_{q_i''=r} \left(\frac{dU_{rq_i}}{dq_i} \right)_{q_i=r} - \left(\frac{dU^*_{rq_i'}}{dq_i'} \right)_{q_i'=r} \left(\frac{dU_{q_i'''r}}{dq_i'''} \right)_{q_i'''=r} \right] \tag{F.68}$$

This may be written in a more compact form. Consider the following expression:

$$\left\{ \frac{d^2 |U_{q_iq_i'}|^2}{dq_i\, dq_i'} \right\}_{q_i = q_i' = r} = \left\{ \frac{d^2 (U^*_{q_iq_i'} U_{q_iq_i'})}{dq_i\, dq_i'} \right\}_{q_i = q_i' = r}$$
$$= \left\{ \frac{d}{dq_i} \left[U_{q_iq_i'} \frac{dU^*_{q_iq_i'}}{dq_i'} + U^*_{q_iq_i'} \frac{dU_{q_iq_i'}}{dq_i'} \right] \right\}_{q_i = q_i' = r}$$
$$= \left\{ U_{q_iq_i'} \frac{d^2 U^*_{q_iq_i'}}{dq_i\, dq_i'} + U^*_{q_iq_i'} \frac{d^2 U_{q_iq_i'}}{dq_i\, dq_i'} \right.$$
$$\left. + \frac{dU_{q_iq_i'}}{dq_i} \frac{dU^*_{q_iq_i'}}{dq_i'} + \frac{dU^*_{q_iq_i'}}{dq_i} \frac{dU_{q_iq_i'}}{dq_i'} \right\}_{q_i = q_i' = r}$$
$$= U_{rr} \left(\frac{d^2 U^*_{q_iq_i'}}{dq_i\, dq_i'} \right)_{q_i = q_i' = r} + U^*_{rr} \left(\frac{d^2 U_{q_iq_i'}}{dq_i\, dq_i'} \right)_{q_i = q_i' = r}$$
$$+ \left(\frac{dU_{q_ir}}{dq_i} \right)_{q_i=r} \left(\frac{dU^*_{rq_i'}}{dq_i'} \right)_{q_i'=r} + \left(\frac{dU^*_{q_ir}}{dq_i} \right)_{q_i=r} \left(\frac{dU_{rq_i'}}{dq_i'} \right)_{q_i'=r} \tag{F.69}$$

We see that the first two terms in the square parenthesis in Eq. (F.68) are identical to the first two terms in Eq. (F.69). The next two terms in the equations are also almost identical, except for the sign. Taking this into account we see that the coordinate representation of the time-correlation function in Eq. (F.58) may finally be written as

$$C_F(t) = \left(\frac{\hbar}{2m} \right)^2 \left\{ \frac{d^2}{dq_i dq_i'} \left| U_{q_iq_i'} \right|^2 - 4 \left| \frac{d}{dq_i} U_{q_iq_i'} \right|^2 \right\}_{q_i = q_i' = r} \tag{F.70}$$

The discussion has so far been restricted to one degree of freedom, that is, one coordinate and its conjugate momentum, to make the notation simple. When we extend our description to a system of N particles with $3N$ degrees of freedom, there will be a modification of the equations in axiom 2. Corresponding to $3N$ Cartesian coordinates $q_1, \ldots, q_{3N}$ describing the classical system, there exist in quantum mechanics $3N$ mutually commuting operators $\hat{q}_1, \ldots, \hat{q}_{3N}$. In the eigenvector basis $|q_1, \ldots, q_{3N}\rangle$ of these operators, called the coordinate basis and normalized as

$$\langle q_1, \ldots, q_{3N}|q'_1, \ldots, q'_{3N}\rangle = \Pi_{j=1}^{3N}\delta(q_j - q'_j) \tag{F.71}$$

we have the following correspondence:

$$\begin{aligned} \langle q_1, \ldots, q_{3N}|\hat{q}_i|q'_1, \ldots, q'_{3N}\rangle &= q_i\Pi_{j=1}^{3N}\delta(q_j - q'_j) \\ \langle q_1, \ldots, q_{3N}|\hat{p}_i|q'_1, \ldots, q'_{3N}\rangle &= -i\hbar\delta'_{q_i}(q_i - q'_i)\Pi_{j=1}^{i-1}\delta(q_j - q'_j)\Pi_{j=i+1}^{3N}\delta(q_j - q'_j) \end{aligned} \tag{F.72}$$

The other axioms remain the same.

Further reading/references

[1] C. Cohen-Tannoudji, B. Diu, and F. Laloë, *Quantum mechanics*, volume one (Wiley, 1977)

G

An integral

The two-dimensional integral in Eq. (11.11) may be reduced to two one-dimensional integrals by a change of variables. This is illustrated in Fig. G.0.1, which shows a Cartesian coordinate system with unit axes (i, j) and associated coordinates τ_1 and τ_2.

We now rotate the (i, j) system by 45° to the (i', j') system. The orthonormal transformation is given by

$$[i' \quad j'] = [i \quad j] \begin{bmatrix} \frac{\sqrt{2}}{2} & \frac{-\sqrt{2}}{2} \\ \frac{\sqrt{2}}{2} & \frac{\sqrt{2}}{2} \end{bmatrix} \tag{G.1}$$

so a vector $\mathbf{R}$ represented in the two coordinate systems is

$$\begin{aligned} \mathbf{R} &= [i' \quad j'] \begin{bmatrix} \tau_1' \\ \tau_2' \end{bmatrix} \\ &= [i \quad j] \begin{bmatrix} \frac{\sqrt{2}}{2} & \frac{-\sqrt{2}}{2} \\ \frac{\sqrt{2}}{2} & \frac{\sqrt{2}}{2} \end{bmatrix} \begin{bmatrix} \tau_1' \\ \tau_2' \end{bmatrix} \\ &\equiv [i \quad j] \begin{bmatrix} \tau_1 \\ \tau_2 \end{bmatrix} \end{aligned} \tag{G.2}$$

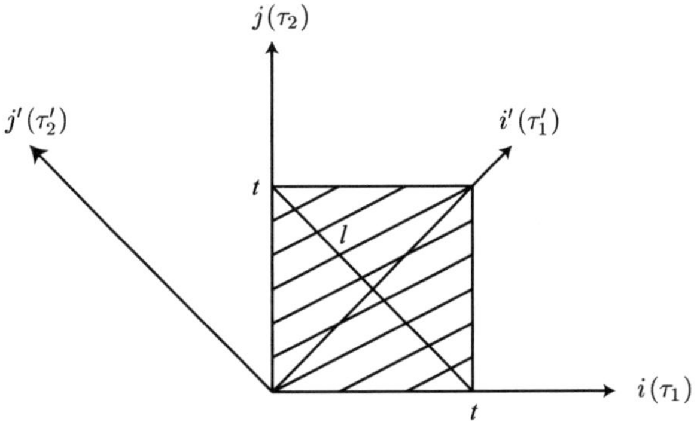

Fig. G.0.1 A sketch of the coordinate transformation for the evaluation of the double integral in Eq. (11.11).

Then the relation between the 'old' and 'new' coordinates is

$$\begin{bmatrix} \tau_1 \\ \tau_2 \end{bmatrix} = \begin{bmatrix} \frac{\sqrt{2}}{2} & \frac{-\sqrt{2}}{2} \\ \frac{\sqrt{2}}{2} & \frac{\sqrt{2}}{2} \end{bmatrix} \begin{bmatrix} \tau_1' \\ \tau_2' \end{bmatrix} \tag{G.3}$$

or the inverse

$$\begin{bmatrix} \tau_1' \\ \tau_2' \end{bmatrix} = \begin{bmatrix} \frac{\sqrt{2}}{2} & \frac{\sqrt{2}}{2} \\ \frac{-\sqrt{2}}{2} & \frac{\sqrt{2}}{2} \end{bmatrix} \begin{bmatrix} \tau_1 \\ \tau_2 \end{bmatrix} \tag{G.4}$$

and, since the Jacobian determinant for the transformation is unity, the area element in the new coordinate system is identical to the one in the old coordinate system, that is, $d\tau_1 d\tau_2 = d\tau_1' d\tau_2'$. The integral spans the 'hatched' square with side t in Fig. G.0.1. To cover the same area in the new coordinate system, the integral is divided into two integrals, one for the triangle below the line l and one for the triangle above this line. This division is naturally introduced in order to set the limits for τ_2' when we integrate over τ_1'. For the lower triangle we see from Eqs (G.3) and (G.4) that the region spanned by τ_2' for a given value of τ_1' is

$$-\sqrt{2}\tau_1' \leqslant \tau_2' \leqslant \sqrt{2}\tau_1' \tag{G.5}$$

and for the upper triangle

$$-2t - \sqrt{2}\tau_1' \leqslant \tau_2' \leqslant 2t - \sqrt{2}\tau_1' \tag{G.6}$$

The integral in Eq. (11.11) then becomes

$$\begin{aligned} &\int_0^{\sqrt{2}t/2} d\tau_1' \int_{-\sqrt{2}\tau_1}^{\sqrt{2}\tau_1} d\tau_2' \exp(\gamma\sqrt{2}\tau_1')\phi(\sqrt{2}\tau_2') \\ &\qquad + \int_{\sqrt{2}t/2}^{\sqrt{2}t} d\tau_1' \int_{-(2t-\sqrt{2}\tau_1)}^{2t-\sqrt{2}\tau_1} d\tau_2' \exp(\gamma\sqrt{2}\tau_1')\phi(\sqrt{2}\tau_2') \end{aligned} \tag{G.7}$$

where we have used Eq. (11.7) for the correlation function of the random force. We may now introduce the substitution

$$\begin{aligned} \psi &= \sqrt{2}\tau_1' \\ \eta &= \sqrt{2}\tau_2' \end{aligned} \tag{G.8}$$

into Eq. (G.7) and find

$$\begin{aligned} \frac{1}{2}\int_0^t d\psi \int_{-\psi}^{\psi} d\eta \exp(\gamma\psi)\phi(\eta) + \frac{1}{2}\int_t^{2t} d\psi \int_{-(2t-\psi)}^{2t-\psi} d\eta \exp(\gamma\psi)\phi(\eta) = \\ \frac{1}{2}\int_0^t d\psi \exp(\gamma\psi) f(\psi) + \frac{1}{2}\int_t^{2t} d\psi \exp(\gamma\psi) f(2t-\psi) \end{aligned} \tag{G.9}$$

where we have used

$$f(z) = \int_{-z}^{z} \phi(\theta)d\theta \sim f \equiv f(\infty) \tag{G.10}$$

Our assumption that $\phi(\theta)$ is sharply peaked around $\theta = 0$ and drops to zero for $\theta > \tau_c$ implies that its integral $f(z)$ reaches a constant value $f = f(\infty)$ for $\tau \sim \tau_c$. Therefore, when we are interested in much longer times, we may replace $f(\psi)$ and $f(2t-\psi)$ by their constant asymptotic value f in the right-hand side of Eq. (G.9). Hence we may write, for $t \gg \tau_c$,

$$\frac{1}{2}f\int_0^t d\psi\,\exp(\gamma\psi) + \frac{1}{2}f\int_t^{2t} d\psi\,\exp(\gamma\psi) = \frac{1}{2}f\int_0^{2t} d\psi\,\exp(\gamma\psi)$$
$$= f\frac{\exp(2\gamma t)-1}{2\gamma} \qquad \text{(G.11)}$$

Substitution of this result for the double integral in Eq. (11.11) finally gives us Eq. (11.12).

H

Dynamics of random processes

In this appendix we will give a review of the general theory of random or stochastic processes.

Let y denote the variable or set of variables in which we are interested. It may be the position or the velocity or both for a Brownian particle, for example. If y is a deterministic quantity, we can construct a function of time $y(t)$ that determines the value of y at every time t given appropriate initial data at $t = 0$. If it is a random variable, such a function does not exist. At every given time, the variable y can have any value whatsoever within its range of variation. To every possible value there is attached a certain probability, which may have any value between zero and one. As we assume y to be a continuous variable, it is easier to speak about probability densities. We say that the value of y has a probability density $P(y;t)$ at time t if there is a probability $P(y;t)dy$ of finding the value of the variable in the infinitesimal interval $(y, y + dy)$. The mere knowledge of the probability density $P(y;t)$ is not sufficient, in general, for the characterization of the process. If we know that the variable has the value y_1 at t_1, then this knowledge will influence the probability of finding the value y_2 at time t_2 because the various values of y are not necessarily independent; there could be a correlation between what happens at t_1 and what happens at t_2. That is, the joint probability density of finding the value y_1 at t_1 and the value y_2 at t_2, $W(y_2;t_2|y_1;t_1)$, cannot necessarily be inferred from a knowledge of $P(y_1;t_1)$. Hence, for a complete characterization of the random process, we must in principle specify all the joint probability densities $P(y_1;t_1)$, $W_2(y_2;t_2|y_1;t_1)$, $W_3(y_3;t_3|y_2;t_2|y_1;t_1)$, and so on. The W_n must satisfy the following obvious conditions:

(a) $W_n \geqslant 0$ because they are probabilities;

(b) $W_n(y_n;t_n|\cdots|y_1;t_1)$ must be symmetric with respect to permutation of the group of variables among each other, because W_n represents a joint probability;

(c) $W_k(y_k;t_k|\cdots|y_1;t_1) = \int dy_{k+1}\cdots dy_n W_n(y_n;t_n|\cdots|y_1;t_1)$ for any value of k in the range $1 \leqslant k \leqslant n-1$, so W_n must be compatible with all the lower-order joint probabilities.

Notice that the arguments for the joint probabilities are ordered such that $t_1 < t_2 \cdots < t_n$, so the 'order of events' should be read from the right to the left. To continue we must somehow truncate the series of higher-order joint probability densities. The simplest case (often referred to as a *purely random process*) is one in which the knowledge of $P(y;t)$ suffices for the solution of the problem. In particular,

$$W_2(y_2;t_2|y_1;t_1) = P(y_2;t_2)P(y_1;t_1) \tag{H.1}$$

so correlations in time are completely absent. This is, however, a very unrealistic assumption in a continuous physical process; indeed, for short enough time intervals there must be a causal relationship between successive events.

The next simplest case is of fundamental importance in statistical physics and is called the *Markov process*. The whole information is now contained in the two functions P and W_2. To help characterize the problem precisely, it is conventional to introduce the concept of a *transition probability* $w_2(y_2;t_2|y_1;t_1)$ defined by

$$W_2(y_2;t_2|y_1;t_1) = w_2(y_2;t_2|y_1;t_1)P(y_1;t_1) \tag{H.2}$$

This relation defines w_2 and tells us that the joint probability density of finding y_1 at t_1 and y_2 at t_2 equals the probability density of finding y_1 at t_1 times the probability of a transition from y_1 to y_2 in time $t_2 - t_1$.

Conditions (a)–(c) above imply the following properties of w_2:

(a′) $w_2(y_2;t_2|y_1;t_1) \geqslant 0$, because it is a transition probability;

(b′) $\int dy_2 w_2(y_2;t_2|y_1;t_1) = 1$, because the system has to go somewhere;

(c′) $P(y_2;t_2) = \int W_2(y_2;t_2|y_1;t_1)dy_1 = \int w_2(y_2;t_2|y_1;t_1)P(y_1;t_1)dy_1$.

The nth-order transition probability $w_n(y_n;t_n|\cdots|y_1;t_1)$ is defined as the conditional probability of finding the value y_n at time t_n given that y had values $y_{n-1},\ldots,y_1$ at the respective times $t_{n-1},\ldots,t_1$. We now define a *Markov process* by the condition

$$w_n(y_n;t_n|\cdots|y_1;t_1) = w_2(y_n;t_n|y_{n-1};t_{n-1}) \tag{H.3}$$

This definition implies that, for a Markov process, the probability of a transition at time t_{n-1} from a value y_{n-1} to a value y_n at time t_n depends only on the value of y at the time t_{n-1} of the transition and not at all on the previous history of the system.

For example, it is easy to see how W_3 can be expressed in terms of P and W_2. When we use Eq. (H.3), W_3 can be written as

$$\begin{aligned} W_3(y_3;t_3|y_2;t_2|y_1;t_1) &= w_2(y_3;t_3|y_2;t_2)w_2(y_2;t_2|y_1;t_1)P(y_1;t_1) \\ &= w_2(y_3;t_3|y_2;t_2)W_2(y_2;t_2|y_1;t_1) \\ &= \frac{W_2(y_3;t_3|y_2;t_2)W_2(y_2;t_2|y_1;t_1)}{P(y_2;t_2)} \end{aligned} \tag{H.4}$$

If we use proposition (c) then we may write

$$\begin{aligned} W_2(y_3;t_3|y_1;t_1) &= \int dy_2 W_3(y_3;t_3|y_2;t_2|y_1;t_1) \\ &= \int dy_2 w_2(y_3;t_3|y_2;t_2)w_2(y_2;t_2|y_1;t_1)P(y_1;t_1) \\ &= w_2(y_3;t_3|y_1;t_1)P(y_1;t_1) \end{aligned} \tag{H.5}$$

From Eq. (H.5) we finally get

$$w_2(y_3;t_3|y_1;t_1) = \int dy_2 w_2(y_3;t_3|y_2;t_2)w_2(y_2;t_2|y_1;t_1) \tag{H.6}$$

This is an important integral equation for the transition probability and is often taken as the definition of a Markov process. It is called the *Chapman–Kolgomorov* equation or sometimes the *Smoluchowski* equation. The physical interpretation of this equation is: the probability of a transition from y_1 at t_1 to y_3 at t_3 can be calculated by taking the product of the probability of a transition to some value y_2 at an intermediate time t_2 and the probability of a transition from that value to the final one at t_3 and summing over all possible intermediate values. Note that nothing is said about the choice of t_2, only that it should be an intermediate time.

In the physical applications that will be of interest to us, the transition probability and probability density do not depend on the times t_1 and t_2 at which transitions occur but only on the time interval $t_2 - t_1$. This is the condition of *stationarity*, which means that the statistics of the process is invariant to a change of the origin of time or to a translation in time. Equation (H.6) may then be written

$$w_2(y_3|y_1;t) = \int dy_2 w_2(y_3|y_2;t-t_2)w_2(y_2|y_1;t_2) \tag{H.7}$$

where we have set $t_1 = 0$.

H.1 The Fokker–Planck equation

Continuous Markov processes occurring in physical systems like the Brownian particle system are characterized by frequent and small changes in the stochastic variable. When the changes are small, a differential equation for the distribution function $P(y;t)$ may be obtained in the following way. We suppose that the changes in the variable occur at intervals of the order of τ_c, while the distribution function changes in times of order τ_r. If the changes of the variables are very small compared to typical values of the variable, and if the changes are very rapid, then we may expect the two time scales to be widely separated; in other words, we may expect that a time τ exists that satisfies the conditions

$$\tau_c \ll \tau \ll \tau_r \tag{H.8}$$

With this in mind, we may write the Markov integral equation in Eq. (H.7) for a stationary process as

$$w_2(y|y_1;t) = \int dy_2 w_2(y_2|y_1;t-\tau)w_2(y|y_2;\tau) \tag{H.9}$$

or in a slightly more convenient form

$$w_2(y|y_1;t+\tau) = \int dy_2 w_2(y_2|y_1;t)w_2(y|y_2;\tau) \tag{H.10}$$

This equation can be said to relate the w_2 function at two slightly separated time instants, both at finite time, with the limiting form of the function for very short times τ. This function is only non-zero for very small changes in y due to the small time period of τ. The assumption of small changes in the variable during τ may be exploited by substituting

$$y - y_2 = \Delta y \tag{H.11}$$

so

$$dy_2 = -d(\Delta y) \tag{H.12}$$

Then Eq. (H.10) may be written

$$w_2(y|y_1; t+\tau) = \int d(\Delta y) w_2(y - \Delta y|y_1; t) w_2(y|y - \Delta y; \tau) \tag{H.13}$$

The minus sign in Eq. (H.12) has been absorbed in an inversion of the limits of integration in Eq. (H.13). Then comes a somewhat 'tricky' point. The second w_2 function in the integrand is the limiting form of the function for short times. It expresses the probability that y will undergo a transition Δy in a time interval τ starting from $y - \Delta y$. It is plausible that w_2 in the short time interval limit is a function of Δy, so in order to emphasize this it is customary to introduce a particular notation, namely

$$w_2(y|y - \Delta y; \tau) \equiv \tilde{w}_2(\Delta y|y - \Delta y; \tau) \tag{H.14}$$

so we may write Eq. (H.13) as

$$w_2(y|y_1; t+\tau) = \int d(\Delta y) w_2(y - \Delta y|y_1; t) \tilde{w}_2(\Delta y|y - \Delta y; \tau) \tag{H.15}$$

One confusion is that $y - \Delta y$ in the argument does not express the Δy dependence but merely the starting point for the transition through Δy in the time interval τ. Now the Δy dependence is underlined rather than a y dependence. This also enables us to distinguish among the two functions, but it is a rather subtle point. $w_2(y|y_1; t+\tau)$ is now expanded in a Taylor series of powers of τ about $w_2(y|y_1; t)$; by virtue of the right-hand inequality in Eq. (H.8) the expansion may be truncated after the first-order term. We also expand the product in the integrand around $w_2(y|y_1; t)\tilde{w}_2(\Delta y|y; \tau) \equiv w_2\tilde{w}_2$ and obtain

$$\begin{aligned} w_2 + \tau\frac{\partial w_2}{\partial t} + \cdots &= \int d\Delta y \left[w_2\tilde{w}_2 - \Delta y \frac{\partial}{\partial y}(w_2\tilde{w}_2) + \frac{1}{2}(\Delta y)^2 \frac{\partial^2}{\partial y^2}(w_2\tilde{w}_2) \right] \\ &= w_2 - \frac{\partial}{\partial y}(w_2\langle \Delta y \rangle) + \frac{1}{2}\frac{\partial^2}{\partial y^2}(w_2\langle (\Delta y)^2 \rangle) + \cdots \end{aligned} \tag{H.16}$$

where we have used the normalization condition

$$\int d(\Delta y) \tilde{w}_2(\Delta y|y; \tau) = 1 \tag{H.17}$$

and $\langle \cdots \rangle$ is the average of the enclosed variable, conditional upon the given initial value of y:

$$\langle (\Delta y)^n \rangle = \int d(\Delta y)(\Delta y)^n \tilde{w}_2(\Delta y|y; \tau) \tag{H.18}$$

Division by τ finally gives the equation

$$\frac{\partial}{\partial t}(w_2(y|y_1;t)) = -\frac{\partial}{\partial y}(A(y)w_2(y|y_1;t)) + \frac{1}{2}\frac{\partial^2}{\partial y^2}(B(y)w_2(y|y_1;t)) \tag{H.19}$$

with

$$\begin{aligned} A(y) &= \lim_{\tau\to 0}\frac{\langle\Delta y\rangle}{\tau} \\ B(y) &= \lim_{\tau\to 0}\frac{\langle(\Delta y)^2\rangle}{\tau} \end{aligned} \tag{H.20}$$

which are the averages of Δy and Δy^2 over the transition probability rates. They will only be meaningful if the averages are proportional to τ, so there will be no explicit τ dependence. This is usually the case. Eq. (H.19) is known as the *Fokker–Planck* equation; the solutions are the transition probability w_2.

It is easy to see that the probability $P(y;t)$ itself also satisfies a Fokker–Planck equation, when we use the relation from proposition (c):

$$P(y;t) = \int dy_1 w_2(y|y_1;t-t_1)P(y_1;t_1) \tag{H.21}$$

which gives the probability at $(y;t)$ when it is known at $(y_1;t_1)$. Then we find

$$\frac{\partial}{\partial t}P(y;t) = \int dy_1 \frac{\partial}{\partial t}(w_2(y|y_1;t-t_1))P(y_1;t_1) \tag{H.22}$$

Substituting the right-hand side of Eq. (H.19) we get

$$\frac{\partial}{\partial t}P(y;t) = \int dy_1 \left[-\frac{\partial}{\partial y}(A(y)w_2(y|y_1;t-t_1)) + \frac{1}{2}\frac{\partial^2}{\partial y^2}(B(y)w_2(y|y_1;t-t_1))\right]P(y_1;t_1) \tag{H.23}$$

Now the integration over y_1 simply gives $P(y;t)$, so the Fokker–Planck equation for P is

$$\frac{\partial}{\partial t}P(y;t) = -\frac{\partial}{\partial y}(A(y)P(y;t)) + \frac{1}{2}\frac{\partial^2}{\partial y^2}(B(y)P(y;t)) \tag{H.24}$$

with the initial condition $P(y;0) = \delta(y-y_1)$.

We now specialize the Fokker–Planck equation to the case of Brownian motion in Section 11.1. In this case, the variable y is the velocity v of the Brownian particle. We also note that the average of a function of the velocity v at time t, given that $v = v_0$ at $t = t_0$, is simply expressed in terms of the transition probability by

$$\langle f(v)\rangle = \int dv f(v) w_2(v|v_0;t-t_0) \tag{H.25}$$

So we can therefore immediately use our results obtained from the Langevin equation in order to evaluate the coefficients $A(v)$ and $B(v)$ in the Fokker–Planck equation. From Eq. (11.9), we obtain

$$\begin{aligned} \langle v - v_0\rangle &= v_0\exp(-\gamma\Delta t) - v_0 \\ &= -\gamma v_0\Delta t + O((\Delta t^2)) \\ &= -\gamma(v(1+\gamma\Delta t))\Delta t + O((\Delta t^2)) \\ &= -\gamma v\Delta t + O((\Delta t^2)) \end{aligned} \tag{H.26}$$

Hence, from Eq. (H.20),

$$A(v) = -\gamma v \tag{H.27}$$

Similarly, we obtain from Eq. (11.15) that

$$\begin{aligned}\langle (v - v_0)^2 \rangle &= \langle v^2 \rangle + v_0^2 - 2\langle v \rangle v_0 \\ &= \frac{k_B T}{M} + \left(v_0^2 - \frac{k_B T}{M} \right) (1 - 2\gamma\Delta t + \cdots) + v_0^2 - 2v_0^2(1 - \gamma\Delta t \cdots) \\ &= \frac{2\gamma k_B T}{M} \Delta t \end{aligned} \tag{H.28}$$

so

$$B(v) = \frac{2\gamma k_B T}{M} \tag{H.29}$$

The Fokker–Planck equation for the Brownian particle system is then

$$\frac{\partial}{\partial t} P(v;t) = -\gamma \frac{\partial}{\partial v} (vP(v;t)) + \gamma \frac{k_B T}{M} \frac{\partial^2}{\partial v^2} P(y;t) \tag{H.30}$$

The physical mechanism described by this equation can be understood by starting at time zero with a velocity distribution sharply peaked at $v = v_0$. As time passes, the maximum of this distribution is shifted toward smaller velocities, as a result of a systematic friction undergone by the particles (first term on the right-hand side of the equation). Furthermore, the peak broadens progressively as a result of diffusion in velocity space (second term on the right-hand side, which is the velocity space equivalent of the similar coordinate space term in Fick's law of diffusion). The final time-independent distribution reached by the Brownian particle is nothing more than the familiar Maxwell distribution:

$$P(v; \infty) = C \exp(-Mv^2/(2k_B T)) \tag{H.31}$$

This is seen by substitution of this distribution into Eq. (H.30), and with $\partial P/\partial t = 0$ at equilibrium the right-hand side should be identical to zero. This is indeed the case; we get

$$\begin{aligned} -\gamma C \frac{Mv^2}{k_B T} \exp(-Mv^2/(2k_B T)) + C\gamma \exp(-Mv^2/(2k_B T)) \\ + \gamma C \frac{Mv^2}{k_B T} \exp(-Mv^2/(2k_B T)) - \gamma C \exp(-Mv^2/(2k_B T)) = 0 \end{aligned} \tag{H.32}$$

The picture offered by the Fokker–Planck equation is, of course, in complete agreement with the Langevin equation and the assumptions made about the process. If we can solve the partial differential equation we can determine the probability density or eventually the transition probabilities at any time, and thereby determine any average value of functions of v by simple quadratures.

H.2 The Chandrasekhar equation

The idea in Kramers theory is to describe the motion in the reaction coordinate as that of a one-dimensional Brownian particle and in that way include the effects of the solvent on the rate constants. Above we have seen how the probability density for the velocity of a Brownian particle satisfies the Fokker–Planck equation that must be solved. Before we do that, it will be useful to generalize the equation slightly to include two variables explicitly, namely both the coordinate r and the velocity v, since both are needed in order to determine the rate constant in transition-state theory.

For the Markovian random process (r, v), the integral equation Eq. (H.15) can be written

$$w_2(r, v|r_1, v_1; t+\tau) = \iint d(\Delta r)d(\Delta v)w_2(r - \Delta r, v - \Delta v|r_1, v_1; t) \times \tilde{w}_2(\Delta r, \Delta v|r - \Delta r, v - \Delta v; \tau) \quad \text{(H.33)}$$

τ is, as before, long compared to the time scale of molecular fluctuations but short compared to the decay time of the particle velocity, so Δv and Δr are small. Indeed, Δr is not independent and may be written as

$$\Delta r = v\tau \quad \text{(H.34)}$$

so the Δr dependence in $\tilde{w}_2$ can be pulled out into a delta function as follows:

$$\tilde{w}_2(\Delta r, \Delta v|r - \Delta r, v - \Delta v; \tau) = \delta(\Delta r - v\tau)\hat{w}_2(\Delta v|v - \Delta v; \tau) \quad \text{(H.35)}$$

Integration of Eq. (H.33) with respect to $d(\Delta r)$ gives

$$w_2(r, v|r_1, v_1; t+\tau) = \int d(\Delta v)w_2(r - v\tau, v - \Delta v|r_1, v_1; t)\hat{w}_2(\Delta v|v - \Delta v; \tau) \quad \text{(H.36)}$$

As before, we now expand the left-hand side about t and the integrand in a Taylor series about $w_2(r, v|r_1, v_1; t)\hat{w}_2(\Delta v|v; \tau)$, and obtain

$$\begin{aligned} w_2 + \tau\frac{\partial}{\partial t}(w_2) &= \int d(\Delta v)\left[w_2\hat{w}_2 - (\Delta v)\frac{\partial}{\partial v}(w_2\hat{w}_2) + \frac{1}{2}(\Delta v)^2\frac{\partial^2}{\partial v^2}(w_2\hat{w}_2) - v\tau\frac{\partial}{\partial r}(w_2\hat{w}_2)\right] \\ &= w_2 - v\tau\frac{\partial}{\partial r}(w_2) - \frac{\partial}{\partial v}(\langle\Delta v\rangle w_2) + \frac{1}{2}\frac{\partial^2}{\partial v^2}(\langle(\Delta v^2)\rangle w_2) \end{aligned} \quad \text{(H.37)}$$

with

$$\langle(\Delta v)^n\rangle = \int d(\Delta v)(\Delta v)^n\hat{w}_2(\Delta v|v; \tau) \quad \text{(H.38)}$$

and

$$\int d(\Delta v)\hat{w}_2(\Delta v|v; \tau) = 1 \quad \text{(H.39)}$$

From this, the version of the Fokker–Planck equation for the transition probability density with two variables r and v is seen to be

$$\frac{\partial}{\partial t}(w_2(r,v|r_1,v_1;t)) + v\frac{\partial}{\partial r}(w_2(r,v|r_1,v_1;t)) = \\ -\frac{\partial}{\partial v}(A(v)w_2(r,v|r_1,v_1;t)) + \frac{1}{2}\frac{\partial^2}{\partial v^2}(B(v)w_2(r,v|r_1,v_1;t)) \tag{H.40}$$

with

$$\begin{aligned} A(v) &= \lim_{\tau\to 0}\frac{\langle \Delta v\rangle}{\tau} \\ B(v) &= \lim_{\tau\to 0}\frac{\langle (\Delta v)^2\rangle}{\tau} \end{aligned} \tag{H.41}$$

The probability density $P(r,v;t)$ also satisfies a Fokker–Planck equation. This was shown in Eq. (H.24) with one stochastic variable, and similarly we find in this case

$$\frac{\partial P(r,v;t)}{\partial t} + v\frac{\partial P(r,v;t)}{\partial r} = -\frac{\partial}{\partial v}(P(r,v;t)A(v)) + \frac{1}{2}\frac{\partial^2}{\partial v^2}(P(r,v;t)B(v)) \tag{H.42}$$

As before, $\langle \Delta v\rangle$ and $\langle (\Delta v)^2\rangle$ may be determined from the Langevin equation. It differs slightly from the Langevin equation in Eq. (11.5), since the motion takes place in an external potential $U(r)$, as sketched in Fig. 11.0.2. This is the potential from the gas phase modified by the interactions from the solvent molecules.

So the Langevin equation for this problem may be written

$$\frac{dv}{dt} = -\gamma v - \frac{1}{M}\left(\frac{\partial U(r)}{\partial r}\right) + F(t) \tag{H.43}$$

From this equation we can determine Δv and $(\Delta v)^2$ using the separation in time scales indicated in Eq. (H.8). We obtain

$$\Delta v = \int_t^{t+\tau} dt'\frac{dv}{dt'} = -\left[\gamma v + \frac{1}{M}\left(\frac{\partial U(r)}{\partial r}\right)\right]\tau + \int_t^{t+\tau} dt' F(t') \tag{H.44}$$

and therefore

$$(\Delta v)^2 = -\left[\gamma v + \frac{1}{M}\left(\frac{\partial U(r)}{\partial r}\right)\right]\tau \int_t^{t+\tau} dt' F(t') \\ + \int_t^{t+\tau} dt_1 \int_t^{t+\tau} dt_2 F(t_1)F(t_2) + O(\tau^2) \tag{H.45}$$

Hence,

$$\langle \Delta v\rangle = -\left[\gamma v + \frac{1}{M}\left(\frac{\partial U(r)}{\partial r}\right)\right]\tau \tag{H.46}$$

and

$$\langle (\Delta v)^2\rangle = f\tau + O(\tau^2) \tag{H.47}$$

where f is given by Eq. (11.14).

These results are now used to determine $A(v)$ and $B(v)$ in Eq. (H.41). We find

$$A(v) = -\gamma v - \frac{1}{M}\left(\frac{\partial U(r)}{\partial r}\right) \tag{H.48}$$

which is different from before, and

$$B(v) = \frac{2\gamma k_B T}{M} \tag{H.49}$$

which is the same as before. These relations are introduced into Eq. (H.42) and we get the Fokker–Planck equation, sometimes also called the *Chandrasekhar equation*, for the position and velocity of a particle:

$$\frac{\partial P(r,v;t)}{\partial t} = -v\frac{\partial P(r,v;t)}{\partial r} + \frac{1}{M}\frac{\partial U}{\partial r}\frac{\partial P(r,v;t)}{\partial v} + \gamma\frac{\partial}{\partial v}(vP(r,v;t)) + \frac{\gamma k_B T}{M}\frac{\partial^2 P(r,v;t)}{\partial v^2} \tag{H.50}$$

It describes the diffusion of a point in phase space for a one-dimensional Brownian particle and is used in Kramers theory.

Further reading/references

[1] N.G. van Kampen, *Stochastic processes in physics and chemistry* (North-Holland, 1981)

I

Multidimensional integrals, Monte Carlo method

The Monte Carlo method is a very powerful numerical technique used to evaluate multidimensional integrals in statistical mechanics and other branches of physics and chemistry. It is also used when initial conditions are chosen in classical reaction dynamics calculations, as we have discussed in Chapter 4. It will therefore be appropriate here to give a brief introduction to the method and to the ideas behind the method.

From Appendix A on statistical mechanics we have seen that the thermodynamic average value of an observable A is given by the expression

$$\langle A \rangle = \frac{\int dp\, dq\, A(p,q) \exp(-H(p,q)/k_B T)}{\int dp\, dq\, \exp(-H(p,q)/k_B T)} \tag{I.1}$$

In a system with N atoms there are $3N$ momenta p and position coordinates q, so the integrals are $6N$-dimensional. $A(p,q)$ is an observable depending on the coordinates and momenta of the atoms and $H(p,q)$ is the Hamiltonian of the system. Since the kinetic energy term in H is quadratic in the momenta, the integration over momenta can be carried out analytically. Hence, averages of functions that depend on momenta only are usually easy to evaluate. The very difficult problem is the computation of averages of functions depending on the positions q. Except for a few special cases, it is impossible to compute the $3N$-dimensional configurational integral analytically, and in all other cases numerical techniques must be used.

The most straightforward approach may appear to be an evaluation of the integrals by numerical quadrature, for instance using Simpson's rule. It is, however, easy to see that such a method quickly becomes hopeless to use. Suppose, for example, that we consider a system with N atoms. Let us assume that we take m equidistant points along each of the $3N$ Cartesian axes. The total number of points at which the integrand must be evaluated is then equal to m^{3N}. For all but the smallest systems this number becomes very large, even for small values of m. For instance, if we take $m = 5$ in a system with 100 atoms, then we need to evaluate the integrand at $5^{300} = 10^{210}$ points! This clearly demonstrates that better numerical techniques are needed to compute thermal averages. One such technique is the Monte Carlo method or, more precisely, the Monte Carlo importance sampling algorithm introduced by Metropolis *et al.* in 1953 [1], when simulating the neutron flux in the core of nuclear reactors.

I.1 Random sampling and importance sampling

The Monte Carlo method includes both a random sampling scheme and an importance sampling scheme. Both sampling schemes have been used in Section 4.1 on classical trajectory calculations.

Let us first look at the simple random sampling scheme. Suppose we want to evaluate the one-dimensional integral

$$S = \int_a^b dx\, f(x) \tag{I.2}$$

A simple quadrature scheme for evaluating this integral may look like

$$\begin{aligned} S &= \int_a^b dx\, f(x) \\ &\simeq \frac{b-a}{L}\sum_{i=1}^{L+1} f(a+(i-1)*(b-a)/L) \\ &= (b-a)\frac{\sum_{i=1}^{L+1} f(a+(i-1)*(b-a)/L)}{L} \equiv (b-a)\langle f\rangle \end{aligned} \tag{I.3}$$

where the x-axis is divided into L equidistant intervals Δx of magnitude $(b-a)/L$. The integral is determined as the product sum of Δx and f evaluated at L x values. Division of the sum by L, as in the last part of Eq. (I.3), gives a simple average $\langle f\rangle$ of $f(x)$ in the interval $[a,b]$, where each point has the same weight, $1/L$.

In brute force Monte Carlo, this average is determined by evaluating $f(x)$ at a large number, say L, of x values randomly distributed in the interval $[a,b]$. This is equivalent to giving each x value the same weight in the summation, just like in the average in Eq. (I.3). No x values are preferred to others, and in the limit as $L \to \infty$ there will be the same number of x values in each interval, no matter where the interval is chosen between a and b. Hence, an estimate of the integral may be found from

$$S \simeq (b-a) * \frac{1}{L}\sum_{i=1}^{L} f(x_i) \tag{I.4}$$

However, like conventional quadrature, this method is of little use for the evaluation of averages such as in Eq. (I.1) because most of the computing is spent at points where the Boltzmann factor is negligible and the integral is therefore zero. Obviously, it would be more preferable to sample many points in regions where the Boltzmann factor is larger than zero and few elsewhere. This is the basic idea behind importance sampling.

That is, instead of sampling the independent variable x in Eq. (I.2), and position coordinates q in Eq. (I.1), as in brute force Monte Carlo, we should choose the independent variables according to some distribution function, that is, some values are chosen more frequently than others.

Let us begin with the one-dimensional case in Eq. (I.2). Suppose we want to compute the definite integral by Monte Carlo importance sampling with sampling points distributed non-uniformly over the interval $[a, b]$, according to some non-negative normalized probability density $w(x)$, that is,

$$\int_a^b dx\, w(x) = 1 \tag{I.5}$$

Clearly, we may multiply and divide the integrand by $w(x)$, and thereby rewrite the integral in the form

$$S = \int_a^b dx\, w(x) \frac{f(x)}{w(x)} \tag{I.6}$$

Let us now assume that there exists a function $u(x)$ satisfying the relation

$$du = w(x)dx \tag{I.7}$$

with $u(b) - u(a) = 1$, such that $w(x)$ is normalized as expressed in Eq. (I.5). It is not always possible to find such a function u, as we shall see below, but for the moment we assume that it is possible. Then the integral can be written as

$$S = \int_{u(a)}^{u(a)+1} du\, \frac{f(x(u))}{w(x(u))} \tag{I.8}$$

In Eq. (I.8) we have written $x(u)$ to indicate that, if we consider u as the integration variable, then x must be expressed as a function of u by inverting the relation $u = u(x)$. The expression in Eq. (I.8) is now similar to the expression in Eq. (I.3) and may be evaluated by generating L random numbers u_i of u in the interval $[u(a), u(a)+1]$. We then obtain the following estimate of the integral:

$$S \simeq \frac{1}{L} \sum_{i=1}^{L} \frac{f(x(u_i))}{w(x(u_i))} \tag{I.9}$$

An obvious question now is what have we gained by rewriting the integral in Eq. (I.2) in the form of Eq. (I.8)? To see this, let us form the variance σ^2 in the value of the estimate of the integral. We find

$$\sigma^2 = \frac{1}{L^2} \sum_{i=1}^{L} \sum_{j=1}^{L} \left\langle \left(\frac{f(x(u_i))}{w(x(u_i))} - \left\langle \frac{f}{w} \right\rangle \right) \left(\frac{f(x(u_j))}{w(x(u_j))} - \left\langle \frac{f}{w} \right\rangle \right) \right\rangle \tag{I.10}$$

Here the bracket $\langle \cdots \rangle$ denotes the true average as obtained for $L \to \infty$. Since the different samples i and j are random and therefore statistically independent, all cross terms between i and j vanish when we evaluate the expression in Eq. (I.10). It may therefore be written

$$\begin{aligned} \sigma^2 &= \frac{1}{L^2} \sum_{i=1}^{L} \left\langle \left(\frac{f(x(u_i))}{w(x(u_i))} - \left\langle \frac{f}{w} \right\rangle \right)^2 \right\rangle \\ &= \frac{1}{L} \left[\left\langle \left(\frac{f}{w} \right)^2 \right\rangle - \left\langle \frac{f}{w} \right\rangle^2 \right] \end{aligned} \tag{I.11}$$

This shows that the variance behaves as $1/L$, independent of the formulation of the integrals, which is clear when we set $w(x) = 1$. However, we may reduce the variance significantly, for a given L, by a proper choice of w. For instance, if $w = f$, so f/w is a constant, then the variance will be zero. This is the ideal situation, but if we choose a w such that the ratio f/w will be a smoothly-varying function, we may still get a very small variance. In contrast, if $w(x)$ is chosen constant, as in brute force Monte Carlo sampling, there may be large fluctuations in the ratio f/w, which is equivalent to a poor determination of the integral.

This becomes even more clear when we consider the multidimensional integrals encountered in statistical mechanics. Then only a very small fraction of the points in phase space are accessible, that is, correspond to states with a non-zero Boltzmann factor. So, even if L is very large, only ρL of the points will be in such a region corresponding to a sampling of the function with only ρL points, and thus a variance which behaves as $1/(\rho L)$ rather than $1/L$. Here ρ is the fraction of points in phase space accessible to the system. The variance in a random sampling may indeed be very large when it is observed that for a liquid of 100 atoms it has been estimated that the Boltzmann factor will be non-zero for 1 out of about 10^{260} points in phase space, that is, $\rho = 10^{-260}$.

Hence, it would clearly be advisable to carry out a non-uniform Monte Carlo importance sampling of configuration space with a w approximately proportional to the Boltzmann factor.

Unfortunately, the simple importance sampling as described above cannot be used to sample multidimensional integrals over configuration space as in Eq. (I.1). The reason is simply that we do not know how to construct the transformation in Eq. (I.7) that will enable us to generate points in configuration space with a probability density as given by the Boltzmann factor. In fact, in order to do so, we must be able to compute analytically the partition function of the system. If we could do that, there would hardly be any need for computer simulations!

In many cases in statistical mechanics, we are not interested in the configurational part of the partition function itself, but in averages of the type in Eq. (I.1), where the ratio between integrals is involved. Metropolis *et al.* [1] showed that it is possible to devise an efficient Monte Carlo scheme to sample such a ratio even when we do not know the probability density $P(q)$ in configuration space:

$$P(q) = \frac{\exp(-H(q)/k_B T)}{\int dq \, \exp(-H(q)/k_B T)} \tag{I.12}$$

but only know the numerator in Eq. (I.12), that is, the relative probability density of the various regions in phase space.

Further reading/references

[1] N. Metropolis, A.W. Rosenbluth, M.N. Rosenbluth, A.N. Teller, and E. Teller, *J. Chem. Phys.* **21**, 1087 (1953)

Index